Ionosphere Monitoring with Remote Sensing

Ionosphere Monitoring with Remote Sensing

Editor

Fabio Giannattasio

MDPI • Basel • Beijing • Wuhan • Barcelona • Belgrade • Manchester • Tokyo • Cluj • Tianjin

Editor
Fabio Giannattasio
Istituto Nazionale di Geofisica e Vulcanologia
Italy

Editorial Office
MDPI
St. Alban-Anlage 66
4052 Basel, Switzerland

This is a reprint of articles from the Special Issue published online in the open access journal *Remote Sensing* (ISSN 2072-4292) (available at: https://www.mdpi.com/journal/remotesensing/special_issues/ionosphere_monitoring).

For citation purposes, cite each article independently as indicated on the article page online and as indicated below:

LastName, A.A.; LastName, B.B.; LastName, C.C. Article Title. *Journal Name* **Year**, *Volume Number*, Page Range.

ISBN 978-3-0365-5903-2 (Hbk)
ISBN 978-3-0365-5904-9 (PDF)

Contents

About the Editor

Fabio Giannattasio

Fabio Giannattasio (Ph.D., II Level Master in Space Science and Technology) is a Researcher at the Istituto Nazionale di Geofisica e Vulcanologia (INGV) in Rome, Italy. His research activity is focused on the study of Sun–Earth interactions and space weather. He has a particular interest magnetosphere-ionosphere coupling, the complex nature of physical processes in the ionosphere, and some aspects inherent to solar physics.

remote sensing

MDPI

Editorial

Ionosphere Monitoring with Remote Sensing

Fabio Giannattasio

Istituto Nazionale di Geofisica e Vulcanologia, Via di Vigna Murata 605, 00143 Rome, Italy; fabio.giannattasio@ingv.it

1. Introduction

Characterising the physical properties of the Earth's ionosphere is fundamental to shed light on the dynamic processes occurring therein on a wide range of both spatial and temporal scales and to understand several phenomena relevant to Space Weather.

In fact, due to the presence of ions and electrons, the ionosphere reacts to the onset, amplification and evolution of magnetic and electric fields.

This response may substantially change the physical properties of the ionosphere and its energetic budget and may be reflected, for example, in the modification of the propagation properties of electromagnetic signals traveling through the ionospheric medium.

Due to the conspicuous amount of high-quality data, these features can be reliably investigated at different scales taking advantage of remote sensing and in situ facilities such as ionosondes, radars, satellites and Global Navigation Satellite Systems (GNSS) receivers.

2. Overview of Contribution and Future Perspectives

In this context, the Special Issue "Ionosphere Monitoring with Remote Sensing" aims at promoting significant advances in our knowledge of the ionosphere through the use of different data from different facilities as well as currently recognized ionospheric models. In fact, the Special Issue focuses on: (1) the investigation of the impact of sunlit, solar and geomagnetic activity on the ionosphere at all latitudes; (2) the investigation of the impact of ionospheric variations on contemporary technology; (3) the improvement of ionospheric models through new instrumental observations, analyses and data-handling techniques; (4) the investigation of magnetosphere–ionosphere coupling through multi-instrumental approaches; and (5) the promotion of new instruments, missions and tools to monitor the ionosphere.

The Special Issue provides 15 original research papers describing results obtained with a wide range of tools, data and analysis techniques and focused on the characterisation of several properties of the ionosphere.

As mentioned above, great attention has been paid to the development of new facilities and analysis techniques to increase our knowledge of the ionosphere. Shindin et al. [1] presented a prototype of a low-cost and good-quality fast ionosonde capable of performing with the unprecedented speed of one second cadence, which allows recording fast quasiperiodic and moving ionospheric disturbances in the F, E and Es layers. An additional strength is that the ionosonde is equipped with cheap, publicly available components, which favours the multi-position registration of ionograms and, as a consequence, the investigation of ionospheric disturbances in a three-dimensional region of space and the possibility to create a network of observation points. A layer of critical importance for ionospheric studies is the transition region between the lower and upper atmosphere, namely, the sporadic E (Es) layer, which consists of a region of enhanced ion plasma at altitudes between 90 and 120 km with a vertical extent of several kilometres and a horizontal extension of tens of kilometres [2]. The existence of this region can be explained by the wind shear theory and the convergence of metal ions and can be influenced by shear instabilities, tidal, planetary or gravity waves, meteors and thunderstorms [3–8]. The vertical structure of the Es layer is

Citation: Giannattasio, F. Ionosphere Monitoring with Remote Sensing. *Remote Sens.* **2022**, *14*, 5325. https://doi.org/10.3390/rs14215325

Received: 13 October 2022
Accepted: 21 October 2022
Published: 25 October 2022

still poorly understood due to its transient and complex nature together with limitations in observation techniques [9–13]. Liu et al. [14] applied the frequency domain interferometry (FDI) technique by using the Es layer measurements near Wuhan, China, on 8 June 2021. They pointed out that this technique allowed them to obtain increased spatial resolution of ionosonde imaging capable of identifying different types of Es and to capture its internal fine structure. Unlike the "classic" vertical sounding mode, for the oblique sounding mode, the transmitter and the receiver are located at stations that can be hundreds or thousands of kilometres apart [15–18]. The resulting oblique ionograms can capture ionospheric properties at the reflection point, which is usually located at the middle point between the transmitter and receiver. However, the problem of how to automatically scale oblique ionograms is still open, and various solutions have been identified over the years [19–23]. Jiang et al. [24] developed a method to carry out the automatic inversion of oblique ionograms to extract the parameters of the ionosphere together with the electron density profile. Their results show that the accuracy of the inferred autoscaled maximum observable frequency and minimum group path of the ordinary trace of the F2 layer is about 91.98% and 86.41%, respectively. Kim et al. [25] used Vertical Incidence Pulsed Ionospheric Radar (VIPIR) to observe the polar ionosphere with Dynasonde analysis software at Jang Bogo Station (74.6°S, 164.2°E), Antarctica, which is located in the polar cap, cusp or auroral oval depending on the local time and the geomagnetic activity conditions. The resulting F2-layer peak electron density (NmF2) and bottomside total electron content (TEC) exhibit an overall good correlation with GPS TEC measurements during quiet conditions. During the daytime and in summer, the bottomside TEC is less correlated with the GPS TEC due to particle precipitation and the onset of large density irregularities in the polar ionosphere. However, the Dynasonde analysis show some limitations and needs to be improved in order to provide accurate density profiles, especially during disturbed geomagnetic conditions. The use of radar imaging and interferometry techniques also provides important information on the physical properties of the ionosphere. In a horizontally structured atmosphere, radar echoes are strongest near the zenith and decrease with the angle off the zenith. In the presence of ionospheric field-aligned plasma irregularities (FAIs), radar echoes are strongest at the beam direction perpendicular to the geomagnetic field, with a fast decrease in the angle off the perpendicular direction. The aspect angle, which is a measure of the aspect sensitivity, i.e., the half width of half power or the standard deviation of Gaussian fitting in the angular power distribution, is of the order of degrees [26,27]. On the contrary, it can be of order 0.1 degrees or less in FAI echoes (see, e.g., Kudeki and Farley [28]). A way to effectively measure an FAI's aspect angle lies in the radar interferometry technique [28,29]. Chen et al. [30] applied the coherent radar imaging (CRI) technique to estimate the aspect angle of mid-latitude E region FAIs. CRI requires the use of separate antennas as independent receiving channels to collect radar echoes [27,31]. The echoes received allow one to retrieve the in-beam angular power distribution. By using the multireceiver and multifrequency capabilities of the 46.5 MHz middle and upper atmosphere radar in Japan, Chen et al. [30] showed that, among the three methods (namely, Fourier, Capon and norm-constrained Capon) used to recover the brightness distribution, the norm-constrained Capon method produces more reliable results and more trustworthy aspect angle values consistent with those obtained with the RI technique. Their results may help to shed light on the spatial and temporal properties of plasma irregularities in the ionosphere.

Karpachev [32] separated and classified ionospheric troughs (regions of anomalously decreased electron density) in the winter ionosphere of the Southern hemisphere by using CHAMP satellite data during high solar activity (between 2000 and 2002). In particular, the authors identified two kinds of high-latitude troughs: (1) a wide trough associated with a region of particle precipitation on the poleward edge of the auroral oval; (2) a narrow trough of ionisation presumably associated with an electric field. Moreover, the main ionospheric trough (MIT) was separated from the ring ionospheric trough (RIT), the latter being formed by the decay of the magnetospheric ring current.

A relevant aspect at the centre of ionospheric investigation concerns plasma density irregularities, which play a key role in the propagation of electromagnetic signals, being a cause of disturbance for the GNSS. In fact, irregularities are responsible for degradation and, eventually, interruptions in the signals received by the system. In the equatorial F region, irregularities are also known as plasma bubbles and develop on the nightside [33] at magnetic latitudes up to 20° in both hemispheres [34], at heights up to 1000–1550 km [35] and on a wide range of spatial scales, from hundreds of kilometres down to a few decametres [36]. Their spatial and temporal distribution depends on solar and geomagnetic activity and exhibits a diurnal and seasonal variation [33,37,38]. The origin of plasma bubbles is recognised to be due to the establishment of density gradients sufficient to trigger a Rayleigh–Taylor instability growth mechanism [34,39]. The irregularities generated in this way expand vertically and then follow the geomagnetic field lines in both directions above and below the magnetic equator. This dynamic is typically overlaid by an eastward drift motion due to polarising electric fields generated by neutral zonal winds. The instability of these structures can, in turn, generate secondary irregularities and trigger a cascading process. Several studies have pointed out the turbulent nature of plasma bubbles [40–45]. In this context, De Michelis et al. [46] focused on the relationship between the spectral features of electron density and magnetic field strength inside plasma bubbles in order to understand whether it is possible to study the dynamical features of plasma bubbles by using either the magnetic field or the electron density measurements. This is motivated by the fact that, in the past, important plasma bubble features have been detected by analysing their magnetic signatures using the diamagnetic effect [47]. However, studying plasma bubbles by using only magnetic field data may not be the correct way, as it implies that the scaling properties of electron density and magnetic fields are equal. To address this point, De Michelis et al. [46] studied the scaling properties of both electron density and magnetic fields associated with plasma bubbles using about two years of Swarm measurements at 1 Hz. Specifically, they applied the local detrended structure function analysis [48] and found that a complex relation may exist between the spectral features of electron density and magnetic field that depends on local time and latitude due to the evolution and turbulent nature of plasma bubbles. A more in-depth study of diamagnetic currents at high latitudes obtained by Swarm measurements has been performed by Lovati et al. [49]. Such weak currents are driven by pressure gradients and produce a magnetic field that is directed opposite to the background geomagnetic field and causes its reduction. The authors used 4 years of electron density, electron temperature and magnetic field data at 1 Hz to investigate the dependence of diamagnetic currents on local time, season, solar and geomagnetic activity and sunlit conditions. They confirmed the enhancement of diamagnetic currents at high latitudes, around the cleft region, during disturbed periods due to the increase in plasma pressure gradients. In the polar cap, currents flow regardless of the geomagnetic activity due to plasma instabilities driving irregularities and pressure gradients. Moreover, during disturbed periods, features in the correspondence of the auroral oval move to lower latitudes. These findings may help to improve current geomagnetic field models and understand the impact of ionospheric irregularities on dynamics at spatial scales of tens of kilometres.

New insights into dynamic processes in the ionosphere are obtained by studying its turbulent nature, which underlies, for example, chaotic plasma behaviour. The turbulent dynamics of ionospheric plasma has long been established, especially at high latitudes, by investigating, for example, fluctuations in magnetic and electric fields and electron density. Such fluctuations are characterised by power-law spectral densities, scaling features and non-Gaussian statistics of increments at all scales (see, e.g., [50–53]) and can affect plasma dynamics via the ExB drift term. At both high and low latitudes, variations in vertical plasma velocity drift plays a key role in the generation of irregularities [54–56]. In light of this, Consolini et al. [57] used electric and magnetic field measurements provided by the Chinese Seismo-Electromagnetic Satellite (CSES-01) to investigate the properties of the plasma ExB drift velocity during a crossing of the Southern auroral F region. Specifically, they analysed the spectral and scaling features of velocity fluctuations and pointed out the turbulent nature of the drift. In more detail, the authors provided evidence of 2D intermittent turbulence at scales from tens of meters to tens of kilometres. This is consistent with filamentary or thin-tube-like features.

One of the most important application issues is the risk assessment of the impact that ionospheric variations may have on technology. A proper risk assessment allows the development of effective mitigation strategies. For example, ionospheric anomalies may result in potential threats for the ground-based augmentation system (GBAS), which is an airport-based augmentation of the GNSS capable of providing advanced civil-aviation services. When GNSS signals travel through ionospheric regions with enhanced gradients, severe errors may be observed and compromise the reliability of the GBAS. Thus, it is fundamental to quickly detect anomalies. In this context, Gao et al. [58] developed a monitor to clearly detect anomalies with an average detection speed improved by more than 16% when dealing with real data instead of simulations. Valdés-Abreu et al. [59] studied the effects of an annular solar eclipse on GNSS position estimation accuracy based on TEC measurements performed by over 2000 stations worldwide, which were validated with measurements by the Swarm satellite mission and four digisondes in Central and South America. In particular, TEC maps pointed out a TEC depletion under the moon's shadow and important variations in both crests of the Equatorial Ionization Anomaly (EIA). Variations typically affect the amplitude of the signal and its delay (see Bravo et al. [60] and references therein) and can affect regions outside the umbra and penumbra of the eclipse [61–63]. With this global coverage, the work of Valdés-Abreu et al. [59] allowed them to find other locations in the world that could be affected by perturbations in the North Pole and infer how that perturbations propagate to those potential locations.

A fundamental physical parameter for studying the impact of sunlit, solar and geomagnetic activity on the upper ionosphere and its coupling with the magnetosphere is the electron temperature. This quantity exhibits distinct features with spatial, diurnal, seasonal and activity variability [64–71]. Pignalberi et al. [72] performed a statistical and global study of the electron temperature in the topside ionosphere derived from seven years of in situ data acquired by the Swarm mission at 1 s cadence. The results obtained with this unprecedented data set were compared to data modelled by the International Reference Ionosphere (IRI) model, as well as data obtained from incoherent scatter radars (ISRs). This also allowed an understanding of the deviation between the IRI model and the measurements and testing the reliability of including Swarm data in the empirical data set layer of the IRI itself. Finally, the authors showed that adding the Lomidze calibration to Swarm data [73] improved their agreement with ISR data and the IRI model, especially at mid-latitudes and during the daytime. Another significant parameter representative of the ionosphere is the equivalent slab thickness (EST), i.e., the ration of the TEC to the NmF2. By definition, this parameter represents an imaginary equivalent depth of the ionosphere and includes information on both the topside and bottomside ionosphere, thus being useful in the study of variations in the upper atmosphere (see, e.g., [74–76]). EST exhibits diurnal, seasonal solar and geomagnetic activity variations with a dependence on the location of the observing station. The greatest variability is observed during periods of geomagnetic storms. Zhang et al. [77] analysed the EST in Guam, at equatorial latitudes, confirming and discussing previous results in the literature. In addition, they obtained some new results pointing out diurnal and seasonal changes and the effect of geomagnetic storms on EST at the magnetic equator. In particular, they found that during positive storms, the penetration electric field increases plasma uplift, causing an increase in TEC accompanied by small increases in NmF2. Moreover, equatorward winds drive plasma into the topside ionosphere at the equator resulting in TEC that does not undergo severe depletion like NmF2 does during negative storms. Thus, geomagnetic storms enhance EST both during positive and negative storms.

The monitoring of the physical properties of the ionosphere and their perturbation also has applications in the study of phenomena that can be considered as precursors of major seismic events. Since the early work of Moore [78] and Davies & Baker [79], the idea was proposed that the processes of earthquake preparation and occurrence could be linked to ionospheric disturbances due to lithosphere–atmosphere–ionosphere coupling. With the increase in available data, this idea has become more and more widespread, and in the last decades, new satellite missions have been conceived to monitor natural disaster activities (QuakeSat, SICH-1M, COMPASS-2, DEMETER, CSES). Satellites with other declared purposes, such as the European Space Agency's Swarm constellation, have also provided important information for ionospheric disturbances. Recently, several works investigated magnetic field anomalies observed by both ground and space facilities to study the lithosphere–atmosphere–ionosphere coupling effects of earthquakes [80–82]. In this context, deep learning techniques are used to carry out statistical studies based on the analysis of large numbers of earthquakes. Xiong et al. [83] proposed a deep learning framework for pre-earthquake ionospheric perturbation identification model called SafeNet, which performs better in identifying possible pre-earthquake ionospheric anomalies the more intense the earthquakes are. Ionospheric scintillations are also used for correlations with the occurrence of earthquakes. Some studies in the literature pointed out that thermal expansion of the atmosphere derived from land surface temperature increase before earthquakes can generate small gravity waves altering the electron density profile and causing changes in the TEC, and, on the other hand, ionospheric perturbations can be detected in the hours after large earthquakes [84,85] (Tsugawa et al. 2011, Pavilidou et al. 2019). Few works in the literature investigated the correlation between the occurrence of earthquakes and ionospheric scintillation (see, e.g., [86]). These studies take advantage of GPS data from ground stations or ionosondes to measure the scintillation index S4 and study its correlation with earthquakes in the same region. By using statistical tools, Molina et al. [87] for the first time used the GNSS reflectometry [88] technique to obtain global oceanic maps of ionospheric scintillation and correlate them to earthquake precursors. Their results point out a small positive correlation for earthquakes with magnitudes above 4, with better results for increasing magnitudes. Correlation was better when positive increments in the S4 index were observed between 6 and 3 days before the earthquakes than the ones observed after them. In the best case, the correct prediction probability is about 32% and the false alarm probability is 16%; however, the probability of detection is small overall. The authors also recognise that the signature of ionospheric scintillation increments as precursors of earthquakes is still small and should not be regarded as an early warning system for earthquakes.

Funding: This research received no external funding.

Conflicts of Interest: The authors declare no conflict of interest.

References

1. Shindin, A.V.; Sergey, P.M.; Vybornov, F.I.; Grechneva, K.K.; Pavlova, V.A.; Khashev, V.R. The Prototype of a Fast Vertical Ionosonde Based on Modern Software-Defined Radio Devices. *Remote Sens.* **2022**, *14*, 547. [CrossRef]
2. Pietrella, M.; Bianchi, C. Occurrence of sporadic-E layer over the ionospheric station of Rome: Analysis of data for thirty-two years. *Adv. Space Res.* **2009**, *44*, 72–81. [CrossRef]
3. Haldoupis, C.; Pancheva, D.; Singer, W.; Meek, C.; Macdougall, J. An explanation for the seasonal dependence of midlatitude sporadic E layers. *J. Geophys. Res.* **2007**, *112*, A6.
4. Whitehead, J.D. Formation of the sporadic E layer in the temperate zones. *J. Atmos. Terr. Phys.* **1961**, *20*, 49–58. [CrossRef]
5. Axford, W.; Cunnold, D. The wind shear theory of temperate zone sporadic E. *Radio Sci.* **1966**, *1*, 191–197. [CrossRef]
6. Nygrén, T.; Lanchester, B.S.; Huuskonen, A.; Jalonen, L.; Eyken, A. Interference of tidal and gravity waves in the ionosphere and an associated sporadic E-layer. *J. Atmos. Sol.-Terr. Phys.* **1990**, *52*, 609–623. [CrossRef]
7. Goldsbrough, P.F.; Ellyett, C.D. Relationship of meteors to sporadic E, 2. statistical evidence for class 1 Em. *J. Geophys. Res. Atmos.* **1976**, *81*, 6135–6140. [CrossRef]
8. Barta, V.; Haldoupis, C.; Sátori, G.; Buresova, D.; Bencze, P. Searching for effects caused by thunderstorms in midlatitude sporadic E layers. *J. Atmos. Sol.-Terr. Phys.* **2017**, *161*, 150–159. [CrossRef]
9. Mori, H.; Oyama, K.I. Sounding rocket observation of sporadic-E layer electron-density irregularities. *Geophys. Res. Lett.* **1998**, *25*, 1785–1788. [CrossRef]

10. Mori, H.; Oyama, K.I. Rocket observation of sporadic-E layers and electron density irregularities over midlatitude. *Adv. Space Res.* **2000**, *26*, 1251–1255. [CrossRef]
11. Bernhardt, P.A.; Selcher, C.A.; Siefring, C.; Wilkens, M.; Compton, C.; Bust, G.; Yamamoto, M.; Fukao, S.; Takayuki, O.; Wakabayashi, M. Radio tomographic imaging of sporadic-E layers during SEEK-2. *Ann. Geophys.* **2005**, *23*, 2357–2368. [CrossRef]
12. Damtie, B.; Nygrén, T.; Lehtinen, M.S.; Huuskonen, A. High resolution observations of sporadic-E layers within the polar cap ionosphere using a new incoherent scatter radar experiment. *Ann. Geophys.* **2003**, *20*, 1429–1438. [CrossRef]
13. Turunen, T.; Nygrén, T.; Huuskonen, A.; Jalonen, L. Incoherent scatter studies of sporadic-E using 300 m resolution. *J. Atmos. Terr. Phys.* **1988**, *50*, 277–287. [CrossRef]
14. Liu, T.; Yang, G.; Zhou, C.; Jiang, C.; Xu, W.; Ni, B.; Zhao, Z. Improved Ionosonde Monitoring of the Sporadic E Layer Using the Frequency Domain Interferometry Technique. *Remote Sens.* **2022**, *14*, 1915. [CrossRef]
15. Smith, M.S. The calculation of ionospheric profiles from data given on oblique incidence ionograms. *J. Atmos. Terr. Phys.* **1970**, *32*, 1047–1056. [CrossRef]
16. Chen, J.; Bennett, J.A.; Dyson, P.L. Synthesis of oblique ionograms from vertical ionograms using quasi-parabolic segment models of the ionosphere. *J. Atmos. Terr. Phys.* **1992**, *54*, 323–331. [CrossRef]
17. Phanivong, B.; Chen, J.; Dyson, P.L.; Bennett, J.A. Inversion of oblique ionograms including the earth's magnetic field. *J. Atmos. Sol. Terr. Phys.* **1995**, *57*, 1715–1721. [CrossRef]
18. Huang, X.; Reinisch, B.W.; Kuklinski, W.S. Mid-point electron density profiles from oblique ionograms. *Ann. Geophys. Italy* **1996**, *49*, 757–761. [CrossRef]
19. Redding, N.J. Image understanding of oblique ionograms: The autoscaling problem. In Proceedings of the IEEE Australian and New Zealand Conference on Intelligent Information Systems, Adelaide, SA, Australia, 18–20 November 1996; IEEE: Piscataway, NJ, USA, 1996; pp. 155–160.
20. Fan, J.; Lu, Z.; Jiao, P. The intelligentized recognition of oblique propagation modes. *Chin. J. Radio Sci.* **2009**, *24*, 528. (In Chinese)
21. Settimi, A.; Pezzopane, M.; Pietrella, M.; Bianchi, C.; Scotto, C.; Zuccheretti, E.; Makris, J. Testing the IONORT-ISP system: A comparison between synthesized and measured oblique ionograms. *Radio Sci.* **2013**, *48*, 167–179. [CrossRef]
22. Ippolito, A.; Scotto, C.; Francis, M.; Settimi, A.; Cesaronl, C. Automatic interpretation of oblique ionograms. *Adv. Space Res.* **2015**, *55*, 1624–1629. [CrossRef]
23. Heitmann, A.J.; Gardiner-Garden, R.S. A robust feature extraction and parameterized fitting algorithm for bottom-side oblique and vertical incidence ionograms. *Radio Sci.* **2019**, *54*, 115–134. [CrossRef]
24. Jiang, C.; Zhao, C.; Zhang, X.; Liu, T.; Chen, Z.; Yang, G.; Zhao, Z. A Method for Automatic Inversion of Oblique Ionograms. *Remote Sens.* **2022**, *14*, 1671. [CrossRef]
25. Kim, E.; Jee, G.; Ham, Y.-B.; Zabotin, N.; Lee, C.; Kwon, H.-J.; Hong, J.; Kim, J.-H.; Bullett, T. Assessment of Polar Ionospheric Observations by VIPIR/Dynasonde at Jang Bogo Station, Antarctica: Part 1—Ionospheric Densities. *Remote Sens.* **2022**, *14*, 2785. [CrossRef]
26. Hocking, W.K.; Fukao, S.; Tsuda, T.; Yamamoto, M.; Sato, T.; Kato, S. Aspect sensitivity of stratospheric VHF radar wave scatterers, particularly above 15-km altitude. *Radio Sci.* **1990**, *25*, 613–627. [CrossRef]
27. Chen, J.-S.; Furumoto, J. Measurement of atmospheric aspect sensitivity using coherent radar imaging after mitigation of radar beam weighting effect. *J. Atmos. Ocean. Technol.* **2013**, *30*, 245–259. [CrossRef]
28. Kudeki, E.; Farley, D. Aspect sensitivity of equatorial electrojet irregularities and theoretical implications. *J. Geophys. Res.* **1989**, *94*, 426–434. [CrossRef]
29. Farley, D.T.; Hysell, D.L. Radar measurement of very small aspect angles in the equatorial ionosphere. *J. Geophys. Res.* **1996**, *101*, 5177–5184. [CrossRef]
30. Chen, J.-S.; Wang, C.-Y.; Chu, Y.-H. Measurement of Aspect Angle of Field-Aligned Plasma Irregularities in Mid-Latitude E Region Using VHF Atmospheric Radar Imaging and Interferometry Techniques. *Remote Sens.* **2022**, *14*, 611. [CrossRef]
31. Palmer, R.D.; Gopalam, S.; Yu, T.-Y.; Fukao, S. Coherent radar imaging using Capon's method. *Radio Sci.* **1998**, *33*, 1585–1598. [CrossRef]
32. Karpachev, A. Advanced Classification of Ionospheric Troughs in the Morning and Evening Conditions. *Remote Sens.* **2022**, *14*, 4072. [CrossRef]
33. Kil, H.; Heelis, R.A. Global distribution of density irregularities in the equatorial ionosphere. *J. Geophys. Res. (Space Phys.)* **1998**, *103*, 407–418. [CrossRef]
34. Kelley, M.C. *The Earth's Ionosphere: Plasma Physics and Electrodynamics*, 2nd ed.; Academic Press: Burlington, MA, USA, 2009.
35. Anderson, D.N.; Mendillo, M. Ionospheric conditions affecting the evolution of equatorial plasma depletions. *Geophys. Res. Lett.* **1983**, *10*, 541–544. [CrossRef]
36. Tsunoda, R.T.; Livingston, R.C.; McClure, J.P.; Hanson, W.B. Equatorial plasma bubbles: Vertically elongated wedges from the bottomside F layer. *J. Geophys. Res. (Space Phys.)* **1982**, *87*, 9171–9180. [CrossRef]
37. Smith, J.; Heelis, R.A. Equatorial plasma bubbles: Variations of occurrence and spatial scale in local time, longitude, season, and solar activity. *J. Geophys. Res. (Space Phys.)* **2017**, *122*, 5743–5755. [CrossRef]
38. Gurram, P.; Kakad, B.; Bhattacharyya, A.; Pant, T.K. Evolution of Freshly Generated Equatorial Spread F (F-ESF) Irregularities on Quiet and Disturbed Days. *J. Geophys. Res. (Space Phys.)* **2018**, *123*, 7710–7725. [CrossRef]
39. Schunk, R.; Nagy, A. *Ionospheres: Physics, Plasma Physics, and Chemistry*; Cambridge University Press: Cambridge, UK, 2009.

40. Kraichnan, R.H. Inertial Ranges in Two-Dimensional Turbulence. *Phys. Fluids* **1967**, *10*, 1417–1423. [CrossRef]
41. Kraichnan, R.H.; Montgomery, D. Two-dimensional turbulence. *Rep. Prog. Phys.* **1980**, *43*, 547–619. [CrossRef]
42. McDaniel, R.D.; Hysell, D.L. Models and DE II observations of inertial-regime irregularities in equatorial spread F. *J. Geophys. Res. (Space Phys.)* **1997**, *102*, 22233–22246. [CrossRef]
43. Yokoyama, T.; Shinagawa, H.; Jin, H. Nonlinear growth, bifurcation, and pinching of equatorial plasma bubble simulated by three-dimensional high-resolution bubble model. *J. Geophys. Res. (Space Phys.)* **2014**, *119*, 10474–10482. [CrossRef]
44. Yokoyama, T. A review on the numerical simulation of equatorial plasma bubbles toward scintillation evaluation and forecasting. *Prog. Earth Planet. Sci.* **2017**, *4*, 37. [CrossRef]
45. Hysell, D.L.; Shume, E.B. Electrostatic plasma turbulence in the topside equatorial F region ionosphere. *J. Geophys. Res. (Space Phys.)* **2002**, *107*, 1269. [CrossRef]
46. De Michelis, P.; Consolini, G.; Alberti, T.; Tozzi, R.; Giannattasio, F.; Coco, I.; Pezzopane, M.; Pignalberi, A. Magnetic Field and Electron Density Scaling Properties in the Equatorial Plasma Bubbles. *Remote Sens.* **2022**, *14*, 918. [CrossRef]
47. Lühr, H.; Rother, M.; Maus, S.; Mai, W.; Cooke, D. The diamagnetic effect of the equatorial Appleton anomaly: Its characteristics and impact on geomagnetic field modeling. *Geophys. Res. Lett.* **2003**, *30*, 1906. [CrossRef]
48. De Michelis, P.; Consolini, G.; Tozzi, R. Magnetic field fluctuation features at Swarm's altitude: A fractal approach. *Geophys. Res. Lett.* **2015**, *42*, 3100–3105. [CrossRef]
49. Lovati, G.; De Michelis, P.; Consolini, G.; Berrilli, F. Pressure-Gradient Current at High Latitude from Swarm Measurements. *Remote Sens.* **2022**, *14*, 1428. [CrossRef]
50. Kintner, P.M., Jr. Observations of velocity shear driven plasma turbulence. *J. Geophys. Res.* **1976**, *A28*, 5114–5122. [CrossRef]
51. Kintner, P.M.; Seyler, C.E. The status of observations and theory of high latitude ionospheric and magnetospheric plasma turbulence. *Space Sci. Rev.* **1985**, *41*, 1572–9672. [CrossRef]
52. Basu, S.; Basu, S.; MacKenzie, E.; Fougere, P.F.; Coley, W.R.; Maynard, N.C.; Winningham, J.D.; Sugiura, M.; Hanson, W.B.; Hoegy, W.R. Simultaneous density and electric field fluctuation spectra associated with velocity shears in the auroral oval. *J. Geophys. Res. Space Phys.* **1988**, *93*, 115–136. [CrossRef]
53. Kozelov, B.V.; Golovchanskaya, I.V.; Ostapenko, A.A.; Fedorenko, Y.V. Wavelet analysis of high-latitude electric and magnetic fluctuations observed by the Dynamic Explorer 2 satellite. *J. Geophys. Res. Space Phys.* **2008**, *113*, A03308. [CrossRef]
54. Fejer, B.G.; Kelley, M.C. Ionospheric irregularities. *Rev. Geophys. Space Phys.* **1980**, *18*, 401–454. [CrossRef]
55. Fejer, B.G. Low latitude electrodynamic plasma drifts—A review. *J. Atmos. Terr. Phys.* **1991**, *53*, 677–693. [CrossRef]
56. Anderson, D.; Anghel, A.; Yumoto, K.; Ishitsuka, M.; Kudeki, E. Estimating daytime vertical ExB drift velocities in the equatorial F-region using ground-based magnetometer observations. *Geophys. Res. Lett.* **2002**, *29*, 1596. [CrossRef]
57. Consolini, G.; Quattrociocchi, V.; Benella, S.; De Michelis, P.; Alberti, T.; Piersanti, M.; Marcucci, M.F. On Turbulent Features of E $\times$ B Plasma Motion in the Auroral Topside Ionosphere: Some Results from CSES-01 Satellite. *Remote Sens.* **2022**, *14*, 1936. [CrossRef]
58. Gao, Z.; Fang, K.; Zhu, Y.; Wang, Z.; Guo, K. An Ionospheric Anomaly Monitor Based on the One Class Support Vector Algorithm for the Ground-Based Augmentation System. *Remote Sens.* **2021**, *13*, 4327. [CrossRef]
59. Valdés-Abreu, J.C.; Díaz, M.A.; Bravo, M.; Báez, J.C.; Stable-Sánchez, Y. Ionospheric Behavior during the 10 June 2021 Annular Solar Eclipse and Its Impact on GNSS Precise Point Positioning. *Remote Sens.* **2022**, *14*, 3119. [CrossRef]
60. Bravo, M.; Martínez-Ledesma, M.; Foppiano, A.; Urra, B.; Ovalle, E.; Villalobos, C.; Souza, J.; Carrasco, E.; Muñoz, P.R.; Tamblay, L.; et al. First Report of an Eclipse from Chilean Ionosonde Observations: Comparison with Total Electron Content Estimations and the Modeled Maximum Electron Concentration and Its Height. *J. Geophys. Res. Space Phys.* **2020**, *125*, e2020JA027923. [CrossRef]
61. Le, H.; Liu, L.; Yue, X.; Wan, W. The ionospheric behavior in conjugate hemispheres during the 3 October 2005 solar eclipse. *Ann. Geophys.* **2009**, *27*, 179–184. [CrossRef]
62. He, L.; Heki, K.; Wu, L. Three-Dimensional and Trans-Hemispheric Changes in Ionospheric Electron Density Caused by the Great Solar Eclipse in North America on 21 August 2017. *Geophys. Res. Lett.* **2018**, *45*, 10933–10940. [CrossRef]
63. Aa, E.; Zhang, S.R.; Shen, H.; Liu, S.; Li, J. Local and conjugate ionospheric total electron content variation during the 21 June 2020 solar eclipse. *Adv. Space Res.* **2021**, *68*, 3435–3454. [CrossRef]
64. Willmore, A.P. Electron and ion temperatures in the ionosphere. *Space Sci. Rev.* **1970**, *11*, 607–670. [CrossRef]
65. Rishbeth, H.; Garriott, O. *Introduction to Ionospheric Physics*; International Geophysics Series v. 14; Academic Press: New York, NY, USA, 1969.
66. Ratcliffe, J.A. *An Introduction to the Ionosphere and Magnetosphere*; Cambridge University Press: Cambridge, UK, 1972.
67. Banks, P.M. Ion temperature in the upper atmosphere. *J. Geophys. Res. Space Phys.* **1967**, *72*, 3365–3385.
68. Roble, R. The calculated and observed diurnal variation of the ionosphere over Millstone Hill on 23–24 March 1970. *Planet. Space Sci.* **1975**, *23*, 1017–1033. [CrossRef]
69. Schunk, R.W.; Nagy, A.F. Electron temperatures in the region of the ionosphere: Theory and observations. *Rev. Geophys.* **1978**, *16*, 355–399. [CrossRef]
70. Bilitza, D. Electron and ion temperature data for ionospheric modelling. *Adv. Space Res.* **1991**, *11*, 139–148. [CrossRef]
71. Evans, J. Theory and practice of ionosphere study by Thomson scatter radar. *Proc. IEEE* **1969**, *57*, 496–530. [CrossRef]

72. Pignalberi, A.; Giannattasio, F.; Truhlik, V.; Coco, I.; Pezzopane, M.; Consolini, G.; De Michelis, P.; Tozzi, R. On the Electron Temperature in the Topside Ionosphere as Seen by Swarm Satellites, Incoherent Scatter Radars, and the International Reference Ionosphere Model. *Remote Sens.* **2021**, *13*, 4077. [CrossRef]
73. Lomidze, L.; Knudsen, D.J.; Burchill, J.; Kouznetsov, A.; Buchert, S.C. Calibration and validation of swarm plasma densities and electron temperatures using ground-based radars and satellite radio occultation measurements. *Radio Sci.* **2018**, *53*, 15–36. [CrossRef]
74. Maltseva, O.A.; Mozhaeva, N.S.; Nikitenko, T.V. Comparison of model and experimental ionospheric parameters at high latitudes. *Adv. Space Res.* **2013**, *51*, 599–609. [CrossRef]
75. Maltseva, O.A.; Mozhaeva, N.S.; Nikitenko, T.V. Validation of the Neustrelitz Global Model according to the low latitude ionosphere. *Adv. Space Res.* **2014**, *54*, 463–472. [CrossRef]
76. Maltseva, O.A.; Mozhaeva, N.S. The Use of the Total Electron Content Measured by Navigation Satellites to Estimate Ionospheric Conditions. *Int. J. Navig. Obs.* **2016**, *2016*, 7016208. [CrossRef]
77. Zhang, Y.; Wu, Z.; Feng, J.; Xu, T.; Deng, Z.; Ou, M.; Xiong, W.; Zhen, W. Statistical Study of Ionospheric Equivalent Slab Thickness at Guam Magnetic Equatorial Location. *Remote Sens.* **2021**, *13*, 5175. [CrossRef]
78. Moore, G.W. Magnetic Disturbances preceding the 1964 Alaska Earthquake. *Nature* **1964**, *203*, 508–509. [CrossRef]
79. Davies, K.; Baker, D.M. Ionospheric effects observed around the time of the Alaskan earthquake of March 28, 1964. *J. Geophys. Res.* **1965**, *70*, 2251–2253. [CrossRef]
80. De Santis, A.; Balasis, G.; Pavón-Carrasco, F.J.; Cianchini, G.; Mandea, M. Potential earthquake precursory pattern from space: The 2015 Nepal event as seen by magnetic Swarm satellites. *Earth Planet. Sci. Lett.* **2017**, *461*, 119–126. [CrossRef]
81. Marchetti, D.; De Santis, A.; D'Arcangelo, S.; Poggio, F.; Piscini, A.; Campuzano, S.A.; De Carvalho, W.V.J.O. Pre-earthquake chain processes detected from ground to satellite altitude in preparation of the 2016–2017 seismic sequence in Central Italy. *Remote Sens. Environ.* **2019**, *229*, 93–99. [CrossRef]
82. Zhu, K.; Fan, M.; He, X.; Marchetti, D.; Li, K.; Yu, Z.; Chi, C.; Sun, H.; Cheng, Y. Analysis of Swarm Satellite Magnetic Field Data Before the 2016 Ecuador (Mw = 7.8) Earthquake Based on Non-negative Matrix Factorization. *Front. Earth Sci.* **2021**, *9*, 1976. [CrossRef]
83. Xiong, P.; Marchetti, D.; De Santis, A.; Zhang, X.; Shen, X. SafeNet: SwArm for Earthquake Perturbations Identification Using Deep Learning Networks. *Remote Sens.* **2021**, *13*, 5033. [CrossRef]
84. Tsugawa, T.; Saito, A.; Otsuka, Y.; Nishioka, M.; Maruyama, T.; Kato, H.; Nagatsuma, T.; Murata, K.T. Ionospheric disturbances detected by GPS total electron content observation after the 2011 off the Pacific coast of Tohoku Earthquake. *Earth Planets Space* **2011**, *63*, 875–879. [CrossRef]
85. Pavlidou, E.; Van der Meijde, M.; Van der Werff, H.; Hecker, C. Time Series Analysis of Land Surface Temperatures in 20 Earthquake Cases Worldwide. *Remote Sens.* **2019**, *11*, 61. [CrossRef]
86. Kandalyan, R.A.; AlQuran, M.K. Ionosphere scintillation and earthquakes. *Jordan J. Phys.* **2010**, *3*, 69–76.
87. Molina, C.; Boudriki Semlali, B.-E.; Park, H.; Camps, A. A Preliminary Study on Ionospheric Scintillation Anomalies Detected Using GNSS-R Data from NASA CYGNSS Mission as Possible Earthquake Precursors. *Remote Sens.* **2022**, *14*, 2555. [CrossRef]
88. Camps, A.; Park, H.; Foti, G.; Gommenginger, C. Ionospheric Effects in GNSS-Reflectometry from Space. *IEEE J. Sel. Top. Appl. Earth Obs. Remote Sens.* **2016**, *9*, 5851–5861. [CrossRef]

remote sensing

MDPI

Technical Note

The Prototype of a Fast Vertical Ionosonde Based on Modern Software-Defined Radio Devices

Alexei V. Shindin [1,2,*], Sergey P. Moiseev [1,2], Fedor I. Vybornov [1,3], Kseniya K. Grechneva [2], Viktoriya A. Pavlova [1] and Vladimir R. Khashev [2]

[1] Radiophysics Research Institute, Lobachevsky State University of Nizhni Novgorod, Bolshaya Pecherskaya St. 25/12a, 603950 Nizhni Novgorod, Russia; moiseev@rf.unn.ru (S.P.M.); vybornov@nirfi.unn.ru (F.I.V.); pavlova@nirfi.unn.ru (V.A.P.)
[2] Department of Radio Propagation and Radio Astronomy, Radiophysics Faculty, Lobachevsky State University of Nizhni Novgorod, Gagarin Av. 23, 603950 Nizhni Novgorod, Russia; grekhneva@rf.unn.ru (K.K.G.); khashev@rf.unn.ru (V.R.K.)
[3] Department of Physics, Volga State University of Water Transport, Nizhni Novgorod, 5 Nesterova Str., 603950 Nizhni Novgorod, Russia
* Correspondence: shindin@rf.unn.ru

Abstract: The description and test results of the prototype of a fast ionosonde for the vertical sounding of the ionosphere, which makes it possible to record ionograms once a second, are presented. Such a high rate of registration of ionograms is required to study the fast processes of redistribution of electron concentration during heating experiments, for registration of fast quasiperiodic and moving ionospheric disturbances in the F, E, and Es layers. The key feature of the presented development is the usage of publicly available radio-electronic components. This provided a significant reduction in the cost of creating the prototype. In the current version, the prototype is based on the software-defined radio (SDR) devices Red Pitaya SDRlab 122-16 and LimeSDR. The test results showed that the quality of the ionograms recorded using the prototype is not worse than the quality of ionograms recorded using the professional CADI ionosonde. The low cost of the components allows providing multi-position registration of ionograms for determination the dynamics of natural and artificial ionospheric disturbances in 3D region of space at a lower expenses rate, as well as to create a network of ionospheric observation points with an increased number of ionosondes.

Keywords: ionosphere; vertical pulse sounding; ionosonde; ionogram; software-defined radio

Citation: Shindin, A.V.; Moiseev, S.P.; Vybornov, F.I.; Grechneva, K.K.; Pavlova, V.A.; Khashev, V.R. The Prototype of a Fast Vertical Ionosonde Based on Modern Software-Defined Radio Devices. *Remote Sens.* **2022**, *14*, 547. https://doi.org/10.3390/rs14030547

Academic Editor: Fabio Giannattasio

Received: 21 December 2021
Accepted: 22 January 2022
Published: 24 January 2022

1. Introduction

Over the past 20 years, software-defined radio (SDR) devices have expanded its applicability from professional to amateur radio (see, for example, [1]). This was facilitated by the following factors: cheaper hardware components and the development of field-programmable gate array (FPGA) technology. The latter factor allowed extreme simplification of the structure of SDR devices for the RF range, factually excluding application-specific integrated circuits (ASIC) chips from it. Moreover, a substantial role was played by the fact that development environments for FPGAs from major vendors became free for individual use. Modern ADC/DAC with a sampling rate of more than 80 MHz allow the entire RF range (3–30 MHz) to be transmitted to the FPGA for processing. All of the above makes such devices extremely convenient for monitoring the passage of HF radio waves through the ionosphere and, in particular, for the implementation of the chirp ionosonde [2,3]. To monitor the current ionospheric situation and structure of the ionosphere, the reconstruction of the electron density profile, the most common technique is the vertical sounding of the ionosphere with short coded pulses with a filling frequency varying within 1–20 MHz. This paper presents the description and test results of the vertical sounding ionosonde prototype based on currently available SDR devices. The developed prototype can be used by scientific groups to create their own devices for monitoring ionospheric conditions.

2. Materials and Methods

2.1. Hardware Part

The vertical sounding ionosonde is an HF radar station (for more details, see [4]). During operation, the ionosonde emits short radio pulses with a filling frequency ranging from 1 to 20 MHz. As a rule, various types of manipulations are applied to the emitted pulses (usually amplitude or phase). The pulses reflected from the ionosphere are recorded and processed using the receiving part. As a result of signal processing a height-frequency characteristic is obtained, so that it is possible to restore the electron concentration profile in the altitude range of 80–700 km.

While developing the ionosonde prototype for vertical sounding, we focused on the technical characteristics of the CADI ionosonde [5] at our disposal: the radiation power is 600 W, the type of pulse encoding is phase shift keying with a 13-bit Barker code with a duration of one bit of 40 µs, the pulse repetition rate is from 25 µs, the range of probing frequencies is from 1 to 20 MHz with a step of 50 kHz in standard mode. In [6], you can find examples of ionograms obtained using the CADI ionosonde. It should be noted that in order to increase the signal-to-noise ratio when registering a signal reflected from the ionosphere, our CADI ionosonde uses averaging over 4 pulses emitted with the same filling frequency. In general, this requires approximately 40 s to emit all the pulses, and then one minute to process the signal and record the ionogram data into the file. The height resolution of ionograms obtained using the CADI ionosonde is 3 km.

The block diagram of the developed prototype of the fast vertical-sounding ionosonde is shown in Figure 1. In the current version, the transmitting and receiving parts of the ionosonde are separate devices. The transmitting part includes (1) DIY HF linear amplifier on LDMOS transistors [7] with a maximum power of 600 W with an operating range of 1.8–72 MHz; (2) 5 W pre-amplifier with an operating range of 100 kHz–40 MHz; (3) programmable attenuator with an operating range of up to 6 GHz and a maximum attenuation of 30 dB; (4) SDR device Red Pitaya SDRlab 122-16 [8]. Figure 2 shows a photo of the transmitting part of the ionosonde mockup without housing on a laboratory bench.

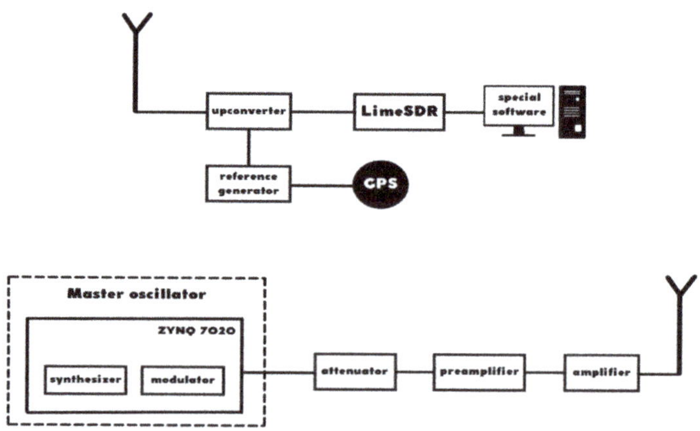

Figure 1. Block diagram of the transmitting (**bottom**) and receiving (**top**) parts of the ionosonde.

The Red Pitaya SDRlab 122-16 is the development board based on a Xilinx Zynq 7020 system that combines FPGAs and a general-purpose dual-core ARM processor. The board is equipped with two-channel 14-bit DAC and 16-bit ADC with a sampling rate of 122.88 MS/s, as well as Ethernet and USB2.0 interfaces for data transfer, communication with other devices (PC), and for connecting additional devices to the board (e.g., WI-FI dongle). This board is used as a master oscillator (probe pulse generator). We plan to use this device as a basis for the receiving part of the ionosonde in the future.

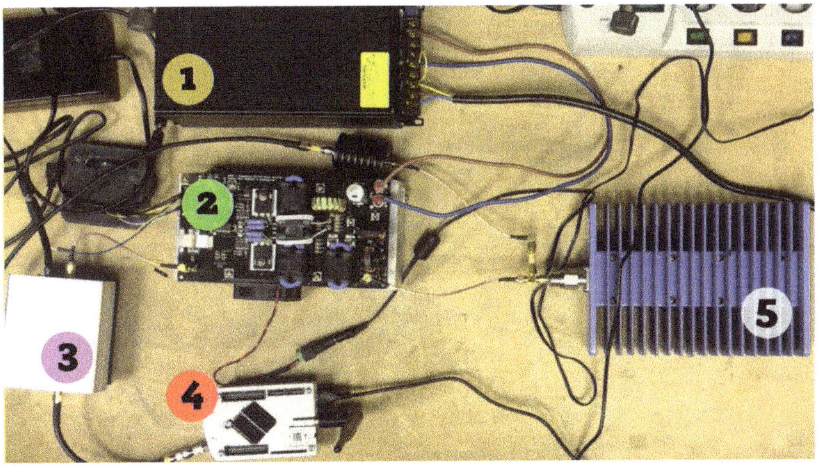

Figure 2. The photo of the assembled model of the transmitting part of the ionosonde without a housing. The numbers indicate 1—amplifier power supply, 2—linear amplifier A600, 3—5 W preamplifier, 4—SDRlab 122-16 in a 3d-printed case as a master oscillator, 5—matched load.

The receiving part of the developed layout (see Figure 3) is based on a two-channel SDR device LimeSDR [9] with a declared operating frequency range from 100 kHz to 3.8 GHz, which is capable of recording a signal in the 61.44 MHz bands. To be able to register the HF signal in the 10 MHz bands modified upconverters were used filtering the HF signal and moving it to the frequency range 120–130 MHz, where the LimeSDR can work more efficiently. Since the operation of the ionosonde assumes precise frequency matching between the transmitting and receiving parts, we used the Leo Bodnar precision GPS reference clock to generate the reference signal for the LimeSDR and upconverters. Two channels of the device allow, in presence of appropriate antennas and a polarizer, to register two polarizations of the reflected signal (O and X modes). A low noise amplifier (+20 dB) was used to amplify the received signal. To record a signal in the 10 MHz bands, the LimeSDR was connected to a PC or laptop via the USB3.0 interface.

Figure 3. Photo of the assembled model of the ionosonde receiving part without a control PC. The numbers indicate: 1—USB hub, 2—precision GPS reference clock, 3—upconverter, 4—LimeSDR.

2.2. Software Part

The SDR concept assumes that the functionality and even the purpose of the device is determined by software components that can be easily changed or upgraded. In our case, the function and operation speed of the vertical sounding ionosonde is provided by the development of firmware for the SDRlab 122-16 FPGA board and software for signal processing and obtaining ionograms.

To achieve the ionogram recording time of the order of 1 s, we: (1) used a pulse repetition period of 5 ms instead of 25 ms (as in the CADI ionosonde); (2) abandoned the averaging of signals over four pulses at one frequency; (3) reduced the frequency range of sounding to 10 MHz; (4) developed software that allows recording ionograms in real-time, i.e., with a delay less than the time required for the emission of all probing pulses. All these measures ensured a sounding time of 0.9 s.

The functional block diagram of the master oscillator firmware on the SDRlab 122-16 board is shown in Figure 4. The master oscillator firmware was implemented by means of the Xilinx Vivado development environment in the Verilog hardware description language using embedded Xilinx IP cores. The project consists of separate modules that have a functional connection with one another: the input clock frequency of 122.88 MHz, coming from the crystal oscillator of the Red Pitaya board, is fed to the input of the clk_wizard module, where it is converted to the frequency of 100 MHz. The clock pulse is fed to the input of the clk_divider module and is divided into two clock frequencies: 25 kHz is the symbol frequency (1 symbol corresponds to 1 bit of a 13-bit Barker code with a duration of 40 μs) and 200 Hz is the frequency of the probing pulses (200 Hz corresponds 5 ms sounding pulse period). These two frequencies provide AM and FM control. In the FM_Modulator module, the phase increment of the harmonic signal changes every 5 ms with a frequency step of 50 kHz (to ensure the ionogram recording time—0.9 s), after which the increment is fed to the input of the DDS digital computational synthesizer module, where the cosine values are generated. The received harmonic signal enters the amplitude modulation module and the already modulated pulse is fed to the input of the digital-to-analog converter control module. Phase shift keying was implemented with a slight upgrade of the AMK firmware, taking advantage of the fact that changing the phase by π is equivalent to multiplying the signal by -1. Note that the selected firmware parameters implement the radiation scheme of the CADI ionosonde, which we were guided by.

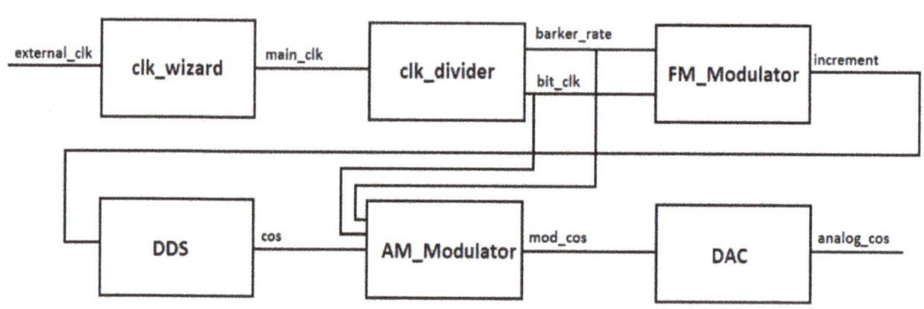

Figure 4. Functional block diagram of the master oscillator firmware on the SDRlab 122-16 board.

The software for recording ionograms was created on the basis of the GNU radio framework [10]. In addition, a trigger module was developed to compensate for lacking synchronization between the transmitting and receiving parts of the model at this stage, as well as possible missing samples. The used flow graph of the GNU Radio Companion tool is shown in Figure 5. An additional program written in python using the libraries numpy [11], scipy [12], and matplotlib [13] was used for autocorrelation analysis and ionograms. The developed software makes it possible to register ionograms with a height resolution of 1.5 km. The computing power of a laptop equipped with a quad-core central processor is sufficient to obtain an ionogram in less than 1 s, which in fact provides real-time monitoring of the ionosphere. A separate application (for example, OBS Studio [14]) can be used to quickly publish ionograms on the Internet video services such as YouTube.

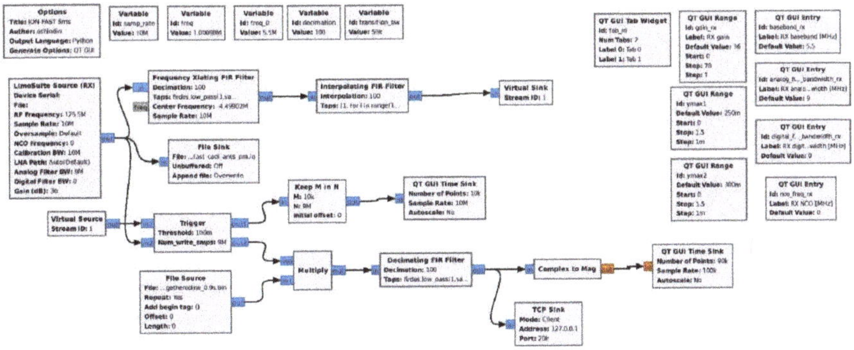

Figure 5. GNU Radio Companion flowgraph used in ionosonde's receiving part.

3. Test Results

Several series of tests were carried out for the developed prototype of the fast ionosonde. The first series of experiments included continuous operation of the ionosonde for 24 h at the maximum ionogram recording rate. Tests have shown that the thermal regime and performance of the components are not disturbed under prolonged loads. We have to mention that the computing capabilities of the hardware used are sufficient to obtain ionograms and their automatic publication on the Internet. The second series of experiments consisted in using several types of receiving antennas as part of the ionosonde. Among them: (1) a large diagnostic transmitting/receiving antenna (in-phase horizontal antenna array 126 × 126 m in size, suspended on 12 masts 16 m high; each of the two linear polarizations has 12 emitters; each emitter consists of three dipoles of different length connected in parallel, due to which the antenna has three resonant frequencies of 2.95, 4.6 and 5.7 MHz), that used to register artificial radio emission of the ionosphere, multifrequency Doppler sounding, diagnostics of the lower layers of the ionosphere and mesosphere, etc. (see Figure 6); standard receiving (two broadband crossed dipoles on four 12 m masts with an operating range of 2–10 MHz) antenna of the CADI ionosonde available at the Vasilsursk experimental base (see Figure 8); the receiving-transmitting Delta type antenna of the chirp ionosonde on a 15 m mast with the northern directional pattern and with an effective operating range of 4–15 MHz. Note that in all the tests, the standard transmitting antenna of the CADI ionosonde was used as the transmitting antenna (vertical delta antenna on a 40 m mast with an operating range of 2–30 MHz). All transceiver equipment was located at a distance of no more than 1 km from each other. It is planned to use the T2FD antenna for the mobile version of the ionosonde. The developed ionosonde prototype can be used together with any broadband HF antenna operating in the range of 1.8–10 MHz and vertical antenna pattern.

The receiving part of the developed ionosonde prototype in various test sessions included either an Acer Predator Helios 500 laptop with the Manjaro Linux 21 operating system installed or an ECS Liva SF110-A320 platform with an AMD Ryzen 5 PRO 2400GE quad-core processor.

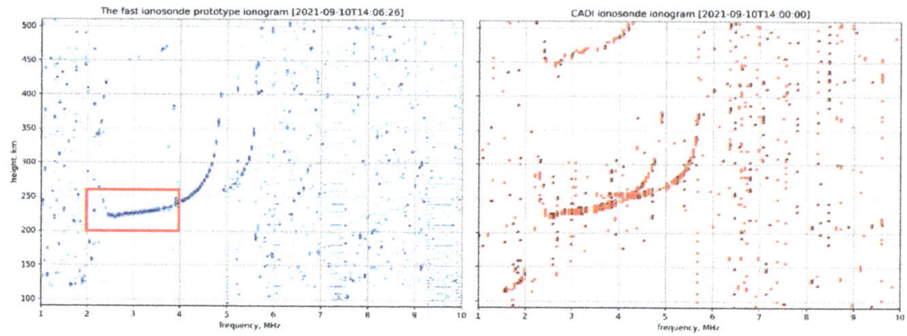

Figure 6. (**Left panel**): an example of an ionogram recorded with the developed ionosonde prototype using a diagnostic antenna (O mode output). Red square corresponds to the ionograms's subarray which presented in detail in Figure 7. (**Right panel**): an example of an ionogram registered with a CADI ionosonde. Ionogram's registration times are shown in UTC at the top of the panels.

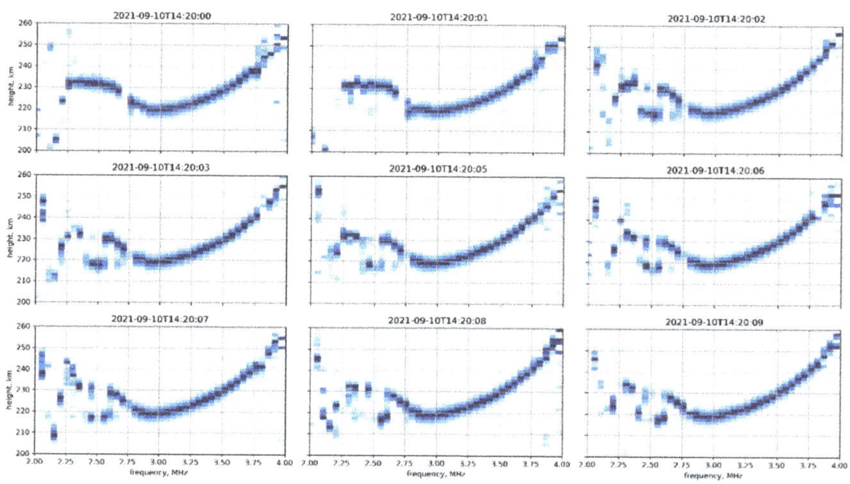

Figure 7. An example of 9 consecutive ionograms recorded at a rate of 1 ionogram per second, illustrating the effect of a rapid decrease in the effective reflection height in the F region of the ionosphere.

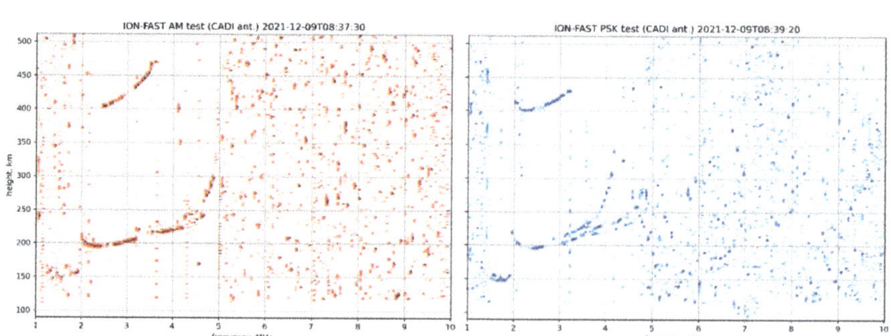

Figure 8. Examples of ionograms recorded using the developed prototype ionosonde using CADI ionosonde antennas. (**Left panel**): Barker code amplitude manipulation. (**Right panel**): Barker code phase shift keying.

Comparative analysis of the obtained ionograms showed that the diagnostic antenna, due to its good parameters, provides ionograms with the highest signal-to-noise ratio. Figure 7 shows 9 consecutive ionograms registered by the developed ionosonde prototype using a diagnostic antenna with a time resolution of 1 s. The figure shows the process of an abrupt change in the reflection height by a value of about 15 km in the ionosphere F layer, which could not be detected at the standard ionogram registration rate. Video files demonstrating the operation of the developed ionosonde prototype can be found in the Supplementary Materials. The standard receiving antenna of the CADI ionosonde required the use of a low-noise amplifier (we used 20 dB 1–30 MHz amplifier) to obtain ionograms of comparable quality. When using an inclined antenna of the chirp ionosonde, it was impossible to obtain ionograms with distinguishable traces of ionospheric layers at this stage of testing. In this case, oblique-sounding ionograms for this antenna are successfully recorded. In the third series of tests, two modes of operation of the ionosonde were compared: with amplitude and phase manipulations of the probe pulses. In both cases, the encoding was carried out with a 13-bit Barker code. Tests have shown that phase shift keying ionograms have a significantly higher signal-to-noise ratio. Apparently, this is due to the higher average sounding power in this mode.

4. Discussion

As can be seen from Figures 6–8, the developed prototype of the vertical ionosonde is capable of registering ionograms comparable in quality to ionograms obtained using the CADI ionosonde. At the same time, the developed prototype uses practically the same temporal radiation pattern and the same transmitter power. The difference in the representation of ionograms by the two instruments is due to many factors. Among the main ones are different characteristics of the receiving parts (the CADI ionosonde has an 8-bit ADC in the receiver) and different approaches to obtaining ionograms (the CADI ionosonde uses fast Fourier transform). The ionograms obtained by the developed ionosonde are quite suitable for further analysis (scaling) in order to reconstruct the electron density profile. The high rate of ionogram registration makes it possible to determine the parameters of fast movements in the ionosphere.

The fast ionosonde can be used to study fast natural variations of electron density profiles in the F-layer of the ionosphere. For example, the use of the fast ionosonde with an ionogram recording time of 2 s allowed detecting local disturbances moving vertically at speeds up to 50 m/s [15]. The horizontal stratification found in this case is capable of performing cyclic vertical movements with an amplitude of up to 5 km and a period of about 90 s.

It is known that traveling ionospheric disturbances (TIDs) with a characteristic spatial scale of up to 100 km are often observed at mid-latitudes in the daytime. Ionosondes operating according to the standard 15-min ionosphere sounding program record them on single ionograms (see, for example, [16]). The rare territorial location of ionosondes does not allow to determine TIDs parameters unambiguously. The relevance of the problem is explained by the difficulty of predicting TIDs and the strong influence on the HF communication channels. The use of linear frequency modulation (LFM) ionosondes and a one-minute sounding cycle made it possible to determine the spatial and dynamic characteristics of the TID from several (usually 5–15) ionograms [17]. As a rule, in this case, a weekly inclined mode of sounding the ionosphere is used by a system of synchronously operating chirp stations [18,19]. Reducing the sounding time to a few seconds would enable tracing the dynamics of the TID propagation processes in detail, especially if several closely located (at a distance of 50–100 km) automatic synchronously operating ionospheric stations were used. The use of several low-cost fast vertical-sounding ionosondes makes it possible to create such promising automatic systems for recording the TID parameters.

The fast vertical-sounding ionosonde can be useful in the development of studies of the processes occurring in the E and Es layers of the ionosphere. It is known that the processes occurring in these layers are characterized by fast dynamics [20].

Experiments on modifying the Earth's ionosphere with powerful short-wavelength radiation [21] have shown the need to develop and use fast vertical-sounding ionosonde. Firstly, the ionosonde in such experiments is used for diagnostic purposes (determination of the height of reflection of a powerful wave) and control of the operation of the heating facility (selection of the frequency of powerful transmitters). As a source of impulse noise, the ionosonde should work periodically but for a short time without interfering with the operation of the diagnostic equipment. Secondly, a high-speed ionosonde is required for research purposes. The interaction of high-power HF radiation of ordinary polarization is accompanied by the excitation of artificial ionospheric turbulence (AIT). There are several stages of its development with characteristic times (the development of striction parametric instability—5–20 ms, the stage of restoration of the level of the reflected signal of the pump wave—0.5–3 s, anomalous attenuation—0.5–10 s, the development of self-focusing non-stability—10–30 s) [21]. Typically, AIT surveys are conducted at fixed frequencies using a probe wave transmitter. The use of a new fast vertical-sounding ionosonde can be useful in investigating the properties of AIT.

5. Conclusions

In this paper, we showed that it is possible now to assemble the prototype of vertical sounding ionosonde using publicly available radio-electronic components with the total cost of approximately 1600 EUR (the prices are of December 2021), which has an ionogram recording rate of 1 ionogram per second.

This cost does not contain the cost of a PC for recording ionograms, as well as transmitting and receiving antennas. Moreover, the transmitting part of the prototype costs 1050 EUR. We plan to implement the receiving part on the SDRlab 122-16 board in the future. It will reduce the total cost by 550 EUR. Unfortunately, the CADI ionosonde is currently not available for order. However, at the time of purchase to equip the Vasilsursk base in 2007, it cost about 50,000 USD.

As we can see from Figure 6 (as well as from the demonstrations in the paper Supplementary File and data set), the developed ionosonde prototype allows obtaining ionograms at an unprecedented speed, comparable in quality to CADI ionosonde ionograms, at a significantly lower cost. The authors are not aware of any cases of demonstration of continuous long-term operation of other ionosondes at a similar speed.

Supplementary Materials: The following supporting information can be downloaded at: https://www.mdpi.com/article/10.3390/rs14030547/s1, Video S1: AM-Large_antenna, Video S2: AM-CADI_antenna, Video S3: PSK_CADI_antenna.

Author Contributions: Conceptualization, A.V.S. and S.P.M.; methodology, A.V.S., S.P.M.; software, A.V.S., S.P.M., K.K.G., V.A.P. and V.R.K.; set up and conducted the experiments, A.V.S., S.P.M., K.K.G. and V.A.P., data processing, A.V.S. and V.A.P.; theoretical analysis, F.I.V.; writing—original draft preparation, A.V.S.; writing—review and editing, S.P.M. and F.I.V. All authors have read and agreed to the published version of the manuscript.

Funding: The work is supported by a Russian Science Foundation grants #21-72-10131 (A.V.S., S.P.M. and V.A.P., Sections 1, 2 and 5), #20-12-00197 (F.I.V., K.K.G. and V.R.K., Sections 3 and 4).

Institutional Review Board Statement: Not applicable.

Informed Consent Statement: Not applicable.

Data Availability Statement: SDRlab 122-16 firmwares for AM and PSK fast ionosonde modes, code for receiving part of the fast ionosonde including the GNU RADIO Out-of-Tree module source code, fast ionogram receiving video demonstrations, fast ionogram examples in npz data and image file format can be found in Shindin, Alexey. (2021). The Prototype of a Fast Vertical Ionosonde Based on Modern SDR Devices-paper dataset [Data set]. Zenodo. https://doi.org/10.5281/zenodo.5795786, accessed on 17 December 2021.

Acknowledgments: Fedor I. Vybornov is grateful to the project No. 0729-2020-0057 within the framework of the basic part of the State assignment of the Ministry of Science and Higher Education of the Russian Federation for the technical feasibility of using CADI stations in Vasilsursk.

Conflicts of Interest: The authors declare no conflict of interest.

References

1. Yamamoto, M. Digital beacon receiver for ionospheric TEC measurement developed with GNU Radio. *Earth Planet Space* **2008**, *60*, e21–e24. [CrossRef]
2. Vierinen, J. GNU Chirp Sounder. Available online: https://www.sgo.fi/~{}j/gnu_chirp_sounder/ (accessed on 21 December 2021).
3. Ivanov, V.A.; Kurkin, V.I.; Nosov, V.E.; Uryadov, V.P.; Shumaev, V.V. Chirp ionosonde and its application in the ionospheric research. *Radiophys. Quantum Electron.* **2003**, *46*, 821–851. [CrossRef]
4. Hunsucker, R.D. *Radio Techniques for Probing the Terrestrial Ionosphere*; Book Series Physics and Chemistry in Space 22 Planetology; Springer: Berlin/Heidelberg, Germany, 1991; pp. 65–93.
5. Canadian High Arctic Ionospheric Network Cadi. Available online: http://chain.physics.unb.ca/chain/pages/cadi/ (accessed on 21 December 2021).
6. CADI at Vasilsursk Site. Available online: https://ip-2-165.unn.ru:3389/cadi/ (accessed on 21 December 2021).
7. A600—HF_6m 600W LDMOS Amplifier QRPblog. Available online: https://qrpblog.com/a600-hf-6m-600w-ldmos-amplifier/ (accessed on 21 December 2021).
8. SDRlab 122-16—Red Pitaya 0.97 Documentation. Available online: https://redpitaya.readthedocs.io/en/latest/developerGuide/hardware/122-16/top.html (accessed on 21 December 2021).
9. LimeSDR—Lime Microsystems. Available online: https://limemicro.com/products/boards/limesdr/ (accessed on 21 December 2021).
10. GNU Radio—The Free & Open Source Radio Ecosystem · GNU Radio. Available online: https://www.gnuradio.org/ (accessed on 21 December 2021).
11. NumPy. Available online: https://numpy.org/ (accessed on 21 December 2021).
12. SciPy. Available online: https://scipy.org/ (accessed on 21 December 2021).
13. Matplotlib—Visualization with Python. Available online: https://matplotlib.org/ (accessed on 21 December 2021).
14. Open Broadcaster Software OBS. Available online: https://obsproject.com (accessed on 21 December 2021).
15. Branitskii, A.V.; Kim, V.Y.; Polimatidi, V.P. Rapid Variations in Electron Density Profiles in the Ionosphere Detected with the Izmiran High-Speed Ionosonde. *Geomagn. Aeron.* **2020**, *60*, 63–79. [CrossRef]
16. Vybornov, F.I.; Mityakova, E.E.; Rakhlin, A.V.; Krupenya, N.D. Analysis of appearance of moving ionospheric disturbances of the "sickle" type at middle latitudes. *Radiophys. Quantum Electron.* **1997**, *40*, 980–986. [CrossRef]
17. Vertogradov, G.G.; Uryadov, V.P.; Vybornov, F.I.; Pershin, A.V. Modeling of Decameter Radio Wave Propagation Under Conditions of a WaveLike Electron-Density Disturbance. *Radiophys. Quantum Electron.* **2018**, *61*, 407–417. [CrossRef]
18. Harris, T.J.; Cervera, M.A.; Meehan, D.H. SpICE: A program to study small-scale disturbances in the ionosphere. *J. Geophys. Res.* **2012**, *117*, A06321. [CrossRef]
19. Kurkin, V.I.; Laryunin, O.A.; Podlesny, A.V.; Pezhemskaya, M.D.; Chistyakova, L.V. Studying morphological characteristics of traveling ionospheric disturbances with the use of near-vertical ionospheric sounding data. *Atmos. Ocean. Opt.* **2014**, *27*, 303–309. [CrossRef]
20. Bakhmet'eva, N.V.; Belikovich, V.V.; Ignat'ev, Y.A.; Ponyatov, A.A. Vertical motions in the lower ionosphere and a sporadic E layer. *Radiophys. Quantum Electron.* **1999**, *42*, 22–30. [CrossRef]
21. Frolov, V.L.; Bakhmet'eva, N.V.; Belikovich, V.V.; Vertogradov, G.G.; Vertogradov, V.G.; Komrakov, G.P.; Kotik, D.S.; Mityakov, N.A.; Polyakov, S.V.; Rapoport, V.O.; et al. Modification of the earth's ionosphere by high-power high-frequency radio waves. *Phys.-Uspekhi* **2007**, *50*, 315–324. [CrossRef]

remote sensing

Technical Note

A Method for Automatic Inversion of Oblique Ionograms

Chunhua Jiang [1,*], Cong Zhao [1], Xuhui Zhang [1], Tongxin Liu [1], Ziwei Chen [2], Guobin Yang [1] and Zhengyu Zhao [1,3]

1 Department of Space Physics, School of Electronic Information, Wuhan University, Wuhan 430072, China; zhaocong@whu.edu.cn (C.Z.); zhangxuhui@whu.edu.cn (X.Z.); tongxin_liu@whu.edu.cn (T.L.); gbyang@whu.edu.cn (G.Y.); zhaozy@whu.edu.cn (Z.Z.)
2 School of Electronic and Information Engineering, Beijing Jiaotong University, Beijing 100044, China; zwchen@bjtu.edu.cn
3 Institute of Space Science and Applied Technology, Harbin Institute of Technology, Shenzhen 518055, China
* Correspondence: chuajiang@whu.edu.cn; Tel.: +86-139-8619-0965

Abstract: In this study, a method is proposed to carry out automatic inversion of oblique ionograms to extract the parameters and electron density profile of the ionosphere. The proposed method adopts the quasi-parabolic segments (QPS) model to represent the ionosphere. Firstly, numerous candidate electron density profiles and corresponding vertical traces were, respectively, calculated and synthesized by adjusting the parameters of the QPS model. Then, the candidate vertical traces were transformed to oblique traces by the secant theorem and Martyn's equivalent path theorem. On the other hand, image processing technology and characteristics of oblique echoes were adopted to automatically scale the key parameters (the maximum observable frequency and minimum group path, etc.) from oblique ionograms. The synthesized oblique traces, whose parameters were close to autoscaled parameters, were selected as the candidate traces to produce a correlation with measured oblique ionograms. Lastly, the proposed algorithm searched the best-fit synthesized oblique trace by comparing the synthesized traces with oblique ionograms. To test its feasibility, oblique ionograms were automatically scaled by the proposed method and these autoscaled parameters were compared with manual scaling results. The preliminary results show that the accuracy of autoscaled maximum observable frequency and minimum group path of the ordinary trace of the F2 layer is, respectively, about 91.98% and 86.41%, which might be accurate enough for space weather specifications. It inspires us to improve the proposed method in future studies.

Keywords: oblique ionogram; automatic inversion; electron density profile; quasi-parabolic segments

Citation: Jiang, C.; Zhao, C.; Zhang, X.; Liu, T.; Chen, Z.; Yang, G.; Zhao, Z. A Method for Automatic Inversion of Oblique Ionograms. *Remote Sens.* **2022**, *14*, 1671. https://doi.org/10.3390/rs14071671

Academic Editor: Fabio Giannattasio

Received: 8 March 2022
Accepted: 29 March 2022
Published: 30 March 2022

1. Introduction

There is a long history of remotely sensing the ionosphere through radio waves as a vertical sounding mode. In this sounding mode, the transmitter and receiver are collocated at the same station. The ionosonde, as the vertical sounding mode, is a widely used tool for monitoring the ionosphere and plays a significant role for studying ionosphere characteristics in the near real-time method. With the development of the modern advanced ionospheric sounders, many notable ionosondes, such as DPS-4D (Digisonde Portable Sounder) [1], Dynasonde [2], CADI (Canadian Advanced Digital Ionosonde) [3], AIS-INGV (Advanced Ionospheric Sounder-Istituto Nazionale di Geofi sica e Vulcanologia) [4], WISS (Wuhan Ionospheric Sounding System) [5], etc., have been developed to carry out the vertical sounding of the ionosphere. Subsequently, many well-established software tools, including ARTIST (Automatic Real-Time Ionogram Scaling True-height) [6], NeXtYZ (pronounced "next wise") [7], UDIDA (Univap Digital Ionosonde Data Analysis) [8], Autoscala [9], and ionoScaler [10], have been equipped with ionosondes to automatically extract parameters and electron density profiles from vertical ionograms.

Unlike the vertical sounding mode, for the oblique sounding mode, the transmitter and the receiver are located at different stations, and can be hundreds or thousands of kilometers apart. It is possible to implement an oblique sounding function with the improvement and additional development of ionosondes. Oblique ionograms recorded by the oblique sounding receiver can represent characteristics of the ionosphere at the reflection point, which mostly is located at the middle point between the transmitter and receiver. We know that the idea of obtaining electron density profiles from oblique ionograms is not new [11–14]; however, how to automatically scale of oblique ionograms is still a challenging task compared with automatic scaling of vertical ionograms for space weather specifications. Similar to autoscaled techniques of vertical ionograms, algorithms are required to be developed to automatically extract parameters and electron density profiles from oblique ionograms. Redding [15] adopted image processing algorithms to extract the traces from oblique ionograms. Fan et al. [16] utilized image processing algorithms and characteristics of oblique ionograms to obtain parameters of the ionosphere. Settimi et al. [17] calculated a 3D ray-tracing algorithm to synthesize oblique ionograms and compared them with measured oblique ionograms to obtain parameters. Ippolito et al. [18] applied the same technique as the Autoscala programs to oblique ionograms for determination of the maximum usable frequency between the transmitter and receiver. Heitmann et al. [19] propose a robust feature extraction and parameterized fitting algorithm for automatic scaling of oblique and vertical ionograms by analytic ray-tracing.

In our previous work, Song et al. [20] proposed a method for obtaining the trace and parameters of the F2 layer from oblique ionograms using the quasi-parabolic model. However, the method proposed by Song et al. [20] is not accurate enough for inversion of oblique ionograms due to only the F2 layer being considered in the model. In this study, the proposed method aims to improve and extend the algorithm developed by Song et al. [20], and further implement automatic inversion of oblique ionograms. The quasi-parabolic segments (QPS) model (nine parameters) was used to represent the ionosphere in this study. Therefore, the proposed method can obtain the parameters of the oblique propagation in the E, F1 and F2 layers and the corresponding electron density profile of the ionosphere; thus, it can improve the accuracy of inversion of oblique ionograms.

2. Methods

The quasi-parabolic segments model [21,22], widely used for inversion of vertical ionograms, was adopted to represent the ionosphere in this study. Then, vertical ionograms could be synthesized by the integral of group refraction index along the propagation path in the ionosphere. Furthermore, synthesized vertical ionograms could be transformed to oblique ionogram by the secant theorem and Martyn's equivalent path theorem [23,24]. The synthesized oblique ionograms were further fitted to the measured oblique ionogram. Similar to automatic inversion of vertical ionograms by Jiang et al. [25,26], the initial nine parameters of QPS were determined from the IRI [27] and NeQuick model [28], and then adjusted to obtain a large amount of candidate oblique traces. Last, the best-fitted trace and parameters could be selected as the best-fitted output of automatic inversion of the oblique ionogram. Figure 1 shows the flowchart of the automatic inversion of oblique ionograms.

2.1. Automatic Scaling Key Parameters from Measured Oblique Ionograms

In the practical application, autoscaled parameters from measured oblique ionograms could be used to reduce the size of candidate traces. Therefore, the proposed method first automatically scaled the key parameters (the maximum observable frequency and minimum group path) from measured oblique ionograms. Similar to automatic scaling of vertical ionograms [10,25,26], the searching window and image projection techniques proposed by Jiang et al. [25,26] could also be utilized to scale oblique ionograms. The image projection technique is similar to the histogram technique by Lynn [29]. At first, the same technique as the ionoScaler software [10] was utilized to automatically extract the maximum observable frequency (MOF) and minimum group path of the E/Es (Sporadic E layer), F1, and F2 layer from oblique ionograms. Figure 2 shows a typical oblique ionogram recorded at 13:07 LT on 3 April 2013 between Beijing (40.3° N, 116.25° E) and Wuhan (30.5° N,

114.37° E). The black lines (on the horizontal and vertical axes) in Figure 2, respectively, represent the projection values at the Frequency and group path. As shown in Figure 2, the measured oblique ionogram could be divided into three regions (please see the back line on the vertical axis in Figure 2) according to the characteristics of the image projection at the group path. The maximum observable frequency could be identified by the projection values on frequency (please see the back line on the horizontal axis in Figure 2). Therefore, the searching window could be used to scale the maximum observable frequency and minimum group path for the E/Es, F1 and F2 layer. Therefore, we first defined the size of the searching window on measured oblique ionograms. Generally, the transmitter and receiver stations are known in the oblique sounding mode of the ionosphere. Then, the ground distance between the transmitter and receiver stations was adopted to determine the searching window in this study.

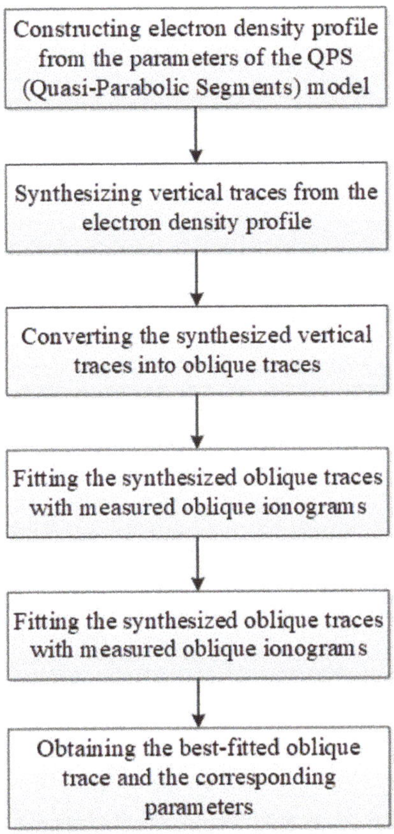

Figure 1. Flowchart of the automatic inversion of oblique ionograms.

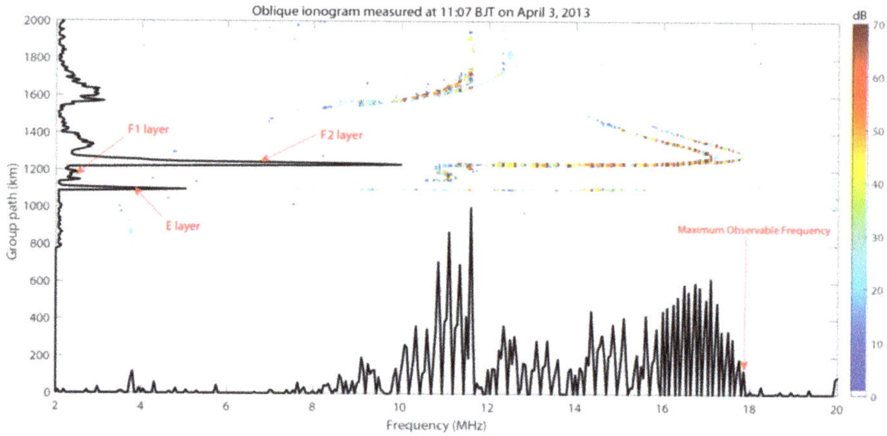

Figure 2. A typical oblique ionogram with projections at the group path and frequency recorded at 13:07 LT on 3 April 2013, between Beijing and Wuhan. The back lines on the vertical and horizontal axes, respectively, represent the projection values at the group path and frequency.

The searching window as $W[M, N]$, proposed by Jiang et al. [10,25,26], was represented by Equation (1):

$$\begin{cases} M = \text{int}[(\Delta f_w)/\Delta f] + 1 \\ N = \text{int}[(P_{\max} - P_{\min})/\Delta P] + 1 \end{cases} \tag{1}$$

where Δf_w is the horizontal size of the searching window, Δf is the resolution of the frequency in oblique ionograms, $P_{\max}$ is the maximum height of the searching window, $P_{\min}$ is the minimum height of the searching window, and ΔP is the resolution of the group path in oblique ionograms.

The present method defined Δf_w as the width of the working frequency (2–15 MHz). The values of $P_{\max}$ and $P_{\min}$ varies depending on the E and F layers.

For the E/Es and F2 layer, the size of the searching window was represented by Equation (2).

$$\begin{cases} P_{\max}E = 2 \cdot \sqrt{(D/2)^2 + (h_m E)^2} - \delta P_E \\ P_{\min}E = 2 \cdot \sqrt{(D/2)^2 + (h_m E)^2} + \delta P_E \end{cases} ,\textit{for E layer}$$
$$\begin{cases} P_{\max}F = P_{\max}E + \delta P_F \\ P_{\min}F = P_{\max}E \end{cases} ,\textit{for F2 layer} \tag{2}$$

where D is the ground distance between the transmitter and receiver stations, $h_m E$ is the peak altitude of the E layer from the IRI model, and δP_E and δP_F, respectively, are the deviation values of the group path of the E layer and F layer in oblique ionograms.

Once the searching window was defined, the image projection values of the searching window [26] were used to calculate the frequency and group path parameters of the E/Es and F layer from measured oblique ionograms. The detail procedure is similar to the methods proposed by Jiang et al. [26] and Song et al. [20]. In this study, we mainly introduce additional routines for identification of the Es and E layers in the present method.

For the Es layer, the proposed method first scaled the maximum observable frequency of the E/Es layer (fMOF_E_Es) from measured ionograms. Then, the fMOF_Emodel was calculated by the secant theorem and Martyn's equivalent path theorem, where the parameters of E layer were estimated from the IRI model. The scaled parameter fMOF_E_Es was further compared with the fMOF_Emodel. If the fMOF_E_Es was larger than the fMOF_Emodel, we suggest that the Es layer existed in measured oblique ionograms. Otherwise, the proposed method suggested that no Es layer occurred in oblique ionograms. Since the altitude of the Es layer was close to the E layer, if the Es layer existed, the group path of the Es layer was suggested to be equal to the E layer in this study. Figure 3 shows the flowchart for estimating the parameters of the E and Es layer from measured oblique ionograms.

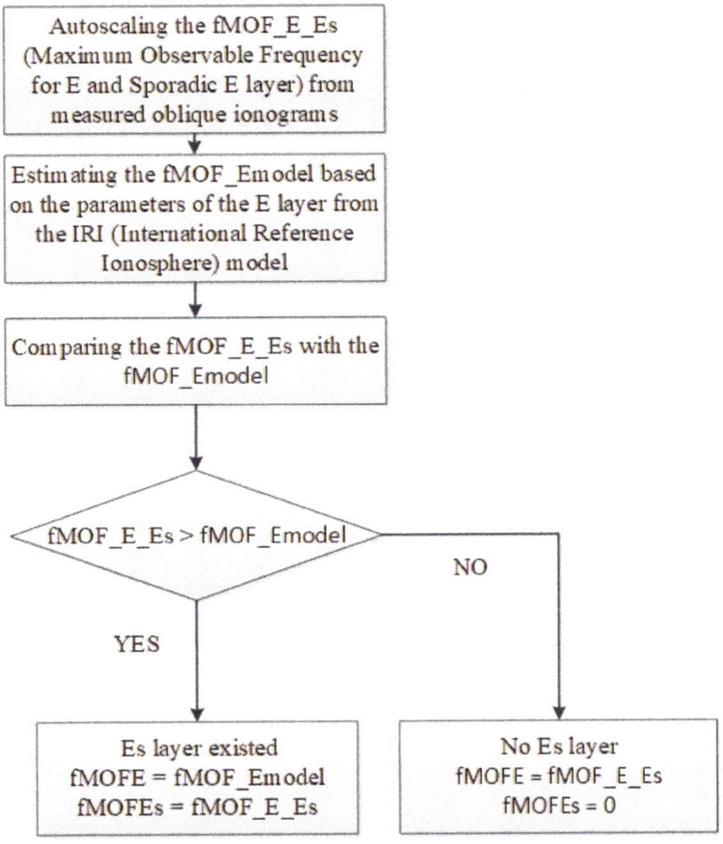

Figure 3. Flowchart of estimating the parameters of E and Es layer from measured oblique ionograms.

In measured oblique ionograms, the echoes of the F1 layer usually do not develop well compared with the E and F2 layers. Thus, the parameters of the F1 layer are estimated from the IRI and NeQuick models, but not from measured ionograms in this study. As a result, the frequency and group path parameters of E, Es, F1 and F2 layers could be estimated from measured oblique ionograms.

2.2. Synthesizing Oblique Ionogram through the QPS Model

The secant theorem and Martyn's equivalent theorem could be used to study the relationship between the vertical ionogram and oblique ionograms. Many studies [20,30] used these theorems to convert oblique ionograms into vertical ionograms. On the contrary, vertical ionograms were required to be converted into oblique ionograms by the secant theorem and Martyn's equivalent theorem in this study. Equation (3) was adopted to convert vertical traces into oblique traces.

$$\begin{cases} f_{ob} = f_{vi} \cdot \sqrt{(D/2)^2 + (h')^2}/h' \\ P_{ob} = 2 \cdot \sqrt{(D/2)^2 + (h')^2} \end{cases} \tag{3}$$

where f_{ob} and P_{ob}, respectively, represent the frequency and group path in synthesized oblique traces, D is the ground distance between the transmitter and receiver stations, h' is the virtual height of vertical traces, and f_{vi} is the frequency of vertical traces.

Figure 4 shows a typical synthesized vertical trace (left) and the converted oblique trace (right). The red line in the left panel of Figure 4 is the electron density profile represented by the QPS model with the parameters by Equation (4). The ground distance between transmitter and receiver stations was set to 1000 km.

$$\begin{cases} foE = 3.5(MHz), y_mE = 20(km), h_mE = 110(km) \\ foF1 = 5.0(MHz), y_mF1 = 80(km), h_mF1 = 180(km) \\ foF2 = 12.0(MHz), y_mF2 = 100(km), h_mF2 = 300(km) \end{cases} \tag{4}$$

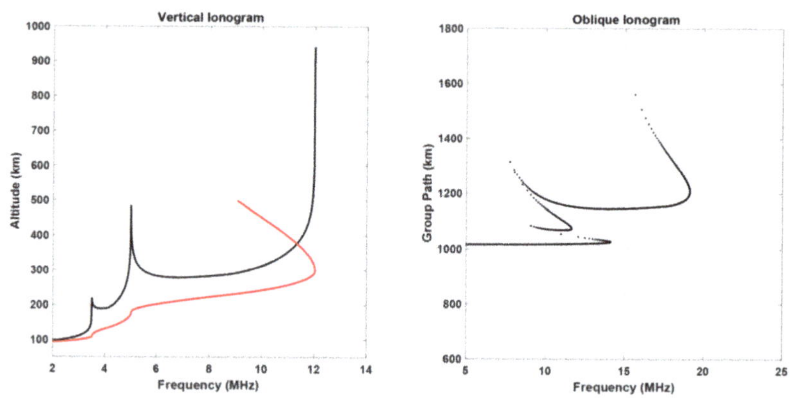

Figure 4. A typical synthesized vertical trace (**left**) and the converted oblique trace (**right**); the red line in the left pane represents the corresponding electron density profile.

2.3. Matching Measured Oblique Ionograms with Synthesized Oblique Traces

Similar to studies of vertical ionograms [26], a large amount of candidate vertical traces could be synthesized by the QPS model in this study. Reasonably, the range of the parameters of the QPS model in this study is similar to the reconstruction of vertical traces [26]. Furthermore, these candidate vertical traces could be transformed to oblique traces. Then, the synthesized oblique traces, whose parameters are close to autoscaled parameters, would be selected as the candidate traces to carry out correlation with measured oblique ionograms. It can reduce the running time of the proposed method to meet the near real-time application associated with inversion of oblique ionograms. Then, the correlation values between synthesized traces and the measured oblique ionogram were compared with the threshold value C_{th}. If there were some correlation values larger than the threshold value C_{th}, the oblique trace with the maximum correlation value and the corresponding parameters would be selected as the best-fitted one. Otherwise, the proposed method would output the autoscaled parameters mentioned above in Section 2.1. Figure 5 shows the flowchart of matching measured oblique ionograms with synthesized oblique traces.

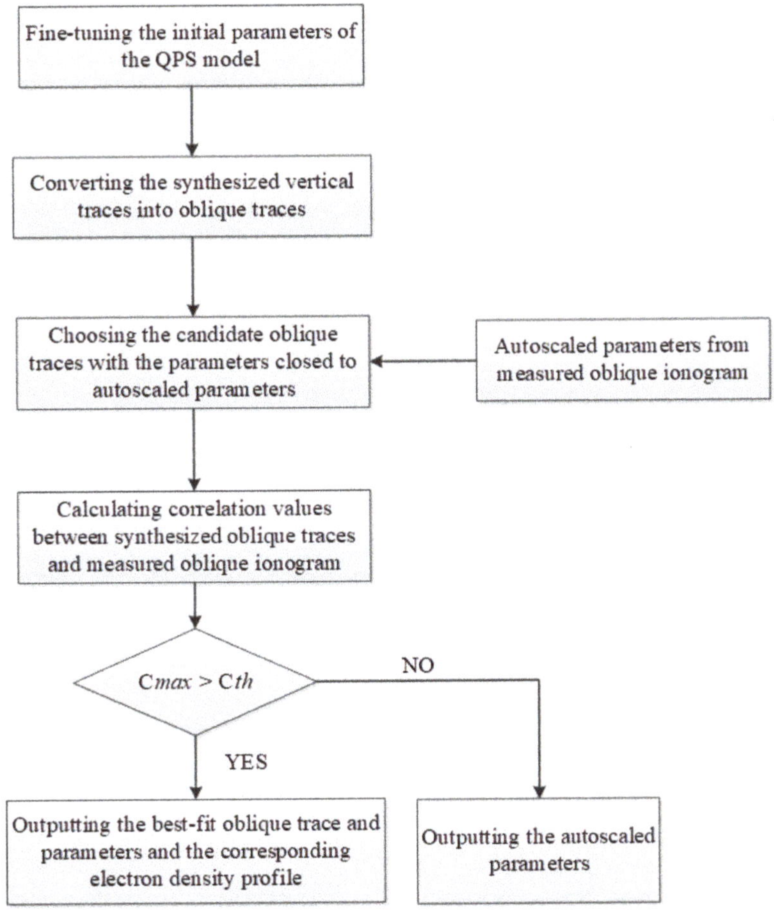

Figure 5. Flowchart of matching measured oblique ionograms with synthesized oblique traces.

3. Results

Figure 6 shows a best-fit synthesized trace (a black line) with the best-fit parameters on a measured oblique ionogram. This typical oblique ionogram includes the echoes of the Es layer. Result shows that the proposed method performed well for automatically scaling the parameters of the E, Es, F1 and F2 layers from oblique ionograms. Because the parameters of the F1 layer were estimated from the model, the synthesized trace did not match well for the echoes of the F1 layer. However, results suggested that it could not affect the performance of the matching trace of the E, Es and F2 layer. The proposed method is inspiring for automatic inversion of measured oblique ionograms, especially for the E/Es layer and F2 layer.

To test the feasibility of this method on different kinds of oblique ionograms, measured oblique ionograms were divided into three categories. The first category is ionograms with the Es layer and F1 layer, the second is without the F1 layer and Es layer, the third is with the F1 layer but without the Es layer. Figure 6 shows the first case. In this section, the second and third cases were tested. Figure 7 shows the best-fitted traces on measured oblique ionograms without the F1 layer and Es layer. Figure 8 shows the third case with the F1 layer but without the Es layer. It can be seen in Figures 7 and 8 that the synthesized oblique traces (black lines) matched well on the echoes of oblique ionograms. The autoscaled parameters were also plotted in Figures 7 and 8. Results matched well on different kinds of measured oblique ionograms, which inspired us to carry out a statistical study of autoscaled parameters on a large amount of oblique ionograms.

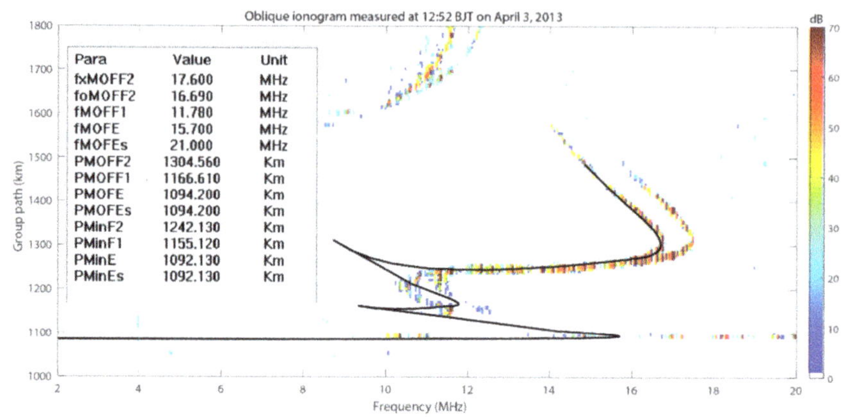

Figure 6. Oblique ionogram measured at 12:52 BJT on 3 April 2013 with the best-fit synthesized trace (a black line) and best-fit parameters.

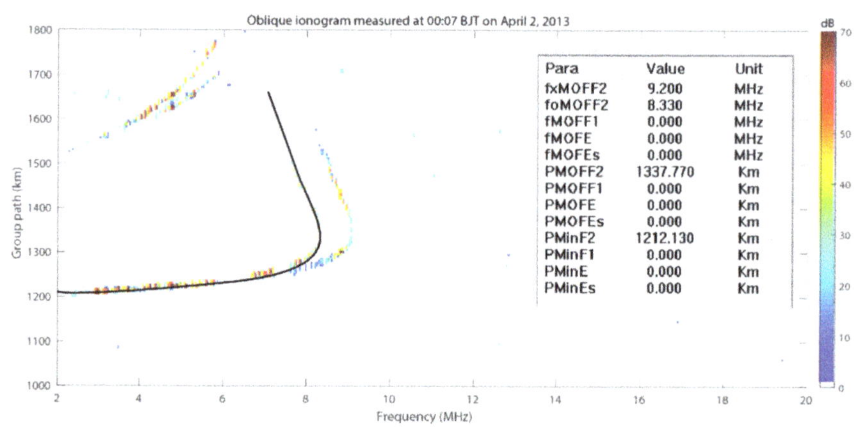

Figure 7. Similar to Figure 6, but the oblique ionogram was measured at 00:07 BJT on 2 April 2013 without the F1 layer.

Because there are no ionosonde installed at the middle point between Wuhan and Beijing stations, the electron density profile inversed from measured oblique ionograms was not used to test the accuracy of the proposed method. It is well known that the parameters of the F2 layer are of great significance for the propagation of radio waves in the ionosphere. Therefore, the maximum observable frequency (maximum frequency of the observed trace) and the minimum group path of the F2 layer were utilized to verify the performance of this method. In this test, the parameters of fxMOFF2, foMOFF2, and PminF2 were adopted. In the presented method, the Es layer was identified by comparison with the E layer from the IRI model. It is difficult to directly identify the Es layer from measured oblique ionograms by operators when the most observed frequency is not large enough. Thus, the maximum observable frequency and the minimum group path of the E or Es layer were used to test the accuracy of autoscaled data. In the case of the E or Es layer, the operator would scale the maximum observable frequency and minimum group path of the echoes of the E or Es layer, but cannot identify that it is the E or Es layer. For autoscaled parameters, if the Es layer was identified, its parameters would be used to compare with manual scaled values. Otherwise, the autoscaled parameters of the E layer would be adopted. As a result, there are five parameters (fxMOFF2, foMOFF2, PminF2, fMOFE, PminE) from oblique ionograms. For measured oblique ionograms, the resolutions of working frequency and the group path are, respectively, 0.05 MHz and 5 km. Thus, an autoscaled value is considered to be acceptable if its error is within ±0.5 MHz for the frequency and ±25 km for the group path, which is in line with the URSI limits of ±5Δ (Δ is the reading accuracy).

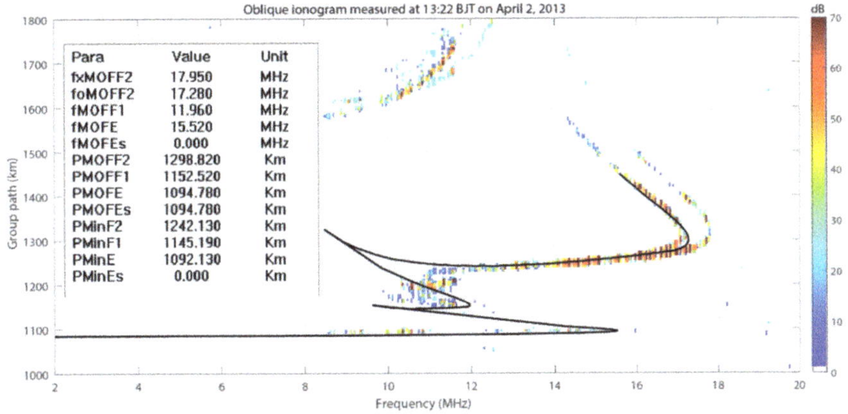

Figure 8. Similar to Figure 6, but oblique ionogram was measured at 13:22 BJT on 2 April 2013 without the Es layer. Oblique ionograms measured during 8–16 April 2013 by the ionosondes between Wuhan and Beijing stations were used to carry out statistical analysis of the performance of the proposed method. The ground distance between Wuhan and Beijing stations is approximately 1040 km. The number of measured oblique ionograms is about 795.

Figures 9 and 10 report the differences between the autoscaled values of the method described here and the standard manual values (Figure 9 for the parameters of the F2 layer, and Figure 10 for the E/Es layer). Table 1 illustrates the percentages of error statistical distributions of parameters for these differences. The accuracy of autoscaled frequency of the F2 layer is above 90%. However, the accuracy of the autoscaled group path is relatively lower (about 86.41%) compared with the maximum observable frequency. Due to the existence of the F1 layer, it is difficult to specify the minimum group path of the F2 layer. On the contrary, the accuracy of the autoscaled minimum group path of the E layer is higher (96.75%) than the autoscaled maximum observable frequency (60.05%). It is noted that the strength of the echoes of the E layer in these measured oblique ionograms are mostly lower, sometimes it is hard to accurately specify the maximum observable frequency of the E layer by the operator (see Figure 8). This will greatly affect the accuracy of autoscaled frequency of the E layer. Therefore, that is the reason that why the accuracy of the maximum observable frequency of the E layer is much less than the F2 layer. For the minimum group path, the lower strength of the echoes of the E layer would not affect the accuracy due to the projection values of the echoes of the E layer at the group path. Thus, we can see that the accuracy of the autoscaled group path for the E layer is high enough. On the contrary, the exact definition of the minimum group path plays a significant role on the accuracy of the autoscaled group path of the F2 layer. In this aspect, the E layer has a greater advantage compared with characteristics of the echoes of the F2 layer. That is the reason why the accuracy of the autoscaled group path of the E layer is greater than the F1 layer. For the lower accuracy of the maximum observable frequency of the E layer, the accuracy of the F2 layer inspires us to believe that it will be improved greatly if the echoes of the E layer are strong enough.

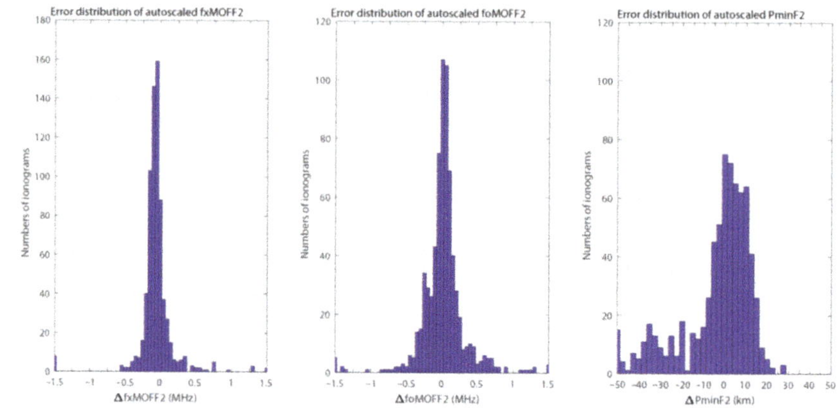

Figure 9. Error distributions of fxMOFF2, foMOFF2, and PminF2 for oblique ionograms between the manual and autoscaled values.

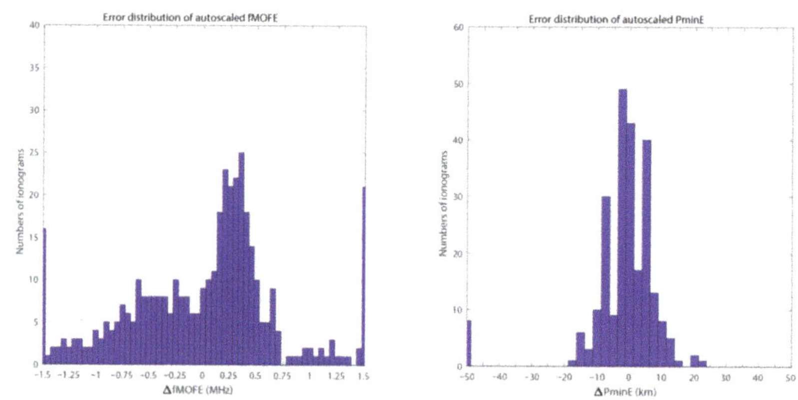

Figure 10. Similar to Figure 9, but for fMOFE and PminE.

Table 1. The percentages of error statistical distributions of autoscaled parameters from oblique ionograms.

Item	fxMOFF2	foMOFF2	PminF2	fMOFE	PminE
Acceptable values	96.34%	91.98%	86.41%	60.05%	96.75%
Total number of oblique ionograms			795		

4. Conclusions

This study describes a method for automatic inversion of oblique ionograms. The proposed method first determined the initial autoscaled parameters using similar technologies of vertical ionograms proposed by Jiang et al. [25,26]. Then, a large number of candidate electron density profiles were constructed by the QPS model, based on the IRI model and the Nequick2 model. Furthermore, the candidate vertical traces, synthesized from electron density profiles, have been converted into oblique traces by the secant theorem and

Martyn's equivalent theorem. Lastly, these candidate oblique traces were used to obtain the best-fitted trace and parameters through matching measured oblique ionograms. Results show that the accuracy of the autoscaled frequency of the F2 layer is above 90% (91.98% for ordinary traces, 96.34% for extraordinary traces). The accuracy of the autoscaled group path is about 86.41%. For the E layer, the accuracy of the autoscaled minimum group path of the E layer can reach up to 96.75%. However, the accuracy of the autoscaled maximum observable frequency is relatively lower (about 60.05%) due to the lower strength of the echoes of the E layer. This indicates that the proposed method might be accurate enough for automatic inversion of oblique ionograms, especially for oblique ionograms with strong echoes (the F2 layer in this study). Results inspire us to develop and improve the performance. The proposed method still requires some adjustments to improve its accuracy and performance, and future studies will focus on the application of the proposed method at different geographic locations.

Author Contributions: Data curation, Z.C. and X.Z.; methodology, C.J.; software, C.J.; validation, C.J., G.Y. and T.L.; investigation, C.Z. and C.J.; writing—original draft preparation, C.J.; writing—review and editing, C.J. and Z.Z.; project administration, C.J. and Z.Z.; funding acquisition, T.L. and C.J. All authors have read and agreed to the published version of the manuscript.

Funding: This research was funded by the National Natural Science Foundation of China (NSFC), grant number 42074184, 41727804, 42104151, 41604133; Youth Foundation of Hubei Provincial Natural (No. 2021CFB134); and the Special Fund for Fundamental Scientific Research Expenses of Central Universities (No. 2042021kf0023).

Institutional Review Board Statement: Not applicable.

Informed Consent Statement: Not applicable.

Data Availability Statement: Oblique ionograms data are available from Chunhua Jiang upon request (chuajiang@whu.edu.cn).

Acknowledgments: We are grateful to the Editor and anonymous reviewers for their assistance in evaluating this paper.

Conflicts of Interest: The authors declare no conflict of interest.

References

1. Reinisch, B.W.; Galkin, I.A.; Khmyrov, G.M.; Kozlov, A.V.; Bibl, K.; Lisysyan, I.A.; Cheney, G.P.; Huang, X.; Kitrosser, D.F.; Paznukhov, V.V.; et al. New Digisonde for research and monitoring applications. *Radio Sci.* **2009**, *44*, RS0A24. [CrossRef]
2. Rietveld, M.T.; Wright, J.W.; Zabotin, N.; Pitteway, M.L.V. The Tromsø dynasonde. *Polar Sci.* **2008**, *2*, 55–71. [CrossRef]
3. MacDougall, J.W.; Grant, I.F.; Shen, X. The Canadian advanced digital ionosonde: Design and results. In *Report UAG-14: Ionospheric Networks and Stations, World Data Center A for Solar-Terrestrial Physics*; UAG-104, Ionosonde Network Advisory Group: Lowell, MA, USA, 1995; pp. 21–27.
4. Zuccheretti, E.; Tutone, G.; Sciacca, U.; Bianchi, C.; Arokiasamy, B.J. The new AIS-INGV digital ionosonde. *Ann. Geophys. Italy* **2003**, *46*, 647–659. [CrossRef]
5. Shi, S.; Yang, G.; Jiang, C.; Zhang, Y.; Zhao, Z. Wuhan Ionospheric Oblique Backscattering Sounding System and Its Applications—A Review. *Sensors* **2017**, *17*, 1430. [CrossRef]
6. Reinisch, B.W.; Huang, X. Automatic calculation of electron density profiles from digital ionograms: 3. Processing of bottomside ionograms. *Radio Sci.* **1983**, *18*, 477–492. [CrossRef]
7. Zabotin, N.A.; Wright, J.W.; Zhbankov, G.A. NeXtYZ: Three-dimensional electron density inversion for dynasonde ionograms. *Radio Sci.* **2006**, *41*, RS6S32. [CrossRef]
8. Pillat, V.G.; Guimaraes, L.N.F.; Fagundes, P.R.; da Silva, J.D.S. A computational tool for ionosonde CADI's ionogram analysis. *Comput. Geosci.* **2013**, *52*, 372–378. [CrossRef]
9. Scotto, C.; Pezzopane, M. A software for automatic scaling of foF2 and MUF(3000)F2 from ionograms. In Proceedings of the URSI 2002, Maastricht, The Netherlands, 17–24 August 2002.
10. Jiang, C.; Yang, G.; Zhou, Y.; Zhu, P.; Lan, T.; Zhao, Z.; Zhang, Y. Software for scaling and analysis of vertical incidence ionograms-ionoScaler. *Adv. Space Res.* **2017**, *59*, 968–979. [CrossRef]
11. Smith, M.S. The calculation of ionospheric profiles from data given on oblique incidence ionograms. *J. Atmos. Terr. Phys.* **1970**, *32*, 1047–1056. [CrossRef]
12. Chen, J.; Bennett, J.A.; Dyson, P.L. Synthesis of oblique ionograms from vertical ionograms using quasi-parabolic segment models of the ionosphere. *J. Atmos. Terr. Phys.* **1992**, *54*, 323–331. [CrossRef]

13. Phanivong, B.; Chen, J.; Dyson, P.L.; Bennett, J.A. Inversion of oblique ionograms including the earth's magnetic field. *J. Atmos. Sol. Terr. Phys.* **1995**, *57*, 1715–1721. [CrossRef]
14. Huang, X.; Reinisch, B.W.; Kuklinski, W.S. Mid-point electron density profiles from oblique ionograms. *Ann. Geophys. Italy* **1996**, *49*, 757–761. [CrossRef]
15. Redding, N.J. Image understanding of oblique ionograms: The autoscaling problem. In *Proceedings of the IEEE Australian and New Zealand Conference on Intelligent Information Systems, Adelaide, SA, Australia, 18–20 November 1996*; IEEE: Piscataway, NJ, USA, 1996; pp. 155–160.
16. Fan, J.; Lu, Z.; Jiao, P. The intelligentized recognition of oblique propagation modes. *Chin. J. Radio Sci.* **2009**, *24*, 528. (In Chinese) [CrossRef]
17. Settimi, A.; Pezzopane, M.; Pietrella, M.; Bianchi, C.; Scotto, C.; Zuccheretti, E.; Makris, J. Testing the IONORT-ISP system: A comparison between synthesized and measured oblique ionograms. *Radio Sci.* **2013**, *48*, 167–179. [CrossRef]
18. Ippolito, A.; Scotto, C.; Francis, M.; Settimi, A.; Cesaronl, C. Automatic interpretation of oblique ionograms. *Adv. Space Res.* **2015**, *55*, 1624–1629. [CrossRef]
19. Heitmann, A.J.; Gardiner-Garden, R.S. A robust feature extraction and parameterized fitting algorithm for bottom-side oblique and vertical incidence ionograms. *Radio Sci.* **2019**, *54*, 115–134. [CrossRef]
20. Song, H.; Hu, Y.; Jiang, C.; Zhou, C.; Zhao, Z.; Zou, X. An automatic scaling method for obtaining the trace and parameters from oblique ionogram based on hybrid genetic algorithm. *Radio Sci.* **2016**, *51*, 1838–1854. [CrossRef]
21. Croft, T.A.; Hoogasian, H. Exact ray calculations in a quasi-parabolic ionosphere with no magnetic field. *Radio Sci.* **1968**, *3*, 69–74. [CrossRef]
22. Dyson, P.L.; Bennett, J.A. A model of the vertical distribution of the electron concentration in the ionosphere and its application to oblique propagation studies. *J. Atmos. Terr. Phys.* **1988**, *50*, 251–262. [CrossRef]
23. Gethinga, P.J.D.; Maliphant, R.G. Unz's application of Schlomilch's integral equation to oblique incidence observations. *J. Atmos. Terr. Phys.* **1967**, *29*, 599–600. [CrossRef]
24. Reilly, M.; Kolesar, J. A method for real height analysis of oblique ionograms. *Radio Sci.* **1989**, *24*, 575–583. [CrossRef]
25. Jiang, C.; Yang, G.; Zhao, Z.; Zhang, Y.; Zhu, P.; Sun, H. An automatic scaling technique for obtaining F2 parameters and F1 critical frequency from vertical incidence ionograms. *Radio Sci.* **2013**, *48*, 739–751. [CrossRef]
26. Jiang, C.; Yang, G.; Zhao, Z.; Zhang, Y.; Zhu, P.; Sun, H.; Zhou, C. A method for the automatic calculation of electron density profiles from vertical incidence ionograms. *J. Atmos. Sol. Terr. Phys.* **2014**, *107*, 20–29. [CrossRef]
27. Bilitza, D.; Altadill, D.; Truhlik, V.; Shubin, V.; Galkin, I.; Reinisch, B.; Huang, X. International Reference Ionosphere 2016: From ionospheric climate to real-time weather predictions. *Soc. Work.* **2017**, *15*, 418–429. [CrossRef]
28. Nava, B.; Coisson, P.; Radicella, S.M. A new version of the NeQuick ionosphere electron density model. *J. Atmos. Sol. Terr. Phys.* **2008**, *70*, 1856–1862. [CrossRef]
29. Lynn, K.J.W. Histogram-based ionogram displays and their application to autoscaling. *Adv. Space Res.* **2018**, *61*, 1220–1229. [CrossRef]
30. Ippolito, A.; Scotto, C.; Sabbagh, D.; Sgrigna, V.; Maher, P. A procedure for the reliability improvement of the oblique ionograms automatic scaling algorithm. *Radio Sci.* **2016**, *51*, 454–460. [CrossRef]

remote sensing

MDPI

Technical Note

Improved Ionosonde Monitoring of the Sporadic E Layer Using the Frequency Domain Interferometry Technique

Tongxin Liu [1], Guobin Yang [1,*], Chen Zhou [1], Chunhua Jiang [1], Wei Xu [1], Binbin Ni [1] and Zhengyu Zhao [1,2]

1 School of Electronic Information, Wuhan University, Wuhan 430072, China; tongxin_liu@whu.edu.cn (T.L.); chenzhou@whu.edu.cn (C.Z.); chuajiang@whu.edu.cn (C.J.); wei.xu@whu.edu.cn (W.X.); bbni@whu.edu.cn (B.N.); zhaozy@whu.edu.cn (Z.Z.)

2 Institute of Space Science and Applied Technology, Harbin Institute of Technology, Shenzhen 518000, China

* Correspondence: gbyang@whu.edu.cn; Tel.: +86-27-6877-8049

Abstract: The sporadic E (Es) layer is a thin layer of ion plasma enhancement in the E-region ionosphere, typically at altitudes of 90–120 km with vertical and horizontal extent of several or several tens of kilometers. As the transition region between the lower and upper atmosphere, this layer is of critical importance for ionospheric studies. The most economical but effective method to observe this layer is using ionosonde, which, however, is incapable of capturing the finer structure or the internal inhomogeneity of the Es layer as the range resolution is on the order of kilometers. To overcome this limitation, we employ the frequency domain interferometry (FDI) technique, a technique that has been successfully applied to the analysis of some radar and sonar measurements. Here, we use the Es layer measurements near Wuhan, China (114°22′E, 30°30′N) on 8 June 2021 as examples to showcase the capability of this technique. Our results show that the spatial resolution of ionosonde imaging is remarkably increased: the complexity of the internal fine structure in the Es layer can be well observed in the FDI-processed ionograms, whereas the intrinsic range resolution is several kilometers. Moreover, by comparing the ionograms obtained with and without the FDI technique, it is found that the FDI-processed ionogram is particularly suitable for the observation of evolutional processes in the Es layer, as well as the identification of different types of Es layer. With this level of spatial resolution, ionosonde, in combination with the FDI technique, opens the possibility for more refined observations of the Es layer.

Keywords: the sporadic E layer; internal fine structure; high-resolution ionosphere imaging; frequency domain interferometry technique

Citation: Liu, T.; Yang, G.; Zhou, C.; Jiang, C.; Xu, W.; Ni, B.; Zhao, Z. Improved Ionosonde Monitoring of the Sporadic E Layer Using the Frequency Domain Interferometry Technique. *Remote Sens.* **2022**, *14*, 1915. https://doi.org/10.3390/rs14081915

Academic Editor: Fabio Giannattasio

Received: 30 March 2022
Accepted: 13 April 2022
Published: 15 April 2022

1. Introduction

The sporadic E (Es) layer is a relatively thinner (compared to other layers of the ionosphere) layer of enhanced ion plasma in the E-region ionosphere, typically at altitudes of 90–120 km with a vertical extent of several kilometers and a horizontal extension of several tens of kilometers [1].

The formation of the Es layer can be well explained by the wind shear theory and the convergence of metal ions [2–4]. Other than these, shear instabilities, tidal, planetary, or gravity waves, meteors, and thunderstorms could somewhat influence the electron density distribution in the Es layer as well [5–7]. For example, the Kelvin–Helmholtz instability (KHI) can lead to a billow structure and a polarization electric field in the Es layer [8,9]. Tidal and planetary waves can give rise to periodic vertical fluctuations in the Es layer [10,11]. Gravity waves can modulate the Es layer and cause distortions along both horizontal and vertical directions, which are ultimately recorded as quasiperiodic backscatter echoes in the very-high-frequency (VHF) range [12,13]. The seasonality of the Es layer is well related with the occurrence of meteors [14], and, as has been reported numerously, the Es formation in the mid-latitudes is also closely related to the sporadic metal layer [15]. The correlation between the spread F-region ionosphere, thunderstorms, and Es layer has implications for the coupling between the upper and lower layers of the Earth's atmosphere as well [16]. Considering these, as the transition region between the lower and upper atmosphere, the morphological structure, composition, and temporal evolution of the Es layer has been the main focus of various theoretical and observational studies, and improving the spatial resolution of Es layer measurements is critical for better understanding the formation and evolution mechanism of the Es layer, as well as the above-mentioned atmospheric processes.

Up to now, although extensive observational efforts have been made, the vertical structure of the Es layer still remains insufficiently investigated due to its transient nature and limitations in observation techniques. By analyzing the in situ data measured during the campaign of Sporadic E Experiment over Kyush (SEEK), Mori and Oyama found that the Es layer can exhibit a complicated multiple layer structure at altitudes with the separation of 10–12 km [17–19]. This multilayer structure was later confirmed by Damtie et al. using the radio sounding data collected by the European Incoherent Scatter (EISCAT) radar [20]. The authors have specifically found that the electrons in the Es layer were distributed at multiple fine layers with spatial intervals of 1–2 km, and during the downward drifting phase, these fine layers were merged into a single layer. Using the EISCAT data, Turunen et al. have further investigated the undulating movement of the Es layer along the vertical direction and revealed that this layer could be compressed by plasma streams and degenerated owing to the fluctuations in the neutral atmosphere [21].

Along the horizontal direction, the Es layer is also found to exhibit many variations, and a "blanketing" or "patchy" structure is found in most cases [22]. The Es layer can be reconstructed via ionosonde using the critical reflection frequency and the direction of arriving waves, which can be utilized to image the embedded structure of field-aligned irregularities (FAIs) in the Es layer [23,24]. The incoherent scatter radar (ISR) images recorded by Hysell et al. have shown that the Es layer at middle latitudes could exhibit both cloud-like and wave-like structures [25]. As for the Es layer at higher latitudes, for example, in Alaska, Hysell et al. have revealed a two-dimensional "patchy" and "stripe" structure [26]. In addition to radar imaging, the Es layer has been extensively studied using measurements from space, for example, the Global Positioning System (GPS) [27], using the total electron content (TEC) anomalies [28].

Compared to other types of Es layer measurements, for example, ISR and GPS, ionosonde has the advantage of providing persistent observation at a relatively low cost. In particular, compared with ISR, large antenna array and great transmission power are not required, and unlike GPS, the bottom-up sounding of the ionosonde is not affected by the F layer, and even weak Es layers can be observed. However, the inhomogeneity of the fine structure of this layer cannot be well captured due to the poor range resolution of the ionosonde, and its applicability in studies of Es layer formation and evolution is limited. To overcome this limitation represents the main goal of this study and, for this purpose, we utilize the frequency domain interferometry (FDI) technique. We show that this technique, which was originally developed for the analysis of radar and sonar measurements [29], is also applicable to ionosonde measurements. In the following, we explain how this technique is utilized to process ionosonde data, and we use several examples of Es layer measurement to showcase the resolution improvement obtained using this technique.

2. Experiment Setup and Methods

2.1. Instruments and Experimental Setup

The ionosonde used in this study is a miniature version of the Wuhan Ionospheric Sounding System (WISS), as developed by the Ionospheric Laboratory of Wuhan University [30]. It specifically uses a 16-bit complementary coded sequence to modulate the high-frequency (HF) signals with a peak power of 500 watts and performs vertical incidence sounding. The transmission duration for any coded chip is 25.6 microseconds (μs), corresponding to an intrinsic range resolution of 3.84 km. In regular mode, each frequency-sweep sounding was performed for a total duration of ~3 min at frequencies from 2 to 20 megahertz (MHz) with steps of 0.05 MHz. The number of accumulations was set to be 32. After a 12 min interruption, the typical sounding period of the frequency-sweep detection was 15 min, which is sufficient for general ionospheric observation missions. The WISS uses an inverted-V-shaped antenna for signal transmission and another 30 m three-wire antenna for signal reception. Therefore, during conventional detection, ordinary (O) and extraordinary (X) modes of ionogram echo traces can be distinguished using image processing [31]. It is worth mentioning that WISS has been widely deployed in China and provides reliable data for various ionosphere studies [32].

To ensure good coherence, i.e., reflection of adjacent frequencies at the same height, the frequency step of the ionosonde was set to be 10 kilohertz (kHz) during this experiment. The number of accumulations at each frequency point was 256 since this number provides sufficient samples for the next-step FDI processing. The resident time of each frequency point is about 4.19 s. We emphasize that the time consumption caused by smaller frequency intervals would not significantly affect the Es detection since the Es layer at low and middle latitudes, in general, can last for at least tens of minutes [14,26]. The sounding frequency range in this experiment was set to be 2–4 MHz, as typically needed for good resolution of the Es layer, also taking into account the acceptability of the time consumption. Thus, the ionosonde sounding period was still approximately 15 min, which is consistent with the conventional ionospheric vertical sounding period. Similarly, also in accordance with the conventional ionospheric vertical measurements, short-term changes in the internal structure of the Es layer during the detection period (15 min) are temporarily ignored. Therefore, this paper is more inclined to reflect the relatively static inhomogeneity of the fine structure of Es layer and the evolution process of a 15 min level. The experiments reported in this study were performed near Wuhan, China (114°22′E, 30°30′N, geomagnetic dip angle: 45°) on 8–9 June 2021. During this period, the Kp indexes are less than 2 [33], which means a geomagnetically quiet day.

2.2. Es Layer Imaging Based on the FDI Technique

The FDI technique has been successfully applied to the analysis of atmospheric radar measurements of ionospheric turbulence and FAIs [34,35]. Good performance has been obtained in general, although with the following drawback: while using a limited number of frequency points to image a small vertical extent, the observation results can only reflect the target response characteristics at a certain frequency band. To remedy this drawback, different from previous radar studies, we first process the frequency-sweep-detected echoes of the Es layer in a wide frequency range. The FDI technique of ionosonde data processing is then applied in the following procedure.

The Es layer is assumed as a target with a slow movement and a narrow height distribution; therefore, when a series of adjacent frequencies is used for sounding and the changes in the Es characteristics in a short duration are ignored, the echo signal $S_p(t)$ of frequency f_p is expressed as

$$S_p(t,r) = A_p s(t - 2r/c)\cdot e^{-j2\pi f_p(2r/c)+\varphi_p} \tag{1}$$

where c represents the speed of light, A_p represents the echo amplitude, $s(\cdot)$ expresses the echo envelope, r is the radial distance, and φ_p is the initial phase. For a coherent radar system such as the WISS, the initial phase of the transmission signal in a small frequency band can be assumed to be the same; thus, for similar working frequencies, the echo phase difference of the same target must meet the following condition:

$$\Delta\phi = 2\pi\cdot\frac{2r}{c}\cdot(\Delta f) \tag{2}$$

where Δf represents the frequency interval. The phase difference is only related to the distance of the target and the frequency interval. The unambiguity range of the signal phase is 2π, and subsequent processing is performed based on the order of the range gate; thus, it is necessary to ensure that no phase ambiguity occurs within a range resolution unit. The frequency step Δf_0 during sounding should satisfy the following condition:

$$\Delta\phi = 2\pi \cdot \frac{2r_0}{c} \cdot (\Delta f_0) \leq 2\pi \Rightarrow \Delta f_0 \leq \frac{c}{2r_0} \tag{3}$$

where r_0 represents the initial range resolution of the ionosonde radial distance. This resolution is 3.84 km for the WISS, indicating that Δf_0 must be smaller than 39.0625 kHz. The frequency step of 10 kHz selected in our experiment clearly meets the requirement.

The signal matrix for the echo signals of a certain frequency f_p and the following k adjacent frequencies $f_p \sim f_{p+k}$, $(f_p \leq 4\ \text{MHz} - k \cdot \Delta f)$ is constructed as follows:

$$S(t,r) = \left[S_p(t,r), S_{p+1}(t,r), \ldots, S_{p+k}(t,r)\right]^T \tag{4}$$

where $[\cdot]^T$ represents the matrix transpose. For each range gate, the echo data of each frequency are extracted, and the covariance matrix R_S is calculated as Equation (5):

$$R_S = SS^H/n \tag{5}$$

where $n = 256$ is the accumulation number in our experiment, and $[\cdot]^H$ represents the conjugate transpose of the matrix. A range-dimensional steering vector is determined based on the required resolution:

$$\begin{cases} a(f_p, r_1) = [e^{-j2\pi f_p(2r_1/c)}, e^{-j2\pi f_{p+1}(2r_1/c)}, \ldots, e^{-j2\pi f_{p+k}(2r_1/c)}]^H \\ a(f_p, r_2) = [e^{-j2\pi f_p(2r_2/c)}, e^{-j2\pi f_{p+1}(2r_2/c)}, \ldots, e^{-j2\pi f_{p+k}(2r_2/c)}]^H \\ \vdots \\ a(f_p, r_m) = [e^{-j2\pi f_p(2r_m/c)}, e^{-j2\pi f_{p+1}(2r_m/c)}, \ldots, e^{-j2\pi f_{p+k}(2r_m/c)}]^H \end{cases} \tag{6}$$

where r_s is defined as the expected resolution, and $m = r_0/r_s$ represents the refined factor.

Therefore, based on the data of each range gate with the initial resolution, the spectrum of the jth refined range unit can be estimated using the Capon method:

$$b(f_p, r_j) = \frac{1}{a^H(f_p, r_j) R_S{}^{-1} a(f_p, r_j)}, j = 1, 2, \cdots, m \tag{7}$$

The sounding range resolution can be increased by m times by scanning the range spectrum. Essentially, it is a type of spectrum estimation method in the range dimension that uses the coherence between the echo signals of the same target in the frequency domain.

In this study, we performed the frequency-sweep sounding of the Es layer using the FDI technique in a frequency steeping mode. Considering the observational frequency range of 2–4 MHz, the electromagnetic environments of each frequency were differed slightly, possibly inducing severe interference during the spectral estimation. For enhanced frequency scanning imaging, we normalized the spectral estimation results and used a Gaussian window to perform range smoothing. Eleven adjacent frequencies were considered in one run of the spectral estimation ($k = 11$). The scanning step of the range spectrum was set to 38.4 m ($m = 100$). Note that because the maximum frequency interval of the signals used in one imaging process was only 100 kHz, the difference in the phase-frequency response of the system between the adjacent frequencies was not considered to induce severe adverse effects to the range spectral estimation.

Inevitably, for a single-channel ionosonde, using a wide beam antenna, it is difficult to have the capabilities of direction-estimating and beam-pointing. Naturally, it is impossible to accurately locate the target position. In spite of this, it should be feasible to use this method to monitor the inhomogeneity and complexity of the internal structure of the Es layer. If the Es layer is dense and uniform, the imaging result should also be a narrow thin line. This is because when the signal is not vertically incident, it will be reflected to other directions and will not return to the ionosonde. On the contrary, if the Es layer is inhomogeneous, due to the unsmooth lower boundary of the Es layer or the presence of embedded irregularities, the diffuse range spectrum should be obtained.

3. Results

3.1. The Inhomogeneous Es Layer

Figure 1a shows the height–intensity ionogram of the Es layer measured near Wuhan, China (114°22′E, 30°30′N) at 17:46 LT (UTC+8) on 8 June 2021. Figure 1b shows the normalized energy at different altitudes and frequencies, while Figure 1c shows the FDI-processed ionogram with a range resolution of 38.4 m. The FDI technique needs more than one frequency point to ensure good coherence, and thus the frequency points close to the upper boundary of present ionogram cannot be well imaged. As such, in this paper, only the part of frequencies below 3.8 MHz were FDI-processed, which is shown in Figure 1b,c for comparison.

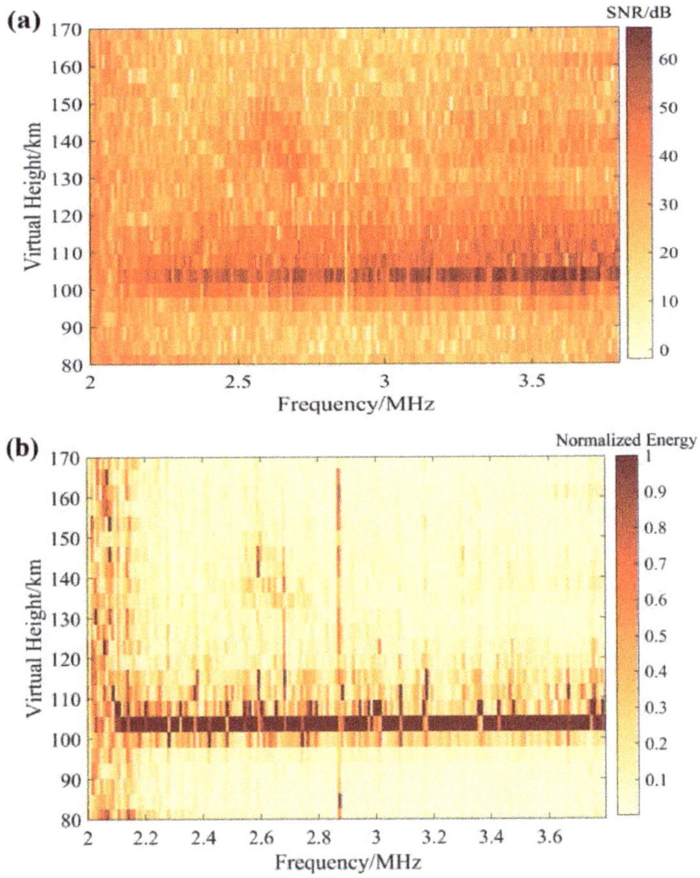

Figure 1. *Cont.*

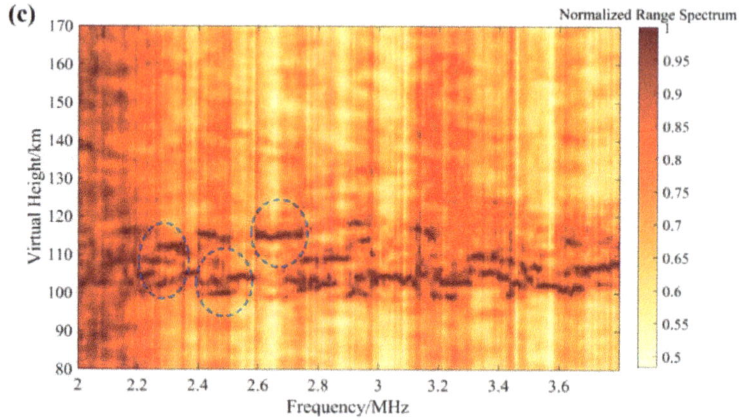

Figure 1. Height–intensity ionogram of the Es layer measured near Wuhan, China at 17:46 LT (UTC+8) on 8 June 2021. (**a**) Conventional ionogram with an intrinsic range resolution of 3.84 km. (**b**) Ionogram showing the normalized energy at different frequencies and altitudes. (**c**) Ionogram with a range resolution of 38.4 m as obtained using the FDI technique.

According to International Reference Ionosphere (IRI) 2016 [36], at this time, the peak height of E layer is 110 km and the critical frequency is 1.7 MHz; therefore, Figure 1 shows a diffuse Es layer, for which imaging results do not focus on a certain range. The reflection height corresponding to the strongest energy was approximately 107 km with the smaller echoes distributed at altitudes of 103–123 km. It is clear from Figure 1a,b that, with the intrinsic resolution of 3.84 km, it is almost impossible to recognize the fine structure and diffusion features. In contrast, the echoes due to smaller-scale electron density irregularities (as circled by the dotted blue line) are clearly shown in Figure 1c. The virtual height of the sounding echoes dramatically changes at varying frequencies, as shown in Figure 1c. A direct comparison between Figure 1a–c shows the improvement of spatial resolution of the FDI technique. At the same time, it can also be observed from Figure 1c that this is a highly inhomogeneous Es layer.

3.2. *Quiet Es Layer*

In addition to the diffuse Es layer shown in Figure 1, we have also examined the FDI method during quiet Es conditions, as shown in Figure 2. Figure 2a shows an example of quiet Es condition measured by WISS near Wuhan, China at 22:34 LT (UTC+8) on 8 June 2021. The echo trace of the Es layer is a clear and uniform straight line, indicating that the Es layer is dense and stable at this time. In this sense, it can be regarded as quiet. Figure 2a,b show the ionograms obtained without and with applying the FDI technique. The corresponding range resolution is 3.84 km and 38.4 m, respectively. From the comparison between these two panels, it is clear that the sounding echoes are almost flat at ~110 km altitude as in the conventional ionogram (Figure 2a), while smaller-scale fluctuations are resolved in the FDI-processed ionogram (Figure 2b).

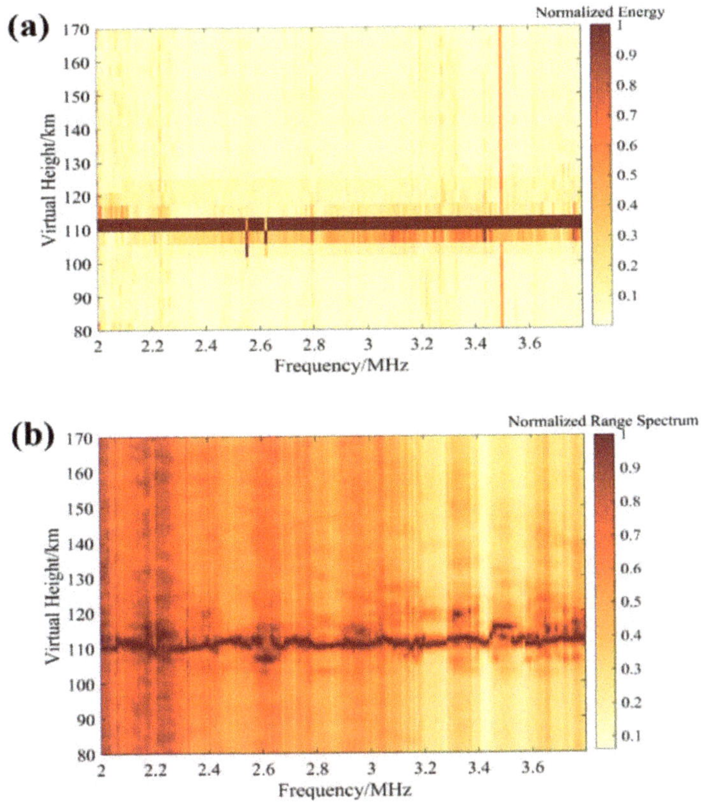

Figure 2. An example of quiet Es layer measured near Wuhan, China at 22:34 LT (UTC+8) on 8 June 2021. (**a**) Ionogram with an intrinsic range resolution of 3.84 km. (**b**) Ionogram with a range resolution of 38.4 m as obtained using the FDI technique.

3.3. Short-Term Evolution of the Es Layer

Figure 3a shows the ionograms measured between 22:06 and 23:18 LT (UTC+8) on 8 June 2021. Figure 3b shows similar results, but obtained using the FDI technique with a range resolution of 38.4 m. During this time interval, the echo trace was first compressed and then expanded, implying that the Es layers are evolving from a thin layer to inhomogeneously distributed irregularities. This example somewhat resembles the event recorded by Hysell et al. using ISR [29]. The evolutional process is hardly recognizable in the conventional ionograms, whereas how it was compressed and expanded is clearly resolved in Figure 3b.

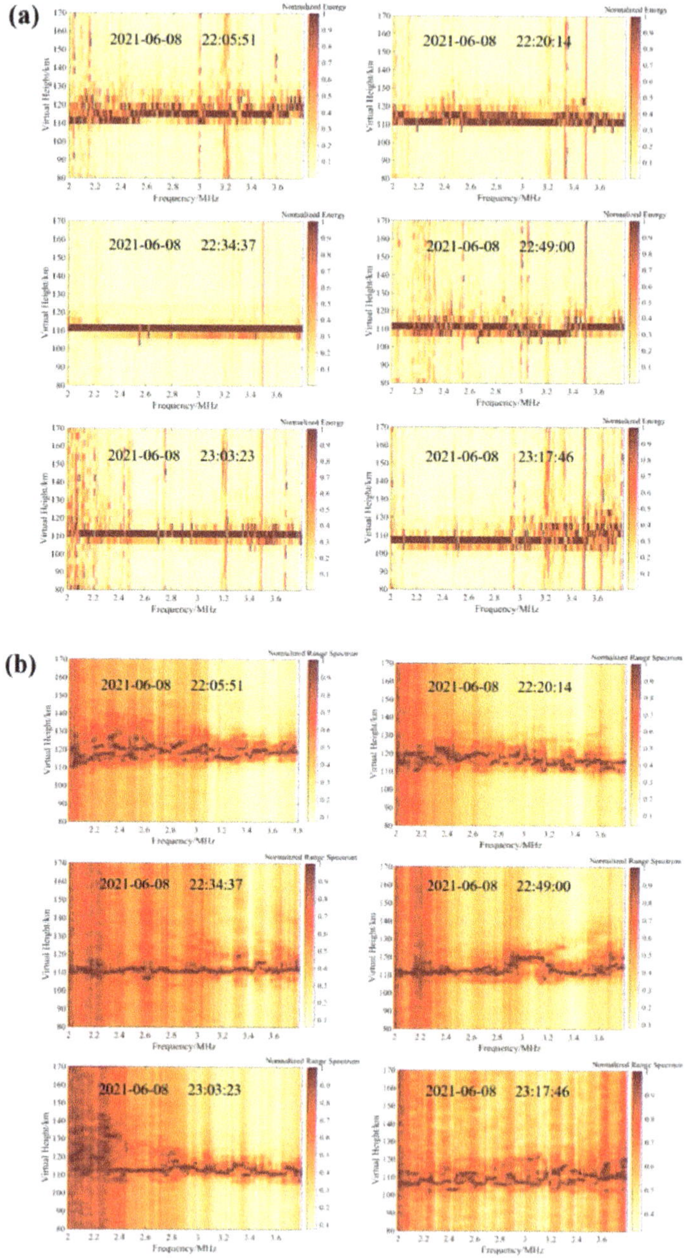

Figure 3. Drastic short-term evolution process of the Es layer within ~1 h. (**a**) Ionograms of the 3.84 km range resolution at ~22:06–23:18 LT (UTC+8) on 8 June 2021, with an interval of 15 min. Although obvious changes are observed in the Es layer, the details can hardly be observed. (**b**) Ionograms of the 38.4 m range resolution. The evolution details are clearly observed. The compression process of the Es layer proceeds at the same speed at each frequency; however, rediffusion starts at the high-frequency band.

3.4. Different Types of Es Layer

The proposed FDI technique is also particularly suitable for the identification of different types of Es layer, especially in a relatively narrow frequency band.

Figure 4a shows the conventional ionogram measured at 19:56 LT on 8 June 2021. Figure 4b shows the same event but obtained using the FDI technique. The echo trace at frequencies of 2.4–2.7 MHz shown in Figure 4a is indicative of a multilayer structure, but it is difficult to determine which type of Es layer was recorded. On the other hand, after applying the FDI technique, it is found that the traces of the Es layers were connected and the traces at 2.4–2.6 MHz exhibited a continuous "spike" shape; both features suggest a c(cusp)-type Es layer according to the manual of ionogram scaling [37].

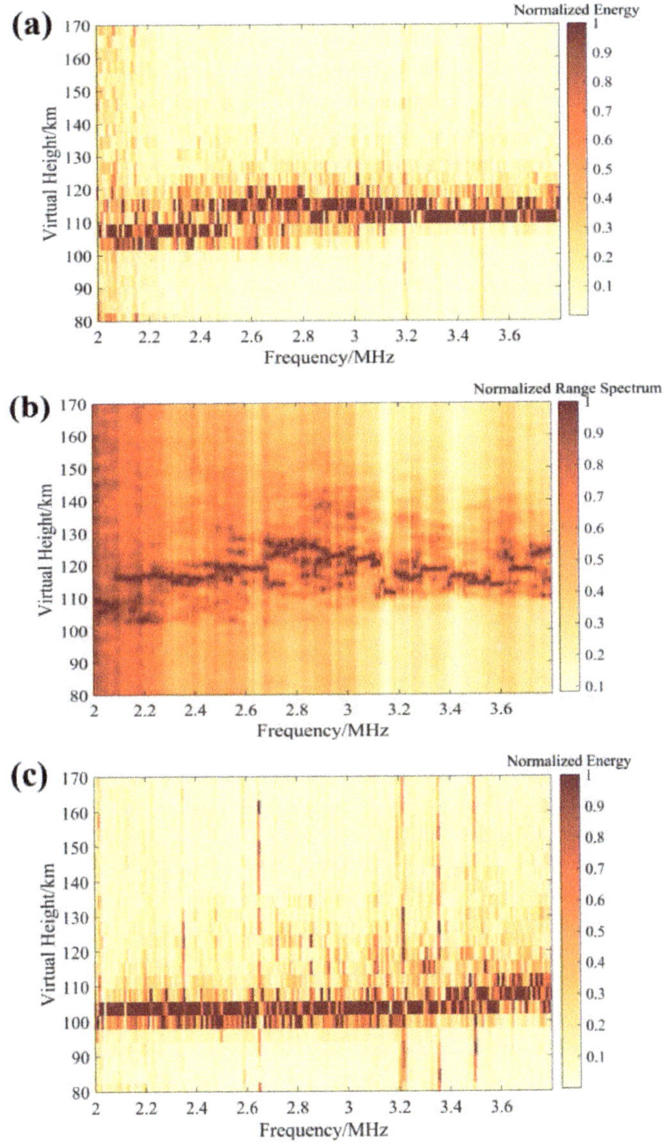

Figure 4. *Cont.*

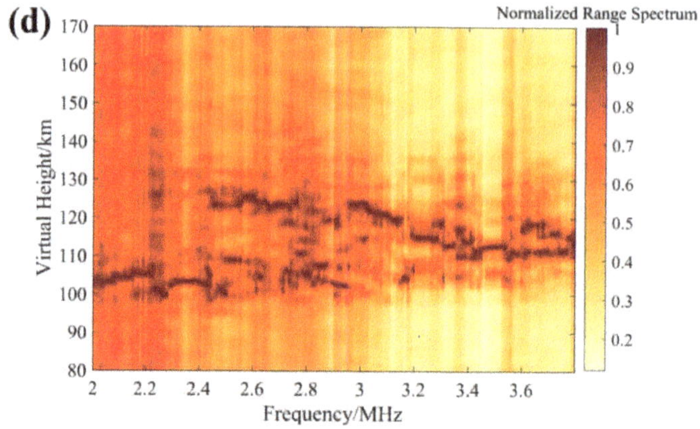

Figure 4. Different types of the Es layer. Ionograms of a c-type Es layer measured near Wuhan, China at ~19:56 LT (UTC+8) on 8 June 2021, with a range resolution of (**a**) 3.84 km and (**b**) 38.4 m. The bottom two panels, (**c**) 3.84 km and (**d**) 38.4 m, show similar results, but for an h-type Es layer measured at 01:12 LT (UTC+8) on 9 June 2021.

Another example is shown in the bottom two panels of Figure 4. These two panels show the Es layer measured at 01:12 LT on 9 June 2021. Based on the conventional ionogram (Figure 4c), the Es layer was likely l (low)-type. Nevertheless, after being processed using the FDI method, this Es layer is actually found to be the h (high)-type (Figure 4d) with no "symmetrical spikes". The upper and lower traces are likely to represent the Xmode and O-mode.

4. Discussion

In Figure 1c, at frequencies lower than 3.2 MHz, diffused range spectra of the echoes were patchily distributed at altitudes of 100–115 km. This reflects the complexity of the internal structure of the Es layer at this time, and the electron density distribution is obviously inhomogeneous. As the spectral peaks are obvious and separated from each other, there may be discontinuous and drastic changes in the spatial distribution of the internal electron density. A reasonable explanation for this phenomenon is, as Whitehead suggested [23], electron density irregularities that are embedded in the Es layer and can considerably scatter the sounding signals. The spectral intensity of the scattered echoes can even suppress the reflected signals at certain frequencies. The inconsistency in terms of the scatterers' position and scale could thus extend the intensity spectra in both range and frequencies. The echo trace between 3.2 and 3.8 MHz is typical of the ionosonde reflection mode and the signal echo height was divided into two layers, indicating an internal multilayer structure with intervals of ~5 km which are likely semi-shielded by each other. Note that the higher layer should not be suspected as caused by the interference due to the reflected waves from nearby objects within the irradiation range of the antenna beam. If this were the case, continuous multilayer echoes would show up in the low-frequency band as well. This feature is more in line with the partial reflection theory and hole structures of the Es layer along the horizontal direction [38,39]. However, this hypothesis needs to be further examined by using a two-dimensional imaging technique or measuring the incoming direction of the signals for positioning with an antenna array.

As for Figure 2b, these fluctuations are reflective of a weak inhomogeneity distribution of electrons in the Es layer. Of special interest is the virtual heights at 2.6 and 3.5 MHz, which are either higher or lower than the overall trend, and not captured by the conventional ionogram (Figure 2a). In other studies, using the ISR data, such phenomena had also been confirmed to exist in the Es layer [29]. This example also demonstrates that the spatial resolution of the ionosonde is largely increased and the FDI technique can be well used to study the physical processes involved in the formation and evolution of the Es layer. Note that this level of spatial resolution can also be achieved using ISR, but with the cost of much higher power consumption and larger antenna array.

Based on Figure 3b, it is worth noting that the compression speed of the echo trace is faster than that of rediffusion. It can be observed that the diffusion range of the spectral virtual height was compressed from 20 km to a thread with slight jitters, continuous echo traces in <15 min, probably indicting that the Es layer evolved from an uneven structure to a dense and uniform layer. For the echoes of different frequencies, the compression process on the imaging ionogram proceeded at basically the same speed. However, rediffusion started from the high-frequency band. At 3.2–3.8 MHz, the traces first showed folds and then extended to the low-frequency band. A reasonable explanation for this could be that the uniform and dense Es layer first yielded small fragments due to the instability or modulation of atmospheric gravity waves and then broke into patches and drifted. This observation indicates an Es variation from "blanketing" to "patchy", and vice versa, which can be temporally dynamic and unstable. These findings obtained within ~1 h indicated a relatively drastic short-term disturbance in the Es layer. Although it remains unclear whether KHI caused by the strong shear of the background neutral wind led to the formation of the unstable patchy Es and FAIs [8] or the wind field of the gravity waves [29] or the internal instability caused by the polarization electric field [9,40–42] is the dominant mechanism, the FDI technique based on ionosondes can provide a new, convenient, and promising way of investigating this open question.

The effect of Figure 4 reveals that there may be some uncertainties in the ionosonde measurements, as caused by the coarse resolution of conventional ionograms. The FDI technique is thus more suitable for Es layer identification. Note also that the virtual heights of the echo traces in Figure 4b at 2.2–2.4 MHz and in Figure 4d at 2.4–2.6 MHz are higher than their corresponding positions in Figure 4a,c, respectively. This is because although the energy of the echoes at the lower position is strong, they mainly originate from different scatters, and the coherence between the signals at different frequencies is not prominent. Alternatively, the higher echoes in Figure 4b,d are due to the reflected signals, suggesting that good coherence leads to an enhanced-range spectral estimation.

5. Conclusions

In this study, we employ the FDI technique to improve the spatial resolution of ionosonde measurements. Using the Es layer measurements near Wuhan as examples, our results show that the spatial resolution of height–intensity ionograms is remarkably increased: the complexity and inhomogeneity of the internal fine structures can be well monitored in the FDI-processed ionograms, compared to the intrinsic range resolution of several kilometers. Moreover, it is found that the detailed evolutional processes and different types of Es layer can be better resolved due to this resolution improvement. Given this level of spatial resolution, ionosonde, in combination with the FDI technique, represents a promising means for more refined observation of the Es layer, as well as the physical processes involved in the formation and evolution of this layer.

Based on the current results, two follow-up studies will be carried out. One is using the antenna array to scan and observe the Es layer with a narrow beam and determine its spatial structure. Another is reducing the sounding period and investigating the evolution process of the Es layer at the second level.

Author Contributions: Conceptualization, C.Z. and Z.Z.; methodology, T.L.; investigation, T.L. and Z.Z.; data curation, G.Y.; writing—original draft preparation, T.L.; writing—review and editing, C.Z., C.J., B.N. and W.X. All authors have read and agreed to the published version of the manuscript.

Funding: This research was funded by the National Natural Science Foundation of China (NSFC No. 42104151, 41774162, 42074187 and 42074184), the Excellent Youth Foundation of Hubei Provincial Natural Science Foundation (No. 2019CFA054), the Youth Foundation of Hubei Provincial Natural

(No. 2021CFB134), and the Special Fund for Fundamental Scientific Research Expenses of Central Universities (No. 2042021kf0023).

Data Availability Statement: Data sharing is not applicable to this article.

Conflicts of Interest: The authors declare no conflict of interest.

References

1. Pietrella, M.; Bianchi, C. Occurrence of sporadic-E layer over the ionospheric station of Rome: Analysis of data for thirty-two years. *Adv. Space Res.* **2009**, *44*, 72–81. [CrossRef]
2. Haldoupis, C.; Pancheva, D.; Singer, W.; Meek, C.; Macdougall, J. An explanation for the seasonal dependence of midlatitude sporadic E layers. *J. Geophys. Res.* **2007**, *112*, A6. [CrossRef]
3. Whitehead, J.D. Formation of the sporadic E layer in the temperate zones. *J. Atmos. Terr. Phys.* **1961**, *20*, 49–58. [CrossRef]
4. Axford, W.; Cunnold, D. The wind shear theory of temperate zone sporadic E. *Radio Sci.* **1966**, *1*, 191–197. [CrossRef]
5. Nygrén, T.; Lanchester, B.S.; Huuskonen, A.; Jalonen, L.; Eyken, A. Interference of tidal and gravity waves in the ionosphere and an associated sporadic E-layer. *J. Atmos. Sol.-Terr. Phys.* **1990**, *52*, 609–623. [CrossRef]
6. Goldsbrough, P.F.; Ellyett, C.D. Relationship of meteors to sporadic E, 2. statistical evidence for class 1 Em. *J. Geophys. Res. Atmos.* **1976**, *81*, 6135–6140. [CrossRef]
7. Barta, V.; Haldoupis, C.; Sátori, G.; Buresova, D.; Bencze, P. Searching for effects caused by thunderstorms in midlatitude sporadic E layers. *J. Atmos. Sol.-Terr. Phys.* **2017**, *161*, 150–159. [CrossRef]
8. Bernhardt, P. The modulation of sporadic-E layers by Kelvin–Helmholtz billows in the neutral atmosphere. *J. Atmos. Sol.-Terr. Phys.* **2002**, *64*, 1487–1504. [CrossRef]
9. Cosgrove, R.; Tsunoda, R. Wind-shear-driven, closed-current dynamos in midlatitude sporadic E. *Geophys. Res. Lett.* **2002**, *29*, 1020. [CrossRef]
10. Haldoupis, C.; Pancheva, D.; Mitchell, N.J. A study of tidal and planetary wave periodicities present in midlatitude sporadic E layers. *J. Geophys. Res. Space Phys.* **2004**, *109*, A02302. [CrossRef]
11. Haldoupis, C.; Pancheva, D. Terdiurnal tidelike variability in sporadic E layers. *J. Geophys. Res. Space Phys.* **2006**, *111*, A07303. [CrossRef]
12. Tsunoda, R.T.; Fukao, S.; Yamamoto, M.; Hamasaki, T. First 24.5-MHz radar measurements of quasi-periodic backscatter from field-aligned irregularities in midlatitude sporadic E. *Geophys. Res. Lett.* **1998**, *25*, 1765–1768. [CrossRef]
13. Woodman, R.F.; Yamamoto, M.; Fukao, S. Gravity wave modulation of gradient drift instabilities in mid-latitude sporadic E irregularities. *Geophys. Res. Lett.* **1991**, *18*, 1197–1200. [CrossRef]
14. Maeda, J.; Suzuki, T.; Furuya, M.; Heki, K. Imaging the midlatitude sporadic E plasma patches with a coordinated observation of spaceborne insar and gps total electron content. *Geophys. Res. Lett.* **2016**, *43*, 1419–1425. [CrossRef]
15. Yuan, T.; Wang, J.; Cai, X.; Sojka, J.; Rice, D.; Oberheide, J.; Criddle, N. Investigation of the seasonal and local time variations of the high-altitude sporadic Na layer (Nas) formation and the associated midlatitude descending E layer (Es) in lower E region. *J. Geophys. Res. Space Phys.* **2014**, *119*, 5985–5999. [CrossRef]
16. Hysell, D.; Larsen, M.; Fritts, D.; Laughman, B.; Sulzer, M. Major upwelling and over-turning in the mid-latitude F region ionosphere. *Nat. Commun.* **2018**, *9*, 3326. [CrossRef]
17. Mori, H.; Oyama, K.I. Sounding rocket observation of sporadic-E layer electron-density irregularities. *Geophys. Res. Lett.* **1998**, *25*, 1785–1788. [CrossRef]
18. Mori, H.; Oyama, K.I. Rocket observation of sporadic-E layers and electron density irregularities over midlatitude. *Adv. Space Res.* **2000**, *26*, 1251–1255. [CrossRef]
19. Bernhardt, P.A.; Selcher, C.A.; Siefring, C.; Wilkens, M.; Compton, C.; Bust, G.; Yamamoto, M.; Fukao, S.; Takayuki, O.; Wakabayashi, M. Radio tomographic imaging of sporadic-E layers during SEEK-2. *Ann. Geophys.* **2005**, *23*, 2357–2368. [CrossRef]
20. Damtie, B.; Nygrén, T.; Lehtinen, M.S.; Huuskonen, A. High resolution observations of sporadic-E layers within the polar cap ionosphere using a new incoherent scatter radar experiment. *Ann. Geophys.* **2003**, *20*, 1429–1438. [CrossRef]
21. Turunen, T.; Nygrén, T.; Huuskonen, A.; Jalonen, L. Incoherent scatter studies of sporadic-E using 300 m resolution. *J. Atmos. Terr. Phys.* **1988**, *50*, 277–287. [CrossRef]
22. Zaalov, N.Y.; Moskaleva, E.V. Statistical analysis and modelling of sporadic E layer over Europe. *Adv. Space Res.* **2019**, *64*, 1243–1255. [CrossRef]
23. Whitehead, J.D. The structure of sporadic e from a radio experiment. *Radio Sci.* **1972**, *7*, 355–358. [CrossRef]
24. Dudeney, J.R.; Rodger, A.S. Spatial structure of high latitude sporadic E. *J. Atmos. Terr. Phys.* **1985**, *47*, 529–535. [CrossRef]
25. Hysell, D.L.; Nossa, E.; Aveiro, H.C.; Larsen, M.F.; Munro, J.; Sulzer, M.P.; Gonzalez, S.A. Fine structure in midlatitude sporadic E layers. *J. Atmos. Sol.-Terr. Phys.* **2013**, *103*, 16–23. [CrossRef]
26. Hysell, D.L.; Munk, J.; Mccarrick, M. Sporadic E ionization layers observed with ra-dar imaging and ionospheric modification. *Geophys. Res. Lett.* **2014**, *41*, 6987–6993. [CrossRef]
27. Hocke, K.; Igarashi, K.; Nakamura, M.; Wilkinson, P.; Wu, J.; Pavelyev, A.; Wickert, J. Global sounding of sporadic E layers by the GPS/MET radio occultation experiment. *J. Atmos. Sol.-Terr. Phys.* **2001**, *63*, 1973–1980. [CrossRef]

28. Maeda, J.; Heki, K. Two-dimensional observations of midlatitude sporadic E irregularities with a dense gps array in japan. *Radio Sci.* **2016**, *49*, 28–35. [CrossRef]
29. Palmer, R.D.; Yu, T.-Y.; Chilson, P.B. Range imaging using frequency diversity. *Radio Sci.* **1999**, *34*, 1485–1496. [CrossRef]
30. Liu, T.; Yang, G.; Hu, Y.; Jiang, C.; Lan, T.; Zhao, Z.; Ni, B. A novel ionospheric sounding network based on complete complementary code and its application. *Sensors* **2019**, *19*, 779. [CrossRef]
31. Jiang, C.; Yang, G.; Zhao, Z.; Zhang, Y.; Chen, Z. A method for the automatic calculation of electron density profiles from vertical incidence ionograms. *J. Atmos. Sol.-Terr. Phys.* **2014**, *107*, 20–29. [CrossRef]
32. Jiang, C.; Chen, Z.; Jing, L.; Lan, T.; Yang, G.; Zhao, Z.; Peng, Z.; Sun, H.; Xiao, C. Comparison of the kriging and neural network methods for modeling foF2 maps over north china region. *Adv. Space Res.* **2015**, *56*, 38–46. [CrossRef]
33. Matzka, J.; Stolle, C.; Yamazaki, Y.; Bronkalla, O.; Morschhauser, A. The geomagnetic Kp index and derived indices of geomagnetic activity. *Space Weather* **2021**, *19*, e2020SW002641. [CrossRef]
34. Chen, J.; Zecha, M. Multiple-frequency range imaging using the oswin VHF radar: Phase calibration and first results. *Radio Sci.* **2009**, *44*, 1–16. [CrossRef]
35. Chen, J.; Chu, Y.; Su, C.; Hashiguchi, H.; Ying, L. Observations of field-aligned irregularities in the ionosphere using multi-frequency range imaging technique. In Proceedings of the Geoscience & Remote Sensing Symposium, Milan, Italy, 26–31 July 2015. [CrossRef]
36. Bilitza, D.; Altadill, D.; Truhlik, V.; Shubin, V.; Galkin, I.; Reinisch, B.; Huang, X. International Reference Ionosphere 2016: From ionospheric climate to real-timeweather predictions. *Space Weather* **2017**, *15*, 418–429. [CrossRef]
37. Wakai, N.; Ohyama, H.; Koizumi, T. *Manual of Ionogram Scaling Third Version*; Radio Research Laboratory Ministry of Posts and Telecommunications: Tokyo, Japan, 1987.
38. Reddy, C.; Rao, M.; Matsushita, S.; Smith, L.G. Rocket observations of electron densities in the night-time auroral E-region at Fort Churchill, Canada. *Planet. Space Sci.* **1969**, *17*, 617–628. [CrossRef]
39. Kagan, L.M.; Bakhmet'Eva, N.V.; Belikovich, V.V.; Tolmacheva, A.V.; Kelley, M.C. Structure and dynamics of sporadic layers of ionization in the ionospheric E region. *Radio Sci.* **2002**, *37*, 1–12. [CrossRef]
40. Cosgrove, R.; Tsunoda, R. Polarization electric fields sustained by closed-current dynamo structures in midlatitude sporadic E. *Geophys. Res. Lett.* **2001**, *28*, 1455–1458. [CrossRef]
41. Cosgrove, R.; Tsunoda, R. A direction-dependent instability of sporadic E layers in the nighttime midlatitude ionosphere. *Geophys. Res. Lett.* **2002**, *29*, 1864. [CrossRef]
42. Cosgrove, R.; Tsunoda, R. Simulation of the nonlinear evolution of the sporadic-E lay-er instability in the nighttime midlatitude ionosphere. *J. Geophys. Res.* **2003**, *108*, 1283. [CrossRef]

remote sensing

MDPI

Article

On the Electron Temperature in the Topside Ionosphere as Seen by Swarm Satellites, Incoherent Scatter Radars, and the International Reference Ionosphere Model

Alessio Pignalberi [1,*], Fabio Giannattasio [1], Vladimir Truhlik [2], Igino Coco [1], Michael Pezzopane [1], Giuseppe Consolini [3], Paola De Michelis [1] and Roberta Tozzi [1]

1 Istituto Nazionale di Geofisica e Vulcanologia, Via di Vigna Murata 605, 00143 Roma, Italy; fabio.giannattasio@ingv.it (F.G.); igino.coco@ingv.it (I.C.); michael.pezzopane@ingv.it (M.P.); paola.demichelis@ingv.it (P.D.M.); roberta.tozzi@ingv.it (R.T.)
2 Institute of Atmospheric Physics of the Czech Academy of Sciences, Boční II 1401, 14100 Prague, Czech Republic; vtr@ufa.cas.cz
3 INAF-Istituto di Astrofisica e Planetologia Spaziali, Via del Fosso del Cavaliere 100, 00133 Roma, Italy; giuseppe.consolini@inaf.it
* Correspondence: alessio.pignalberi@ingv.it

Citation: Pignalberi, A.; Giannattasio, F.; Truhlik, V.; Coco, I.; Pezzopane, M.; Consolini, G.; De Michelis, P.; Tozzi, R. On the Electron Temperature in the Topside Ionosphere as Seen by Swarm Satellites, Incoherent Scatter Radars, and the International Reference Ionosphere Model. *Remote Sens.* **2021**, *13*, 4077. https://doi.org/10.3390/rs13204077

Academic Editor: Yunbin Yuan

Received: 30 August 2021
Accepted: 8 October 2021
Published: 12 October 2021

Abstract: The global statistical median behavior of the electron temperature (T_e) in the topside ionosphere was investigated through in-situ data collected by Langmuir Probes on-board the European Space Agency Swarm satellites constellation from the beginning of 2014 to the end of 2020. This is the first time that such an analysis, based on such a large time window, has been carried out globally, encompassing more than half a solar cycle, from the activity peak of 2014 to the minimum of 2020. The results show that Swarm data can help in understanding the main features of T_e in the topside ionosphere in a way never achieved before. T_e data measured by Swarm satellites were also compared to data modeled by the empirical climatological International Reference Ionosphere (IRI) model and data measured by Jicamarca (12.0°S, 76.8°W), Arecibo (18.2°N, 66.4°W), and Millstone Hill (42.6°N, 71.5°W) Incoherent Scatter Radars (ISRs). Moreover, the correction of Swarm T_e data recently proposed by Lomidze was applied and evaluated. These analyses were performed for two main reasons: (1) to understand how the IRI model deviates from the measurements; and (2) to test the reliability of the Swarm dataset as a new possible dataset to be included in the underlying empirical dataset layer of the IRI model. The results show that the application of the Lomidze correction improved the agreement with ISR data above all at mid latitudes and during daytime, and it was effective in reducing the mismatch between Swarm and IRI T_e values. This suggests that future developments of the IRI T_e model should include the Swarm dataset with the Lomidze correction. However, the existence of a quasi-linear relation between measured and modeled T_e values was well verified only below about 2200 K, while for higher values it was completely lost. This is an important result that IRI T_e model developers should properly consider when using the Swarm dataset.

Keywords: electron temperature; topside ionosphere; ESA Swarm satellites; International Reference Ionosphere model; Langmuir Probes in-situ data; Incoherent Scatter Radar data

1. Introduction

The ionosphere is a plasma medium consisting of ions and electrons with also a neutral component that, for most of the cases, are not in thermal equilibrium. In fact, since the early satellites and rockets missions in the 1950s and 1960s [1], a bulk of evidence has been accumulated demonstrating the non-thermal equilibrium between ions, electrons, and neutrals in the ionosphere. Plasma temperature (ions plus electrons) usually exceeds the neutral one by a factor strongly dependent on altitude, time, and location; moreover, large differences among ion and electron temperatures are also observed. As a consequence, when talking about the temperature of the ionosphere, we need to differentiate between ion (T_i), electron (T_e), and neutral (T_n) temperature.

The non-thermal equilibrium of the ionosphere has a number of physical and chemical implications for the ionospheric dynamics and composition. For example, the temperature of charged particles deeply influences the distribution of electron density by driving the recombination rate in the bottomside ionosphere, and affects the diffusive equilibrium state in the topside ionosphere [2,3]. The main source of energy heating the ionosphere is the Sun: either directly, through the EUV and X-ray radiations illuminating the Earth, or indirectly, at high latitudes, through the precipitation of charged particles induced by the interactions between the solar wind and the magnetosphere-ionosphere system and the consequent Joule heating [1,4–6]. In the ionization processes occurring in the ionosphere, a certain amount of energy is released as an excess of kinetic energy of photoelectrons. Such photoelectrons dissipate their energy through collisions with charged and neutral particles in a cascade process that first raises the temperature of the electrons (because the transfer of energy via elastic collisions is more efficient between particles of similar mass). In a second stage, the established temperature difference drives the transfer of energy between electrons and the surrounding medium. Above the F2 region, most of the energy transfer is from electrons to ions because ion-electron Coulomb collisions are more efficient than neutral-electron ones; then, the temperature of the ions also increases above the neutrals one. As a further step, the ions lose their excess energy through collisions with neutrals and other ions, thereby raising the temperature of neutrals. As a result of these complex interactions, the entire photoelectron energy dissipates in the neutrals and it is finally transferred down in the lower layers of the atmosphere by conduction and radiation.

Among the three mentioned temperatures, T_e is the most variable because of the very low heat capacity of electrons [4]; moreover, electrons are the most affected by the solar radiation energy input and by the interaction with charged particles entering the Earth's ionosphere following the geomagnetic field lines at auroral latitudes. The close coupling with the geomagnetic field lines and the plasmasphere produces very distinct spatial patterns in the T_e distribution [6]. As a consequence, T_e exhibits very distinctive diurnal, seasonal, spatial, and solar activity trends. This large variability makes it difficult to obtain a precise and reliable description of T_e through in-situ or remote sensing measurements. In fact, in-situ measurements taken by Langmuir Probes (LPs) and Retarding Potential Analyzers (RPAs) [1,7] on-board Low-Earth-Orbit (LEO) satellites provide a global spatial description for different local times, but they are highly dependent on the satellite orbital configuration and altitude and are limited to the satellite mission lifetime. Conversely, ground-based remote sensing techniques such as Incoherent Scatter Radars (ISRs) [7,8] can better describe the altitudinal, diurnal, and seasonal variability but for a fixed geographic location and for the very few existing ISR facilities. Moreover, the solar activity variability can be described when long time series become available. Furthermore, the cost of this measurement technique strongly limits the number and distribution of ISRs worldwide. As a consequence, for modeling purposes, both in-situ satellites and remote sensing ISRs observations are very important and complement each other.

An outstanding example is represented by the International Reference Ionosphere (IRI) model [9], which is the empirical climatological model of the ionosphere used as reference by the ionospheric community, and recently recognized as the official ISO (International Organization for Standardization) standard for the ionosphere [10]. Through the exploitation of different measurement datasets of ionospheric parameters collected by ionosondes, ISRs, LEO satellites, Global Positioning System (GPS) radio occultation, and rockets, IRI describes the hourly monthly median behavior of the electron density, electron and ion temperature, and ion composition in the ionospheric altitude range, on a global basis, for different levels of solar and magnetic activity. Due to its empirical nature, IRI evolves and improves when new datasets become available and are used to derive the empirical coefficients of its analytical formulations. In this framework, the ability to model the changes of the ionospheric ground state as a function of the different climatological conditions of both the interplanetary medium and the Earth's seasons is also fundamental in relation to space weather science.

Despite a long series of measurements acquired by both satellites and ISRs, the major current gap in modeling and studying ionospheric T_e consists in the lack of an accurate description of its dependence on the most important geographic and geophysical parameters. There is still insufficient knowledge of the global distribution and relevant description of T_e longitudinal variations as well as of its seasonal and hemispheric asymmetries, despite some attempts [11–13]. A difference between long-term and short-term T_e variations with F10.7 has also been noted [14], which further complicates the modeling of T_e. Empirical models still struggle to describe small-scale structures such as the morning overshoot [15–17] and its dependence on solar and magnetic activity or the mid-latitude T_e enhancement [18,19]. The dependence on the magnetic activity and the choice of an appropriate magnetic activity proxy for T_e modeling is still an open problem; for example, Brace and Theis [20] found no clear dependence of T_e on the *ap* index, with exception of perhaps a slight systematic increase of T_e with *ap* at polar latitudes. Apart from modeling, there are still gaps in the fundamental theory. For instance, Pavlov et al. [21] introduced an anomalous conductivity to obtain an agreement between theory and EXOS-D data in the plasmasphere. Even very advanced physics-based models still encounter problems with reproducing observational data (e.g., the post-sunset T_e enhancement) and must be driven by measured data [22]. Another tricky point concerns the existing problems associated with the frequent T_e differences among satellite data from different missions, and between in-situ data and ground-based ISR measurements [23,24], which require in many cases a cross calibration and/or a correction of these data. Because of this, studying and global modeling of T_e has become a very challenging problem.

From this point of view, T_e observations collected by LPs on-board the European Space Agency (ESA) Swarm satellites [25] are very promising for a future inclusion in the IRI T_e model [26]. In fact, the three Swarm satellites launched at the end of 2013 and still operating provide T_e data in the topside ionosphere region on a global basis, for different local times, at a 2-Hz rate. As a consequence, the present dataset of Swarm LP T_e measurements is that large to cover different latitudes, local times, seasons, and also solar activity levels. Contrary to the electron density, T_e data have not been extensively exploited so far; only a few and rather narrowly focused studies have been published, such as [17,24]. In this context, we aimed to investigate the main climatological trends of T_e in the topside ionosphere by taking advantage of seven full years (from the beginning of 2014 to the end of 2020) of Swarm T_e observations by comparing the corresponding spatial, diurnal, and seasonal trends with those modeled by IRI. Recently, to improve the accuracy of Swarm T_e data, Lomidze et al. [24] proposed a correction to observations, based on the comparison with three ISRs located at different latitudes. In this study, we applied the Lomidze correction to Swarm data to verify its performance when compared against IRI-modeled T_e data and T_e data observed by Jicamarca (12.0°S, 76.8°W), Arecibo (18.2°N, 66.4°W), and Millstone Hill (42.6°N, 71.5°W) ISRs over the last decades.

The paper is organized as follows. In Section 2, the Swarm and ISRs T_e datasets, along with the IRI T_e model and the Lomidze correction, are discussed. In Section 3, an overall comparison between Swarm-measured (original and after applying the Lomidze correction) and IRI-modeled T_e values is first given based on the entire dataset composed by the three Swarm satellites; in addition, Swarm B and IRI T_e data are climatologically compared with the corresponding values recorded by the Jicamarca, Arecibo, and Millstone Hill ISRs. In the same section, we also describe and discuss the geographic, diurnal, and seasonal trends shown by both Swarm B and IRI data. Section 4 is devoted to the final discussion of our results and is supported by the conditioned probability density function analysis of Swarm-measured and IRI-modeled data. The conclusions and the future developments are the subject of Section 5.

2. Materials and Methods

2.1. ESA Swarm In-Situ Electron Temperature Data and Application of the Lomidze Correction

The ESA Swarm mission is a constellation of three LEO satellites launched at the end of 2013, and still operating, with the aim of studying the geomagnetic field, the electric currents in the magnetosphere and ionosphere, and the impact of the solar wind on the dynamics of the upper atmosphere [25]. The three Swarm satellites are labeled A, B, and C, and share the same design and on-board instrumentation. All three satellites were put in a circular near-polar orbit. Swarm A and C have the same orbit configuration (inclination of 87.35°, initial altitude of about 460 km, east-west separation of about 1–1.5° in longitude), while Swarm B has a different one (inclination of 87.75°, initial altitude of about 510 km). As a consequence, Swarm A and C fly in tandem, while Swarm B moves away from the couple A and C by covering different local times. Each satellite takes about 130–140 days to cover all the local times.

Among the instruments carried by Swarm satellites, we focused on LPs data [27]. LPs provide measurements of in-situ electron density (N_e) and T_e along the satellite orbit with an original sampling frequency of 2 Hz in the harmonic mode [24,28]. Moreover, 1-Hz data are also provided by interpolation of the 2-Hz dataset; it is this dataset containing time series of in-situ N_e and T_e data at 1-Hz rate that is used in this work. Specifically, LP data collected by Swarm A, B, and C from 1 January 2014 to 31 December 2020 (seven full years) are taken into account. Swarm's data are freely downloadable at ftp://swarm-diss.eo.esa.int (accessed on 24 September 2021).

Swarm's LP data are provided with three flags [28]: *Flags_LP* contains information on the source of LP measurements, and *Flags_Ne* and *Flags_Te* characterize N_e and T_e measurements, respectively. Detailed information about Swarm data quality is available at https://earth.esa.int/web/guest/swarm/data-access/quality-of-swarm-l1b-l2cat2-products (accessed on 24 September 2021). In this study, only the most reliable data were considered; specifically, those recorded by LPs in *High-Gain* mode. LP data were then selected with the following choice of the flags: *Flags_LP* = 1, *Flags_Ne* $\leq$ 29, and *Flags_Te* = 10 or 20.

Lomidze et al. [24] developed a correction of both N_e and T_e Swarm LP data based on the analysis of Swarm measurements matching coincident values retrieved by ISRs, COSMIC/FORMOSAT-3 GPS radio occultation, and ionosondes. For what concerns the Swarm's T_e data correction, Lomidze et al. [24] separately considered the measurements collected when the LPs are in *High-Gain* mode (HG in the following) from those collected with the LPs in *Low-Gain* mode. Since only HG data were considered in this study, we applied only the HG Lomidze correction to T_e data. As a consequence, the Equations (4)–(6) in Lomidze et al. [24] were applied on the Swarm A, B, and C T_e datasets, namely:

$$T_{e,Lom} = 1.2815 \cdot T_{e,LP} - 1167 + 7.293 \cdot N_{e,LP}/10^4 \text{ Swarm A, HG} \quad (1)$$

$$T_{e,Lom} = 1.2248 \cdot T_{e,LP} - 1047 + 8.548 \cdot N_{e,LP}/10^4 \text{ Swarm B, HG} \quad (2)$$

$$T_{e,Lom} = 1.1334 \cdot T_{e,LP} - 762 + 4.088 \cdot N_{e,LP}/10^4 \text{ Swarm C, HG} \quad (3)$$

In Equations (1)–(3) $T_{e,Lom}$ is the Lomidze-corrected T_e, $T_{e,LP}$ is the original Swarm-measured value, $N_{e,LP}$ is the corresponding N_e value measured by Swarm LP (i.e., unadjusted). In Equations (1)–(3), and throughout the paper, T_e values are in K and N_e values are in cm^{-3}. Equation (1) is valid for the dataset of HG Swarm A LP T_e dataset, Equations (2) and (3) for the corresponding Swarm B and C datasets, respectively.

As outlined by Equations (1)–(3), the Lomidze correction is based on a regression analysis of Swarm-measured $T_{e,LP}$ with data retrieved by the three ISRs located at Jicamarca, Millstone Hill, and Arecibo from December 2013 to June 2016 (for more information on the dataset selection on which Equations (1)–(3) are based see [24]). The shortness of the dataset did not allow a characterization of possible diurnal, seasonal, geographic, and solar activity dependences of the corrections introduced in Equations (1)–(3). As a consequence, Equations (1)–(3) were applied on the entire Swarm datasets although some residual diurnal, seasonal, geographic, and solar activity dependences were somewhat expected. The application of these ISRs-based corrections allowed greatly improving the accuracy of Swarm T_e data when compared to ISRs ones (stepping from median differences between 300 and 400 K to near zero) without affecting the corresponding precision and correlation [24]. As outlined by Lomidze et al. [24], the inclusion of the term $N_{e,LP}$ in Equations (1)–(3) produced a worsening in the precision of the correction (less than 60 K) but with the advantage of removing some certain non-physical patterns from Swarm T_e data.

In this work, both Swarm $T_{e,LP}$ original data and Lomidze-corrected $T_{e,Lom}$ data were employed. In Section 3, a statistical comparison between T_e data measured by Swarm, calibrated after applying the Lomidze correction, and the corresponding ones modeled by IRI is presented. In the same section, a statistical comparison with T_e data recorded by Jicamarca, Arecibo, and Millstone Hill ISRs is also performed.

2.2. Electron Temperature Observations by Incoherent Scatter Radars

ISRs exploit the Thomson backscatter from ionospheric electrons to retrieve the value of several ionospheric parameters including T_e [8]. Specifically, T_e profiles are recorded by ISRs in the ionospheric altitude range from about 100 to 1000 km.

T_e observations collected by Jicamarca, Arecibo, and Millstone Hill ISRs were used in this work for validation purposes. Jicamarca ISR is located right above the magnetic equator at a Quasi-Dipole (QD, [29]) latitude of 0.2°N, Arecibo (QD latitude 27.0° N) is a low-latitude station, while Millstone Hill (QD latitude 51.8°N) is a mid-latitude/sub-auroral station. Jicamarca data are from 1996 to 2020, Arecibo ones from 1974 to 2015, and Millstone Hill ones from 1976 to 2020. ISRs data are not continuous in time but the length of the datasets here used allow for a quite uniform representation of the different diurnal and seasonal trends.

The ISRs T_e data (including corresponding T_e error) were downloaded from the http://cedar.openmadrigal.org website (accessed on 24 September 2021). The altitudes of Swarm satellites fall into standard remote sensing altitude ranges of all three ISRs, which provide reliable data for decades. A similar dataset referring to T_i, for Millstone Hill was recently employed in Pignalberi et al. [30].

To perform a comparison with Swarm B-measured and IRI-modeled values, we selected T_e values recorded by ISRs in the range 510 ± 20 km. Only the most reliable ISRs' data were selected, i.e., those with a percentage error lower than 10%. This filtering mostly affected data recorded during the night and early-morning, which are those characterized by the largest dispersion.

2.3. Electron Temperature Description by the IRI Model

In the last version of IRI (IRI-2016, [9]), two different models for T_e are implemented; they are conventionally named BIL-1995, and TBT-2012 (http://irimodel.org/, accessed on 24 September 2021). The BIL-1995 is the oldest option and is based on the works by Bilitza [31–33] who, through the application of the Booker approach [34] to the Brace and Theis [20] and Spenner and Pluge [35] models, describes the T_e vertical profile for the ionospheric altitude range between 60 and 2000 km. Since the IRI-2001 version [36], IRI has included an additional option based on Intercosmos data [37], that since the IRI-2012 version [38] was updated with the most recent TBT-2012 model [26] which incorporates all available satellite T_e data. The solar activity dependence, also described in Truhlik et al. [26], is mainly based on results from Truhlik et al. [39]. In the current IRI version, this additional option is called TBT-2012+SA and represents the option related to the TBT-2012 model with the solar activity term switched on. Since the TBT-2012+SA is the current default option for the T_e modeling in IRI, it has been applied in this work to perform a comparison with Swarm and ISRs measured data and will be briefly described here.

In the TBT-2012 model [26], T_e values at five fixed anchor points at 350, 550, 850, 1400, and 2000 km of altitude are modeled through a spherical harmonic expansion in a system of associated Legendre polynomials (up to the 8th order) in terms of the magnetic local time and a latitudinal coordinate based on a combination of invariant and dip latitude (invdip, [40]). Through the use of T_e measurements from LPs on-board several satellite missions (for a detailed description of the dataset see Bilitza et al. [23] and Truhlik et al. [26]), the coefficients of the spherical harmonic expansions were calculated for summer and winter solstices, and for the combined equinoxes (independently of the hemisphere); then, interhemispheric differences in magnetic coordinates were not considered in the model. The solar activity dependence was then included by selecting three solar activity ranges and describing the corresponding variability as a function of the $PF_{10.7}$ solar index [26].

The complete vertical T_e profile was obtained by applying the Booker profile function formalism [33,34]. First, a piece-wise function was built making a linear interpolation of modeled T_e values between anchor points, which produced a vertical profile divided in regions with a constant T_e gradient. Second, Epstein-step functions were applied to realize the transition from regions with different gradients and obtain the skeleton profile, which produced a continuous analytical representation of the T_e vertical profile. Finally, the skeleton profile was integrated to obtain the Booker profile function of T_e, which output a continuous and smooth description of the entire modeled T_e vertical profile. For a detailed mathematical description of the Booker profile function formalism, refer to the Appendix section of Pignalberi et al. [30].

In IRI, T_e is constrained to be equal to T_n described by the NRLMSISE-00 model [41] for altitudes lower than 120 km. An example of IRI (TBT-2012+SA default option) modeled vertical T_e profile is given in Figure 1 (green curve) where the anchor points that IRI uses to describe the entire profile are highlighted. An anchor point at a non-fixed altitude around 270 km ($T_{e,m}$ in Figure 1) was introduced on the basis of Jicamarca and Arecibo ISRs data to describe a local daytime maximum in T_e characterizing mostly the low latitudes [33].

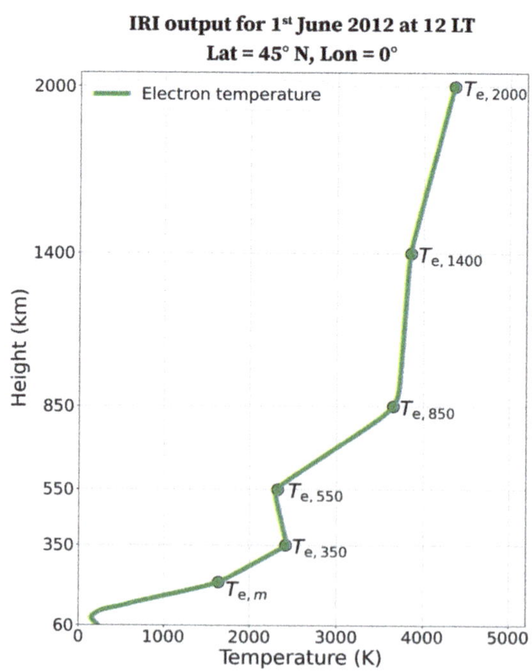

Figure 1. IRI modeled (TBT-2012+SA default option) vertical profile of T_e (green curve) at 45° N of latitude and 0° of longitude for 1 June 2012 at 12 local time (LT). The five fixed anchor points at 350, 550, 850, 1400, and 2000 km of altitude are highlighted along with the additional anchor point $T_{e,m}$.

In this work, the IRI model (TBT-2012+SA option) was run on the entire Swarm satellites dataset as if it were collocated and synchronized with LPs on-board Swarm satellites. As a consequence, 1-Hz time series of IRI-modeled T_e values were produced for the time and locations probed by Swarm satellites. In this way, a one-to-one comparison between Swarm-measured and IRI-modeled T_e values was guaranteed.

3. Results

3.1. Overall Statistical Comparison between Measured and Modeled Electron Temperature Values

First of all, we aimed to investigate the effect of the application of the Lomidze correction on Swarm T_e measurements through a comparison with IRI-modeled values for the period 2014–2020 for all the three Swarm satellites.

An overall comparison between Swarm-measured (original and Lomidze-corrected) and IRI-modeled T_e values is given in Figures 2 and 3 as joint histograms and distributions of residuals between measured and modeled data. From the residuals between measured and modeled values, the following statistical metrics were calculated: mean and standard deviation of the residuals, Root Mean Square Error (RMSE, Equation (4)), and Relative Root Mean Square Error (RRMSE, Equation (5)).

$$\text{RMSE [K]} = \sqrt{\frac{\sum_{i=1}^{N}\left(T_{e,\text{measured},i} - T_{e,\text{modeled},i}\right)^2}{N}} \quad (4)$$

$$\text{RRMSE [\%]} = \sqrt{\frac{\sum_{i=1}^{N}\left(\frac{T_{e,\text{measured},i} - T_{e,\text{modeled},i}}{T_{e,\text{modeled},i}} \cdot 100\right)^2}{N}} \quad (5)$$

In Equations (4) and (5), $T_{e,\text{measured}}$ is the T_e measured by Swarm (i.e., $T_{e,\text{LP}}$) or corrected according to Lomidze et al. [24] (i.e., $T_{e,\text{Lom}}$), while $T_{e,\text{modeled}}$ is the corresponding value modeled by IRI. Temperatures are in K. N is the number of values on which the statistics are calculated.

Figure 2 represents the joint histograms between T_e values measured by Swarm and modeled by IRI and gives information on the distribution of measured and modeled data as a function of the corresponding magnitude. Most of the values cluster in the reddish region, i.e., where more than 10^4 values are counted inside each bin (the scale is logarithmic). It is interesting to note that IRI-modeled data never go beyond about 4300 K, while Swarm data (both original and Lomidze-corrected ones) show a long tail well beyond 5000 K (not shown in the figure), even though the application of the Lomidze correction reduces their magnitude. To characterize the accordance between modeled and measured values and the global effect of the application of the Lomidze correction, corresponding residuals were studied.

Figure 3 represents the statistical distributions of the residuals between values measured by Swarm and modeled by IRI. By looking at the statistical metrics reported in the upper left corner of each panel, it emerges how the application of the Lomidze correction to Swarm data improved the accuracy when compared to IRI-modeled data. In fact, the mean of the residuals decreased from about 560–650 K to about 150–250 K, with a 400 K average improvement. Conversely, the application of the Lomidze correction did not affect the precision of Swarm data, as highlighted by the standard deviation values that remained in the range 380–470 K. Nevertheless, the application of the Lomidze correction generally improved the agreement with IRI data, as demonstrated by the RMSE and RRMSE values that decreased, respectively, from about 690–780 K to about 440–530 K, and from about 27–29% to about 20–23%.

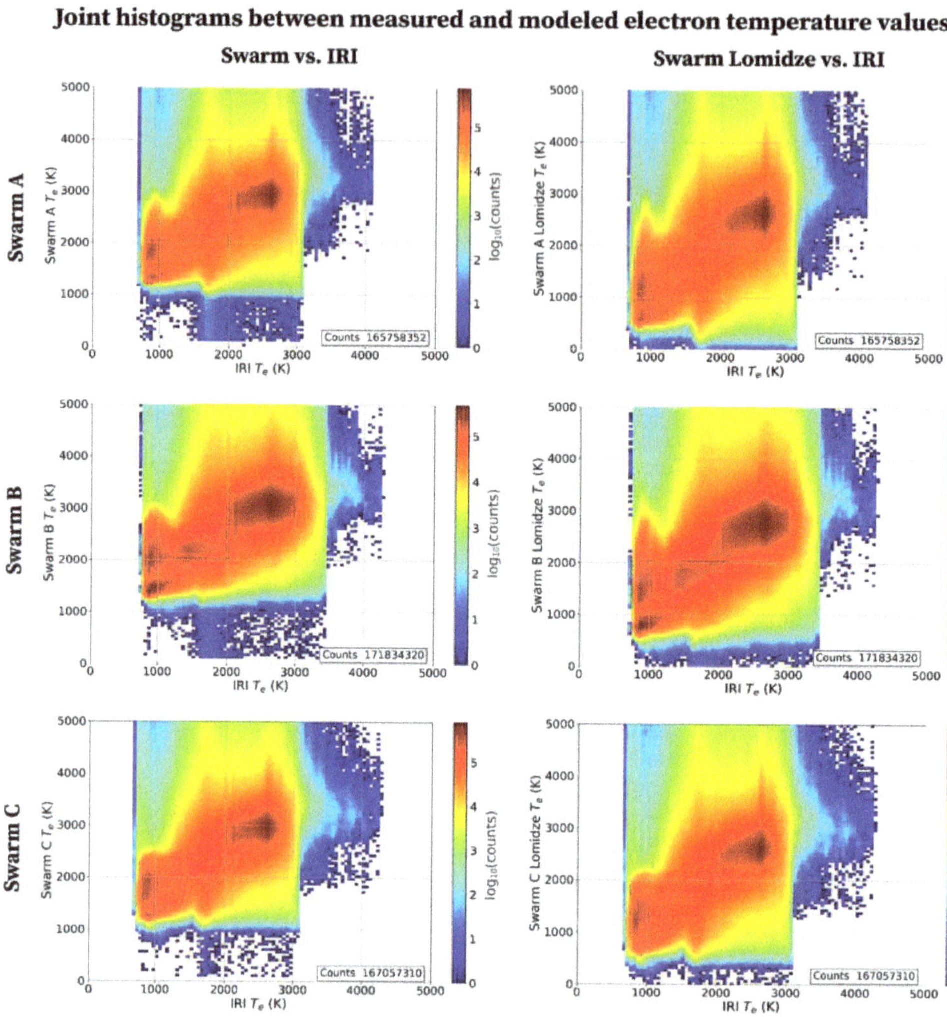

Figure 2. Joint histograms between T_e values measured by Swarm and modeled by IRI (TBT-2012+SA default option): (left panels) original Swarm-recorded values, (right panels) Swarm values corrected with Lomidze et al. [24]. From top to bottom, the analysis refers to Swarm A, B, and C. In each panel, the number of total counts is reported in the lower right corner.

To sum up, the results shown in Figure 3 demonstrate how the application of the Lomidze correction to Swarm T_e data can improve the agreement with IRI-modeled values, without impacting the corresponding precision. However, we are aware that the comparison with a model, IRI, cannot guarantee the reliability and efficacy of the Lomidze correction; the only comparison that can truly tell whether this correction improves real T_e observations is that with other independent measurements. This is why we performed a comparison against T_e data observed by ISRs.

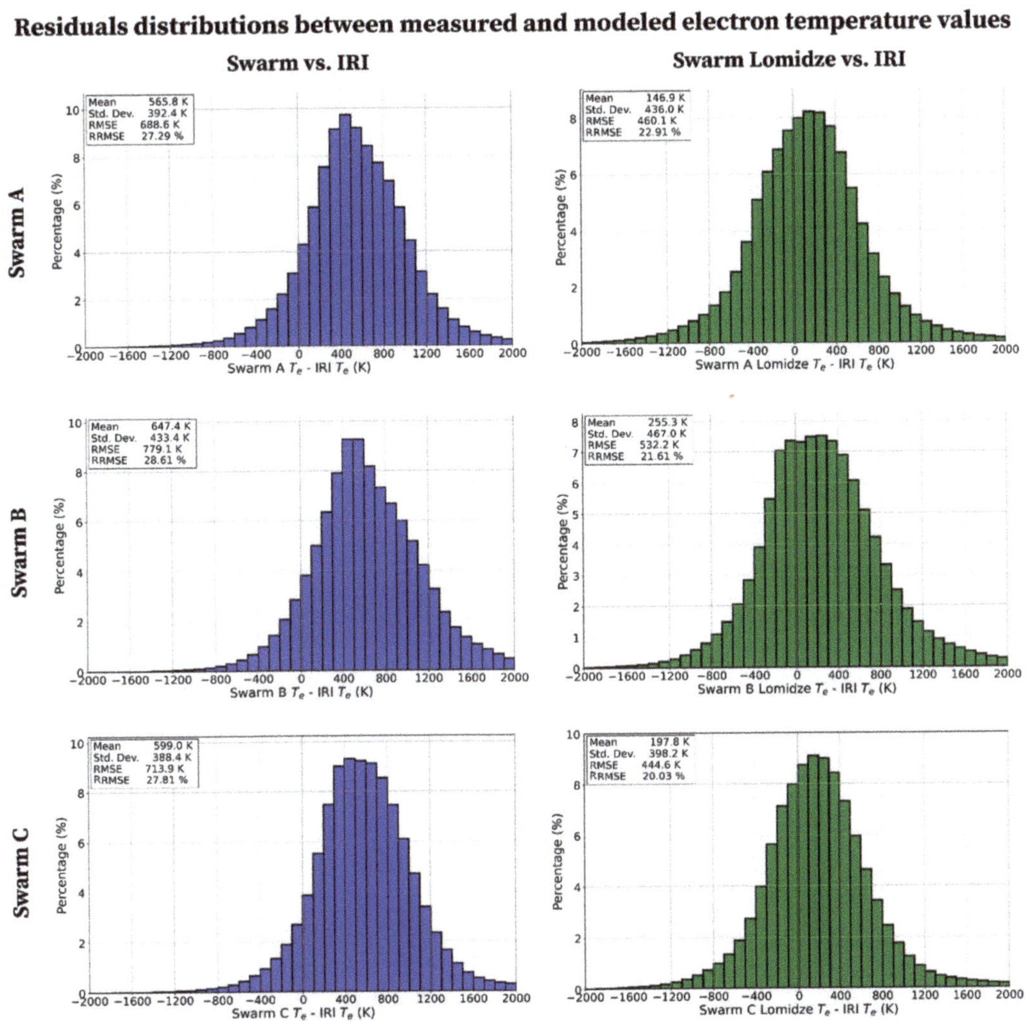

Figure 3. Statistical distributions of the residuals between T_e values measured by Swarm and modeled by IRI (TBT-2012+SA default option): (left panels, in blue) original Swarm-recorded values, (right panels, in green) Swarm values corrected with Lomidze et al. [24]. From top to bottom, the analysis refers to Swarm A, B, and C. In each panel, some statistical metrics are reported in the upper left corner.

3.2. Statistical Comparison against ISRs Data

We implemented a statistical comparison among Swarm (original and corrected values), IRI, and ISRs T_e data for the Jicamarca, Arecibo, and Millstone Hill locations to verify the ability of Swarm and IRI to describe the diurnal and seasonal trends of T_e shown by ISRs data, and their accuracy.

Unlike Lomidze et al. [24], we did not look for correspondences (in both time and space) between ISRs and Swarm data because of the limited number of correspondences achievable (of the order of some hundreds from 2014 to 2020, for each satellite). Instead, we applied a statistical approach by comparing the diurnal and seasonal trends exhibited by both T_e values collected by ISRs and by Swarm B for the ISRs locations. In this way, we could investigate how much Swarm, and also the IRI model, can describe the diurnal and seasonal T_e trends at equatorial, low, and mid latitudes by making a comparison with Jicamarca, Arecibo, and Millstone Hill data, respectively. This comparison and those of the following section were limited to the Swarm B satellite because it is the nearest to the IRI vertical profile anchor point at 550 km of altitude. Moreover, as follows from Figures 2 and 3, the statistical analysis for all three satellites was very similar, with slightly larger differences for Swarm B.

Specifically, T_e values collected at Jicamarca, Arecibo, and Millstone Hill in the range (510 ± 20) km were binned as a function of the magnetic local time (MLT, in QD coordinates) and season, and corresponding boxplots were generated. The binning in 15-minutes wide (in MLT) bins allowed describing the diurnal trend of T_e at ISRs location. The seasonal trend was studied by binning data in four periods (expressed in day of the year, doy) around solstices and equinoxes and covering the entire year:

- March Equinox: $35 \leq$ doy ≤ 125;
- June Solstice: $126 \leq$ doy ≤ 217;
- September Equinox: $218 \leq$ doy ≤ 309;
- December Solstice: doy ≤ 34 OR doy ≥ 310.

No solar activity data selection was performed because the quantity of both ISRs and Swarm data does not allow further split, and because the investigation of T_e dependence on solar activity is outside of the scope of this study. As a consequence, the results in Figures 4–6 are representative of average solar activity conditions of the years covered by ISRs datasets (see Section 2.2), and of the years 2014–2020 for Swarm and IRI.

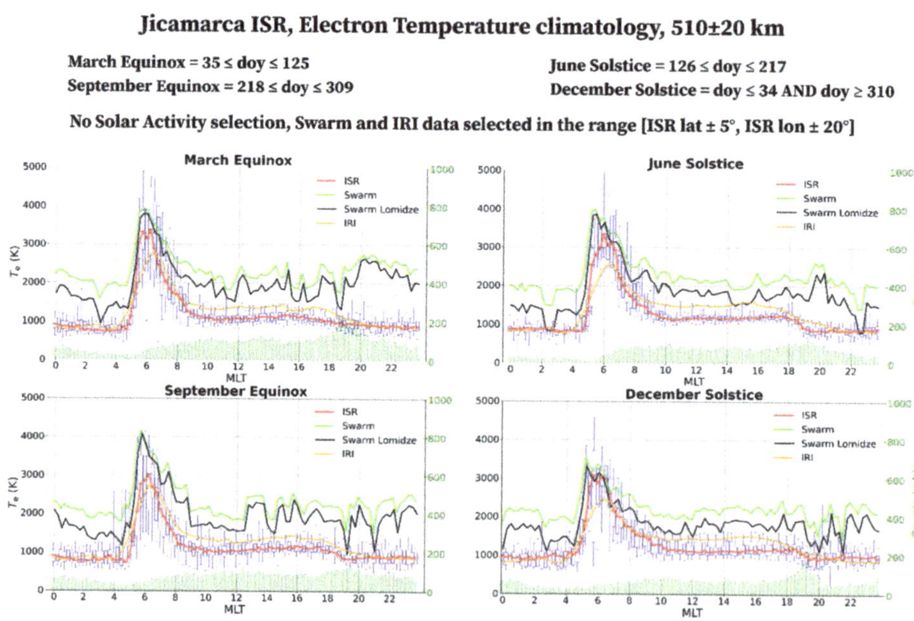

Figure 4. Comparison between T_e data observed by Jicamarca ISR (12.0°S, 76.8°W, QD latitude 0.2°N) at around 510 km of altitude (boxplots), Swarm B satellite (green line), Swarm B corrected with Lomidze (black line), and those modeled by IRI (TBT-2012+SA default option) (orange line). Both measured and modeled data are binned as a function of MLT (x-axis) in bins 15-minutes wide, and of the season by selecting data around the equinoxes and solstices (each panel represents a different season). ISR data are represented as boxplots in which the red horizontal line is the median; the 25th and 75th percentiles are represented as the lower and upper limits of each box; the 5th and 95th percentiles are shown as lines extending below and above each box (whiskers). Green shaded bars at the bottom of each panel represent the number of ISR data falling in that bin. Conversely, for Swarm and IRI, only the median values are represented as solid curves.

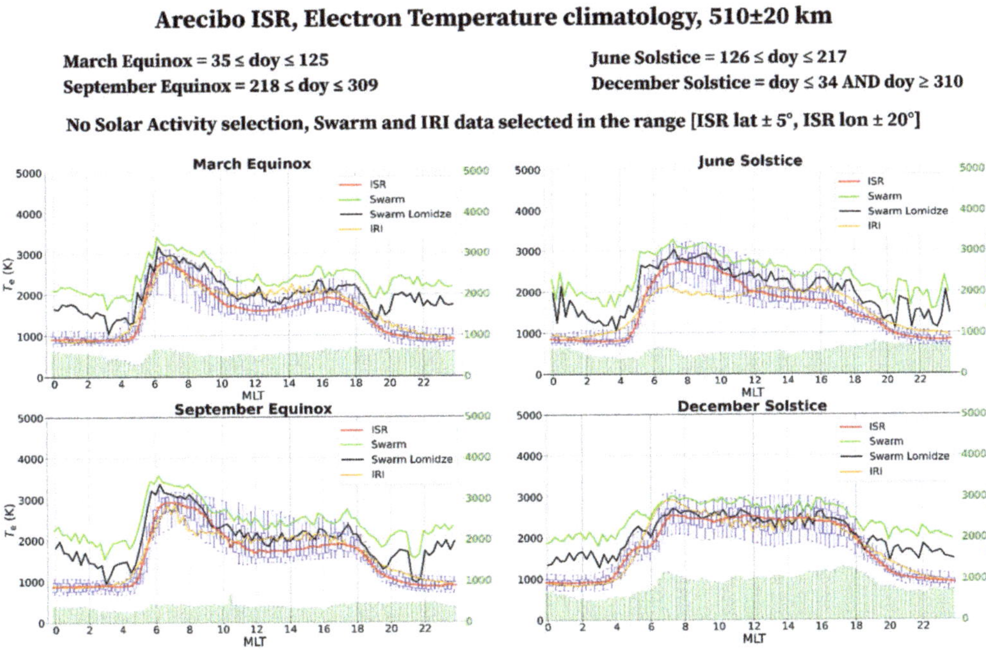

Figure 5. Same as Figure 4 but for Arecibo ISR (18.2°N, 66.4°W, QD latitude 27.0°N).

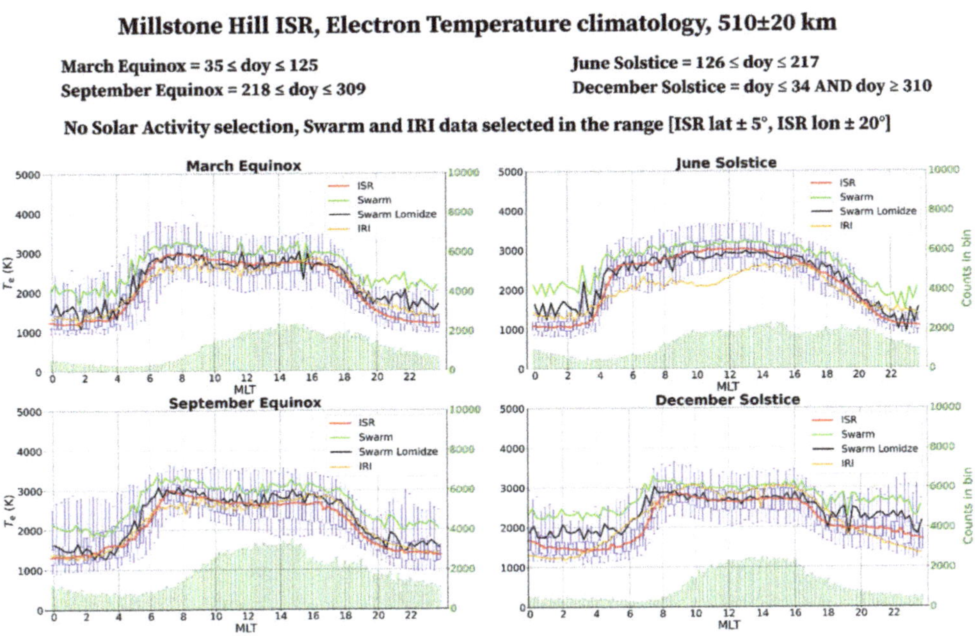

Figure 6. Same as Figure 4 but for Millstone Hill ISR (42.6°N, 71.5°W, QD latitude 51.8°N).

In summary, every ISR dataset was split in 384 (96 diurnal x 4 seasonal) bins and the following statistical metrics were calculated for the boxplot representation:

- 5th percentile, i.e., the lower whisker;
- 25th percentile, i.e., the first quartile;
- 50th percentile, i.e., the second quartile, representative of the median;
- 75th percentile i.e., the third quartile;
- 95th percentile, i.e., the upper whisker.

For this kind of comparison, the median is the more appropriate statistical metric because it cuts off the outliers of the distribution. The difference between 75th and 25th percentiles defines the inter-quartile range (IQR), which provides a measure of the data dispersion around the median value, while the whiskers highlight the tails of the distribution.

Figures 4–6 show the boxplots for the Jicamarca (Figure 4), Arecibo (Figure 5), and Millstone Hill (Figure 6) sites. For the comparison with ISRs results, both Swarm B-measured (original and corrected with Lomidze) and IRI-modeled T_e data were selected and binned as done for ISRs data. Specifically, from the Swarm B dataset, we selected those measurements falling inside a region centered at ISRs' geographic coordinates and extending $\pm 5^{\circ}$ in latitude and $\pm 20^{\circ}$ in longitude. We tested regions of different extension in both latitude and longitude around the ISRs' location. The choice we made is the one balancing the amount of data needed for the statistical comparison and the preservation of the characteristic trends at the ISRs locations. Then, the selected Swarm B (original and corrected with Lomidze) and IRI data were binned as ISRs data and the corresponding median values are superposed to ISRs' boxplots in Figures 4–6.

Figure 4 shows how equatorial values at Jicamarca are characterized by the presence of a remarkable morning peak between 5 and 8 MLT, with median values from ISR exceeding 3000 K; this phenomenon is often referred as morning overshoot [15–17,42]. Overall, daytime values stand at about 1000–1300 K, while nighttime ones at about 700–1000 K, in a quite steady fashion. Differently from the dawn solar terminator hours, the ones related to the dusk solar terminator do not exhibit a peak but only a small ramp with T_e smoothly stepping from the daytime to the nighttime behavior. Of course, the MLTs of both the morning overshoot and the dusk ramp slightly change according to the season. The MLTs characterized by the morning overshoot exhibit the highest dispersion of data as described by the IQRs and whiskers values. The seasonal variability is really weak at Jicamarca due to the illumination conditions that do not change significantly over the year. Both Swarm and IRI can catch the main diurnal features of T_e for each season but with remarkable differences in the absolute values. IRI represents reliably nighttime values, while it slightly overestimates daytime values and underestimates values in the morning sector when the overshoot occurs. Differently, Swarm (both original and corrected with Lomidze) performs a general overestimation independently of the hour. Moreover, Swarm values do not exhibit differences between day and night, with the exception of the morning peak. The application of the Lomidze correction reduces the original overestimation of Swarm measurements, but not enough to make them match with ISR data. When comparing with ISRs data, we have to take into account that the IRI TBT-2012+SA option is based on data recorded from satellites [26], that are often limited in both space and time. In fact, the differences between IRI and Jicamarca ISR values might be related to the limited spatial and temporal resolution of the model (20 degrees in latitude and 80 minutes in MLT), resulting in the inability of the model to accurately describe the equatorial daytime dip in T_e and thus resulting in an overestimation of daytime T_e at the magnetic equator by IRI. At MLTs of the narrow and high morning peak, this limited resolution is most likely at the base of both the underestimation and some mismatches of the position of the morning peak maximum made by the IRI model. Another reason that could explain the observed differences is the fact that in-situ probes can produce higher T_e values in a non-equilibrium plasma [43] or if some sort of probe contamination occurs [44]. A possible combination of both effects thus could result in both an overestimation of about 30% during daytime and an underestimation of about 20% of the peak maximum made by the IRI model (as evidenced by Figure 4). During nighttime, when plasma is in thermal equilibrium and the limited spatial and temporal resolution of the model is not so important, the agreement between IRI and ISR is very good.

Figure 5 shows the results for Arecibo. Compared to Jicamarca, at Arecibo the seasonal variability is more marked above all for MLT hours from noon to dusk. Like Jicamarca, Arecibo values also show a remarkable morning increase with temperatures abruptly increasing from about 1000 to 2500–3000 K. However, Arecibo daytime values do not decrease as for Jicamarca but remain at around 2000 K (except for the December solstice) with significant differences in the sector 14–19 MLT, depending on the season. In fact, March and September equinoxes show a very similar trend where the first hump corresponding to the morning increase is followed by a small second hump before sunset, while for the June solstice (summer), the second hump does not appear and values steadily decrease from daytime to nighttime. The December solstice (winter) is instead characterized by quite constant values (at about 2500 K like the morning peak one) during daytime and a very distinct decrease at sunset. IRI can reliably represent the nighttime T_e at Arecibo, while some departures are evident during daytime, above all in the morning for the June solstice (underestimation) and in the central hours of the day at equinoxes (overestimation). Compared to Jicamarca, at Arecibo, the accordance between Swarm and ISR data is better, especially in the daytime sector, while nighttime values are still affected by a remarkable overestimation (even after applying the Lomidze correction). A very good accordance is found for the winter season (December solstice) during daytime. Swarm and IRI can describe both the diurnal and seasonal trends of T_e at Arecibo.

Figure 6 summarizes the results for Millstone Hill, where T_e manifests a very distinct diurnal behavior characterized by a sharp division between nighttime and daytime values. Nighttime values are of the order of 1100–1500 K, while the daytime values range between 2500 and 3000 K. Differently from Arecibo, in this case, daytime values are always quite constant with a slight tendency to maximize at noon during summer (June solstice). Winter (December solstice) nighttime values are remarkably higher than summer ones due to conjugate heating [23,45]. Except for June solstice, IRI can reliably describe the diurnal behavior of T_e at Millstone Hill. It is the same for Swarm after applying the Lomidze correction; in this case the agreement with ISR's data is remarkable for both the representation of the diurnal and seasonal trends and the accuracy of the absolute values; only a slight overestimation is visible during nighttime particularly at December solstice.

It has to be considered that the Lomidze correction, aside from being based on the same three ISRs here studied, is heavily biased toward Millstone Hill ISR data that were more represented in the Lomidze et al. [24] dataset than Arecibo data and above all Jicamarca. As a consequence, since Lomidze et al. [24] cumulated the data of the three ISRs to derive a single calibration curve, it should be expected that the correction is more effective at Millstone Hill than at Arecibo and Jicamarca. This agrees very well with the results shown in Figures 4–6. Moreover, the Lomidze correction does not discriminate between daytime and nighttime or between different seasons. Our results show that the Lomidze correction has the best performance for daytime hours and the worst for nighttime hours, with no distinct seasonal difference. This suggests that a latitudinal and day/night dependence should be included in the Lomidze correction when enough data become available in the future.

As we can see from Figures 4–6, the daily T_e profile varies with the ISR geographic location. For example, at Jicamarca, which is at the magnetic equator, T_e shows a single sharp maximum around dawn that can be associated with the well-known morning overshoot. This feature is characteristic of low latitudes, and in fact is present also in the daily profiles acquired at Arecibo. Here, and at Millstone Hill (middle and sub-auroral latitudes), the increase of T_e in the dayside is due to the increased sunlight and a second hump at dusk marks the decrease of T_e in passing from day to night. At Millstone Hill, an increased T_e during nighttime is recorded at the December solstice, mainly due to the decrease of plasma density in the F layer and the following reduction of the collisional cooling effect, and also to the conjugate heating. To better highlight and describe all of these different spatial features exhibited by the topside T_e, in the next section, the main spatial trends of T_e will also be investigated, in both geographic and magnetic coordinates.

3.3. Statistical Trends of Swarm-Measured Electron Temperature Values and Comparison with IRI-Modeled Ones

The results of Section 3.1 and 3.2 demonstrate that the application of the Lomidze correction improves the agreement between Swarm data and those modeled by IRI and measured by different ISRs. This is why, in the following analyses, we focus only on Swarm B data corrected according to Lomidze et al. [24].

Swarm B data from 2014 to 2020, and corresponding IRI-modeled values, were binned as a function of the geographic coordinates with bins 2.5°-wide in latitude and 5°-wide in longitude. For each bin, the median value was considered, as representative of the main statistical trend, after discarding the outliers through the Median Absolute Deviation statistical procedure (see Section 3.3 by Pignalberi et al. [46] for a detailed description). The binning was first performed on the entire dataset and then by selecting MLT (in QD coordinates) sectors representative of the dawn (5–7 MLT), daytime (12–15 MLT), and nighttime (0–3 MLT) conditions. This was done because at a fixed altitude, the T_e diurnal variations are known to be the most important along with the latitudinal ones [20,26,39].

Figure 7 shows the corresponding results considering all the MLTs and the different MLT sectors, for both Swarm-measured and IRI-modeled data, and the corresponding percentage of the normalized residuals between them. The figure highlights how the T_e spatial distribution follows the configuration of the Earth's magnetic field lines. In fact, the characteristic shape of the magnetic equator and of the auroral regions stand out, and this is somewhat expected because the plasma in the topside ionosphere is generally constrained by the geomagnetic field [47]. This behavior has also been recently confirmed by Giannattasio et al. [48,49] that, by using Swarm data, identified patterns in the parallel electrical conductivity (which is crucially affected by T_e) related to features of the geomagnetic field such as regions R1 and R2 [50] or the magnetic cusp [51]. They also identified variations in the parallel electrical conductivity due to MLT, QD latitude, season, solar, and geomagnetic activity. The MLTs around dawn highlight the fact that the morning peak is a phenomenon affecting only the low latitudes, which follows the magnetic equator shape and reinforces in the longitudinal sector between 80°W and 80°E (including South America and Africa) where the departure between the magnetic equator and the geographic one is the largest. At the morning peak, Swarm B values exceed 3000 K in many longitudinal sectors, well above those recorded at mid latitudes and even higher than high latitude ones. In contrast, during daytime and nighttime, the high latitudes experience the highest values; in both cases, the increase of T_e from mid to high latitudes is limited to a quite narrow latitudinal range where the spatial gradient is concentrated: at mid latitudes during daytime and at the sub-auroral and auroral boundaries during nighttime. When compared to IRI-modeled values, the largest differences emerge at magnetic equator latitudes at dawn, and at low and mid latitudes during daytime and nighttime, respectively. Overall, Swarm measurements exceed IRI values for most cases and this is in line with the results of Figures 2 and 3. The excess is particularly clear at the morning peak and at low and mid latitudes at nighttime. If we recall the analysis shown in the previous section, based on the comparison with ISRs data (Figures 4–6), we can say that Swarm overestimates the nighttime values; at the morning peak, Jicamarca ISR results (Figure 4) point out that the difference between Swarm and IRI is mainly due to a joint action of both the Swarm overestimation and the IRI underestimation.

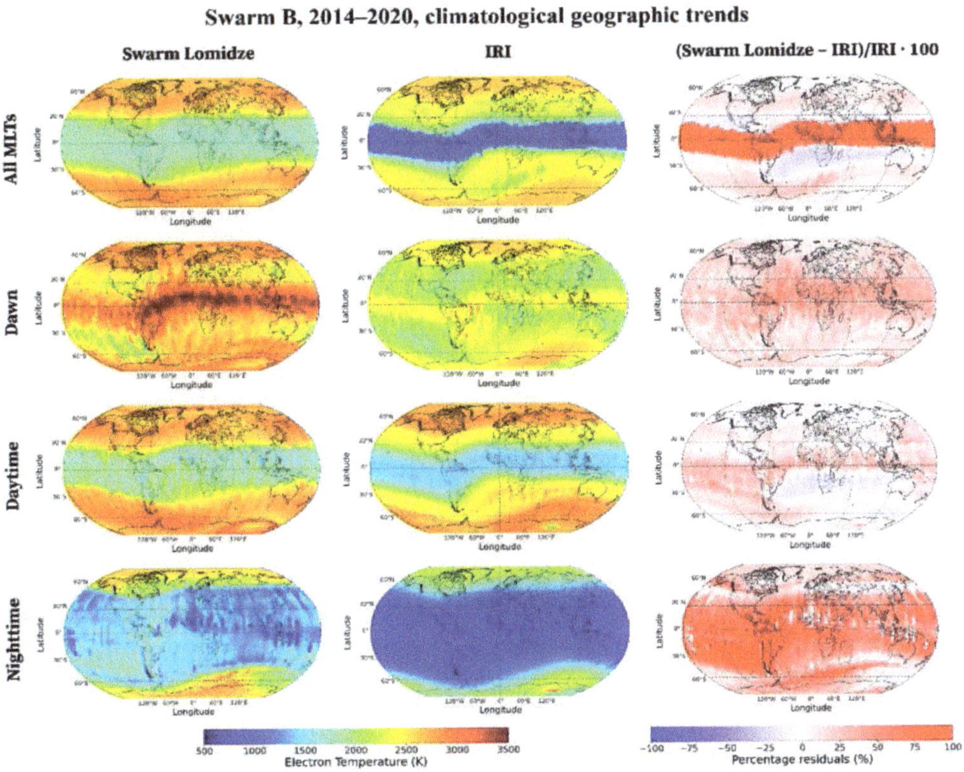

Figure 7. Statistical geographic trends of T_e measured by Swarm B after applying the Lomidze et al. [24] correction (first column), of the corresponding values modeled by IRI (second column), and of the percentage of normalized residuals between measured and modeled values (third column). Those represented are median values binned in geographic coordinates; 2.5° in latitude, 5° in longitude. The first row pertains to all MLTs (i.e., to the entire Swarm B dataset), second row to MLTs around dawn, third row to daytime MLTs, and fourth row to nighttime MLTs.

As highlighted by Figure 7, the Earth's magnetic field has a strong influence on the diurnal trend of Swarm and IRI values. Therefore, we binned T_e data also as a function of QD latitude and MLT with bins 2.5°-wide and 15 minutes-wide, respectively. The binning was first performed on the entire dataset and then by selecting different days of the year representative of the March equinox, June solstice, September equinox, and December solstice conditions, as done in Section 3.2. for ISRs data.

Figure 8 shows that the overall diurnal trend of T_e is characterized by different well-defined features at different latitudes. At low latitudes, the morning peak is the main feature and small day–night differences are visible, which confirms the results from Jicamarca ISR (Figure 4). At mid latitudes (30–60° QD latitude), no distinct morning peak is present and the main feature is the remarkable difference of absolute values between daytime and nighttime, which confirms the results from Millstone Hill ISR (Figure 6). At high latitudes, the diurnal trend is less pronounced with high values throughout the day that intensify during daytime at the auroral boundaries. The agreement between Swarm and IRI is very good for what concerns the overall diurnal pattern. The main differences characterize the morning peak hours and the nighttime hours for low and mid latitudes. It should be noted that IRI cannot describe the very well-located morning peak maximum just over the magnetic equator, and the T_e increase at the auroral boundaries during daytime. For both cases the reason has to be found in the limited spatial resolution of the model.

As expected, no remarkable differences characterize the equinoxes, while the comparison between the solstices highlights seasonal differences and a different description of these by Swarm and IRI. In fact, while IRI shows a clear seasonal dependence for daytime hours, with higher T_e values in the winter hemisphere than in the summer one, Swarm data show a less marked seasonal difference.

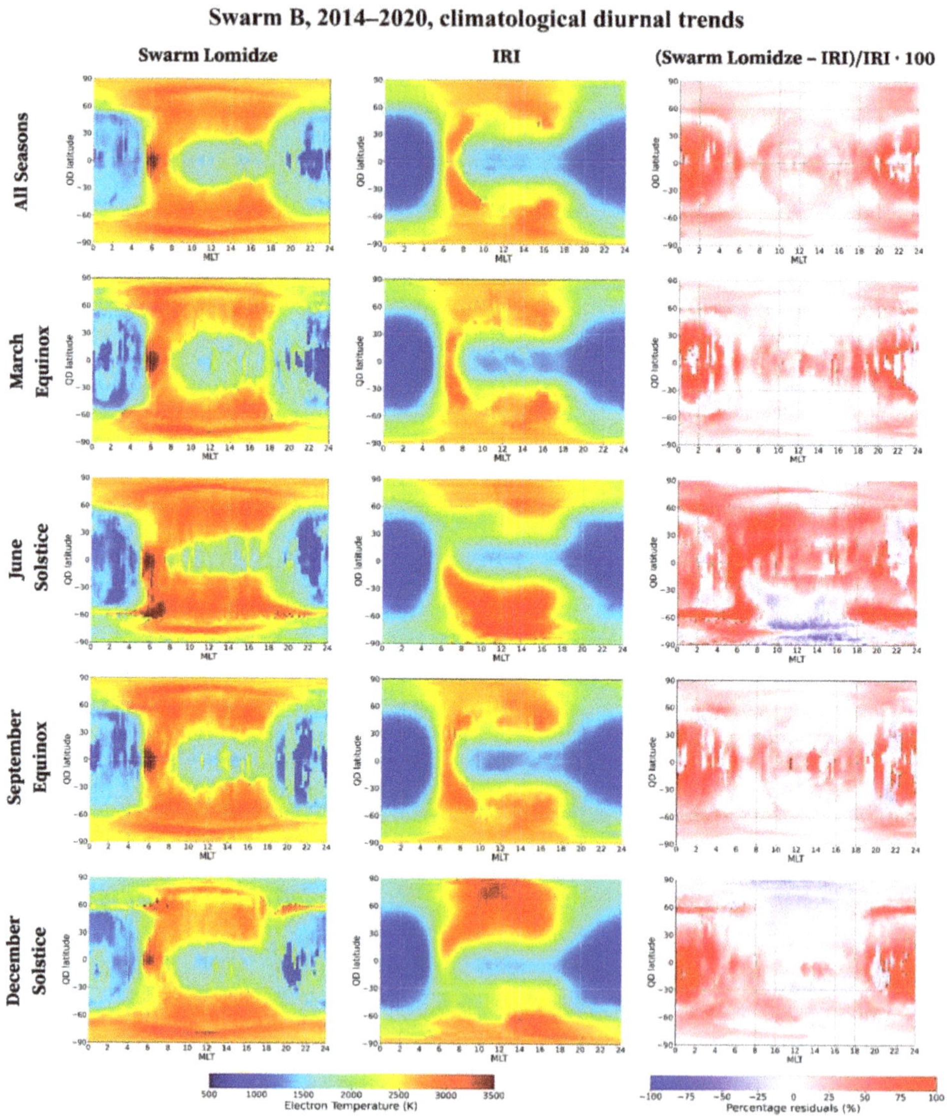

Figure 8. Statistical diurnal trends of T_e measured by Swarm B after applying the Lomidze et al. [24] correction (first column), of the corresponding values modeled by IRI (second column), and of the percentage residuals between measured and modeled values (third column). Those represented are median values binned in QD magnetic coordinates; (x-axis) 15 minutes in MLT, (y-axis) 2.5° in QD latitude. First row independently of the season (i.e., for the entire Swarm B dataset), second row for the March equinox, third row for the June solstice, fourth row for the September equinox, and fifth row for the December solstice.

The diurnal behavior of T_e shown in Figures 4–8 is also consistent with previous results in the literature, where the observed trend was recognized to be mainly due to solar UV radiation (in the dayside) and the precipitation of particles (both in the dayside and nightside) caused by magnetosphere-ionosphere coupling [52]. Apart from the obvious contribution due to the excess of energy of photoelectrons in the dayside especially at mid latitudes, electrons, protons and alpha particles of solar origin may be injected into the ionosphere during open magnetosphere conditions, i.e., when reconnection at the magnetopause occurs, at QD latitudes of about 80° and around 12:00 MLT [53,54]. The precipitation of such particles is expected to enhance T_e in the cusp regions [51] and at lower latitudes in the pre-noon sector [55]. T_e also increases on the nightside due to particle precipitation. In more detail, plasma acceleration occurs in response to reconnection phenomena in the far-away geomagnetic tail regions, generating a three-dimensional current system modeled by the current wedge paradigm [56] that connects the tail to the nightside ionosphere in correspondence of the R2 region [50]. Interestingly, the peak of such enhancement is observed in correspondence with the boundary between R2 and the main ionospheric trough in the pre-dawn sector due to the joint action of particle precipitation and the decrease of N_e [49,57].

The diurnal variation of T_e also shows a clear dependence on the local season, which emerges from Figure 8 when considering the six-months shift of local seasons in Northern and Southern hemispheres in correspondence with solstices and equinoxes. Specifically, the increase of T_e is remarkable above all in the nightside winter, from about 20 MLT to pre-dawn at QD latitudes of about 60° and especially in the Southern hemisphere, and in the dayside summer. Such a behavior can be easily interpreted in light of the previous literature. In fact, previous studies concluded that field-aligned currents density changes from winter to summer by about a factor of 2 in the dayside due to the increase of photoelectron flux and the following increase of T_e [58–62]. In this case, the increase of T_e is driven by both the increase of EUV photoionization and, at very high latitudes (~80°), by precipitations in the cusp region around noon. In the nightside, particle precipitation is the dominant process at auroral and sub-auroral latitudes and becomes more and more important when proceeding from summer to winter. In particular, during winter the enhancement of T_e reaches the maximum expansion from dusk to about 08:00 MLT, especially in the Southern hemisphere. The explanation of this behavior is the simultaneous collapse of both electron and ion density in the ionospheric F layer, and the following collapse of the contribution of collisional cooling to T_e. Therefore, in the absence of mechanisms that reduce the energy of precipitating particles T_e is free to increase due, e.g., to energy exchanges with the nightside magnetosphere. Not surprisingly, there is an anticorrelation between N_e and T_e, and T_e is significantly enhanced in regions of depleted N_e, such as between region R2 and the main trough in the nightside winter; while, in contrast, it decreases in regions of enhanced N_e due to energy loss to the ions [19,48,49,57,63]. Some studies in the past tried to establish a link between N_e and T_e, since N_e is a major factor in electron energy loss due to the ionospheric plasma quasi-neutrality (density of ions is nearly equal to N_e). Generally, an anti-correlation was found. Brace and Theis [20] established an interrelationship between N_e and T_e using Atmosphere Explorer C data. Kakinami et al. [64] found a U shape in the T_e–N_e dependence using Hinotori satellite data. Their results suggest that an additional heat source has to be considered, even though the mechanism is unknown. Su et al. [65] dealt with the investigation of T_e–N_e correlation using data from the DEMETER satellite, by highlighting the variability of this correlation depending on the latitude, local time, and season. Hu et al. [66] used N_e measured from COSMIC as a driver for T_e in their neural network T_e model. Generally, they found a better agreement between their model and ISR data than between ISR data and IRI values. Thus, it is evident that the T_e–N_e relationship could help in the investigation of T_e behavior in the ionosphere but it is a complex problem itself. This is why, in this study, we did not investigate the T_e–N_e correlation using the Swarm data and we focus only on T_e.

Compared to the spatial and diurnal trends, the interpretation of the seasonal trends exhibited by Swarm data is more difficult and the differences with IRI are significant. Overall, the main seasonal differences concern the summer and winter seasons for specific MLT sectors. Nighttime T_e values at high latitudes have been found to be deeply influenced by the solar illumination conditions [67]; in fact, high values are found in the summer hemisphere lit by the Sun. In contrast, low latitudes are not affected by seasonal changes at nighttime. T_e at mid latitudes during local winter can be affected by strong heating due to conjugate photoelectrons if the opposite hemisphere is already sunlit.

4. Discussion

The results of Section 3.1. highlighted how the application of the Lomidze correction to Swarm-measured data reduces the mismatch with IRI-modeled data but, as pointed out by Section 3.3. results and the comparison with ISRs data (Section 3.2), several differences between measured and modeled data are still present for what concerns both T_e magnitude and the corresponding spatial, diurnal, and seasonal variations.

To better highlight the relation between measured and modeled data, the conditioned probability density function between them has been calculated. Specifically, we calculated the probability density of measured T_e values, both original and Lomidze-corrected, conditioned by IRI-modeled ones in 50 K-wide range around a given value T_0 (ranging from 25 to 4975 K in steps of 50 K):

$$p(T_{e,\text{measured}}|T_{e,\text{modeled}})|T_{e,\text{modeled}} \in [T_0 - 25; T_0 + 25)\ \text{K} \tag{6}$$

In Equation (6), $T_{e,\text{measured}}$ is the T_e measured by Swarm (i.e., $T_{e,\text{LP}}$) or corrected according to Lomidze et al. [24] (i.e., $T_{e,\text{Lom}}$), while $T_{e,\text{modeled}}$ is the corresponding one modeled by IRI and p is the conditioned probability density. Equation (6) was calculated for each T_0 equal to the $T_{e,\text{modeled}}$ values used to define the ranges for the conditioned probability density calculation. Conditioned probability density values are represented in Figure 9 as colored bins, while bins with $p < 10^{-6}$ are in white. For each $[T_0 - 25; T_0 + 25)$ K range, the median is calculated and represented as a black circle, and corresponding error bars are the median absolute deviation values [46].

Median values of Figure 9 were used to perform a linear regression analysis and to obtain the corresponding slope, intercept, and Pearson correlation coefficient, namely:

$$R = \frac{\text{cov}(T_{e,\text{measured}}, T_{e,\text{modeled}})}{\sigma_{T_{e,\text{measured}}}\sigma_{T_{e,\text{modeled}}}} \in [-1, 1] \tag{7}$$

where cov is the covariance matrix and σ the variance.

Linear regression best fit lines are represented in Figure 9 in magenta, with corresponding slope, intercept, and R values reported in each panel. The analysis was restricted to $T_{e,\text{modeled}} \leq 2200$ K because the linear relation between measured and modeled data fully breaks above this threshold, as clearly highlighted by the median values of Figure 9.

The results of Figure 9 confirm and complement those shown by Figures 2 and 3. Specifically, the slopes of the linear best fit, which range between 0.80 and 0.89 for the original Swarm data, after applying the Lomidze correction range between 0.97 and 1.00, while the intercept values decrease from 920–990 K to 140–320 K. This confirms the accuracy improvement provided by the Lomidze correction. Differently, the correlation coefficient (R) was slightly affected by the Lomidze correction because this correction is substantially linear, apart from the term dependent on $N_{e,\text{LP}}$ (see Equations (1)–(3)); anyway, the correlation is very high in both cases ($R > 0.94$). However, we have to take into account that these results are valid only in the range $T_{e,\text{modeled}} \leq 2200$ K, where a linear dependence with a slope near 1 between measured and modeled data is evident. A non-linear dependence is clearly visible for $T_{e,\text{modeled}} > 2200$ K and in principle could be indifferently due to the Swarm LPs instrumentation, possible shortcomings of the Lomidze correction, and the IRI representation of the highest T_e values.

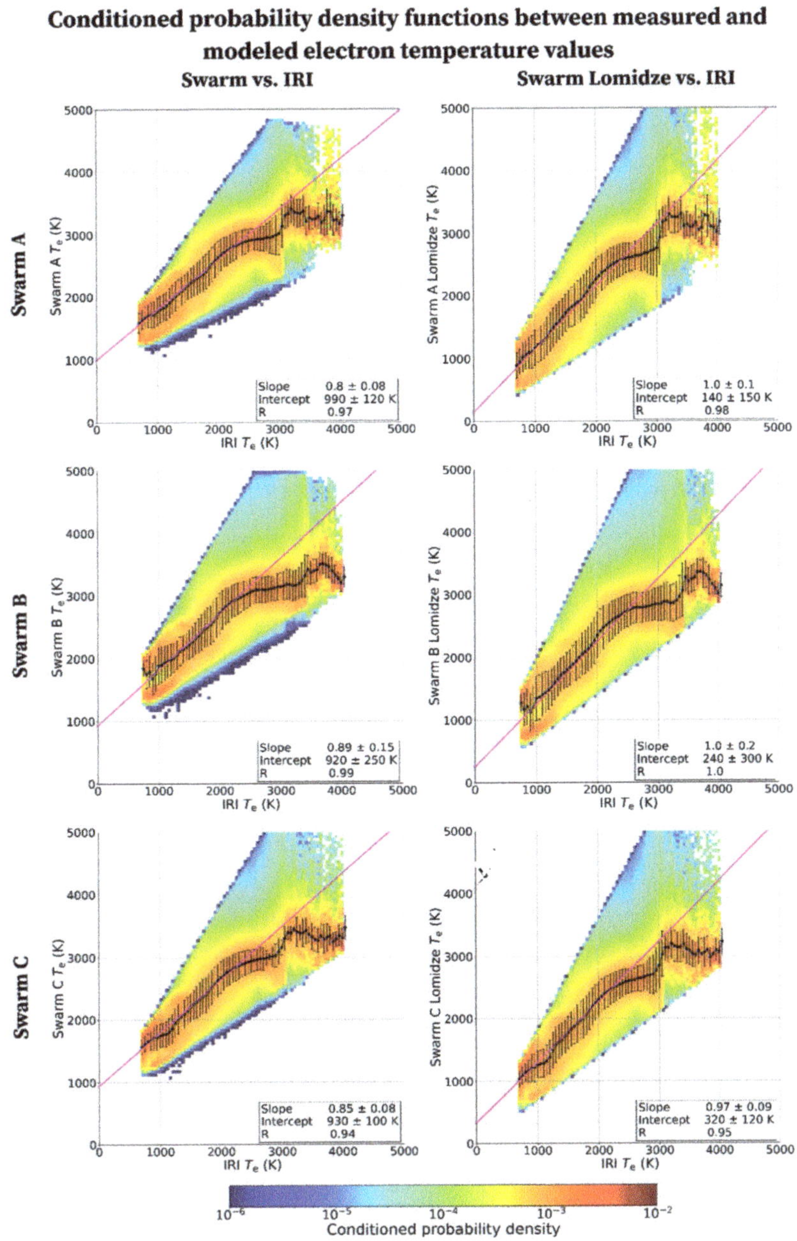

Figure 9. Conditioned probability density function of T_e values measured by Swarm satellites (left panels), and after applying the Lomidze et al. [24] correction (right panels), at fixed IRI-modeled T_e values. From top to bottom, the analysis refers to Swarm A, B, and C. The black circles refer to the median values of measured and corrected Swarm T_e at fixed IRI-modeled T_e values. Error bars are the median absolute deviation. The magenta lines are linear regression fits for IRI T_e below 2200 K, with corresponding parameters shown in the lower right boxes.

Recently, Pignalberi et al. [68] studied the occurrence of very high Swarm-measured T_e values ($T_{e,measured} \geq 6000$ K), and corresponding spatial, diurnal, and seasonal variations. As expected, very high T_e values were recorded above all at high latitudes but with well-defined trends. Specifically, the occurrence of very high $T_{e,measured}$ values maximizes in winter and is minimum in summer, with intermediate values at both equinoxes. Moreover, the occurrence is asymmetric between hemispheres, being highest in the Southern hemisphere, where it can reach 40% of the total observations in winter. These extremely high T_e values are not represented at all by IRI, which is limited to values below about 4300 K (see Figures 2 and 9). As a consequence, the departure from the linear dependence between $T_{e,measured}$ and $T_{e,modeled}$ for the highest values cannot be related to the application of the Lomidze correction. The fact that the occurrence of $T_{e,measured} \geq 6000$ K values is quite remarkable at high latitudes and that it usually happens at small spatial and temporal scales [68] drives us to the conclusion that the median climatological representation given by the IRI model cannot properly represent such very fast and steep T_e variations. This is mainly due to the limited spatial (20 degrees in invdip latitude) and temporal (80 minutes in MLT) resolution of its current spherical harmonic expansion approach [26]. Nevertheless, we point out that the Lomidze correction is based on a dataset of T_e measurements from Arecibo, Jicamarca, and Millstone Hill ISRs, which are all low- and mid-latitude stations. This means that no high-latitude measurement was considered and, most importantly, only values in the range $T_{e,measured} < 3500$ K (in HG mode) were used to derive the calibration curves (see Figure 7 in [24]). With regard to this, it would be of great benefit to include high-latitude ISR stations (e.g., Resolute Bay (74.7°N, 94.9°W), Poker Flat (65.1°N, 147.5°W,), Irkutsk (52.9°N, 103.3°E), EISCAT Tromsø (69.3°N, 19.1°E), EISCAT Kiruna (67.5°N, 20.3°E), EISCAT Sodankylä (67.2°N, 26.4°E), EISCAT Longyearbyen (78.1°N, 16.0°E)) in the Lomidze dataset to also characterize the high T_e values usually recorded at auroral latitudes. Finally, the possible overestimation of T_e values measured by Swarm LPs due to instrumental/environmental effects cannot be completely ignored. This is why, a more in-depth analysis, based on different datasets and measurement techniques, is needed to clarify the origin and reliability of the very high T_e values recorded by Swarm satellites.

5. Conclusions

In this study, the main statistical median features of the topside ionosphere T_e were investigated on a global basis through data collected by LPs on-board Swarm satellites from the beginning of 2014 to the end of 2020. Swarm T_e measured data were compared with corresponding values modeled by IRI, and with data measured by Jicamarca, Arecibo, and Millstone Hill ISRs, to evaluate the reliability of both Swarm data and IRI model in the description of the spatial, diurnal, and seasonal trends of T_e in the topside ionosphere.

We also investigated the effectiveness of the correction to Swarm T_e values proposed by Lomidze et al. [24] through comparison with IRI-modeled data and ISRs-measured ones. Specifically, we found that the application of the Lomidze correction:

- improves the agreement between Swarm data and corresponding ones modeled by IRI when the entire dataset is considered, for every Swarm satellite. This is attested by the average 400 K improvement in the mean residuals between Swarm and IRI after the Lomidze correction application;
- does not alter either the dispersion of Swarm data around the mean and the correlation between Swarm and IRI, due to the linear character of the correction;
- reduces the Swarm data RMSE from about 690–780 K to about 440–530 K, and RRMSE from about 27–29% to about 20–23%, when compared to IRI data;
- generally improves the agreement between Swarm and ISR data. The improvement is particularly evident at Millstone Hill while it is lower at Arecibo and even lower at Jicamarca. Moreover, the correction is more effective during daytime than nighttime;
- does not alter the linear relation trend between measured and modeled T_e values in the range below 2200 K, but it improves the corresponding slope (closer to 1) and intercept (closer to 0) values.

The results obtained with ISRs data suggest that the calibration of Swarm T_e data made by Lomidze et al. [24] is heavily biased towards mid latitudes and daytime hours due to the ISRs dataset they used, and to the fact that they cumulated the data of the three ISRs to derive a single calibration curve. As a consequence, the accuracy of Swarm T_e data would benefit from the inclusion of the latitudinal and diurnal dependence in the calibration procedure. Swarm T_e data are very valuable to study the relative changes of T_e, but a more detailed correction is needed to obtain more reliable absolute values of T_e especially at low and equatorial latitudes. Moreover, to better describe the high T_e values, the inclusion of data recorded by high-latitude ISRs in the Lomidze dataset is advisable.

The orbital configuration of Swarm satellites and the extension of the dataset are very tailored for the study of the global features of T_e with a spatial and time resolution hardly accessible in the past. Because of this, we were able to study the median spatial, diurnal, and seasonal trends of T_e in the topside ionosphere in a very accurate fashion. Moreover, this allowed implementing an important statistical comparison between measured data and corresponding ones modeled by IRI. This comparison is beneficial both to validate the IRI model and to unveil the most important departures between measured and modeled data for a future inclusion of the Swarm T_e dataset in the IRI model. The statistical comparison between Swarm and IRI evidenced that:

- the largest differences emerge at magnetic equator latitudes at dawn, and at low and mid latitudes during daytime and nighttime, respectively;
- IRI needs to be improved in the description of the morning peak at low latitudes. This can be achieved by increasing the order of spherical harmonics underlying the IRI Te description;
- IRI needs to be improved in summer daytime for both hemispheres, and in the description of the high Te values characteristic of daytime auroral oval latitudes;
- IRI data never go beyond about 4300 K, while Swarm data show values well beyond 5000 K, and in general Swarm values are higher than IRI ones.

The main statistical features of T_e in the topside ionosphere have been pointed out and discussed in the paper and linked to the physical phenomena affecting the T_e behavior at these altitudes, and to the existing literature on the topic. One of the most relevant results of this study is the characterization of the high T_e values recorded by Swarm LPs, which are customary above all at high latitudes. These high T_e values are not represented by the climatological description given by the IRI model. However, IRI has been developed to represent the median behavior of T_e at climatological scales, and extreme (off-median) values or very fast and steep T_e variations are out of the scope of the model. Nevertheless, a careful characterization of the high T_e values recorded by Swarm LPs deserves particular attention for both modeling and scientific purposes.

In this paper, we did not deal with T_e variations with the solar and magnetic activity. Since Swarm satellites are still in orbit and the mission will be supported for the next years, the continuous growth of the dataset will allow also studying the effects of the solar and magnetic activity variations on T_e in the years to come. Moreover, we focused on the large-scale spatial and temporal variations of T_e in order to make a fair comparison with the IRI model that describes the climatological behavior of the ionosphere. However, the very-high resolution of Swarm LPs data also allows the study of small-scale T_e features [68,69] which complement the results of this study and can be of help in the comprehension of the T_e behavior in the topside ionosphere.

Author Contributions: Conceptualization, A.P.; methodology, A.P., F.G., and V.T.; software, A.P.; data curation, A.P. and V.T.; investigation, A.P., F.G., V.T., I.C., M.P., G.C., P.D.M. and R.T.; validation, A.P., F.G., V.T., I.C., M.P., G.C., P.D.M. and R.T.; formal analysis, A.P.; writing—original draft preparation, A.P.; writing—review and editing, A.P., F.G., V.T., I.C., M.P., G.C., P.D.M. and R.T.; funding acquisition, F.G. and P.D.M. All authors have read and agreed to the published version of the manuscript.

Funding: This research is partially supported by the Italian MIUR-PRIN grant 2017APKP7T on Circumterrestrial Environment: Impact of Sun–Earth Interaction, and by European Space Agency

(ESA contract N. 4000125663/18/I-NB-"EO Science for Society Permanently Open Call for Proposals EOEP-5 BLOCK4" INTENS). V. Truhlik was supported by grant LTAUSA17100 of the Ministry of Education, Youth and Sports of the Czech Republic.

Data Availability Statement: ESA Swarm data are publicly available at ftp://swarmdiss.eo.esa.int (accessed on 24 September 2021). Jicamarca, Arecibo, and Millstone Hill ISR data are available via the public access portal at http://cedar.openmadrigal.org (accessed on 24 September 2021).

Acknowledgments: The IRI team is acknowledged for developing and maintaining the IRI model and for giving access to the corresponding Fortran code via the IRI website (http://irimodel.org/, accessed on 24 September 2021). Thanks to the European Space Agency for making Swarm data publicly available via ftp://swarmdiss.eo.esa.int (accessed on 24 September 2021), and for the considerable efforts made for the Langmuir probes data calibration and maintenance. This publication uses data from Jicamarca, Arecibo, and Millstone Hill ISR made available via the public access portal at http://cedar.openmadrigal.org (accessed on 24 September 2021). The authors are indebted to the observatory directors and operators for the significant investments of their time, effort, expertise, and funds needed to acquire and provide measurement data to academic research over the years.

Conflicts of Interest: The authors declare no conflict of interest.

Abbreviations

The following abbreviations are frequently used in this manuscript:

BIL-1995	Bilitza—1995 model
doy	Day of the year
ESA	European Space Agency
EUV	Extreme ultra-violet
GPS	Global positioning system
HG	High gain
IQR	Inter-quartile range
IRI	International Reference Ionosphere
ISR	Incoherent Scatter Radar
LEO	Low-Earth-orbit
LP	Langmuir Probe
LT	Local time
MLT	Magnetic local time
N_e	Electron density
QD	Quasi dipole
R	Pearson correlation coefficient
RMSE	Root mean square error
RRMSE	Relative root mean square error
TBT-2012	Truhlik Bilitza Triskova—2012 model
TBT-2012+SA	Truhlik Bilitza Triskova—2012 + Solar Activity model
T_e	Electron temperature
T_i	Ion temperature
T_n	Neutral temperature

References

1. Willmore, A.P. Electron and ion temperatures in the ionosphere. *Space Sci. Rev.* **1970**, *11*, 607–670. [CrossRef]
2. Rishbeth, H.; Garriott, O. *Introduction to Ionospheric Physics*; International Geophysics Series v. 14; Academic Press: New York, NY, USA, 1969.
3. Ratcliffe, J.A. *An Introduction to the Ionosphere and Magnetosphere*; Cambridge University Press: Cambridge, UK, 1972.
4. Banks, P.M. Ion temperature in the upper atmosphere. *J. Geophys. Res. Space Phys.* **1967**, *72*, 3365–3385. [CrossRef]
5. Roble, R. The calculated and observed diurnal variation of the ionosphere over Millstone Hill on 23–24 March 1970. *Planet. Space Sci.* **1975**, *23*, 1017–1033. [CrossRef]
6. Schunk, R.W.; Nagy, A.F. Electron temperatures in the region of the ionosphere: Theory and observations. *Rev. Geophys.* **1978**, *16*, 355–399. [CrossRef]
7. Bilitza, D. Electron and ion temperature data for ionospheric modelling. *Adv. Space Res.* **1991**, *11*, 139–148. [CrossRef]
8. Evans, J. Theory and practice of ionosphere study by Thomson scatter radar. *Proc. IEEE* **1969**, *57*, 496–530. [CrossRef]

9. Bilitza, D.; Altadill, D.; Truhlik, V.; Shubin, V.; Galkin, I.; Reinisch, B.; Huang, X. International reference ionosphere 2016: From ionospheric climate to real-time weather predictions. *Space Weather* **2017**, *15*, 418–429. [CrossRef]
10. Bilitza, D. IRI the International Standard for the Ionosphere. *Adv. Radio Sci.* **2018**, *16*, 1–11. [CrossRef]
11. Kakinami, Y.; Lin, C.C.H.; Liu, J.Y.; Kamogawa, M.; Watanabe, S.; Parrot, M. Daytime longitudinal structures of electron density and temperature in the topside ionosphere observed by the Hinotori and DEMETER satellites. *J. Geophys. Res. Space Phys.* **2011**, *116*, A05316. [CrossRef]
12. Ma, H.; Liu, L.; Chen, Y.; Le, H.; Li, Q.; Zhang, H. Longitudinal differences in electron temperature on both sides of zero declination line in the mid-latitude topside ionosphere. *J. Geophys. Res. Space Phys.* **2021**, *126*, e2020JA028471. [CrossRef]
13. Slominska, E.; Rothkaehl, H. Mapping seasonal trends of electron temperature in the topside ionosphere based on DEMETER data. *Adv. Space Res.* **2013**, *52*, 192–204. [CrossRef]
14. Brace, L.; Theis, R.; Hoegy, W. Ionospheric electron temperature at solar maximum. *Adv. Space Res.* **1987**, *7*, 99–106. [CrossRef]
15. Oyama, K.-I.; Balan, N.; Watanabe, S.; Takahashi, T.; Isoda, F.; Bailey, G.J.; Oya, H. Morning overshoot of Te enhanced by downward plasma drift in the equatorial topside ionosphere. *J. Geomagn. Geoelectr.* **1996**, *48*, 959–966. [CrossRef]
16. Stolle, C.; Liu, H.; Truhlík, V.; Luehr, H.; Richards, P.G. Solar flux variation of the electron temperature morning overshoot in the equatorialFregion. *J. Geophys. Res. Space Phys.* **2011**, *116*, A04308. [CrossRef]
17. Yang, T.; Park, J.; Kwak, Y.; Oyama, K.; Minow, J.I.; Lee, J. Morning overshoot of electron temperature as observed by the swarm constellation and the international space station. *J. Geophys. Res. Space Phys.* **2020**, *125*, e2019JA027299. [CrossRef]
18. Brace, L.H. Solar cycle variations in F-region Te in the vicinity of the midlatitude trough based on AE-C measurements at solar minimum and DE-2 measurements at solar maximum. *Adv. Space Res.* **1990**, *10*, 83–88. [CrossRef]
19. Prölss, G.W. Subauroral electron temperature enhancement in the nighttime ionosphere. *Ann. Geophys.* **2006**, *24*, 1871–1885. [CrossRef]
20. Brace, L.; Theis, R. Global empirical models of ionospheric electron temperature in the upper F-region and plasmasphere based on in situ measurements from the atmosphere explorer-c, isis-1 and isis-2 satellites. *J. Atmos. Terr. Phys.* **1981**, *43*, 1317–1343. [CrossRef]
21. Pavlov, A.V.; Abe, T.; Oyama, K.-I. Comparison of the measured and modelled electron densities and temperatures in the ionosphere and plasmasphere during 20-30 January 1993. *Ann. Geophys.* **2000**, *18*, 1257–1262. [CrossRef]
22. Richards, P.G.; Buonsanto, M.J.; Reinisch, B.W.; Holt, J.; Fennelly, J.A.; Scali, J.L.; Comfort, R.H.; Germany, G.A.; Spann, J.; Brittnacher, M.; et al. On the relative importance of convection and temperature to the behavior of the ionosphere in North America during 6–12 January 1997. *J. Geophys. Res. Space Phys.* **2000**, *105*, 12763–12776. [CrossRef]
23. Bilitza, D.; Truhlik, V.; Richards, P.; Abe, T.; Triskova, L. Solar cycle variations of mid-latitude electron density and temperature: Satellite measurements and model calculations. *Adv. Space Res.* **2007**, *39*, 779–789. [CrossRef]
24. Lomidze, L.; Knudsen, D.J.; Burchill, J.; Kouznetsov, A.; Buchert, S.C. Calibration and validation of swarm plasma densities and electron temperatures using ground-based radars and satellite radio occultation measurements. *Radio Sci.* **2018**, *53*, 15–36. [CrossRef]
25. Friis-Christensen, E.; Lühr, H.; Hulot, G. Swarm: A constellation to study the Earth's magnetic field. *Earth Planets Space* **2006**, *58*, 351–358. [CrossRef]
26. Truhlik, V.; Bilitza, D.; Triskova, L. A new global empirical model of the electron temperature with the inclusion of the solar activity variations for IRI. *Earth Planets Space* **2012**, *64*, 531–543. [CrossRef]
27. Knudsen, D.J.; Burchill, J.K.; Buchert, S.C.; Eriksson, A.I.; Gill, R.; Wahlund, J.; Åhlen, L.; Smith, M.; Moffat, B. Thermal ion imagers and Langmuir probes in the Swarm electric field instruments. *J. Geophys. Res. Space Phys.* **2017**, *122*, 2655–2673. [CrossRef]
28. National Space Institute, Technical University of Denmark. *Swarm L1b Product Definition*; SW-RS-DSC-SY-0007; National Space Institute, Technical University of Denmark: Kongens Lyngby, Denmark, 2018. Available online: https://earth.esa.int/documents/10174/1514862/Swarm_L1b_Product_Definition (accessed on 10 October 2021).
29. Laundal, K.M.; Richmond, A.D. Magnetic coordinate systems. *Space Sci. Rev.* **2017**, *206*, 27–59. [CrossRef]
30. Pignalberi, A.; Aksonova, K.D.; Zhang, S.-R.; Truhlik, V.; Gurram, P.; Pavlou, C. Climatological study of the ion temperature in the ionosphere as recorded by Millstone Hill incoherent scatter radar and comparison with the IRI model. *Adv. Space Res.* **2021**, *68*, 2186–2203. [CrossRef]
31. Bilitza, D. Models for ionospheric electron and ion temperature. In *International Reference Ionosphere-IRI 79*; Report UAG-82; Rawer, K., Lincoln, J.V., Conkright, R.O., Eds.; World Data Center A for Solar-Terrestrial Physics: Boulder, CO, USA, 1981; p. 245. Available online: http://www.irimodel.org/ (accessed on 10 October 2021).
32. Bilitza, D.; Brace, L.; Theis, R. Modelling of ionospheric temperature profiles. *Adv. Space Res.* **1985**, *5*, 53–58. [CrossRef]
33. Bilitza, D. *International Reference Ionosphere*; Report 90-22; National Space Science Data Center: Greenbelt, MD, USA, 1990. Available online: http://irimodel.org/docs/IRI1990pp0-84.pdf (accessed on 10 October 2021).
34. Booker, H.G. Fitting of multi-region ionospheric profiles of electron density by a single analytic function of height. *J. Atmos. Terr. Phys.* **1977**, *39*, 619–623. [CrossRef]
35. Spenner, K.; and Plugge, R. Empirical model of global electron temperature distribution between 300 and 700 Km based on data from Aeros-A. *J. Geophys.* **1979**, *46*, 43–56. Available online: https://journalgeophysicsjournalcom/JofG/article/view/286 (accessed on 10 October 2021).
36. Bilitza, D. International reference ionosphere 2000. *Radio Sci.* **2001**, *36*, 261–275. [CrossRef]

37. Truhlík, V.; Třísková, L.; Šmilauer, J.; Afonin, V. Global empirical model of electron temperature in the outer ionosphere for period of high solar activity based on data of three Intercosmos satellites. *Adv. Space Res.* **2000**, *25*, 163–169. [CrossRef]
38. Bilitza, D.; Altadill, D.; Zhang, Y.; Mertens, C.J.; Truhlik, V.; Richards, P.; McKinnell, L.-A.; Reinisch, B.W. The international reference ionosphere 2012—A model of international collaboration. *J. Space Weather Space Clim.* **2014**, *4*, A07. [CrossRef]
39. Truhlik, V.; Bilitza, D.; Triskova, L. Latitudinal variation of the topside electron temperature at different levels of solar activity. *Adv. Space Res.* **2009**, *44*, 693–700. [CrossRef]
40. Truhlík, V.; Třísková, L.; Smilauer, J. Improved electron temperature model and comparison with satellite data. *Adv. Space Res.* **2001**, *27*, 101–109. [CrossRef]
41. Picone, J.M.; Hedin, A.E.; Drob, D.; Aikin, A.C. NRLMSISE-00 empirical model of the atmosphere: Statistical comparisons and scientific issues. *J. Geophys. Res. Space Phys.* **2002**, *107*, 1468. [CrossRef]
42. Oyama, K.-I.; Watanabe, S.; Su, Y.; Takahashi, T.; Hirao, K. Season, local time, and longitude variations of electron temperature at the height of ∼600 km in the low latitude region. *Adv. Space Res.* **1996**, *18*, 269–278. [CrossRef]
43. Hoegy, W.R. Probe and radar electron temperatures in an isotropic nonequilibrium plasma. *J. Geophys. Res. Space Phys.* **1971**, *76*, 8333–8340. [CrossRef]
44. Oyama, K.-I.; Lee, C.H.; Fang, H.K.; Cheng, C.Z. Means to remove electrode contamination effect of Langmuir probe measurement in space. *Rev. Sci. Instrum.* **2012**, *83*, 55113. [CrossRef]
45. Zhang, S.-R.; Holt, J.M. Ionospheric plasma temperatures during 1976–2001 over Millstone Hill. *Adv. Space Res.* **2004**, *33*, 963–969. [CrossRef]
46. Pignalberi, A.; Habarulema, J.; Pezzopane, M.; Rizzi, R. On the development of a method for updating an empirical climatological ionospheric model by means of assimilated vTEC measurements from a GNSS receiver network. *Space Weather* **2019**, *17*, 1131–1164. [CrossRef]
47. Schunk, R.; Nagy, A.F. *Ionospheres: Physics, Plasma Physics, and Chemistry*, 2nd ed.; Cambridge University Press: Cambridge, UK, 2009.
48. Giannattasio, F.; De Michelis, P.; Pignalberi, A.; Coco, I.; Consolini, G.; Pezzopane, M.; Tozzi, R. Parallel electrical conductivity in the topside ionosphere derived from swarm measurements. *J. Geophys. Res. Space Phys.* **2021**, *126*, e2020JA028452. [CrossRef]
49. Giannattasio, F.; Pignalberi, A.; De Michelis, P.; Coco, I.; Consolini, G.; Pezzopane, M.; Tozzi, R. Dependence of parallel electrical conductivity in the topside ionosphere on solar and geomagnetic activity. *J. Geophys. Res. Space Phys.* **2021**, *126*, e2021JA029138. [CrossRef]
50. Iijima, T.; Potemra, T.A. Large-scale characteristics of field-aligned currents associated with substorms. *J. Geophys. Res. Space Phys.* **1978**, *83*, 599–615. [CrossRef]
51. Milan, S.E.; Clausen, L.B.N.; Coxon, J.; Carter, J.; Walach, M.-T.; Laundal, K.M.; Østgaard, N.; Tenfjord, P.; Reistad, J.P.; Snekvik, K.; et al. Overview of solar wind-magnetosphere-ionosphere-atmosphere coupling and the generation of magnetospheric currents. *Space Sci. Rev.* **2017**, *206*, 547–573. [CrossRef]
52. Zmuda, A.J.; Martin, J.H.; Heuring, F.T. Transverse magnetic disturbances at 1100 km in the auroral region. *J. Geophys. Res. Space Phys.* **1966**, *71*, 5033–5045. [CrossRef]
53. Brinton, H.C.; Grebowsky, J.M.; Brace, L.H. The high-latitude winter region at 300 km: Thermal plasma observations from AE-C. *J. Geophys. Res. Space Phys.* **1978**, *83*, 4767–4776. [CrossRef]
54. Foster, J.C. An empirical electric field model derived from Chatanika radar data. *J. Geophys. Res. Space Phys.* **1983**, *88*, 981–987. [CrossRef]
55. Dyson, P.L.; Winningham, J.D. Top side ionospheric spread and particle precipitation in the day side magnetospheric clefts. *J. Geophys. Res. Space Phys.* **1974**, *79*, 5219–5230. [CrossRef]
56. McPherron, R.; Russell, C.T.; Aubry, M.P. Satellite studies of magnetospheric substorms on 15 August 1968: Phenomenological model for substorms. *J. Geophys. Res. Space Phys.* **1973**, *78*, 3131–3149. [CrossRef]
57. Wang, W.; Burns, A.G.; Killeen, T.L. A numerical study of the response of ionospheric electron temperature to geomagnetic activity. *J. Geophys. Res. Space Phys.* **2006**, *111*, A11301. [CrossRef]
58. Fujii, R.; Iijima, T.; Potemra, T.A.; Sugiura, M. Seasonal dependence of large-scale Birkeland currents. *Geophys. Res. Lett.* **1981**, *8*, 1103–1106. [CrossRef]
59. Liou, K.; Newell, P.T.; Meng, C.-I. Seasonal effects on auroral particle acceleration and precipitation. *J. Geophys. Res. Space Phys.* **2001**, *106*, 5531–5542. [CrossRef]
60. Papitashvili, V.; Christiansen, F.; Neubert, T. A new model of field-aligned currents derived from high-precision satellite magnetic field data. *Geophys. Res. Lett.* **2002**, *29*, 28-1–28-4. [CrossRef]
61. Christiansen, F.; Papitashvili, V.; Neubert, T. Seasonal variations of high-latitude field-aligned currents inferred from Ørsted and Magsat observations. *J. Geophys. Res. Space Phys.* **2002**, *107*, SMP5-1–SMP5-13. [CrossRef]
62. Wang, H.; Lühr, H.; Ma, S.Y. Solar zenith angle and merging electric field control of field-aligned currents: A statistical study of the Southern Hemisphere. *J. Geophys. Res. Space Phys.* **2005**, *110*, 306. [CrossRef]
63. McDonald, J.; Williams, P. The relationship between ionospheric temperature, electron density and solar activity. *J. Atmos. Terr. Phys.* **1980**, *42*, 41–44. [CrossRef]
64. Kakinami, Y.; Watanabe, S.; Liu, J.-Y.; Balan, N. Correlation between electron density and temperature in the topside ionosphere. *J. Geophys. Res. Space Phys.* **2011**, *116*, A12331. [CrossRef]

65. Su, F.; Wang, W.; Burns, A.G.; Yue, X.; Zhu, F. The correlation between electron temperature and density in the topside ionosphere during 2006-2009. *J. Geophys. Res. Space Phys.* **2015**, *120*, 10724–10739. [CrossRef]
66. Hu, A.; Carter, B.; Currie, J.; Norman, R.; Wu, S.; Zhang, K. A deep neural network model of global topside electron temperature using incoherent scatter radars and its application to GNSS radio occultation. *J. Geophys. Res. Space Phys.* **2020**, *125*, e2019JA027263. [CrossRef]
67. Kitamura, N.; Ogawa, Y.; Nishimura, Y.; Terada, N.; Ono, T.; Shinbori, A.; Kumamoto, A.; Truhlík, V.; Smilauer, J. Solar zenith angle dependence of plasma density and temperature in the polar cap ionosphere and low-altitude magnetosphere during geomagnetically quiet periods at solar maximum. *J. Geophys. Res. Space Phys.* **2011**, *116*, A08227. [CrossRef]
68. Pignalberi, A.; Coco, I.; Giannattasio, F.; Pezzopane, M.; De Michelis, P.; Consolini, G.; Tozzi, R. A new ionospheric index to investigate electron temperature small-scale variations in the topside ionosphere. *Universe* **2021**, *7*, 290. [CrossRef]
69. Pignalberi, A. TITIPy: A Python tool for the calculation and mapping of topside ionosphere turbulence indices. *Comput. Geosci.* **2021**, *148*, 104675. [CrossRef]

remote sensing

Article

An Ionospheric Anomaly Monitor Based on the One Class Support Vector Algorithm for the Ground-Based Augmentation System

Zhen Gao [1], Kun Fang [2], Yanbo Zhu [1], Zhipeng Wang [1,*] and Kai Guo [3]

[1] National Key Laboratory of CNS/ATM, School of Electronics and Information Engineering, Beihang University, Beijing 100191, China; gaozhen@buaa.edu.cn (Z.G.); zhuyanbo@buaa.edu.cn (Y.Z.)
[2] Research Institute for Frontier Science, Beihang University, Beijing 100191, China; fangkun@buaa.edu.cn
[3] Nottingham Geospatial Institute, University of Nottingham, Nottingham NG7 2RD, UK; Kai.guo@nottingham.ac.uk
* Correspondence: wangzhipeng@buaa.edu.cn; Tel.: +86-10-8231-6977

Abstract: An ionospheric anomaly is the irregular change of the ionosphere. It may result in potential threats for the ground-based augmentation system (GBAS) supporting the high-level precision approach. To counter the hazardous anomalies caused by the steep gradient in ionospheric delays, customized monitors are equipped in GBAS architectures. A major challenge is to rapidly detect the ionospheric gradient anomaly from environmental noise to meet the safety-critical requirements. A one-class support vector machine (OCSVM)-based monitor is developed to clearly detect ionospheric anomalies and to improve the robust detection speed. An offline-online framework based on the OCSVM is proposed to extract useful information related to anomalous characteristics in the presence of noise. To validate the effectiveness of the proposed framework, the influence of noise is fully considered and analyzed based on synthetic, semi-simulated, and real data from a typical ionospheric anomaly event. Synthetic results show that the OCSVM-based monitor can identify the anomaly that cannot be detected by other commonly-used monitors, such as the CCD-1OF, CCD-2OF and KLD-1OF. Semi-simulation results show that compared with other monitors, the newly proposed monitor can improve the average detection speed by more than 40% and decrease the minimum detectable gradient change rate to 0.002 m/s. Furthermore, in the real ionospheric anomaly event experiment, compared with other monitors, the OCSVM-based monitor can improve the detection speed by 16%. The result indicates that the proposed monitor has encouraging potential to ensure integrity of the GBAS.

Keywords: GNSS; GBAS; ionospheric gradient anomaly; one class support vector machine

Citation: Gao, Z.; Fang, K.; Zhu, Y.; Wang, Z.; Guo, K. An Ionospheric Anomaly Monitor Based on the One Class Support Vector Algorithm for the Ground-Based Augmentation System. *Remote Sens.* **2021**, *13*, 4327. https://doi.org/10.3390/rs13214327

Academic Editor: Fabio Giannattasio

Received: 3 October 2021
Accepted: 26 October 2021
Published: 28 October 2021

1. Introduction

The Ground-Based Augmentation System (GBAS) is a short-baseline, airport-based augmentation of the Global Navigation Satellite Systems (GNSS). It can provide advanced civil-aviation services concerning accuracy, integrity, continuity, and availability. Integrity is one of the key aspects to indicate the safety of precision approach at airports around the world. It refers to the capability to alert when system outputs cannot be trusted. The GBAS ensures integrity in three ways [1], i.e., providing local-area differential corrections and removing common-mode errors; providing the authorized users with customized monitors for rare system faults; allowing users to establish a bound on residual errors to perform navigation operations.

Among all the identified hazards, the severest one is the undetected ionospheric anomaly [2]. Ionospheric delays because of free electrons along the path of the GNSS signal are always uniformly distributed under normal conditions. However, when GNSS signals travel through the disturbed ionosphere, severe extreme errors have been observed, which potentially compromise integrity of the GBAS. A model is thus developed to predict the maximum errors that GBAS users can suffer from this threat. The threat of ionospheric anomaly in mid-latitude regions is modeled as a spatially linear semi-infinite and wave-front wedge moving at a constant speed. A predigested model of the ionospheric anomaly is shown in Figure 1a [3]. This piecewise and linear curve is parameterized by the slope of the ramp, gradient width and front speed. In the presence of the ionospheric anomaly, the ionospheric delay changes monotonously. The spatial change rate of the ionospheric delay is the ionospheric gradient. It is the "slope" with the dimension of mm/km in Figure 1a measured by ground stations, namely the maximum ionospheric delays (with the dimension of mm) divided by the gradient width (with the dimension of km). We can directly estimate the ionospheric correction residual through the distance between the user and the reference station. The ionospheric gradient is simplified to a constant in small regions, although seasonal variances play a significant role on the dynamics of the ionosphere [4]. This model is consistent with the study of ionospheric delays in the case of ionospheric anomalies in the mid-latitude region. Compared with the ionospheric delay anomaly, an extreme steep gradient in the ionosphere caused by intense ionospheric storms is generally considered as a dominant ionospheric threat. A predigested model of the ionospheric gradient anomaly is shown in Figure 1b.

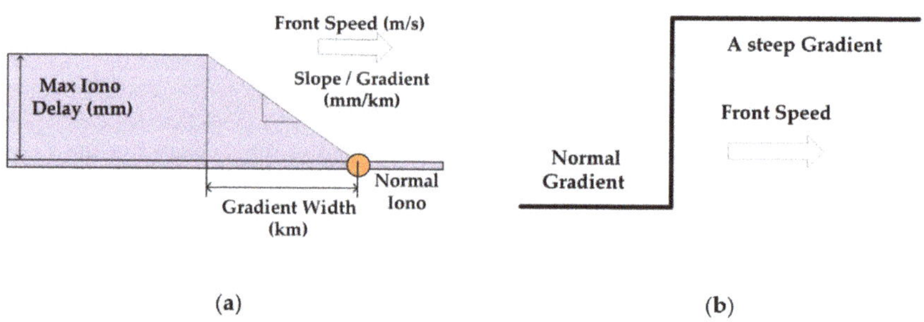

Figure 1. Illustrations of predigested ionospheric anomaly (**a**) and ionospheric gradient anomaly (**b**) models (Figure based on [3]).

Ionospheric gradients can be estimated using code and carrier measurements in practice, including the long-term ionospheric anomaly monitoring and the single-frequency carrier-based and code-aided method [5,6]. Severe extreme gradients have been identified on rare occasions [7]. The largest slant gradient observed in the ionosphere was up to 850 mm/km in Brazil [8], which means measurement errors as large as 850 mm per kilometer were generated between the airborne user and the ground reference station. These errors cannot be accurately corrected using the precise ionospheric model, thus causing inaccurate positioning results and catastrophic consequences for civil aviation. This is unacceptable for the safety-critical GNSS application. An ideal way is to apply a combination of dual-frequency measurements, which can theoretically eliminate the first-order effect of the ionosphere [9,10]. However, this way is not yet available for civil aviation applications, because the correlative GBAS Approach Service Type F (GAST F) is still in the stage of standardization. Therefore, rapid and accurate detection of ionospheric anomalies for single-frequency-based GBAS is still crucial for its performance.

Monitors deployed in the GBAS to detect the ionospheric anomaly mainly refer to the code-carrier divergence (CCD) monitor [11], the dual solution pseudo-range ionospheric gradient monitor [12], and the ground ionospheric gradient monitor [2,13]. Among all the monitors, the CCD monitor is the only one capable of detecting ionospheric gradients observable to both the user and ground subsystem for GAST D. CCD is the anomalous phenomenon caused by the GNSS signal propagation in the disturbed ionosphere, where code phase measurements are delayed and carrier phase measurements are advanced. The CCD monitor is a filter operating on the difference of the code-minus-carrier (CMC) to estimate ionospheric divergences, i.e., the change rate of ionospheric delay over time. If anomalies occur, the divergence increases and can be tested by the customized monitor. The normal ionospheric data with well-defined patterns and the data with unexpected profiles can be separated. In addition, the divergence can also be triggered by potential payloads occurring at the ranging source. Thus, the monitor performs dual tasks, deployed to counter both satellite-induced faults and ionospheric anomalies by detecting divergences.

Two kinds of traditional moving-average-based filters are implemented in the GBAS to monitor the divergence [14,15], i.e., one first-order low-pass auto-regressive and moving average (ARMA) filter, named CCD-1OF, and two first-order cascade ARMA filters, named CCD-2OF. Generally, the CCD-2OF outperforms the CCD-1OF in monitoring divergences due to the smaller smoothing constant. However, both monitors cannot detect divergence rapidly and suffer from a serious delay [16]. These monitors are statistical based, which extract information in the form of simple one-dimensional time series. We cannot understand which process can generate such phase time series. Additionally, the inherent averaging characteristics of filters lead to the weak anomaly-related components being easily overwhelmed by normal noise. Consequently, it is often difficult for traditional divergence monitors to quickly detect the ionospheric anomaly. These monitors even fail to alert in noisy environment. These limitations increase the probability of missed detection, thus leading to misleading information [1]. Therefore, high-level precision approaching requirements are difficult to meet. Developing a novel monitor is highly necessary for future GBAS constructions.

To improve detection performance, a variety of optimization strategies have been studied. Instances of creative research include the monitors based on generalized least squares [17], a two-step method based on an adaptive Kalman filter [3], and a Kullback–Leibler divergence metric [18]. These approaches prove the feasibility of capturing abnormal ionospheric behaviors. However, further work needs to be carried out to overcome the shortcomings, such as the insufficient estimation accuracy, low generalization ability caused by miscellaneous assumptions and poor performance under weak noise.

To overcome these issues, an option is to exploit machine learning algorithms rather than purely statistical-based algorithms. The support vector machine (SVM) is a supervised algorithm designed for binary classification problems [19]. It is famous and appealing, as it can simplify the complex task of classification in an explainable way. It has been extensively utilized in the GNSS community such as ionospheric scintillation detection [20], GBAS availability prediction [21], and high precision positioning [22]. A one class support vector machine (OCSVM) algorithm extends the SVM algorithm to solve a single classification problem [23]. It learns the underlying characteristics of the existing normal samples to judge whether the new samples come from this distribution, thus well suitable for anomaly detection. It has complete mathematical derivation and a meaningful interpretation. As anomalies occur rarely, it is reasonable to assume that they only occur in the tails of normal distributions, thus the goal of anomaly detection is to estimate the density level sets of the normal distributions. This assumption is typical for the OCSVM algorithm. When only a few anomalies occur and the knowledge of new information is limited or unavailable, the OCSVM is suitable, as it does not require any anomalous data. The OCSVM has been found successful in accurately sensing the abnormal information in many domains, such as materials science [24], computer vision [25], power system [26], and agriculture [27]. However, it has not been applied in the GBAS. It is of great significance to investigate the feasibility of using the OCSVM for ionospheric anomaly detection in the GBAS.

The remaining sections are arranged as follows. Section 2 introduces the principle of the OCSVM, followed by the methodology proposed in this work, i.e., an offline-online ionospheric anomaly monitor based on the OCSVM. In Section 3, the synthetic, semi-simulated and real anomalous events are discussed in detail. Finally, conclusions and perspectives of this work are summarized in Section 4.

2. Ionospheric Anomalies Detection Based on the OCSVM

As previously described, a new monitor is required to meet the requirements of the high-level precision approach. The major goal is to perform ionospheric anomaly detection in a time-efficient way. To achieve this purpose, we present an analysis of the statistics of OCSVM metrics and establish a framework of the monitor in this section.

2.1. The OCSVM Algorithm

The principles of the OCSVM are described in this section. As a special form of the SVM, the OCSVM is a domain-based detection algorithm and can tackle the one-class classification problem. It learns the training set with identical labels and determines whether the samples from the new test set are subordinate to the training set [23].

Suppose that we have vectors $\mathbf{X} = \{\mathbf{x}_i, i = 1, \ldots, K\}$. K is the number of vectors in training sets and $\mathbf{x}_i$ comes from the input space $\boldsymbol{R}$. $\phi(\mathbf{x}_i)$ represents the projection of the original training sample $\mathbf{x}_i$ in $\boldsymbol{R}$ to the huge dimensional feature space $\boldsymbol{F}$. A hyper-plane f is used to separate the projected vectors from the original which is assigned as the only anomaly in $\boldsymbol{F}$. The hyper-plane is given by

$$f(\mathbf{x}) = \omega^T \phi(\mathbf{x}) - \rho \tag{1}$$

where ω and ρ are the normal phase vectors and compensation values of the hyperplane in $\boldsymbol{F}$, respectively. These parameters can be calculated by the following quadratic programming issues

$$\begin{array}{ll} \min\limits_{\omega,\rho,\xi} & \frac{1}{2}\|\omega\|^2 + \frac{1}{\gamma K}\sum\limits_{i=1}^{K} \xi_i - \rho \\ s.t. & \omega \cdot \phi(\mathbf{x}_i) \geq \rho - \xi_i \\ & \xi_i \geq 0, i = 1, \ldots, K \end{array} \tag{2}$$

where $\gamma \in (0,1]$ is the trade-off constant value used to determine the proportion of the normal and anomalous data. ξ_i is the slack variable, referring to the extent where the samples are misclassified. $\|\Delta\|$ denotes the 2-norm of the vector Δ. The Lagrange factor $[\alpha_1, \ldots, \alpha_K]^T$ is introduced for each vector $\mathbf{x}_i$ to obtain the dual problem of (2). Solving the dual problem gives rise to

$$\omega = \sum_{i=1}^{K} \alpha_i \phi(\mathbf{x}_i) \tag{3}$$

where $0 \leq \alpha_i \leq 1/\gamma K$. The famous kernel trick can replace the inner product of two vectors in $\boldsymbol{F}$ with a kernel operation in $\boldsymbol{R}$. This projection transforms a non-linearly separable problem into a linearly separable problem. Thus (1) combined with (3) becomes a function as follows

$$f(\mathbf{x}) = \sum_{i=1}^{K} \alpha_i G(\mathbf{x}_i, \mathbf{x}) - \rho \tag{4}$$

where G($\mathbf{x}_i$,$\mathbf{x}$) is the inner product of two projecting functions, i.e., $G(\mathbf{x}_i, \mathbf{x}) = (\phi(\mathbf{x}_i), \phi(\mathbf{x}))$. A popular radial basis function (RBF) kernel function is utilized here to expand vectors into the high dimension. The RBF is given by

$$G(\mathbf{x}_i, \mathbf{x}) = \exp\left(-\frac{\|\mathbf{x}_i - \mathbf{x}\|^2}{\sigma}\right) \tag{5}$$

where σ is the hyper-parameter controlling the width of the kernel. The kernel provides the OCSVM the ability to detect abnormal samples in the high dimension feature space $\boldsymbol{F}$. Moreover, it can be proved that only for the vector on the boundary, namely the support vector, the non-zero α imposes effects on the construction of the boundary.

Based on the derivation above, the OCSVM can be constructed and used to detect new samples. If the new samples fit well with the trained model, they are labeled as normal samples. Otherwise, they are adjudged as anomaly samples. $f(\mathbf{x})$ serves as a score function to determine the classification quantitatively. A vector x is classified to be normal with $f(\mathbf{x}) > 0$ and anomalous with $f(\mathbf{x}) < 0$. The smaller the $f(\mathbf{x})$ is, the more abnormal the vector is viewed.

2.2. Ionospheric Anomaly Detection with the OCSVM-Based Monitor

This section applies the OCSVM algorithm to the monitor. Based on the analysis above, a new offline-modeling-online-monitoring monitor is proposed. It has the potential to meet the high-level approaching requirements. Figure 2 shows the overall schematic diagram of the proposed integrity monitoring algorithm illustrating the process. Statistical information can be extracted to characterize the normal states offline. A classifier can then be trained to detect the potential ionospheric anomaly. The GBAS can monitor the ionospheric anomaly more effectively through the OCSVM-based detection strategy. Detailed analyses are given next.

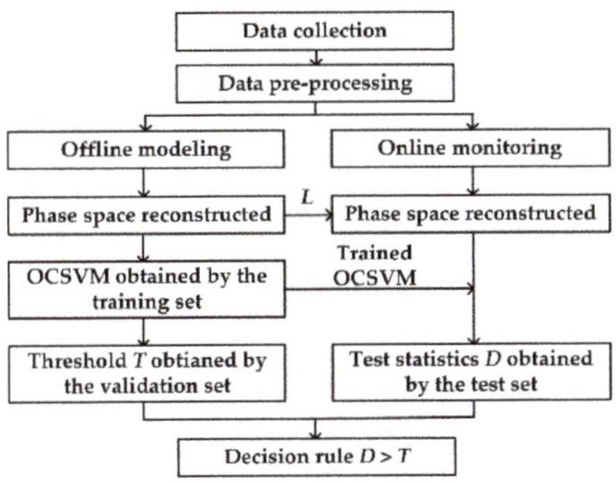

Figure 2. Overall schematic diagram of the ionospheric anomaly monitor based on the OCSVM.

In data collection, we collected both the code-phase and carrier-phase measurements. To eliminate the same item in the code and carrier phase measurements and emphasize the differences, a CMC is introduced, given by

$$\mathrm{CMC} = P - \Phi = 2I + N\lambda + M \tag{6}$$

where P and Φ represent the code-phase measurement and carrier-phase measurement between the receivers and satellites, respectively. I is the ionospheric delay. N is the ambiguity of whole cycles. λ is the carrier wavelength. M includes noise including multipath effects, thermal noise and other error sources.

Before the introduction of the monitor design, it is necessary to understand which original measurements can be dealt with. A pre-processing is performed to process GNSS raw measurements and to perform various corrections to get the true CMC as in [28]. Recall that our aim is to detect the ionospheric anomaly and separate it from the other errors. The first term in the right side of (6) is what we are interested in. The presence of the double ionospheric delay errors is induced by the dispersive nature of the ionosphere. Although ionospheric delays are time-variant during normal ionospheric days, we can use a neural network and single-frequency GNSS signals to estimate the delays [29]. Therefore, the delays during normal ionospheric days can be considered as a time-variant but known bias. The second term in the right side of (6) is the ambiguity term. It does not directly affect the results when cycle slips are absent in measurements. It is always fixed or easily repaired under the circumstance of relatively weak ionospheric perturbations using the Kalman filter [30]. Therefore, the cycle-relevant term is treated as a pure bias. The third term in the right side of (6) indicates ground-related noise. The noise cannot be simply modeled by a zero-mean Gaussian distribution. It is considered the major obstacle in the ionospheric anomaly detection, as the ionospheric anomaly-related components are easily overwhelmed. Based on the above analysis, we understand what we can cope with before stepping into the monitor design. Compared with traditional CCD monitors, we focus on the original CMC instead of the difference of the CMC. Namely, we concern about the accumulation errors induced by the ionospheric gradient anomalies over a period of time.

Due to the fact that only vector inputs can be imported into the OCSVM, one-dimension time series cannot be applied directly [26]. Therefore, the CMC sequence must be converted to a high-dimension vector first. We need to ensure that we can construct a sound mathematical model from a single time series. The method of nonlinear analysis can be used, provided that the CMC is a dynamic, chaotic, and irregular sequence. The CMC is assumed as an observable variable of a multivariate system. Additional knowledge obtained from the collection of real data is required to establish the system. The state phase space of the original system can be reconstructed by the time-delay embedding to maintain the system dynamics. Specifically, let $\mathrm{CMC}_1, \ldots, \mathrm{CMC}_k$ be a time sequence, it can be unfolded into the phase space Q, where $Q \subseteq \boldsymbol{R}^L$ and L refers to embedding dimension (ED). We set the embedding delay as the sampling interval to consider more information [31]. The reconstructed vector can be then obtained by

$$\mathrm{CMC}_{L,k} = [\mathrm{CMC}_{k-L+1}\ \mathrm{CMC}_{k-L+2}\ \ldots\ \mathrm{CMC}_k] \quad (7)$$

where $\mathrm{CMC}_k \subseteq Q$. Thus, a one-dimensional time series can be transferred to a string of vector $V_L(N)$, given by

$$V_L(N) = \{\mathrm{CMC}_{L,k}, k = L, \ldots, N\} \quad (8)$$

A continuous track formed by vectors connected by lines in the reconstructed phase space is called a phase diagram, whose geometry reflects the characteristics of unknown systems. The reconstructed phase space is topologically identical to the original system, provided that the embedding dimension is sufficiently large.

After converting the time series to a set of vectors in the phase space, the OCSVM can be used to monitor the ionospheric anomaly. We can model the collected data offline and monitor the real-time data online [32]. An offline modeling needs to be built by training tests. We first pay attention to the parameter selection. The RBF is chosen as the kernel function of the OCSVM. The selection of the hyper-parameter σ affects the final detection results. A larger σ results in a greater generalization capability but a lower discernibility and vice versa. Note that both σ and L affect the final test results. They will be discussed in the next section. Because the probability of serious ionospheric anomaly in the actual environment is extraordinarily low, it is advisable to set the trade-off parameter γ to 1, indicating that there is no outlier in the training set. Thus, the OCSVM model can be obtained by the offline data. Additionally, the OCSVM operating on the RBF kernel performs poorly with the number of samples in training sets. It does not perform well in the case of big data because of the increased computational load, while hardware auxiliaries can be used to overcome this difficulty [33]. Considering the actual situation, the offline OCSVM is feasible to be trained quickly. At the same time, the offline modeling part is not required to be real-time, and the appropriate computational load is acceptable in practice.

Once a model reflecting the normal state is established, detecting any departure from the normal operation is vital. Because of the complexity of the real environment, we cannot directly determine whether the new sample is an ionospheric anomaly by the sign of the score value. A threshold is set to determine the decision result. A validation test is required to predict new scores. It acts as cross-validation to judge the extent of how the trained OCSVM matches the underlying distribution. Then, the threshold needs to be obtained by the scores. Because the distribution of the OCSVM scores does not follow the Gaussian distribution, the traditional Gaussian overbounding to construct a threshold is a little conservative. Instead, we use non-parametric methods to estimate the threshold. A famous kernel density estimation (KDE) is employed to approach the probability density function in a non-parametric approach [34]. A simple Gaussian kernel is chosen in this work. Since KDE is to estimate the values of unknown random variables from known samples, it is reasonable to assume that the weight of known samples on unknown variables decreases as a Gaussian distribution gradually for the central limit theorem. The Gaussian kernel is widely employed to estimate the underlying distribution of the scores [35]. Note that different from the traditional monitors, the OCSVM monitor uses test statistics to describe the degree of misclassification. When the value is very large, the distribution of the test sample is very close to the real sample, and there is no abnormality. The ionospheric anomaly occurs only when the statistic is negative and its absolute value is large. Therefore, the threshold is unilateral. Then the threshold T_{KDE} can be determined as the unilaterally lower bound of the estimated cumulative distribution function with respect to the corresponding false-alarm risk. As a result, the real threshold T is the minimum of zero and T_{KDE}. For real samples, the errors are strongly correlated with the elevation angle due to the multipath, thus the threshold is determined related to the elevation bin [36]. It is also noted that because the ambiguity term is included, each satellite channel in each elevation bin needs to set a different threshold, which makes a higher request for the GNSS signal channel.

For the online-monitoring part, the parameter is the same as the offline model. The calculated test statistic is the score function as D. This statistic is compared with the threshold values calculated offline to find whether the sample patterns deviate from the majority. If the test statistics are less than the detection threshold T, timely warnings are provided and broadcast. Note that the OCSVM only parse the information in normal conditions without any presupposition for any ionospheric anomaly points, thus any ionospheric anomaly causing CMC changes will be detected. Although this unpredictable and unclassified fault mode may cause troubles in areas with high fault probabilities, any threat identified should be informed for the GBAS. Moreover, the linear characteristics of the online-monitoring ensure that the computational cost is comparable to that of the traditional monitors.

The OCSVM-based monitor has three benefits in monitoring ionospheric anomalies: (1) Using the phase-space reconstruction technique, one-dimensional CMC time series can be transformed into high-dimensional vectors. The full dynamics of the CMC system accessible in the phase space can be passed through the OCSVM; (2) The OCSVM is a domain-based detection algorithm that can distinguish between normal and anomaly classes with a redefined boundary or "domain" as the test statistics. It offers better flexibility compared with the Gaussian statistics; (3) The OCSVM has certain robustness to noise and generalization ability to ionospheric anomalies. Therefore, superior performance can be expected with the proposed OCSVM-based monitor to exhibit high-level precision approach.

3. Experiment Analysis

In this section, three examples are illustrated to show the anomalous monitoring capability of the proposed OCSVM-based monitor. In the first example, synthetic data are used to simulate the ionospheric gradient-free and non-negligible ionospheric gradient anomaly CMC, respectively. In the second example, we verified performance of ionospheric anomaly detection with semi-simulation data. The raw BeiDou Navigation Satellite System (BDS) data collected at Dongying Shengli Airport with an artificial anomaly are used to demonstrate the effectiveness of the proposed monitor. In the third example, a real anomaly event are taken into account to evaluate the performance under real circumstance. The OCSVM-based monitor is compared with traditional CCD monitors and KLD-1OF monitor with respect to the ionospheric anomaly detection performance.

3.1. Ionospheric Anomaly Detection with Synthetic Data

The aim of the synthetic data analysis is to show the effect of the different noise levels on the required time of detecting anomalous events, which cannot be realized using real data. Because the CMC depends solely on the difference in code and carrier phase rather than individual values in the single-frequency mode, code phase and carrier phase anomalies are equivalent [10]. Thus, only one fault mode analysis is required. To demonstrate the ionospheric anomaly detection performance, 2000 samples are generated to simulate the gradient changes induced by the ionosphere. The first 1000 samples are ionospheric gradient-free. To simulate the ionospheric anomaly, a constant steep gradient change, as shown in Figure 1, is added. The ramp-type change starts at the 1001st sample at a velocity of 0.01 m per epoch [3,18]. The parameters of the samples are written as

$$I_k = \begin{cases} 4 + n_k & 0 < k \leq 1000 \\ 4 + 0.01(k - 1000) + n_k & 1000 < k \leq 2000 \end{cases} \tag{9}$$

where n_k is the independent measured noise at the kth epoch. Because the noise follows a heavy-tailed distribution, we choose distributions that can characterize upper and lower bounds of the real distribution to conduct the simulation. The tail distribution of real noise lies between Gaussian and Laplacian distributions [36]. Thus, the noise is assumed to be zero-mean, and it includes components following the Gaussian distribution with the variance ranging from 0.1 to 2.5, and components following the Laplacian distribution with the scale ranging from 0.1 to 2.5, respectively [3,18]. We choose them to simulate the noise of different levels. When the variance and scale are 0.1, the noise is weak. With the increase of the variance and scale, the noise amplitude in the simulated data increases gradually. When the variance and scale approach 2.5, the noise is strong enough to conservatively simulate hostile environment [37]. According to our experience, noise beyond this range seldom appears in the real environment. Figure 3 shows the variation of the simulated data amplitude when both the variance of Gaussian noise and the scale of the Laplacian noise are set to 0.5 and 2.5, respectively.

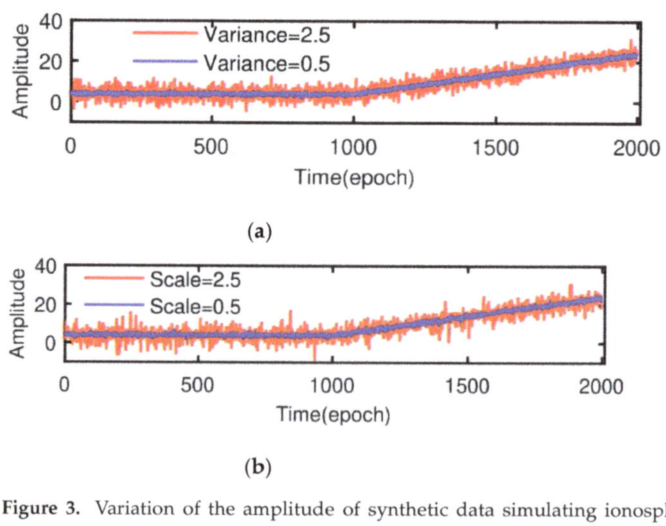

Figure 3. Variation of the amplitude of synthetic data simulating ionospheric anomalies with Gaussian noise (**a**) and Laplacian noise (**b**). The ionospheric gradient anomalies are simulated from the 1001st epoch by adding ramp-type errors. The absolute maximum values of noise in (**a**) are 1.7 m (variance = 0.5) and 6.9 m (variance = 2.5). The absolute maximum values in (**b**) are 1.9 m (scale = 0.5) and 9.6 m (scale = 2.5).

To detect whether anomalies occur, (9) is first reconstructed to the form of (7). Figure 4 takes the $L = 3$ as an example to show the constructed phase state of Gaussian and Laplacian noise. The distributions in the constructed phase state are spindle-shaped, indicating that the distributions of normal and anomalous vectors are different. The thick middle part where points are brought together represents normal vectors, while the sharp head part where points spread far away represents abnormal vectors. These two parts are easily separated.

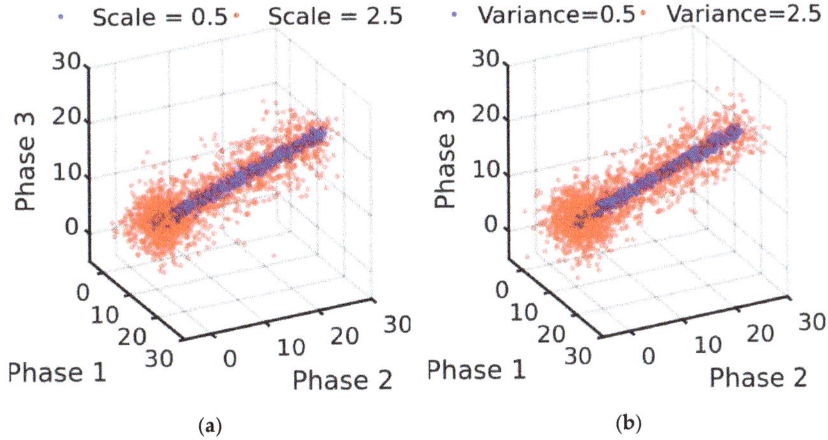

Figure 4. Vector sets of the time series in a phase space (L = 3) under (**a**) Gaussian noise and (**b**) Laplacian noise. Both distributions are spindle-shaped, indicating that ionospheric anomalies, denoted by the sharp head part, can be easily distinguished from the normal points denoted by the thick middle part.

The first 500 epochs of the time sequences are considered as the training set to construct the OCSVM, while the 501st to 1000th epochs are the validation set to obtain the threshold [3,18]. Figure 5 is the visualization of the 3D trained OCSVM decision bound and the corresponding support vectors in the input space. The vectors are constructed under Gaussian noise with the variance set to 2.5 in Figure 5a, and under Laplacian noise with the scale set to 2.5 in Figure 5b. The red plane is the 3D decision boundary. The corresponding hyperplane maximizes the distance between the original points and vectors in the high-dimensional space. Thus, the plane separates the normal vectors from the anomaly ones. The case when the variance and scale parameters are 0.5 is basically the same as that of 2.5, thus it is not shown.

During the monitoring process, the parameter determination of the dimension and the RBF kernel is important for the determination of an appropriate OCSVM-based monitor. It is noticeable that both of them have an impact on the OCSVM-based monitor; thus, they need to be estimated simultaneously. Optimal parameters are chosen to get the lowest average detection time under different levels of noise. By adjusting the amplitude of noise from 0.5 to 2.5 at a step of 0.5 to characterize the main changes in noise [18], the corresponding parameter is obtained. Taking Gaussian noise as an example, the average detection time with different dimensions and kernel scales is shown in Figure 6. There is an obvious variation of the detection time with the increase of the dimension L. The minimum average time of 86 epochs occurs when $L = 10$. As for the RBF kernel scale σ, a larger scale generally results in a lower average time. When it reaches a critical point, the average detection time tends to converge and maintains unchanged even keeping increasing the kernel scale continuously. When $L = 10$, the critical point appears when σ is 20, which is considered the optimal parameter of the RBF kernel scale. As a result, $\sigma = 20$ and $L = 10$ are chosen as the optimal values. A similar method is used to determine the optimal parameter under the Laplacian noise. In this case, $\sigma = 18$ and $L = 60$ are chosen as the optimal values.

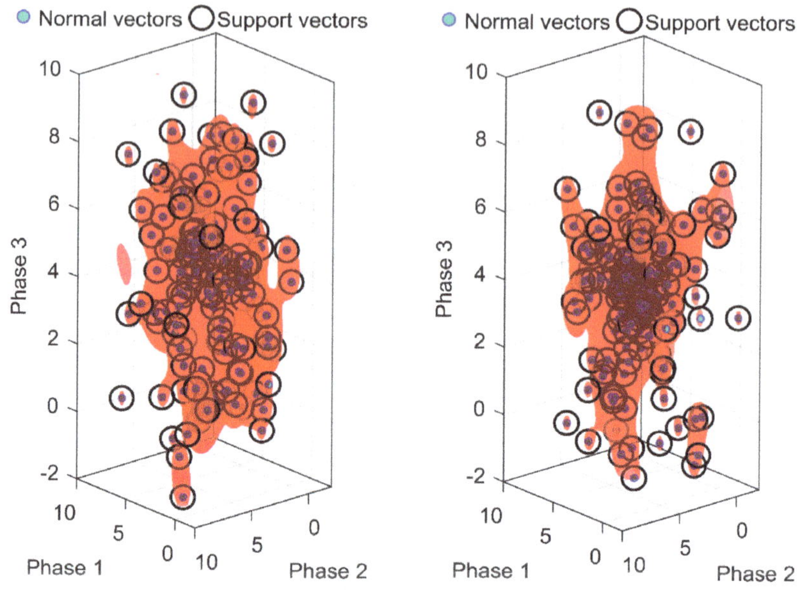

Figure 5. Trained OCSVM and the vectors in an input phase space (L = 3) with (**a**) Gaussian noise and (**b**) Laplacian noise. Red plane is the 3D contour plane.

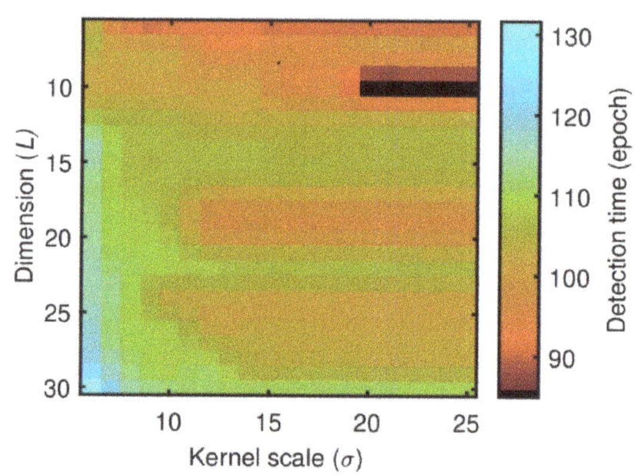

Figure 6. Variation of the average detection time in relation to L and σ with synthetic data. The minimum time of 86 epochs occurs when σ = 20 and L = 10.

To compare the ionospheric anomaly detection performance, three monitors are used as the control group, i.e., the traditional CCD-1OF with τ_1 = 100 epochs [14], the traditional CCD-2OF with $\tau_1 = \tau_2$ = 30 epochs [15], the Kullback-Leibler metric-based monitor KLD-1OF with τ = 150 epochs and L = 50 [18]. The proposed OCSVM-based monitor is selected with the parameters and threshold estimated through the process described previously. The probability of fault detection allocated for the monitor is 10^{-8}.

Figure 7 illustrates the test statistics and the corresponding thresholds of the CCD-1OF, CCD-2OF, KLD-1OF and the OCSVM-based monitors, under Gaussian noise with the variance set as 0.5 and 2.5, respectively. In the presence of weak noise when the noise variance is set to 0.5 in Figure 7a, the test statistic of the OCSVM-based monitor takes the shortest time to cross the threshold, thus the OCSVM-based monitor can perform faster than others. In the presence of strong noise when the noise variance is set to 2.5 in Figure 7b, traditional CCD monitors cannot detect any divergence, thus the detection of the ionospheric anomaly is missed. Therefore, such noise, which may be significant for high-level precision approach, can be missed without timely alarming with CCD-1OF and CCD-2OF. Although the KLD-1OF can also detect anomalies, the OCSVM-based monitor outperforms it in the detection speed.

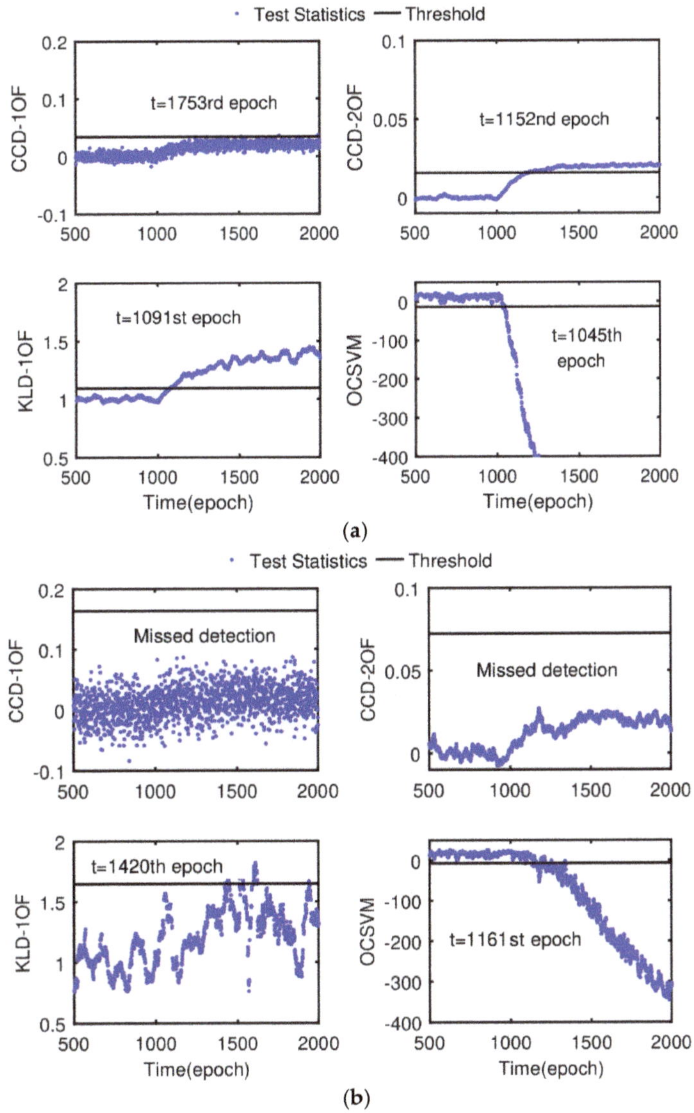

Figure 7. Variation of the test statistics and thresholds of the CCD-1OF, CCD-2OF, KLD-1OF and the proposed OCSVM-based monitors under Gaussian noise, with the noise variance set to 0.5 in (**a**) and 2.5 in (**b**). If the ionospheric anomaly is successfully detected, the detection time is included. Note that the ionospheric anomaly is inserted from the 1000th epoch.

Figure 8 illustrates the test statistics and the corresponding thresholds under Laplacian noise with the scale set to 0.5 and 2.5, respectively. When the amplitude of the noise is small, the OCSVM-based monitor outperforms all the other monitors. When the noise variance is 2.5, only the test statistic of the OCSVM-based monitor crosses the threshold, which means that only this monitor successfully detects anomalies. This result again indicates the effectiveness of the OCSVM-based monitor under all noise levels.

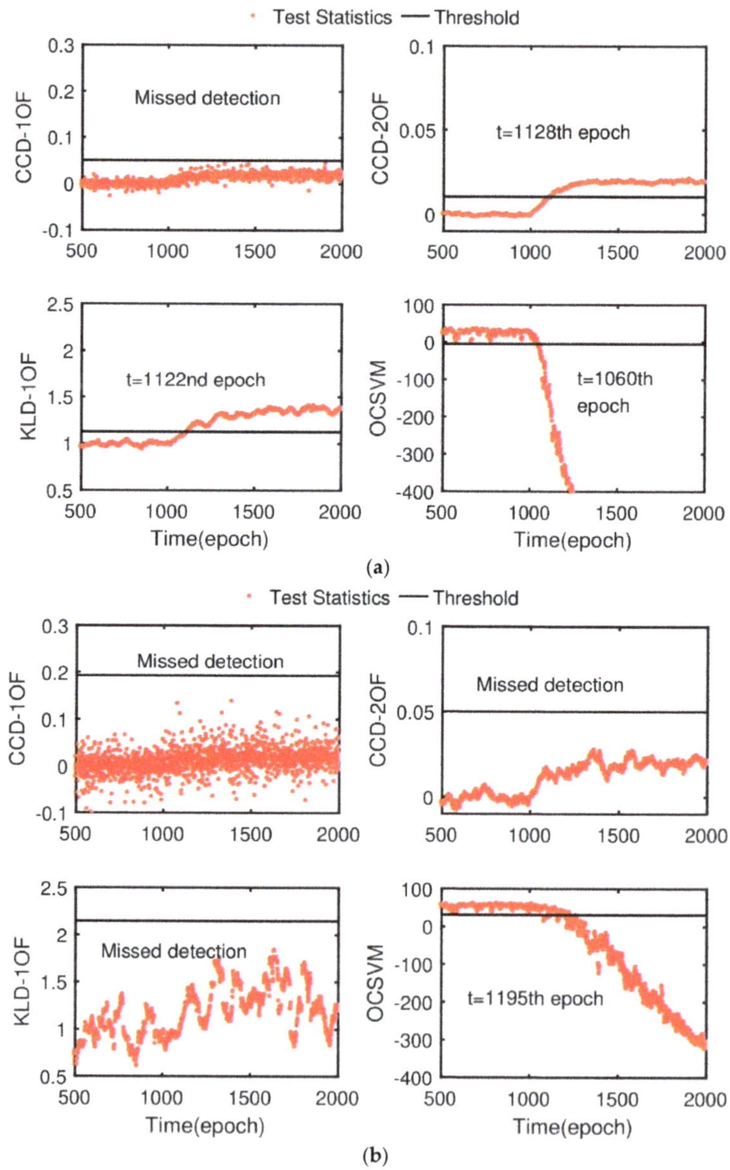

Figure 8. Variation of the test statistics and thresholds of the CCD-1OF, CCD-2OF, KLD-1OF and the proposed OCSVM-based monitors under Laplacian noise, with the noise variance set to 0.5 in (**a**) and 2.5 in (**b**). If the ionospheric anomaly is successfully detected, the detection time is included. Note that the ionospheric anomaly is inserted from the 1000th epoch.

Figure 9 presents the average detection time of the four monitors in the presence of varied noise from 50 repeated tests. The results are not shown when monitors fail to detect any ionospheric anomaly. The proposed OCSVM-based monitor is always faster than the others in detecting the ionospheric anomaly. When the noise amplitude is getting larger, the CCD-1OF and CCD-2OF monitors fail to sense the anomaly and the KLD-1OF detects the anomaly at a lower speed. Only the OCSVM-based monitor can detect the anomaly at a relatively high speed. In conclusion, the OCSVM-based monitor is more suitable for ionospheric anomalous detection compared with the traditional CCD monitors in the GBAS regardless of the noise levels and types.

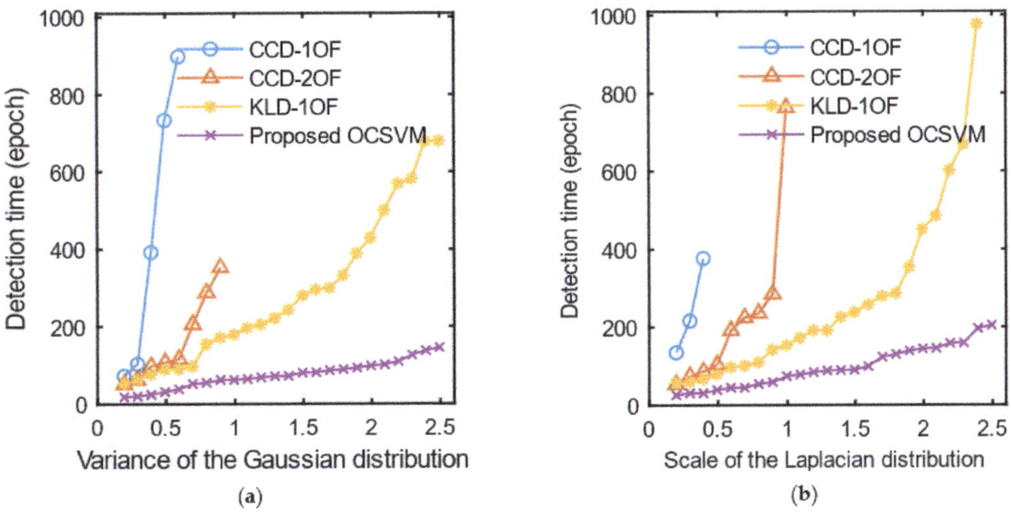

Figure 9. Variations of the ionospheric anomaly time in relation to Gaussian (**a**) and Laplacian (**b**) noise levels when the CCD-1OF, CCD-2OF, KLD-1OF and OCSVM-based monitors are applied. Ionospheric anomalies that cannot be detected are not shown in the figure.

3.2. Ionospheric Anomaly Detection with Semi-Simulation Data

To further evaluate the performance of the proposed ionospheric anomaly monitor, the real observed data are processed. In this work, the real data are collected from the GBAS deployed at the Dongying Airport in Shandong, China [38]. The B1I signals of BDS-2 are captured by four reference receivers, each equipped with the same choke ring antenna as shown in Figure 10. The BDS code and carrier phases measurements are recorded continuously at a sampling rate of 1 Hz. Previous literature shows that one day's data is sufficient to construct a model to perform an evaluation of the detection time [3,5,18]. The antenna of this receiver is installed beside the runway of the airport. The elevation mask is set to 15°. Thus, 24-h data collected by the 1st receiver on 2 September 2020 are used in this analysis. In Figure 11, the ground-related ranging noise of PRN 7 is drawn as an example. It is obtained as shown in [29]. The amplitude lies between -1.5 m and 1.5 m. The data are then binned at an interval of 5° to illustrate the performance of monitors under different noise levels. In addition, because the probability of ionospheric gradient anomaly in practice is very small, which is in the magnitude of about 10^{-7}, we artificially add the simulated anomalies rather than process the real anomalies. As the airport is located in the mid-latitude region, we use the ramp-type error model to simulate the anomaly as shown in Figure 1.

Figure 10. BDS GBAS antenna mounted on the ground near the runway at the Dongying Airport in Shandong, China.

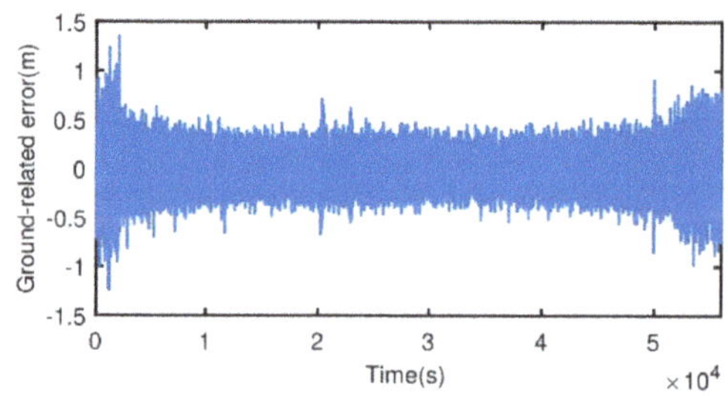

Figure 11. Processed BDS ranging noise of PRN 7.

To test the performance of the OCSVM-based monitor, a parameter determination process for real data is required first, which is the same as that for the synthetic data. L and σ are estimated in the light of the average detection time of the satellite PRN 7 with the ionospheric gradient changing at a rate of 0.018 m/s [3,15]. The satellite is chosen as it covers a larger range of elevations. According to Figure 12, values of $L = 20$ and $\sigma = 9.6$ are chosen as the optimal parameters.

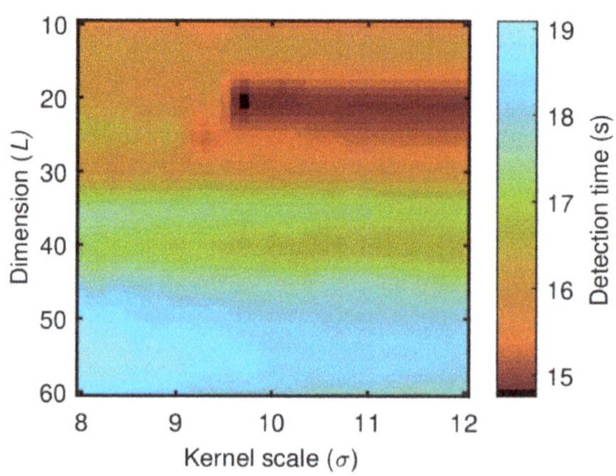

Figure 12. Variation of the average detection time for determining the optimal L and σ with real data. The minimum time of 14.75 s occurs when $\sigma = 9.6$ and $L = 20$.

After determining the parameters of the proposed monitor, the threshold in every elevation bin is calculated with the similar method described in Section 2.2. Figure 13 shows the estimated metric and the determined thresholds as a function of elevations.

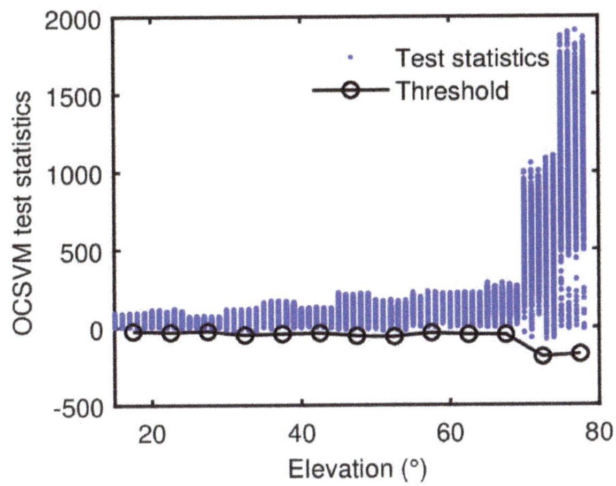

Figure 13. Variation of the metric and threshold of the proposed OCSVM-based monitor in terms of satellite elevations.

The impact of severe ionospheric gradients is then simulated. Two different ionospheric gradients, with the change rate respectively set to 0.01 and 0.02 m/s, are inserted to illustrate the effectiveness of the proposed monitor. The duration of the inserted gradients is 200 s. The experiment is repeated 50 times to compare the stability of different monitors. Figure 14 shows the average detection time in each elevation bin when the four different monitors are implemented in Figure 14a,c. Standard deviations of the detection time are also shown in Figure 14b,d. A greater deviation indicates a lower stability of the detection performance. Because the signal at a lower elevation angle tends to suffer more noise, such as multipath, the detection time decreases as the elevation increases. The standard deviation at lower elevation bin is missing in the figure due to the detection failure. Compared with the other three monitors, the OCSVM-based monitor has the shortest detection time and the greatest detection speed. Especially, compared with the KLD-1OF, the OCSVM-based monitor can decrease the average detection time respectively by about 55% and 44% when the change rate is 0.01 m/s and 0.02 m/s. Additionally, the CCD-2OF and KLD-1OF perform more steadily than the OCSVM-based monitor. The reason is that the averaging process is absent in the OCSVM-based monitor, thus the OCSVM-based monitor has more instabilities in detecting the ionospheric anomalies. However, the shortest detection time of the traditional monitor is mostly greater than the longest detection time of the OCSVM-based monitor. Thus, the instability of the OCSVM-based monitor can be tolerated. In addition, there is no false alarm event in the simulation.

Because the GBAS involves the field of life safety, small ionospheric anomalies are non-negligible for the reliable navigation [18]. Being able to detect small gradient changes, such as at a rate of 0.001m/s is also a clear advantage compared with the traditional monitors. The capabilities of the four different monitors are compared for detecting the ionospheric anomaly occurring on PRN 7. The detection capability refers to the minimum gradient change rate that can be detected during the ionospheric anomaly occurrence. In this example, the rate of the gradient change is varied from 0.001 to 0.025 m/s. Figure 15 shows the minimum gradient changes that can be detected for the four monitors. As it can be seen, the OCSVM-based monitor presents a better detection capability than the other monitors. When it is applied, the minimum gradient changes detected are lower than

0.005 m/s over all elevations. Thus, the OCSVM-based monitor can reduce the integrity risk by alarming quickly and accurately.

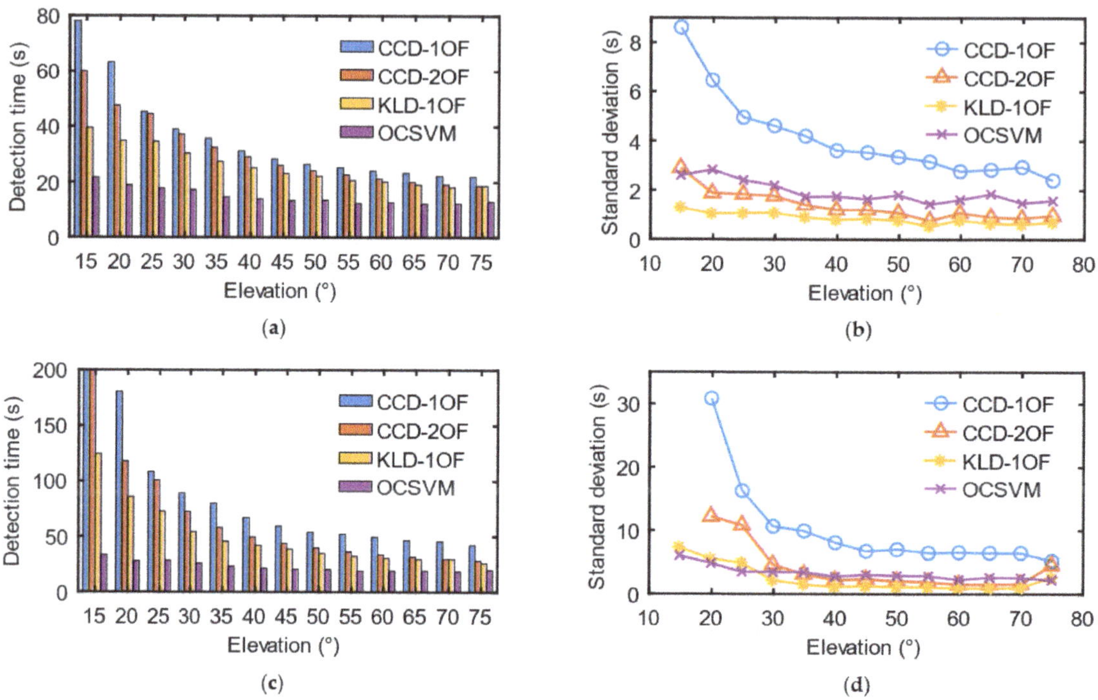

Figure 14. Comparison of the average and the standard deviations of the detection time when applying the CCD-1OF, CCD-2OF, KLD-1OF and the proposed OCSVM-based monitors to detect the ionospheric anomaly on PRN7. The ionospheric gradients, with the change rate set to 0.01 m/s in (**a**,**b**) and 0.02 m/s in (**c**,**d**), are inserted.

To further verify the effectiveness of the OCSVM-based monitor, all visible satellites are considered in the ionospheric anomaly detection. The ionospheric gradient change, with the rate of 0.02 m/s, is added to all the satellites. The detection time of ionospheric anomaly on each satellite in each elevation bin is counted through 50 experiments when the traditional and proposed monitors are implemented. The average and the standard deviation of the detection time in bin are calculated. Additionally, because the signal qualities of satellites are different, the signals with large noise may be excluded when using the traditional monitors [12]. By contrast, the signal quality does not affect the OCSVM-based monitor, since the OCSVM-based monitor can detect ionospheric anomalies in the environment with large noise. The detection time is shown in Figure 16. It can be seen that the detection time maintains the lowest when the OCSVM-based monitor is used. The OCSVM-based monitor decreases the average detection time by about 43% compared with the KLD-1OF monitor. When the elevations are high, the CCD-2OF and KLD-1OF monitors slightly outperform in terms of the standard deviations, i.e., the stability. However, considering the significant contribution of the OCSVM-based monitor to the detection speed, it is acceptable with more instability for the proposed monitor. It can be seen that the test results of all visible satellites are quite similar to the results of PRN7 alone, regardless of the average or the standard deviation of the detection time. This proves that the OCSVM-based monitor performs well under all circumstances. Therefore, the proposed monitor can help to ensure the integrity of the GBAS by accelerating the detection. Thus, the proposed monitor can help to ensure integrity of the GBAS by accelerating the detection.

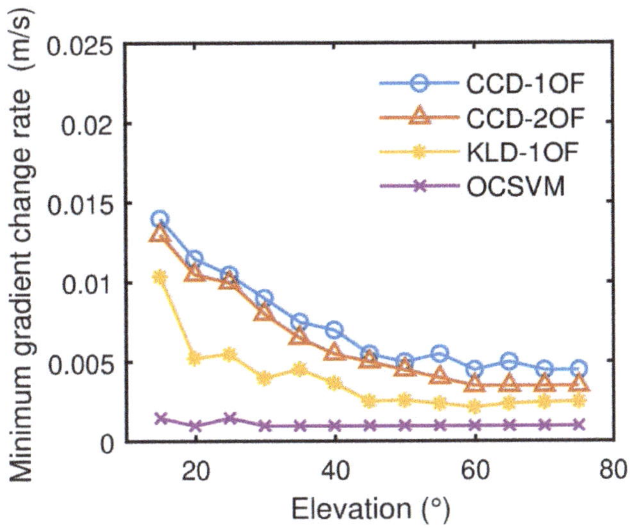

Figure 15. Variations of the minimum gradient changes that can be detected by different monitors in terms of elevations.

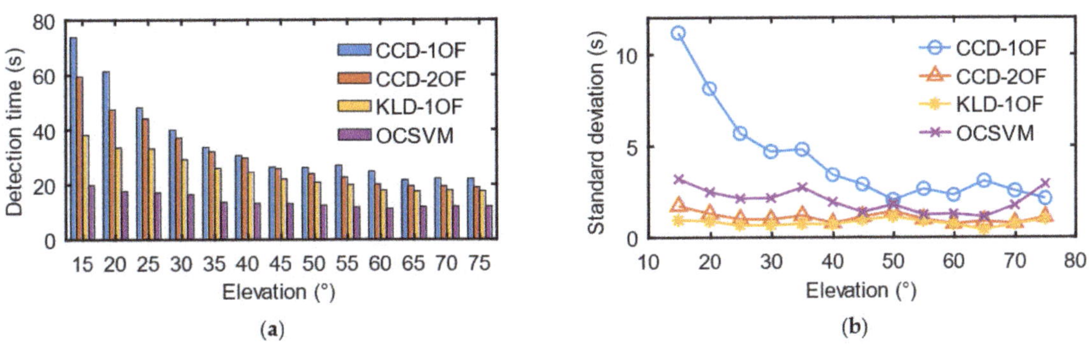

Figure 16. Comparison of the average (**a**) and the standard deviations (**b**) of the detection time of all BDS-2 satellites in relation to satellite elevations using the CCD-1OF, CCD-2OF, KLD-1OF and the proposed OCSVM-based monitors. The ionospheric gradient change with a rate of 0.02 m/s is added to all visible satellites.

3.3. Real Ionospheric Anomaly Event Experiment

The performance of the monitors is evaluated in the presence of real ionospheric anomaly in this section. Global Positioning System (GPS) data on L1 measurements collected from the Crustal Movement Observation Network of China from 2011 to 2014 were processed in [39]. During this period, the largest detected gradient as 71 mm/km in northern China occurred on 23 May 2012, which was identified using the data from two stations, i.e., BJSH (Shisanling, Beijing, 40.3° N, 116.2° E) and BJYQ (Yanqing, Beijing, 40.4° N, 116.0° E), located in middle latitude region with a baseline of 25.5 km. The satellite affected by the ionospheric anomaly was PRN 10. The geomagnetic storm class of this day is moderate (Kp = 4.7, Dst = −30 nT). When the GPS signal passed through the ionosphere, it was affected by the irregular geomagnetic storm, resulting in an ionospheric anomaly on this day [40].

The ionospheric slant delays for PRN 10 captured at the BJSH and BJYQ stations from 15 to 21 Universal Time (UT) are shown in the Figure 17a. It can be seen that ionospheric delay calculated at the two stations shares a similar tendency, while it is slightly larger from 15.6 to 15.7 UT at BJSH. This difference is caused by a spatial ionospheric gradient. Figure 17b shows the gradient levels measured along the BJSH-BJYQ baseline. The maximum value of 71 mm/km occurs at 15.7 UT. This spatial gradient is not as extreme as those reported in [7], but it is still a severe threat for the safety of civil aviation. Figure 17c shows the elevation of PRN 10 viewed at BJSH and BJYQ stations. It can be seen that the anomaly occurred when the satellite was about 30°. Then we pay attention to the measurements at BJSH station. In Figure 17d, the observation ranging noise and ionospheric delay of PRN 10 calculated at BJSH station are shown. During the period of ionospheric anomalies lasting for 420 s, the anomalous ionospheric gradient increased by about 2.1 m; thus, the average change rate is approximately 0.005 m/s. Additionally, the predicted normal ionospheric delay is opposite to the actual abnormal ionospheric delay [29], so the net ionospheric delay change rate exceeds 0.005 m/s and is up to a constant as 0.008 m/s.

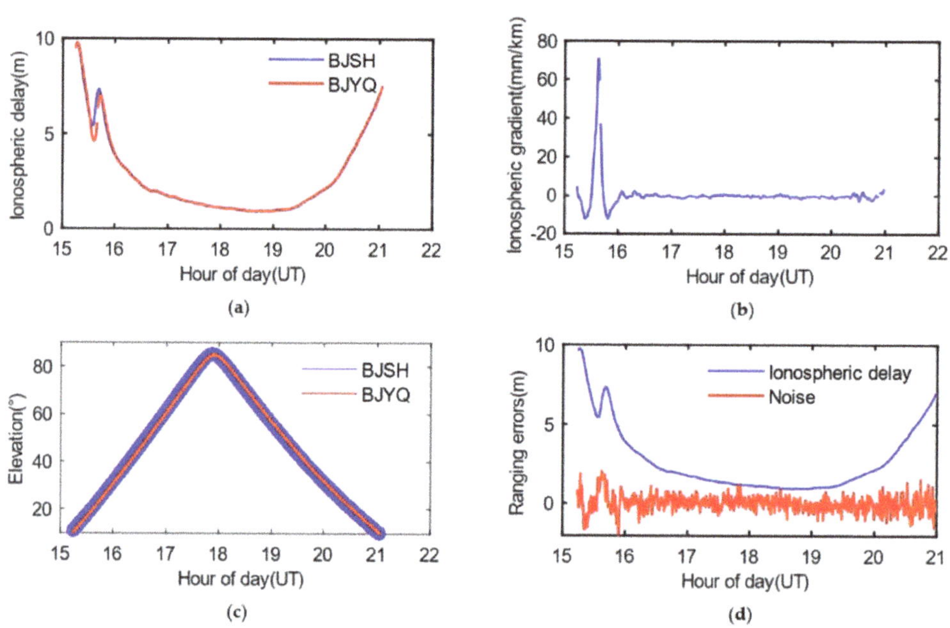

Figure 17. A typical ionospheric anomaly caused by geomagnetic storms in Beijing, China on 23 May 2012. Panels are: (**a**) The ionospheric delay measured on GPS PRN 10 at BJSH and BJYQ stations; (**b**) the ionospheric gradient calculated along the baseline between the two stations; (**c**) the satellite elevations of PRN 10; (**d**) the observation ranging errors of PRN 10 at BJSH station. The amplitude of noise is briefly limited to 1.5 m.

Under the circumstance of the ionospheric anomaly shown in Figure 17, the performance of the proposed ionospheric anomaly monitor is evaluated. Parameters of traditional CCD monitors are the same as that presented in Section 3.2. $\tau = 30$ epochs and $L = 25$ are chosen for KLD-1OF [18], and $L = 21$ and $\sigma = 15$ are chosen for the OCSVM-based monitor. Due to the fact that the time of anomalies is only 420 s, the threshold is set to a fixed value during this period. Figure 18 illustrates the test statistics and the corresponding thresholds of four monitors. The CCD-1OF monitor fails to detect the ionospheric anomaly in this real case. On the other hand, the CCD-2OF, KLD-1OF and the OCSVM-based monitors can detect the ionospheric anomalies, while the OCSVM-based monitor is at a higher speed. Compared with the KLD-1OF monitor, the detection time can be decreased by 16%. There are two possible reasons to explain why the improvement in the real event is not as obvious as that in the simulation data. First, there may be cycle slips that are not fully repaired, resulting in slightly different distributions between the training set and test set. In addition, the distribution of noise in the irregular environment may be more complex than that of normal noise. However, in any case, the OCSVM-based monitor still performs well in detecting the real anomalous ionospheric gradient.

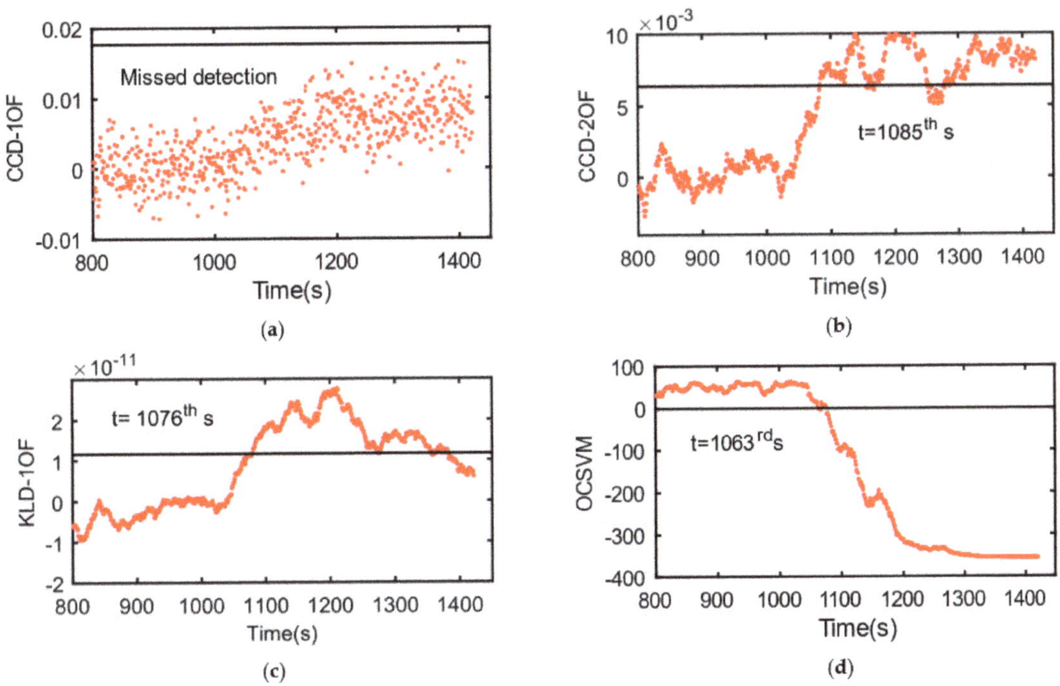

Figure 18. Variation of the test statistics and thresholds of the CCD-1OF (**a**), CCD-2OF (**b**), KLD-1OF (**c**) and the proposed OCSVM-based (**d**) monitors in the presence of an real ionospheric anomaly event. If the ionospheric anomaly is successfully detected, the detection time is included. Note that the ionospheric anomaly starts at the 1000th epoch.

4. Conclusions and Perspectives

Anomalous ionospheric gradient is a challenging risk source during the high-level precision approach in the GBAS. Several traditional monitors such as CCD-1OF, CCD-2OF, KLD-1OF have been proposed to detect the ionospheric anomalies based on the statistics in terms of the divergence. However, they do not guarantee a rapid identification or a prompt warning to users in noisy environment. This issue may dissatisfy the requirements of the precision approach and result in potential risks.

To overcome this issue, a novel OCSVM-based monitor was proposed in this work to more accurately and rapidly detect ionospheric anomalies in the GBAS. By combining phase-space reconstitution, the one-dimensional CMC time series can be transformed into high-dimensional vectors. Then, the trained OCSVM model was determined using the normal vectors offline to effectively extract the characteristics of the normal measurements and the threshold. Then, current scores can be obtained on the basis of the online measurements passing through the trained model. The distinction between the online score and the offline threshold was sensed to determine whether the ionospheric anomalies occurred. On the basis of the synthetic and BDS real data, the detection speed for the single-frequency GBAS under different levels of noise was analyzed. In addition, the performance evaluation based on real ionospheric anomalous event environment is also carried out. The proposed monitor showed superiority over the CCD-1OF, CCD-2OF and

KLD-1OF monitors regardless of the noise levels. Results showed that the efficiency can be increased by more than 40% compared with the KLD-1OF monitor. Additionally, the OCSVM-based monitor mainly has the ability to detect small anomalies. In real abnormal environment, the OCSVM-based monitor performs better than the KLD-1OF monitor for a reduction of detection time by 16%. As a result, the OCSVM-based monitor can provide real-time protection against the ionospheric anomaly, which provides a better safeguard for future civil aviation.

Follow-up work can be carried out in three ways. First, the OCSVM-based monitor can only be utilized when the cycle slip is absent. Thus, this monitor is used under the implicit assumption that the ionospheric disturbance is slight. Under strong ionospheric disturbance, the possibility to improve the OCSVM-based monitor ability will be investigated in future work. Second, limited by hardware, only the ionospheric anomaly monitoring capability at Dongying airport is tested. Further verification of different scenarios is necessary before the OCSVM algorithm can be leveraged in practice in the safety-critical systems. Finally, the proposed OCSVM-based monitor could still be further applied in the other GNSS argumentation systems. A further analysis of how the proposed monitor can be utilized in other systems will be carried out.

Author Contributions: Conceptualization, Z.G., Z.W.; methodology, Z.G., K.F.; validation, Z.G.; formal analysis, Z.G.; writing—original draft preparation, Z.G.; writing—review and editing, K.F., Z.W. and K.G.; visualization, Z.G., K.G.; supervision, Y.Z.; project administration, Z.W.; funding acquisition, Z.W. All authors have read and agreed to the published version of the manuscript.

Funding: The work was carried out with financial support from the National Key Research and Development Program of China (grant no. 2020YFB0505602), the National Natural Science Foundation of China (grant nos. 61871012 and 62022012), the Civil Aviation Security Capacity Building Fund Project (grant nos. CAAC Contract 2021(77) and CAAC Contract 2020(123)), and the Beijing Nova Program of Science and Technology (grant no. Z191100001119134).

Data Availability Statement: The data presented in this study are available upon request from the corresponding author.

Acknowledgments: The authors would like to thank many people at the National Key Laboratory of CNS/ATM for their advice and interests.

Conflicts of Interest: The authors declare no conflict of interest.

Abbreviations

ARMA	Auto-regressive and moving average
BDS	BeiDou Navigation Satellite System
CCD	Code-carrier divergence
CCD-1OF	CCD monitor with one first-order low-pass ARMA filter
CCD-2OF	CCD monitor with two first-order cascade ARMA filters
GBAS	Ground Based Augmentation System
CMC	Code minus carrier
ED	Embedding dimension
GAST	GBAS Approach Service Type
GNSS	Global Navigation Satellite System
GPS	Global Positioning System
KDE	Kernel density estimation
KLD-1OF	Kullback–Leibler divergence metric using one first order ARMA filter
OCSVM	One class support vector machine
RBF	Radial basis function
SVM	Support vector machine
UT	Universal time

References

1. Rife, J.; Pullen, S.; Enge, P. Evaluating fault-mode protection levels at the aircraft in category III LAAS. In Proceedings of the ION AM 2007 of the Institute of Navigation, Cambridge, MA, USA, 23–25 April 2007; pp. 356–371.
2. Cheng, J.; Li, J.; Li, L.; Jiang, C.; Qi, B. Carrier Phase-Based Ionospheric Gradient Monitor Under the Mixed Gaussian Distribution. *Remote Sens.* **2020**, *12*, 3915. [CrossRef]
3. Zhao, L.; Yang, F.; Li, L.; Ding, J.; Zhao, Y. GBAS Ionospheric Anomaly Monitoring Based on a Two-Step Approach. *Sensors* **2016**, *16*, 763. [CrossRef]
4. Giannattasio, F.; De Michelis, P.; Pignalberi, A.; Coco, I.; Consolini, G.; Pezzopane, M.; Tozzi, R. Parallel Electrical Conductivity in the Topside Ionosphere Derived from Swarm Measurements. *J. Geophys. Res. Space Phys.* **2021**, *126*, e2020JA028452. [CrossRef]
5. Lee, J.; Dattabarua, S.; Zhang, G.; Pullen, S.; Enge, P. Observations of low-elevation ionospheric anomalies for ground-based augmentation of GNSS. *Radio Sci.* **2011**, *46*, 1–11. [CrossRef]
6. Fujita, S.; Yoshihara, T.; Saito, S. Determination of Ionospheric Gradients in Short Baselines by Using Single Frequency Measurements. *J. Aero. Astroon. Aviat. A.* **2010**, *42*, 269–275. [CrossRef]
7. Wang, Z.; Li, T.; Li, Q.; Fang, K. Impact of Anomalous Ionospheric Gradients on GBAS in the Low-Latitude Region of China. *GPS Solut.* **2021**, *25*, 2. [CrossRef]
8. Lee, J.Y.; Yoon, M.S.; Sam, P.; Gillespie, J.; Mather, N.; Cole, R.; Souza, J.R.; Doherty, P.; Pradipta, R. Preliminary results from ionospheric threat model development to support GBAS operations in the Brazilian region. Proceedings of ION GNSS+ 2015, Tampa, FL, USA, 14–18 September 2015; pp. 1500–1506.
9. Schlüter, S.; Hoque, M.M. An SBAS Integrity Model to Overbound Residuals of Higher-Order Ionospheric Effects in the Ionosphere-Free Linear Combination. *Remote Sens.* **2020**, *12*, 2467. [CrossRef]
10. Jiang, Y.; Milner, C.; Macabiau, C. Code Carrier Divergence Monitoring for Dual-Frequency GBAS. *GPS Solut.* **2017**, *21*, 769–781. [CrossRef]
11. Kawalec, A.; Mielnik, P. Code and carrier divergence technique to detect ionosphere anomalies. In Proceedings of the 2008 International Radar Symposium, Wrocław, Poland, 21–23 May 2008; pp. 1–4.
12. Wang, Z.; Yin, Y.; Song, D.; Fang, K.; Li, Q.; Li, X. Dual Smoothing Ionospheric Gradient Monitoring Algorithm for Dual-Frequency BDS GBAS. *Chin. J. Aeronaut.* **2020**, *33*, 3395–3404. [CrossRef]
13. Khanafseh, S.; Pullen, S.; Warburton, J. Carrier phase ionospheric gradient ground monitor for GBAS with experimental validation. *Navigation* **2012**, *59*, 51–60. [CrossRef]
14. Xie, G. Optimal On-Airport Monitoring of the Integrity of GPS-Based Landing Systems. Ph.D. Thesis, Department of Aeronautical and Astronautical Engineering, Stanford University, Stanford, CA, USA, 2004.
15. Simili, D.V.; Pervan, B. Code-carrier divergence monitoring for the GPS local area augmentation system. In Proceedings of the 2006 IEEE/ION Position, Location and Navigation Symposium, San Diego, CA, USA, 25–27 April 2006; pp. 483–493.
16. Yun, Y.; Cho, J.; Heo, M. Sensitivity Analysis of Code Carrier Divergence Monitor Algorithms. In Proceedings of the ION GNSS 2012, Institute of Navigation, Nashville, TN, USA, 17–21 September 2012; pp. 3505–3511.
17. Kim, E.; Walter, T.; Powell, J.D. Adaptive carrier smoothing using code and carrier divergence. In Proceedings of the ION NTM, Institute of Navigation, San Diego, CA, USA, 22–24 January 2007; pp. 141–152.
18. Cho, J.; Yun, Y.; Heo, M. GBAS Ionospheric Anomaly Monitoring Strategy Using Kullback-Leibler Divergence Metric. *IEEE Trans. Aerosp. Electron. Syst.* **2015**, *51*, 565–574. [CrossRef]
19. Savas, C.; Dovis, F. Comparative Performance Study of Linear and Gaussian Kernel SVM Implementations for Phase Scintillation Detection. In Proceedings of the 2019 International Conference on Localization and GNSS (ICL-GNSS), Nuremberg, Germany, 4–6 June 2019; pp. 1–6.
20. Liu, Y.; Morton, Y. Automatic detection of ionospheric scintillation-like GNSS satellite oscillator anomaly using a machine-learning algorithm. *Navigation* **2020**, *67*, 651–662. [CrossRef]
21. Wang, Z.; Du, J.; Zhi, W.; Zhu, Y.; Liu, Y.; Sun, Q. An Availability Prediction Method of Ground-Based Augmentation System Based on Support Vector Machine Algorithm. In Proceedings of the ION ITM 2018 of the Institute of Navigation, Reston, VA, USA, 29 January–1 February 2018; pp. 322–338.
22. Lyu, Z.; Gao, Y. An SVM Based Weight Scheme for Improving Kinematic GNSS Positioning Accuracy with Low-Cost GNSS Receiver in Urban Environments. *Sensors* **2020**, *20*, 7265. [CrossRef]
23. Ma, J.; Perkins, S. Time-Series Novelty Detection Using One-Class Support Vector Machines. In Proceedings of the International Joint Conference on Neural Networks, Portland, OR, USA, 20–24 July 2003; Volume 3, pp. 1741–1745.
24. Goryaeva, A.M.; Lapointe, C.; Dai, C.; Dérès, J.; Maillet, J.; Marinica, M. Reinforcing Materials Modelling by Encoding the Structures of Defects in Crystalline Solids into Distortion Scores. *Nat. Commun.* **2020**, *11*, 4691. [CrossRef] [PubMed]
25. Ionescu, R.T.; Smeureanu, S.; Popescu, M.; Alexe, B. Detecting Abnormal Events in Video Using Narrowed Normality Clusters. In Proceedings of the IEEE Winter Conference on Applications of Computer Vision (WACV), Waikoloa, HI, USA, 7–11 January 2019; pp. 1951–1960.
26. Wang, Z.; Fu, Y.; Song, C.; Zeng, P.; Qiao, L. Power System Anomaly Detection Based on OCSVM Optimized by Improved Particle Swarm Optimization. *IEEE Access* **2019**, *7*, 181580–181588. [CrossRef]
27. Clauss, K.; Yan, H.; Kuenzer, C. Mapping Paddy Rice in China in 2002, 2005, 2010 and 2014 with MODIS Time Series. *Remote Sens.* **2016**, *8*, 434. [CrossRef]

28. Robert, E.; Jonas, P.; Vuillaume, J.; Salos, D.; Hecker, L.; Yaya, P. Development of a European Ionosphere Threat Model in Support of GBAS Deployment. In Proceedings of the 2018 IEEE/ION Position, Location and Navigation Symposium (PLANS), 23–26 April 2018; IEEE: Monterey, CA, USA, 2018; pp. 1181–1190.
29. Sayim, I.; Lang, H.; Zhang, D. Ionospheric Delay Prediction and Code-Carrier Divergence Testing for GBAS Using Neural Network and GPS L1. *Aerosp. Sci. Technol.* **2017**, *70*, 66–75. [CrossRef]
30. Ouzeau, C.; Frédéric, B.; Macabiau, C.; Roturier, B. Ionospheric Code Delay Estimation in a Single Frequency Case for Civil Aviation. In Proceedings of the ION GNSS 2006 of the Institute of Navigation, Fort Worth, TX, USA, 26–29 September 2006; pp. 3059–3069.
31. Matilla-García, M.; Morales, I.; Rodríguez, J.M.; Marín, M.R. Selection of Embedding Dimension and Delay Time in Phase Space Reconstruction via Symbolic Dynamics. *Entropy* **2021**, *23*, 221. [CrossRef]
32. Kim, D.; Cho, J. Improvement of Anomalous Behavior Detection of GNSS Signal Based on TDNN for Augmentation Systems. *Sensors* **2018**, *18*, 3800. [CrossRef]
33. Xia, G.; Liu, Y.; Wei, T.; Wang, Z.; Huang, W.; Du, Z.; Zhang, Z.; Wang, X.; Zhou, C. Ionospheric TEC forecast model based on support vector machine with GPU acceleration in the China region. *Adv. Space Res.* **2021**, *68*, 1377–1389. [CrossRef]
34. Botev, Z.I.; Kroese, J. Kernel density estimation via diffusion. *Ann. Stat.* **2010**, *38*, 2916–2957. [CrossRef]
35. Yin, S.; Ding, S.X.; Abandan Sari, A.H.; Hao, H. Data-driven monitoring for stochastic systems and its application on batch process. *Int. J. Syst. Sci.* **2013**, *44*, 1366–1376. [CrossRef]
36. Shively, C.A.; Braff, R. An Overbound Concept for Pseudorange Error from the LAAS Ground Facility. In Proceedings of the ION AM 2000 of the Institute of Navigation, San Diego, CA, USA, 26–28 June 2000; pp. 661–671.
37. Larson, J.D.; Gebre-Egziabher, D.; Rife, J.H. Gaussian-Pareto Overbounding of DGNSS Pseudoranges from CORS. *Navigation* **2019**, *66*, 139–150. [CrossRef]
38. Zhu, Y.; Liu, Y.; Wang, Z.; Li, Q. Evaluation of GBAS Flight Trials Based on BDS and GPS. *IET Radar Sonar Navig.* **2020**, *14*, 233–241. [CrossRef]
39. Wang, Z.; Li, T.; Li, Q.; Zhu, Y. Ionospheric Delay Gradient Threat Model for GBAS based on GPS Data in China. In Proceedings of the ION GNSS+ 2019 of the Institute of Navigation, Nashville, TN, USA, 16–20 September 2012; pp. 3505–3511.
40. Song, X.; Yang, R.; Zhan, X. An analysis of global ionospheric disturbances and scintillations during the strong magnetic storm in September 2017. *Aerosp. Syst.* **2020**, *3*, 255–263. [CrossRef]

remote sensing

MDPI

Article

SafeNet: SwArm for Earthquake Perturbations Identification Using Deep Learning Networks

Pan Xiong [1], Dedalo Marchetti [2,3], Angelo De Santis [3], Xuemin Zhang [1] and Xuhui Shen [4,*]

1 Institute of Earthquake Forecasting, China Earthquake Administration, Beijing 100036, China; xiongpan@ief.ac.cn (P.X.); zxm@ief.ac.cn (X.Z.)
2 College of Instrumentation and Electrical Engineering, Jilin University, Changchun 130061, China; dedalomarchetti@jlu.edu.cn
3 Istituto Nazionale di Geofisica e Vulcanologia, Via di Vigna Murata 605, 00143 Rome, Italy; angelo.desantis@ingv.it
4 National Institute of Natural Hazards, Ministry of Emergency Management of China, Beijing 100085, China
* Correspondence: shenxh@seis.ac.cn

Abstract: Low Earth orbit satellites collect and study information on changes in the ionosphere, which contributes to the identification of earthquake precursors. Swarm, the European Space Agency three-satellite mission, has been launched to monitor the Earth geomagnetic field, and has successfully shown that in some cases it is able to observe many several ionospheric perturbations that occurred as a result of large earthquake activity. This paper proposes the SafeNet deep learning framework for detecting pre-earthquake ionospheric perturbations. We trained the proposed model using 9017 recent (2014–2020) independent earthquakes of magnitude 4.8 or greater, as well as the corresponding 7-year plasma and magnetic field data from the Swarm A satellite, and excellent performance has been achieved. In addition, the influence of different model inputs and spatial window sizes, earthquake magnitudes, and daytime or nighttime was explored. The results showed that for electromagnetic pre-earthquake data collected within a circular region of the epicenter and with a Dobrovolsky-defined radius and input window size of 70 consecutive data points, nighttime data provided the highest performance in discriminating pre-earthquake perturbations, yielding an F1 score of 0.846 and a Matthews correlation coefficient of 0.717. Moreover, SafeNet performed well in identifying pre-seismic ionospheric anomalies with increasing earthquake magnitude and unbalanced datasets. Hypotheses on the physical causes of earthquake-induced ionospheric perturbations are also provided. Our results suggest that the performance of pre-earthquake ionospheric perturbation identification can be significantly improved by utilizing SafeNet, which is capable of detecting precursor effects within electromagnetic satellite data.

Keywords: earthquake; pre-earthquake anomalies; swarm satellites; ionospheric plasma; deep learning; physical mechanisms

Citation: Xiong, P.; Marchetti, D.; De Santis, A.; Zhang, X.; Shen, X. SafeNet: SwArm for Earthquake Perturbations Identification Using Deep Learning Networks. *Remote Sens.* **2021**, *13*, 5033. https://doi.org/10.3390/rs13245033

Academic Editor: Fabio Giannattasio

Received: 2 November 2021
Accepted: 8 December 2021
Published: 10 December 2021

1. Introduction

The ionosphere is an important layer of the solar-terrestrial space observation environment, and the process of earthquake preparation and occurrence can also cause anomalous changes in ionospheric parameters over the preparation zone, known as seismic ionospheric disturbances. These phenomena are a manifestation of earthquakes in the ionosphere, a result of lithosphere–atmosphere–ionosphere coupling, and are considered to be one of the more promising ideas for detecting short-term earthquake signals. Decades ago, Moore [1] and Davies and Baker [2] first reported anomalous ionospheric perturbations associated with the 1964 Alaska earthquake in the USA, and studies on seismic ionospheric phenomena have been rapidly developed. With the accumulation of available observational resources, seismic ionospheric phenomena have been detected, reported, and confirmed [3]. Currently, these phenomena are an issue of great concern and have become a hot topic at the intersection of seismology and space physics. In the 21st century, with the development of space satellite observation technology, many countries launched satellites dedicated to monitoring space environment changes and natural disaster activities such as earthquakes and volcanoes, for example the QuakeSat (USA), SICH-1M (Ukraine), COMPASS-2 (Russia), DEMETER (France), and China Seismo Electromagnetic Satellite (CSES). Furthermore, even satellites dedicated to other scientific purposes have been demonstrated to provide important observations for seismic ionospheric disturbances, such as the Swarm constellation of the European Space Agency (ESA). The Swarm constellation consists of three identical satellites that carry the same payloads. The combined observation of multiple satellites in the Swarm constellation offers significant advantages over single-satellite observations, allowing for better detection efficiency, better global coverage in a day, higher spatial resolution, and improved capabilities for analyzing anomalies. In this way, Swarm represents a new, successful approach to the study of seismic ionospheric phenomena.

Studies of these phenomena from the Swarm satellites are currently ongoing, yielding continuous investigation and reporting of the valuable description of their mechanisms. A review of previous studies on seismic ionospheric phenomena based on Swarm satellites shows that the work is generally divided into two types: earthquake case studies and statistical investigations, both of which focus on the analysis of ionospheric disturbance phenomena associated with the preparation phase of an earthquake before its occurrence. Both types explore the precursor and provide related criteria potentially useful for earthquake prediction and forecasting.

Various perturbations have been observed using Swarm data before the occurrence of large earthquakes. De Santis et al. [4] investigated magnetic field anomalies from Swarm data for one month before and after the magnitude 7.8 Nepal earthquake that occurred on 25 April 2015 at 06:26 UTC, and found that the cumulative number of anomalies followed the same typical power-law behavior of a critical system close to its critical time, thereafter returning to normal after the earthquake. In another study, Marchetti et al. [5] used multi-parameters from ground and space, that is, Earth geomagnetic field data (magnetic data from the ESA Swarm constellation and from L'Aquila and Duronia ground observatories of the INGV (Italian National Institute of Geophysics and Vulcanology)) along with the MERRA-2 climatological dataset to study the lithosphere–atmosphere–ionosphere coupling effects of the 2016–2017 central Italy earthquake sequence. They revealed anomalies in the ground-based geomagnetic observations 275 and 85 days before the earthquake sequence, anomalies from satellite observations 240 days and 3 days before the start of the earthquake sequence, and two perturbations in the chemical/physical composition of the atmosphere 200 and 150 days prior to the earthquake sequence. In a further study, Marchetti et al. [6] analyzed the Swarm satellite magnetic data prior to the magnitude 7.1 California Ridgecrest earthquake that occurred on 6 July 2019 and found some increase in anomalies in the Y (eastern) component of the magnetic field around 200 days before the earthquake; moreover, 15 min before the earthquake, the Swarm Bravo satellite passed right over the epicentral region, and its Y component presented interesting anomalies. Zhu et al. [7] analyzed ionospheric magnetic field data from the Swarm Alpha satellite before the 16 April 2016 Ecuador earthquake (magnitude 7.8) based on the non-negative matrix factorization method and found that the energy and entropy of one of the weighting components were more concentrated within the preparation region of the seismic event. In that study, the cumulative number of orbits with anomalies inside that region showed an acceleration before the Ecuador earthquake and recovered to a linear (i.e., standard) trend after the earthquake. In addition to the aforementioned examples, numerous studies have confirmed that the Swarm ionospheric perturbations are sensitive enough and useful for detecting earthquake-related anomalies, such as for the 2014 Ludian earthquake [8], 2017 Sarpol-e Zahab (Iran) earthquake [9], 2017 Mexico earthquake [10], and the above mentioned 2016 Ecuador earthquake [11,12].

Statistical analysis is a common technique for investigating ionospheric anomalies that occur before earthquakes using satellite data [13–15]. De Santis et al. [16] statistically analyzed the Swarm electron density and magnetic field data observed by the three Swarm satellites over 4.7 years using a superposed epoch approach and found that some electron density and magnetic anomalies were significantly concentrated from more than two months before the earthquake to a few days prior to the earthquake. This confirmed the well-known Rikitake empirical law between the time of the precursors and the magnitude of the earthquake by studying different magnitude ranges. In another statistical analysis of 5.3 years of magnetic field data from the Swarm satellite by Marchetti et al. [17], the distance from the satellite to the earthquake epicenter matched the measured distance arrival time of the coseismic disturbance from the surface to the ionosphere, confirming that observed anomalies were likely to be caused by seismic events due to their occurrence with a mixed transmission mechanism, i.e., by acoustic gravity waves and electromagnetic propagation in the ionosphere.

However, most existing pre-earthquake anomaly studies are lacking in consistent analysis methods and anomaly evaluation metrics; thus, the analysis results of anomalies lack universality and often lead to various or even contradictory interpretations for the same earthquake. Moreover, statistical studies of seismic anomalies often do not consider the influence of non-seismic anomalies.

Deep learning, which has been widely used in recent earthquake research [18–22], could perform consistent analysis and assessment of a large number of earthquakes, and could take into account the influence of non-seismic anomalies, which are an effective tool to solve the above issue. By investigating different DEMETER satellite datasets, Xiong et al. [21] confirmed some frequency bands with low-frequency electric and magnetic fields to be the main features for pre-seismic electromagnetic perturbation identification using deep learning. Xiong et al. [23] also proposed a deep learning framework termed SeqNetQuake by training whole life cycle dataset from the DEMETER satellites and transferring the well-trained model to the CSES satellite to form a new identification model which achieved a 12% improvement in classification performance. Based on the classical AdaBoost machine learning algorithm and the feature of satellite remote sensing products such as infrared and hyperspectral gases, Xiong et al. [22] proposed a novel earthquake prediction framework based on inverse boosting pruning trees (IBPT), and achieved promising forecasting results in the validation of global earthquake cases retrospectively.

In this study, we use deep learning techniques, combined with multi-year accumulated Swarm satellite data for pre-seismic ionospheric disturbance identification. The proposed method, known as SafeNet (SwArm for Earthquake study using deep learning networks), is a deep-learning method based on a sequence-based classification neural network for pre-earthquake perturbation identification. The suggested model was trained using 9017 recent independent 4.8+ magnitude earthquakes and 7-year plasma and magnetic field data from the Swarm A satellite. The results indicated that nighttime data provided the best performance in distinguishing pre-earthquake perturbations, with an F1 score of 0.846 and a Matthews correlation coefficient of 0.717. It also worked effectively in detecting pre-seismic ionospheric abnormalities as earthquake magnitude increased. Furthermore, the findings of this study enabled us to propose a hypothesis regarding the physical mechanism behind earthquake-induced ionospheric disturbances. In general, SafeNet could considerably enhance the effectiveness of pre-earthquake ionospheric perturbation identification.

This paper is organized as follows. In Section 2, the used data from the Swarm satellite and its data pre-processing are discussed, together with the observation cases of the anomalies before two actual earthquakes. Section 3 describes the network structure design and performance evaluation metrics for the proposed deep learning model. The results are analyzed and discussed in Section 4, and a hypothesis of the earthquake mechanism is presented. In Section 5, we provide a conclusion and further possible orientation for future work.

2. Datasets and Observations

2.1. The Swarm Satellites

The Swarm constellation is the first ESA constellation program for geomagnetic observation, whose main scientific objective is to study the delicate structure of the Earth's magnetic field, its dynamics, and its interaction with Earth systems [24,25]. The constellation consists of three identically equipped satellites, Swarm A (Alpha), B (Bravo), and C (Charlie), which were launched together into a near-polar orbit on 22 November 2013, and finalized their commissioning on 17 April 2014. For the main purpose of the Swarm constellation, each satellite is equipped with a high-precision magnetometer in addition to several other sensors to increment the scientific satellites' capabilities. Among them, the charged particle sensor (Langmuir detector) provides a new avenue for the study of the seismic ionosphere phenomena.

2.2. Earthquake Case Study

2.2.1. 2016 Sumatra Earthquake

Figure 1 shows the Earth magnetic field and electron density residuals as obtained by applying the MASS (MAgnetic Swarm anomaly detection by Spline analysis) method (De Santis et al. [4]) to the data measured by the Swarm Alpha satellite on 15 February 2016, i.e., 16 days before the Mw = 7.8 Sumatra 2016 earthquake localized at 4.952°S, 94.330°E, and 24 km depth. The magnetic track shows a clear anomaly highlighted by a red circle in panel E only in the Y-East component; this is compatible for signals coming from below (i.e., the internal ones, Pinheiro et al. [26]). The FFT (fast Fourier transform) spectrum shows a signature highlighted by a red circle in panel B at about 0.4 Hz (i.e., a period of 2.5 s) that could correspond to the frequency of the anomaly. Such a signature is not visible in the track without evident anomalies (the amplitude at this frequency is normally lower than the level in this figure as visible in the other examples in Text S1, Supplementary Materials). The electron density shows the clear EIA diurnal profile, but it is quite unusual that it appeared during nighttime (local time of about 1:30 a.m.). Furthermore, the electron density shows some disturbances at +5° geographic latitude and even they do not coincide with the magnetic anomaly; both phenomena could be produced by the preparation phase of this large earthquake. Geomagnetic indices Dst = 6 nT and ap = 4 nT depict a very quiet geomagnetic condition. Thus, all the investigated measurements show an unusual status of the ionosphere in the nighttime of 15 February 2016, without known external perturbations; furthermore, the longitude of the satellite matched with the one of the future epicenters, so finally we can consider this track a good candidate for an earthquake precursor.

A complete investigation of nighttime Swarm Alpha from this track until earthquake occurrence is presented in Text S1 of the Supplementary Materials. Some days are affected by perturbed geomagnetic conditions and thus do not permit searching for possible seismo-induced ionospheric disturbances. Instead, on at least other 3 days (21, 25, and 26 February 2016), there are interesting signals: all of them present higher signal content in the Y-East component with respect to the vertical and North ones. These anomalies, detected inside Dobrovolsky's area and a few days before earthquake occurrence, can in principle be good candidates for precursors.

2.2.2. The Ecuador Earthquake Occurred on 16 April 2016

A similar example to the previous one was recorded before the Ecuador earthquake occurred on 16 April 2016 at 0.371° N, 79.940° W, and 20.6 km depth by the Swarm Alpha satellite on 19 January 2019 (see Figure 2). Akhoondzadeh et al. [12] found a pattern, in terms of the chain of phenomena, in the lithosphere–atmosphere and ionosphere in preparation of the Ecuador earthquake. In particular, they detected an increase of Swarm magnetic anomalies about 9 days before the event and during geomagnetic quiet conditions; these were probably related to the preparation phase of this seismic event. The track shown in Figure 2 is one example acquired during such an increase of anomalies and it shows an anomaly in three components of the magnetic field highlighted by red circles in panels

D, E, and F. We notice that in the Y-East component the anomaly seems to have a longer duration with respect to the other components and a northern anomalous signal (even if this second North disturbance is formally out of the Dobrovolsky area). From the frequency point of view, in this case the highest intensity in the Y-East FFT spectrum (see panel B) seems to be located around 0.2 Hz (period of 5 s) with some frequency spread, and in particular it seems that there is a double peak at lower frequencies (0.182 Hz and 0.197 Hz) not present in Z FFT (see panel C), while the peak at 0.23 Hz is present also in the Z component as highlighted by the red circle in panel C. Future investigations are necessary to check by a systematic approach if a particular frequency is more prone to identifying possible seismo-induced phenomena. From the multiparametric and multi-instruments investigation, it is possible to note that there is a depletion of electron density which mostly coincides in latitude with the magnetic disturbance. A common alteration of the magnetic field with a decrease of electron density is probably a sign of the crossing of a "plasma bubble". This feature, which is produced for the standard behavior of the ionosphere, can be also induced by air ionization in the preparation of an earthquake [27]. The earthquake was recently re-investigated by another method by Zhu et al. [7], confirming the previous result and offering new insights.

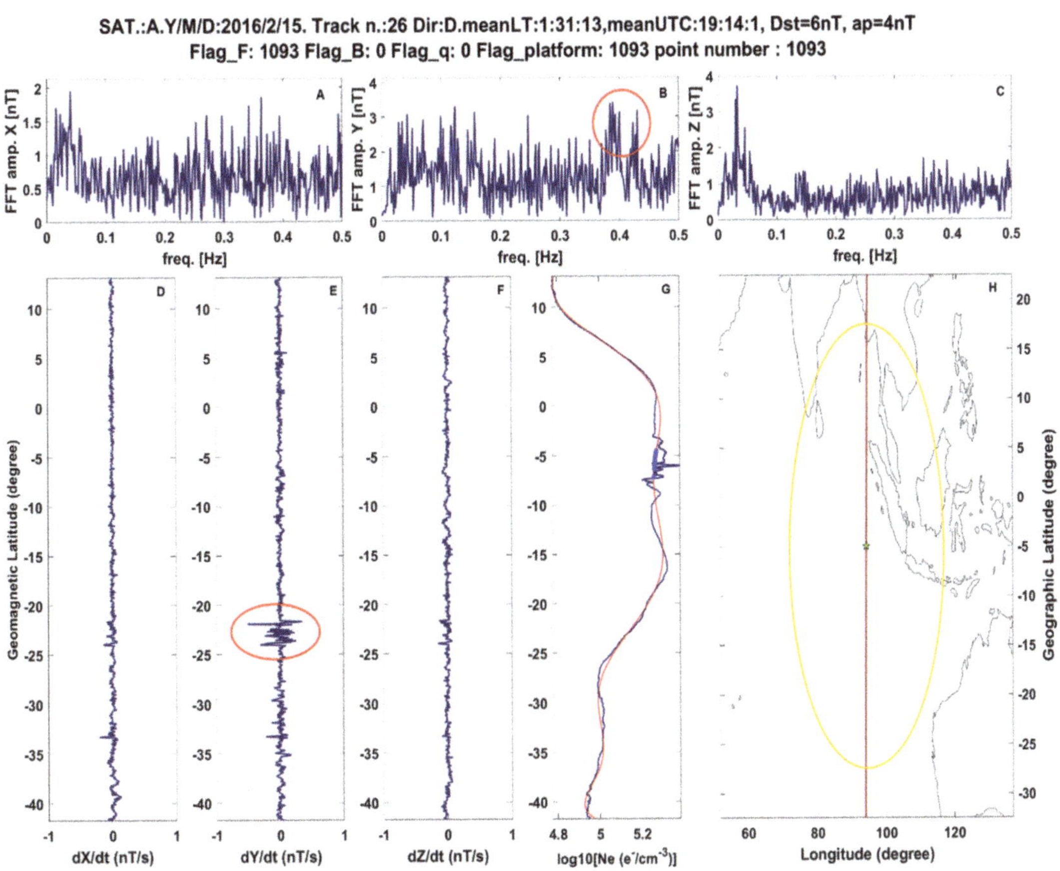

Figure 1. Swarm Alpha satellite nighttime track 26 of 15 February 2016 acquired 16 days before the M7.9 Sumatra 2016 earthquake. Panels D, E, and F show magnetic X, Y, and Z components residual after derivate and cubic-spline removal (MASS method; De Santis et al. [4]). Panels A, B, and C provide the FAST Fourier Transform of the residual of X, Y, and Z,

respectively. Panel G shows the logarithm of electron density, with a 10-degree polynomial fit as a red line. The pixels that present Ne values that significantly deviate from the fit are identified as blue stars and they are potential anomalies as defined by NeLOG in De Santis et al. [28]. The map in panel H shows the epicenter of the earthquake by a green star and its preparation area defined by Dobrovolsky et al. [29] as a yellow circle. In the title of the figure, the satellite (A for Alpha, B for Bravo, and C for Charlie), date, track number (counted daily), and time in local and UTC times of the center of the track are indicated. The values of the geomagnetic indexes Dst and a_p at the track acquisition time are also provided, and in the second line of title the number of flagged samples are provided together with the total number of samples in the track.

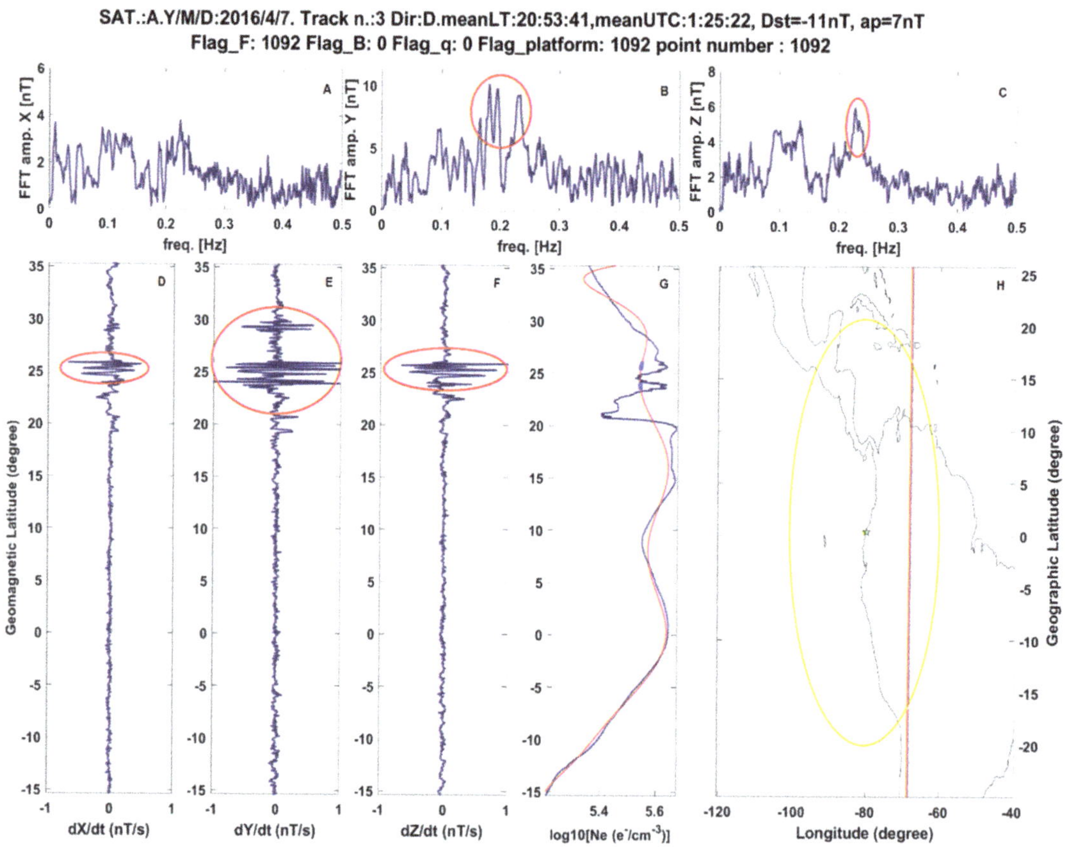

Figure 2. Swarm Alpha satellite nighttime track 3 of 7 April 2016 acquired 9 days before the M7.8 Ecuador 2016 earthquake. The description of the subfigures is the same as in Figure 1.

Also, for the case study in Text S2 (Supplementary Materials), from the Swarm Alpha nighttime data, one track for each day was shown until the earthquake's occurrence. In this case, most of the other detected anomalies were probably associated with external geomagnetic activity.

2.3. Dataset and Preprocessing

All three satellites in the Swarm constellation carry the same scientific payloads to detect ionospheric electromagnetic parameters. This study focused on the use of magnetic field and plasma parameter data from the Swarm A satellite and space weather data. The types of sensors and data utilized in the study are described as follows:

(1) The vector magnetometer (VFM) is a fluxgate magnetometer from which the on-ground processor provides both high-frequency (50 Hz) and low-frequency (1 Hz) signals.

The magnetic field intensity is available in an instrumental reference system as well as in the Earth one (which is used in this study), consisting of three components, X (North), Y (East), and Z (Vertical), and is measured in nT, while time is measured in Universal Time. The VFM measures field components with an accuracy of 0.1 nT every 3 months for signals at global scales within a space resolution of 20 km [25]. In this study, we focus on the X, Y, and Z components of the low-frequency (1 Hz) data.

(2) The Langmuir probe (LP) measures the electron temperature (Te), electron density (Ne), and other parameters of plasma by measuring the current generated by electrons and ions at a sampling rate of 1 Hz. This study used plasma Ne data from a level 1b product [24,25].

(3) To avoid effects caused by space weather events, we collected the Kp index, an indicator of global geomagnetic activity, to be used as an auxiliary means of discriminating between solar (or geomagnetic activity) and seismic ionospheric disturbance phenomena. It should be noted that data corresponding to periods with a Kp index greater than 3.0 were not analyzed in this study.

The data used for this study were from Swarm A and included the parameters mentioned above, collected from 1 April 2014 to 30 April 2020, i.e., 7 years of data. According to the United States Geological Survey, 18,621 earthquakes with magnitudes greater than or equal to 4.8 were reported during this period. Thus, we used the same technique as reported by Yan et al. [14] to exclude aftershocks from the list of earthquakes in order to prevent mixing pre- and post-seismic effects. After this process, the final list comprised 9017 independent earthquakes. We also removed data that corresponded to the aftershocks.

To test the reliability of the machine learning methods and improve their robustness, we created the same number of artificial non-seismic events as actual earthquakes, while randomizing and changing the timings and locations to avoid overlapping with real earthquakes. Within the selected spatio-temporal range, we sampled latitude, longitude, and time at random, adhering to the following constraints: (1) the latitude or longitude is not within 10° of the latitude or longitude of a real earthquake and (2) the time is not within 15 days of the occurrence time of a real earthquake.

3. Methodology

Figure S25 (Supplementary Materials) depicts a flowchart diagram of the methodology used in this study. To begin, a total of 9017 earthquakes with magnitudes of 4.8 or higher were extracted from seismic catalogues considering those that occurred all over the world for the study. Thirteen datasets were built by combining different magnitudes of earthquakes and features. Each dataset was divided into two parts: training data and test data. After that, we used the "sliding window" technique for data preprocessing, and lastly, we generated time series-based features.

In our work, we explored the effect of different model input and spatial window sizes, earthquake magnitudes, and whether the earthquake occurred during the day or at night using different datasets. We also performed a comparison of five state-of-the-art approaches. Considering these approaches are highly sensitive to parameter selection, we preferred to choose those configurations that would allow us to achieve the highest performance in the tests. The performance of each approach was then evaluated after the parameter selection. Finally, we evaluated the performance of each technique using ROC curves and other performance metrics.

3.1. Data Preprocessing

A variety of reasons, including satellite payload interference, the space environment, and other factors, may produce inaccuracies in constantly observed satellite data. As a precaution against such mistakes, we divided continuous observation data into fixed-length sliding windows (commonly known as "sequences"), which we then utilized as inputs to our proposed model. We also divided the data into sliding windows that were continuous but did not overlap because time series data are highly autocorrelated sequences. As a result, we carefully examined the difference between the first and final data points ordered by time in each time window to verify that the data were continuous. Time series windows with unreasonable time differences (i.e., gaps) were removed from the analysis.

Subsequently, we reformulated the pre-earthquake ionospheric perturbation discrimination task as a multiclass multivariate time-series classification problem with the data marked as follows (Figure 3): the non-seismic-related data were marked as 0, seismic-related data were marked as 1, and data with a Kp index greater than 3.0, regardless of whether the data were related to an earthquake, were marked as 2, indicating density perturbations due to solar and magnetic activity [30]. It is known that the Earth's magnetic field undergoes temporal fluctuations and exhibits recognized patterns related to the movement of the poles, and time series data exhibit significant autocorrelation characteristics, which indicates that the field is highly variable. Therefore, to ensure that the overall Swarm dataset could be utilized effectively, it was carefully divided into two contiguous parts: the first 80 percent (in chronological order) of the data was used for model training (the training set), and the last 20 percent was used for model testing and final assessment (the test set).

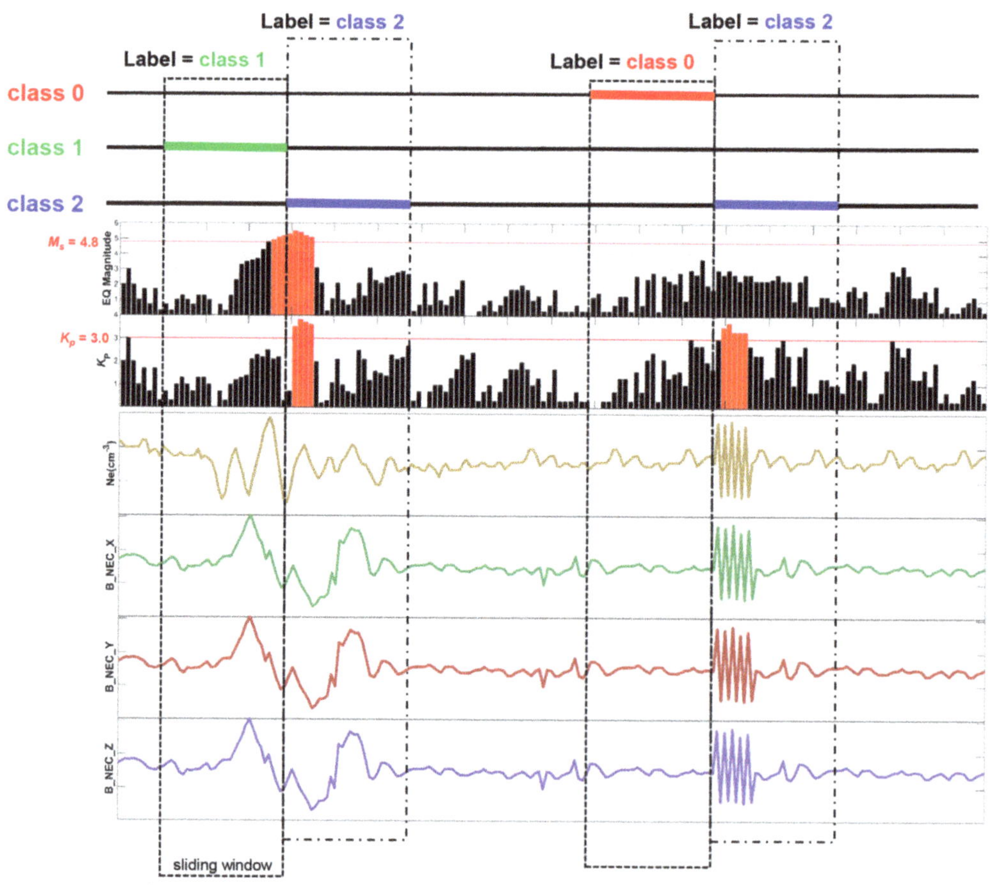

Figure 3. Sequence labeling after data segmentation using a sliding window. Consecutive observations of used parameters are segmented by non-overlapping sliding windows. T is the window's length. Non-seismic data are marked as 0 (class 0), seismic data are marked as 1 (class 1), and data with a Kp index greater than 3 are marked as 2 (class 2), regardless of whether they are associated with an earthquake.

3.2. Deep Learning Network Architecture

In our research, we used a combination of a convolutional neural network (CNN) and bi-directional long short-term memory (Bi-LSTM) to train our proposed SafeNet model (Figure 4). The model's architecture comprises one-dimensional (1D) convolutional layers, a 1D Bi-LSTM structural layer, and a fully connected (FC) block. For feature extraction from the input data, the SafeNet model employs CNN layers, which are combined with Bi-LSTMs to facilitate sequence prediction. SafeNet accesses subsequences of main sequence as blocks, collects features from each block, and then transfers extracted features to the LSTM layer for interpretation. To enable the CNN model to read each subsequence in the window, the entire CNN model is wrapped in a Time Distributed layer. The extracted features are then flattened and provided to the Bi-LSTM layer for reading, and further features are extracted. The 1D convolutional layers are utilized to extract data features in concept, and then Bi-LSTM structures are employed to optimize feature extraction in sequential data. Finally, the classification probability is calculated using an FC layer. The loss function is categorical cross-entropy, and the optimization is performed using the Adam method [31]. For more details on the SafeNet network architecture, please refer to Text S4 (Supplementary Materials).

The proposed model was developed in TensorFlow 2.0 with the Keras (v. 2.3.0) interface [32]. To facilitate fast training, all models were trained on a server equipped with two Intel Xeon E5-2650 v4 CPUs, 128 GB of RAM, and an NVIDIA GeForce RTX 2080 Ti graphics processing unit (GPU) [33]. Owing to the sensitivity of the proposed method to the chosen parameters, Bayesian hyperparameter tuning was utilized to determine the optimal settings [34] and was developed using the Hyperopt Python package [35]. The negative of the F-measure (F1) was utilized as the objective function's return value (loss) in this procedure. The procedure optimized hyperparameters based on their capacity to minimize an objective function by constructing a probability model based on the results of previous evaluations. Consequently, this model can be expected to perform better with fewer iterations than the random or grid searches would require. Table S1 (Supplementary Materials) summarizes the search space for SafeNet's important parameters. Each model was assigned a maximum of 100 iterations. DataSet S1 (Supplementary Materials) provides the hyperparametric optimization trial results for all the datasets used to train the SafeNet model and other benchmarking classifiers.

3.3. Performance Evaluation

Datasets utilized in this study were often class unbalanced, with the number of samples representing the non-seismic class being much greater than the number of samples representing the other classes [36]. In this situation, a simple classifier that predicted each sample as the majority class could achieve a high level of accuracy; thus, the total classification accuracy was insufficient to assess performance. As a result, we used the F-measure (F1) to evaluate model performance, which considers the correct classification of each class to be equally important. The F1 score is a metric that considers precision and recall. This is often referred to as the harmonic mean of both. Consequently, class imbalance was addressed by weighting the various classes according to their sample proportions. The Matthews correlation coefficient (MCC) [37], which emphasizes positives in samples, was also employed in this study. The specific formulas for the performance metrics such as F1 score and MCC are defined in Text S3 (Supplementary Materials).

Furthermore, in this study, receiver operating characteristic (ROC) curves, which are plots of the true positive rate versus the false positive rate, were employed to assess the output quality of the classifier's performance. The ROC curve is often used in binary classification settings to assess the output of a classifier. To extend the ROC curve and ROC area for multiclass classification, the output was binarized, and one ROC curve was drawn and used to evaluate classifier quality per class. In addition, we computed the area under the ROC curve, known as the AUC, which was used to compare various models. In this study, higher AUC values were regarded as better methods for identifying ionospheric perturbations prior to earthquakes.

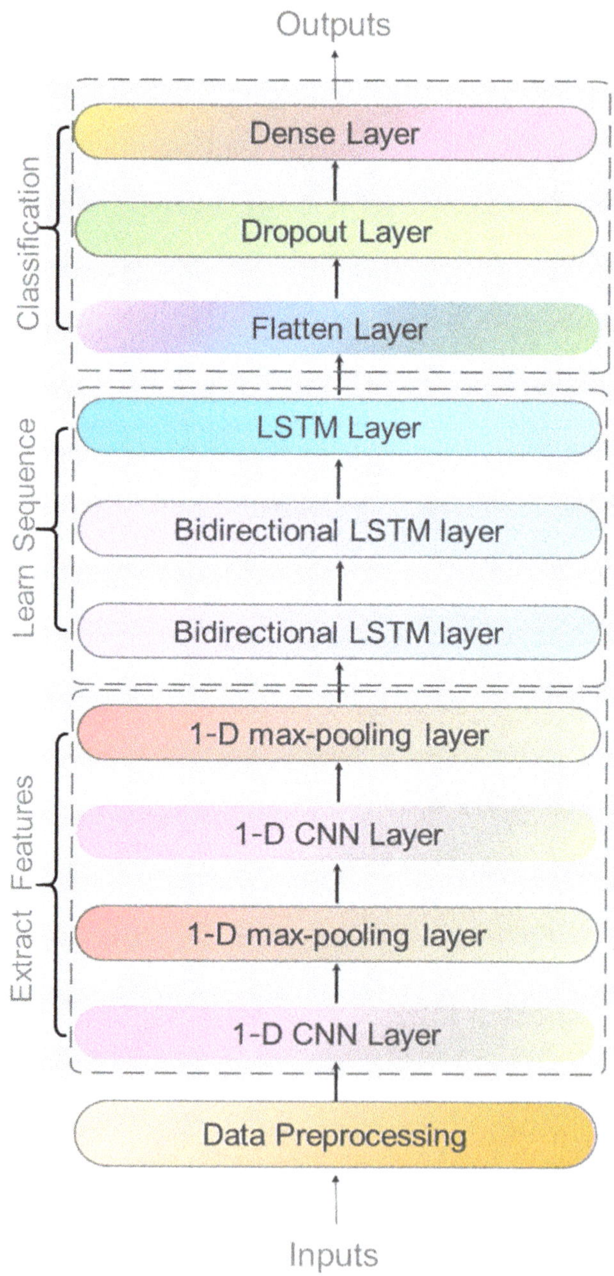

Figure 4. The bottom-up network framework architecture of the SafeNet model. 1-D CNN: One-dimensional convolutional neural network; Dropout: drop-out layer; Bidirectional LSTM: bidirectional long short-term memory layer. "Flatten" and "Dense" are the names of the functional layers.

Finally, to visually illustrate the classification performance of each class, ternary probability diagrams and confusion matrices were used to depict the probability distributions for each input class in the test data, as well as the distribution of predicted and actual values.

Five state-of-the-art machine learning models were benchmarked for the study task: gradient boosting machine [38], deep neural network (DNN) [39], random forest [40], CNN [41], and LSTM [42] models. These methods were implemented in Python (v. 3.6) with scikit-learn (v. 0.20.0) and Keras (v. 2.3.0). Because the explored models are sensitive to parameter selection, we chose parameters that yielded the best performance using Bayesian hyperparameter tuning, as described above. After the optimal parameters were determined for each method, the performances of the different methods were compared.

4. Results

We used the Swarm dataset to train the proposed SafeNet model directly, which had been split into the training and test sets. Initially, we configured the data with 60 consecutive observations as the input sequence length, a spatial window centered at the epicenter, a deviation of the Dobrovolsky radius [29], and nighttime data in the initial configuration, because there is no universal standard for lengths of the input sequence and the spatial window (DataSet 01 in Table 1). In our study, we consistently considered data from 15 days before to 5 days after every earthquake and set it as the temporal window.

Table 1. Datasets with different features generated using Swarm data.

DataSet	Night/Daytime	Spatial Feature	Input Sequence Length	Earthquake Magnitude/No. of Real Earthquakes/Positive to Negative Ratio
DataSet 01	Nighttime	with its center at the epicenter and the Dobrovolsky radius	60 continuous points	above 4.8/9017/1:1
DataSet 02	Nighttime	with its center at the epicenter and the Dobrovolsky radius	80 continuous points	above 4.8/9017/1:1
DataSet 03	Nighttime	with its center at the epicenter and the Dobrovolsky radius	70 continuous points	above 4.8/9017/1:1
DataSet 04	Nighttime	with its center at the epicenter and the Dobrovolsky radius	50 continuous points	above 4.8/9017/1:1
DataSet 05	Nighttime	with its center at the epicenter and the Dobrovolsky radius	40 continuous points	above 4.8/9017/1:1
DataSet 06	Daytime	with its center at the epicenter and the Dobrovolsky radius	70 continuous points	above 4.8/9017/1:1
DataSet 07	Nighttime	with its center at the epicenter and a deviation of 3°	70 continuous points	above 4.8/9017/1:1
DataSet 08	Nighttime	with its center at the epicenter and a deviation of 5°	70 continuous points	above 4.8/9017/1:1
DataSet 09	Nighttime	with its center at the epicenter and a deviation of 7°	70 continuous points	above 4.8/9017/1:1
DataSet 10	Nighttime	with its center at the epicenter and a deviation of 10°	70 continuous points	above 4.8/9017/1:1
DataSet 11	Nighttime	with its center at the epicenter and the Dobrovolsky radius	70 continuous points	4.8~5.2/5136/1:1
DataSet 12	Nighttime	with its center at the epicenter and the Dobrovolsky radius	70 continuous points	5.2~5.8/2793/1:1
DataSet 13	Nighttime	with its center at the epicenter and the Dobrovolsky radius	70 continuous points	5.8~7.5/853/1:1
DataSet 14	Nighttime	with its center at the epicenter and the Dobrovolsky radius	70 continuous points	above 4.8/9017/1:2
DataSet 15	Nighttime	with its center at the epicenter and the Dobrovolsky radius	70 continuous points	above 4.8/9017/1:3
DataSet 16	Nighttime	with its center at the epicenter and the Dobrovolsky radius	70 continuous points	above 4.8/9017/1:4
DataSet 17	Nighttime	with its center at the epicenter and the Dobrovolsky radius	70 continuous points	above 4.8/9017/1:5

As illustrated in Figure 5, the ROC curves were used as a performance measure because they represent relative trade-offs between true positives (benefits) and false positives (costs) for each class, and the performance of the model utilizing nighttime data is shown. The AUC values of classes 0 and 1 were both greater than 0.9 (Figure 5A,B), indicating that the model can roughly distinguish time series related to earthquakes and non-seismic events, but the AUC of class 2 was only 0.50 (Figure 5C), indicating that the accuracy of the model in identifying space weather such as magnetic storms was low. This may be due to the fact that class 2 was trained with a lower number of samples, causing the model to fail to extract the features of the class. Figure 5D depicts the MCC, F1 score, and accuracy bar plot curves, which represent the overall performance of the model. The findings were similar to those implied by the ROC curves, showing that the model could distinguish earthquakes from non-seismic and space events to some degree. In general, the performance of the model based on the initial setup was reasonable, but not remarkable. As a result, we investigated whether combining datasets with various temporal and geographic characteristics, as well as alternative models, might provide an improved performance.

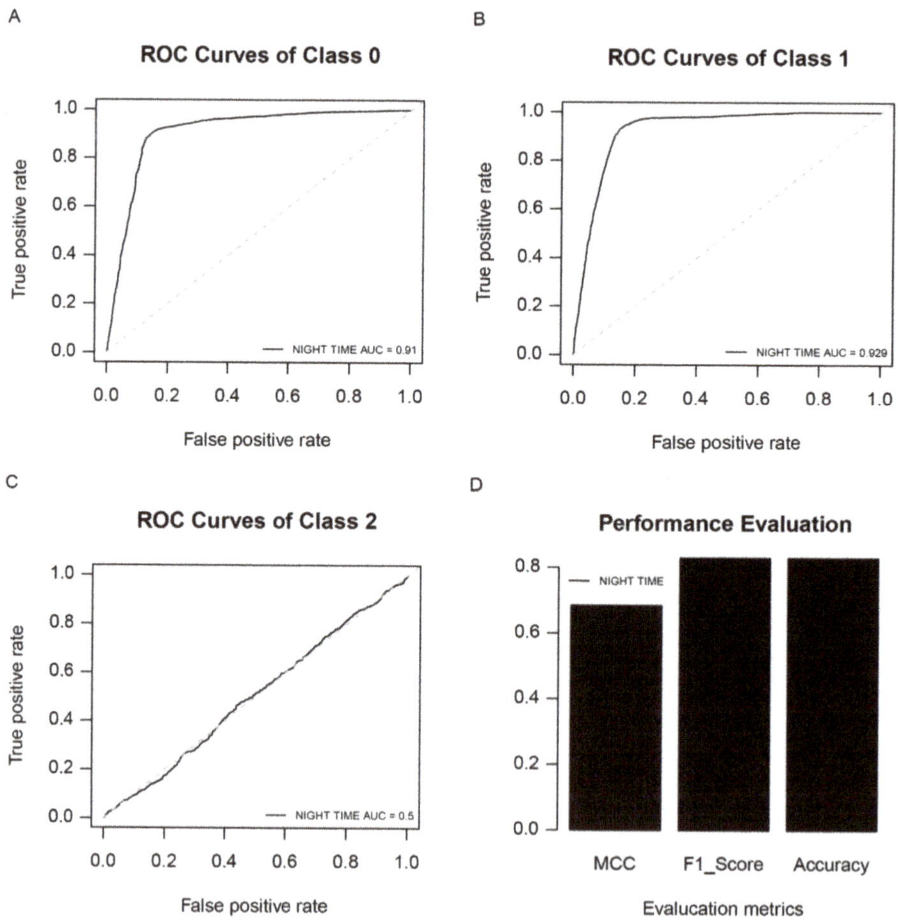

Figure 5. Receiver operating characteristic (ROC) curves showing SafeNet's performance for (**A**) class 0, (**B**) class 1, and (**C**) class 2 utilizing nighttime data. (**D**) Matthews correlation coefficient (MCC), F1 score, and accuracy bar plot curves illustrating the model's performance with nighttime data.

4.1. Considering Various Input Sequence Lengths

To further investigate whether the SafeNet method can identify pre-earthquake disturbances with varying input sequence lengths and whether it can improve performance, datasets with the following input sequence length of consecutive observations were created (Table 1): 80 (DataSet 02), 70 (DataSet 03), 50 (DataSet 04), and 40 (DataSet 05) consecutive observations.

Figure 6 depicts the ROC curves and MCC, F1 score, and accuracy bar plot curves for the datasets with varying input sequence lengths. Table 2 lists the classification performance metrics achieved using the SafeNet. In Table 2, it is revealed that the overall F1 scores vary from 0.812 to 0.846, and the MCC varies from 0.654 to 0.717 for various datasets; these values are also shown in Figure 6D, which shows a performance comparison of the results. It was illustrated that as the input sequence length fluctuated, the performance of the model changed as well, and the optimal performance was achieved using the dataset with an input sequence length of 70 consecutive observations (DataSet 03). According to the ROC curves shown in Figure 6A–C, SafeNet also offered a reasonable performance for each class when DataSet 03 was used. Based on these results, we could conclude that the length of the input sequence had an influence on the performance of the SafeNet model, and that the best performance was achieved with an input sequence of 70 consecutive observations.

Table 2. Performance comparison of SafeNet and benchmark classifiers on different datasets.

Method	DataSet	MCC	F1	Accuracy	AUC of Class 0	AUC of Class 1	AUC of Class 2
SafeNet	DataSet 01	0.684	0.830	0.830	0.910	0.929	0.500
	DataSet 02	0.654	0.825	0.825	0.894	0.927	0.539
	DataSet 03	0.717	0.846	0.846	0.931	0.946	0.545
	DataSet 04	0.690	0.829	0.829	0.899	0.907	0.500
	DataSet 05	0.662	0.812	0.812	0.920	0.907	0.515
	DataSet 06	0.653	0.805	0.805	0.881	0.871	0.534
	DataSet 07	0.665	0.812	0.812	0.909	0.917	0.521
	DataSet 08	0.659	0.809	0.809	0.912	0.919	0.500
	DataSet 09	0.644	0.801	0.801	0.909	0.917	0.531
	DataSet 10	0.657	0.809	0.809	0.913	0.921	0.505
	DataSet 11	0.510	0.697	0.697	0.869	0.898	0.537
	DataSet 12	0.517	0.706	0.706	0.860	0.883	0.518
	DataSet 13	0.656	0.812	0.812	0.896	0.915	0.539
	DataSet 14	0.661	0.835	0.835	0.875	0.916	0.522
	DataSet 15	0.687	0.830	0.830	0.911	0.907	0.530
	DataSet 16	0.665	0.819	0.819	0.911	0.925	0.523
	DataSet 17	0.657	0.814	0.814	0.908	0.926	0.545
CNN	DataSet 03	0.635	0.825	0.825	0.859	0.908	0.520
LSTM	DataSet 03	0.643	0.824	0.824	0.880	0.923	0.520
DNN	DataSet 03	0.660	0.834	0.834	0.890	0.930	0.509
GBM	DataSet 03	0.613	0.813	0.813	0.882	0.923	0.538
RF	DataSet 03	0.450	0.742	0.742	0.836	0.876	0.519

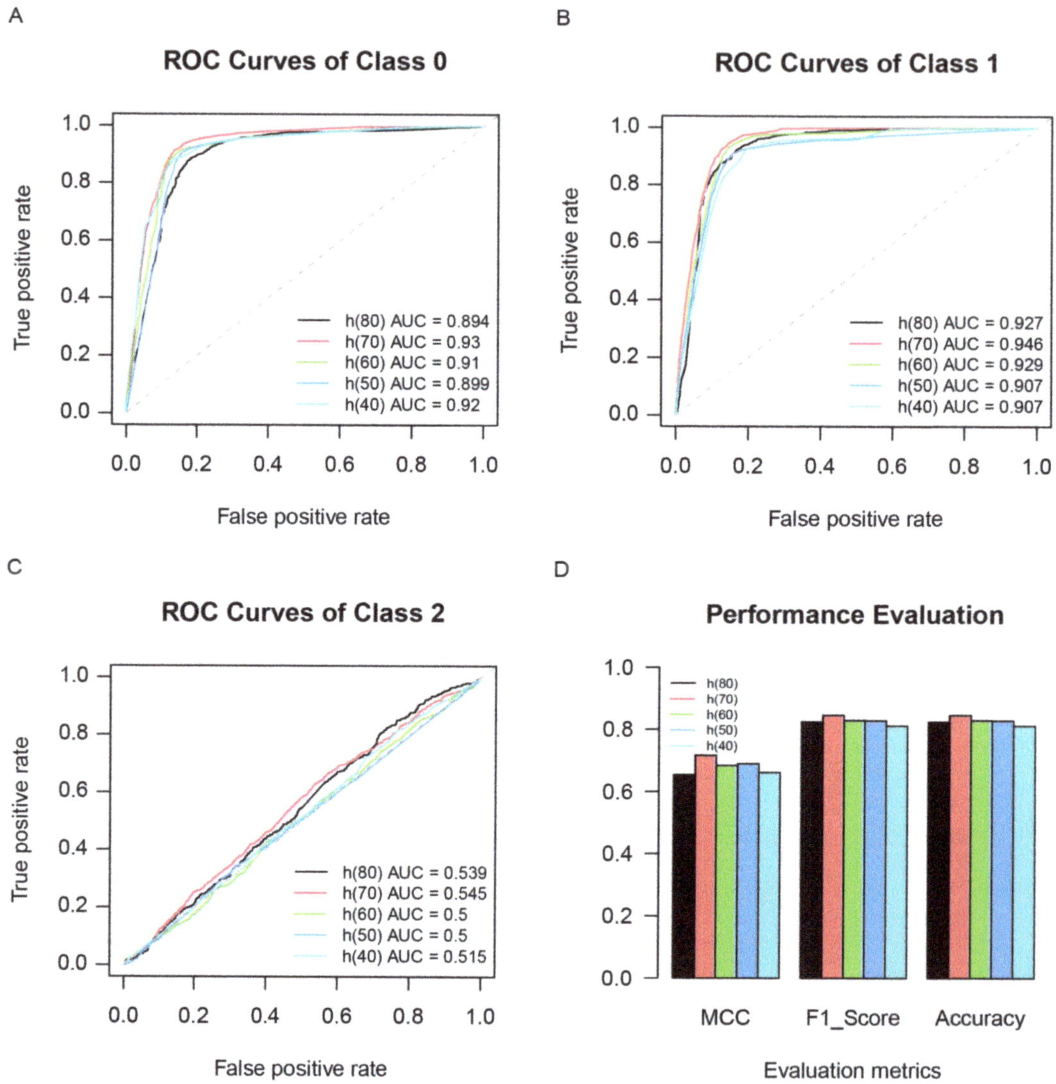

Figure 6. Comparing model performance using ROC curves for (**A**) class 0, (**B**) class 1, and (**C**) class 2 with window sizes of 40, 50, 60, 70, and 80. (**D**) Bar plots of the MCC, F1 score, and accuracy at various window sizes. We use the letter h to define the window size, and the numbers in brackets represent the specific size.

4.2. Data Comparing Nighttime Versus Daytime

The data acquisition time may affect the identification of pre-earthquake electromagnetic perturbations. To illustrate the effect of data collection time, a daytime dataset (Dataset 06 in Table 1) was created. As shown in Figure 7 and Table 2, SafeNet's performance was compared using benchmark datasets generated during the day and night. We evaluated the model's performance using the AUC, MCC, F1, and accuracy indicators.

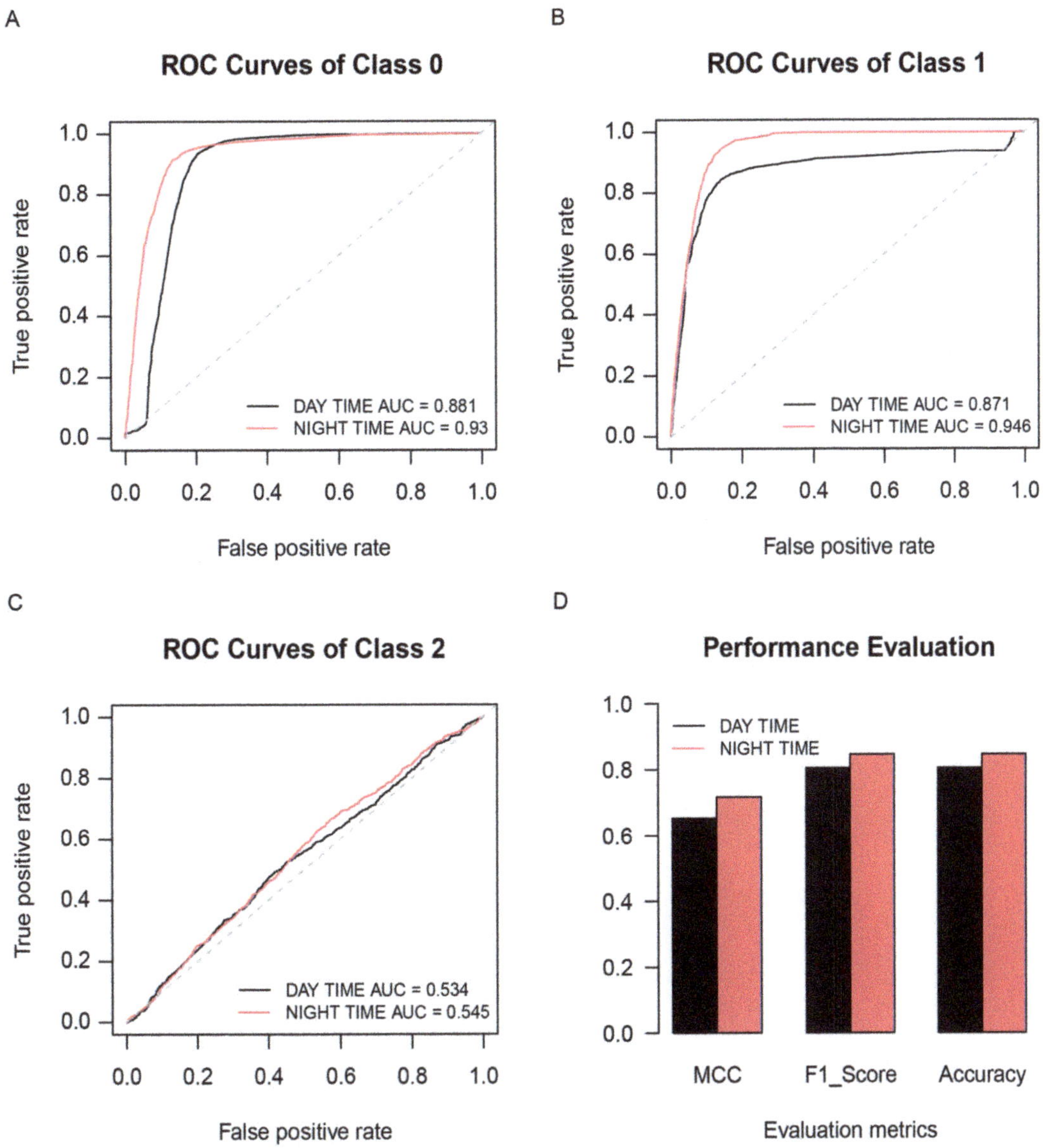

Figure 7. The ROC curves comparing model performance in nighttime vs. daytime data for (**A**) class 0, (**B**) class 1, and (**C**) class 2. (**D**) Comparison of model performance for nighttime and daytime data using bar plots of MCC, F1 score, and accuracy.

For the same spatial and temporal features, we found that using the nighttime datasets (Dataset 03 in Table 1) resulted in a higher classification performance than using the daytime dataset (Table 2). SafeNet's ROC curves for both datasets are given in Figure 7A–C, and we can observe that the AUC curve for nighttime data is somewhat higher than that for daytime data, with approximately 5.7%, 8.6%, and 2.1% increases in AUC for the three classes, respectively. When all classes were taken into account, SafeNet's F1 score improved from 0.805 to 0.846 when nighttime data were used, compared to daytime data, and MCC improved by 9.8% (Table 2).

Figure 7D compares the MCC, F1 score, and accuracy values between the daytime and nighttime data, revealing that the nighttime data performed slightly better than the daytime data. One reason for this may be that, because ionospheric conditions are typically more disturbed during the day, identifying seismic electromagnetic effects is more difficult, which may reflect a small number of significant changes in daytime data. This finding is supported by other research on statistical results of electromagnetic disturbances caused by earthquakes [16,43–46].

4.3. Considering Various Spatial Windows

SafeNet worked effectively for a circular region centered at the epicenter with a Dobrovolsky radius (DataSet 03). To further investigate the impact of various spatial windows on the performance of the model, satellite datasets with spatial windows of 3° (DataSet 07), 5° (DataSet 08), 7° (DataSet 09), and 10° (DataSet 10) were created (Table 1).

Table 2 details the SafeNet's performance on the five datasets, and Figure 8 illustrates the ROC and MCC curves, F1 score, and accuracy bar plot curves. SafeNet performed best when the dataset was used with the spatial window radius given by Dobrovolsky's formula (DataSet 03), with an F1 score of 0.846 and an MCC of 0.717. Comparing the results from Figure 8D and Table 2 shows that an improvement in the overall performance was not achieved with larger spatial windows. This tendency is also evident in the ROC curves in Figure 8A–C, where DataSet 03 had the highest AUC value of the datasets. Though the cause for these findings is unclear, it could be that a disturbance moving upward from the Earth's surface alters the ionosphere's properties geometrically, and the radius of the affected area matches the radius calculated using Dobrovolsky's formula. In addition, among the cases with different sizes of the preparation area (DataSet 07–10), it is the smaller ones that show the best performance. This could be due to the fact that there are more earthquakes with a smaller magnitude (i.e., 4.8–5.0) than those with a larger magnitude.

4.4. Considering the Magnitude of the Earthquake

It is well known that earthquake magnitude may play an active role in the identification of pre-earthquake perturbations. Therefore, to demonstrate the influence of magnitudes, we divided earthquakes into groups of three—5136 earthquakes with magnitudes between 4.8 and 5.2, 2793 earthquakes with magnitudes between 5.2 and 5.8, and 853 earthquakes with magnitudes between 5.8 and 7.5, and the corresponding datasets were created: DataSet 11, DataSet 12, and DataSet 13 (Table 1).

Table 2 illustrates the performance of the SafeNet model for the three datasets. As shown in Figure 9, the model had similar performance over the three datasets; for example, the AUC of class 0 and class 1 on all three datasets using the SafeNet method was above 0.86, and the F1 score ranged from 0.697 to 0.812, which suggests that the SafeNet model provides a satisfactory performance in discriminating electromagnetic pre-earthquake perturbations on the datasets with different magnitudes. Moreover, it can be concluded from Table 2 and Figure 9 that the larger the magnitude, the better the classification performance. This is also in line with the reality.

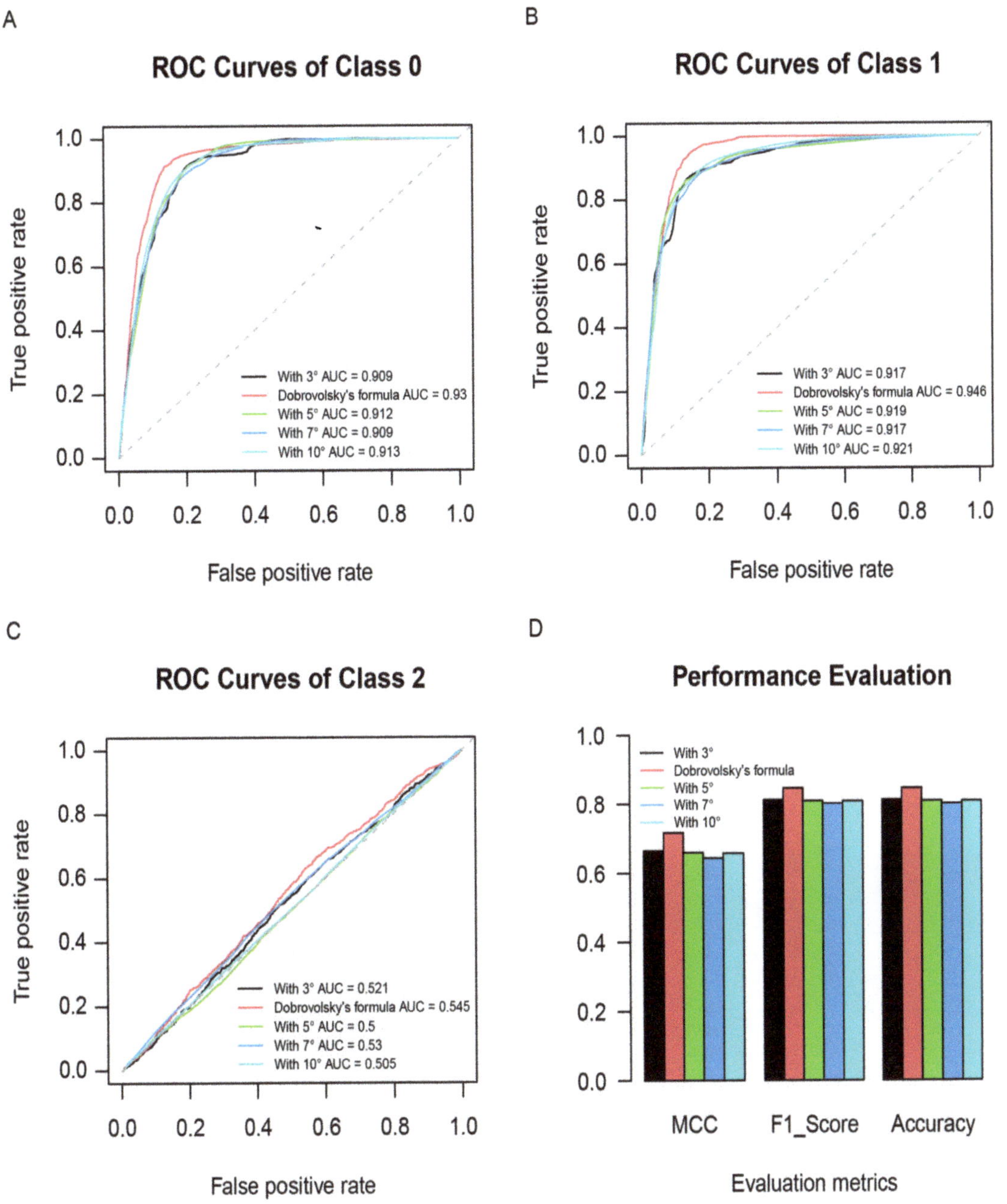

Figure 8. The ROC curves comparing model performance for (**A**) class 0, (**B**) class 1, and (**C**) class 2 with various spatial window radii of 3°, Dobrovolsky's formula, 5°, 7°, and 10°. (**D**) Bar plots showing the MCC, F1 score, and accuracy of the results.

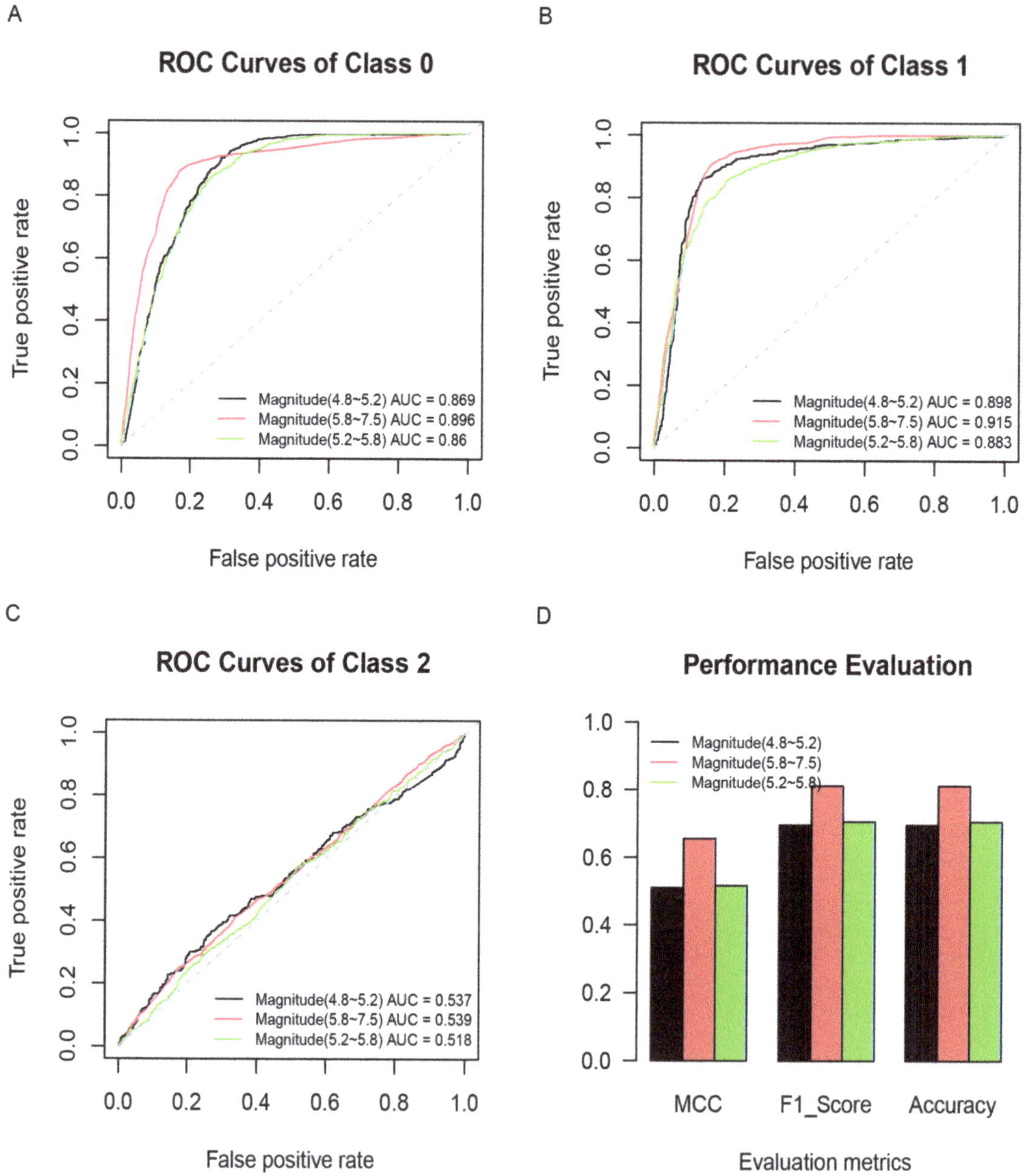

Figure 9. The ROC curves comparing model performance for (**A**) class 0, (**B**) class 1, and (**C**) class 2 of earthquakes with magnitudes between 4.8 and 5.2, earthquakes with magnitudes between 5.2 and 5.8, and earthquakes with magnitudes between 5.8 and 7.5. (**D**) Bar graphs displaying the MCC, F1 score, and accuracy.

4.5. Considering Unbalanced Datasets

The actual situation of earthquake issues is usually highly unbalanced, and it is obvious that non-earthquake datasets are always significantly larger than earthquake datasets. To test the real performance of the SafeNet model on unbalanced datasets as well as to investigate whether our proposed method could be applied to earthquake anomaly identification on unbalanced datasets, datasets with the positive to negative ratio of 1:2 (DataSet 14), 1:3 (DataSet 15), 1:4 (DataSet 16), and 1:5 (DataSet 17) were generated (shown in Table 1).

Table 2 illustrates the performance of the SafeNet model on five datasets (including DataSet 03). Although the method performs most effectively on the dataset with a positive to negative ratio of 1:1 (DataSet 03), the overall performance is similar on the five datasets; for instance, the proposed method has F1 scores around 0.83 (ranging from 0.814 to 0.846) as well as MCC values ranging from 0.661 to 0.717. Figure 10 shows the ROC curves for all three classes, together with a comparison of the performance metrics; we also observed a similar trend of our proposed method's performance on the five datasets, which suggests that the SafeNet model achieves a good performance for pre-seismic perturbation identification on the unbalanced dataset. Although the five unbalanced datasets are different, these results show that our method is less sensitive on the positive to negative ratio, and our method can be used to identify possible electromagnetic preseismic perturbations on an unbalanced dataset, which suggests that it could provide a good realistic performance.

4.6. Comparative Analysis of Other Classifiers

Table 2 and Figure 11 report the performance of our SafeNet model with five other benchmarking classifiers for DataSet 03. The performance of the existing methods ranged from F1 = 0.742 to 0.846 and MCC = 0.450 to 0.717. However, SafeNet had the best performance, improving MCC by 8.6% over that of the next-best DNN model. Figure 11A–C compares the ROC curves obtained for the SafeNet model with those of the five other classifiers, and SafeNet again demonstrated the best performance with a 4.6% improvement in AUC for class 0, a 1.7% improvement for class 1, and a 7.1% improvement for class 2 over the second-best DNN model.

To further confirm the performance of SafeNet, ternary probability diagrams and a confusion matrix were used to indicate the distribution of the predicted and true values and to allow for more profound insight into the classification performance of the model. Figure 12 shows the ternary probability diagrams and confusion matrix for the three classes obtained from the SafeNet model. Ternary probability diagrams allow for a qualitative evaluation of classification results. From Figure 12A–C as a whole, most of the samples in class 0 and class 1 were correctly predicted, and classes predicted by SafeNet could be classified into the correct classes, while the performance of class 2 was slightly worse, with some samples having a prediction probability concentrated around 0.5. The confusion matrices shown in Figure 12D quantitatively present the prediction accuracy for each class, with the correct prediction accuracy for class 0, class 1, and class 2 of 89.5%, 85.5%, and 87.9%, respectively. In general, SafeNet provides a good classification performance for identifying seismic signals and space weather events from Swarm satellite data.

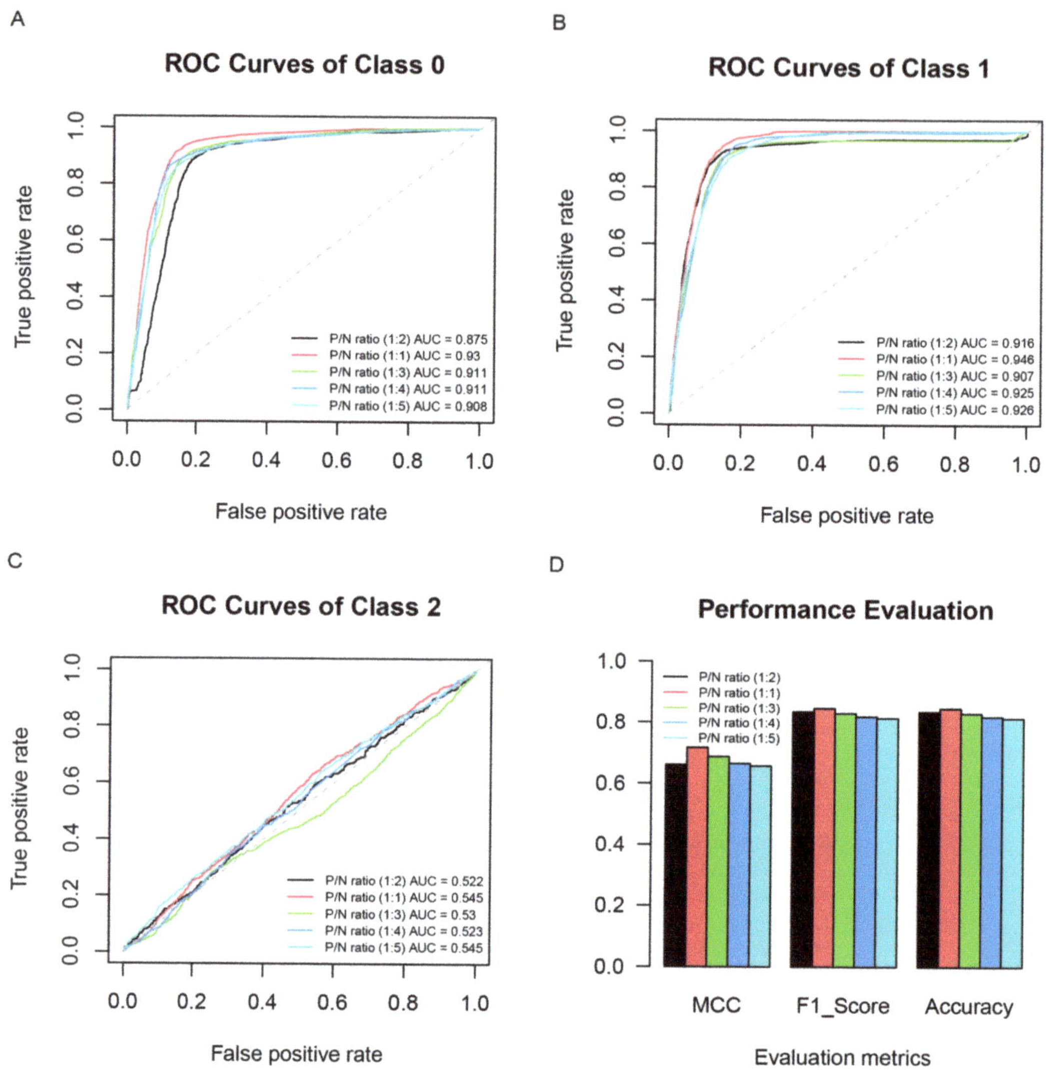

Figure 10. The ROC curves comparing model performance for (**A**) class 0, (**B**) class 1, and (**C**) class 2 of datasets with the positive to negative ratio of 1:1 (DataSet 3), 1:2 (DataSet 14), 1:3 (DataSet 15), 1:4 (DataSet 16), and 1:5 (DataSet 17). (**D**) Bar graphs displaying the MCC, F1 score, and accuracy.

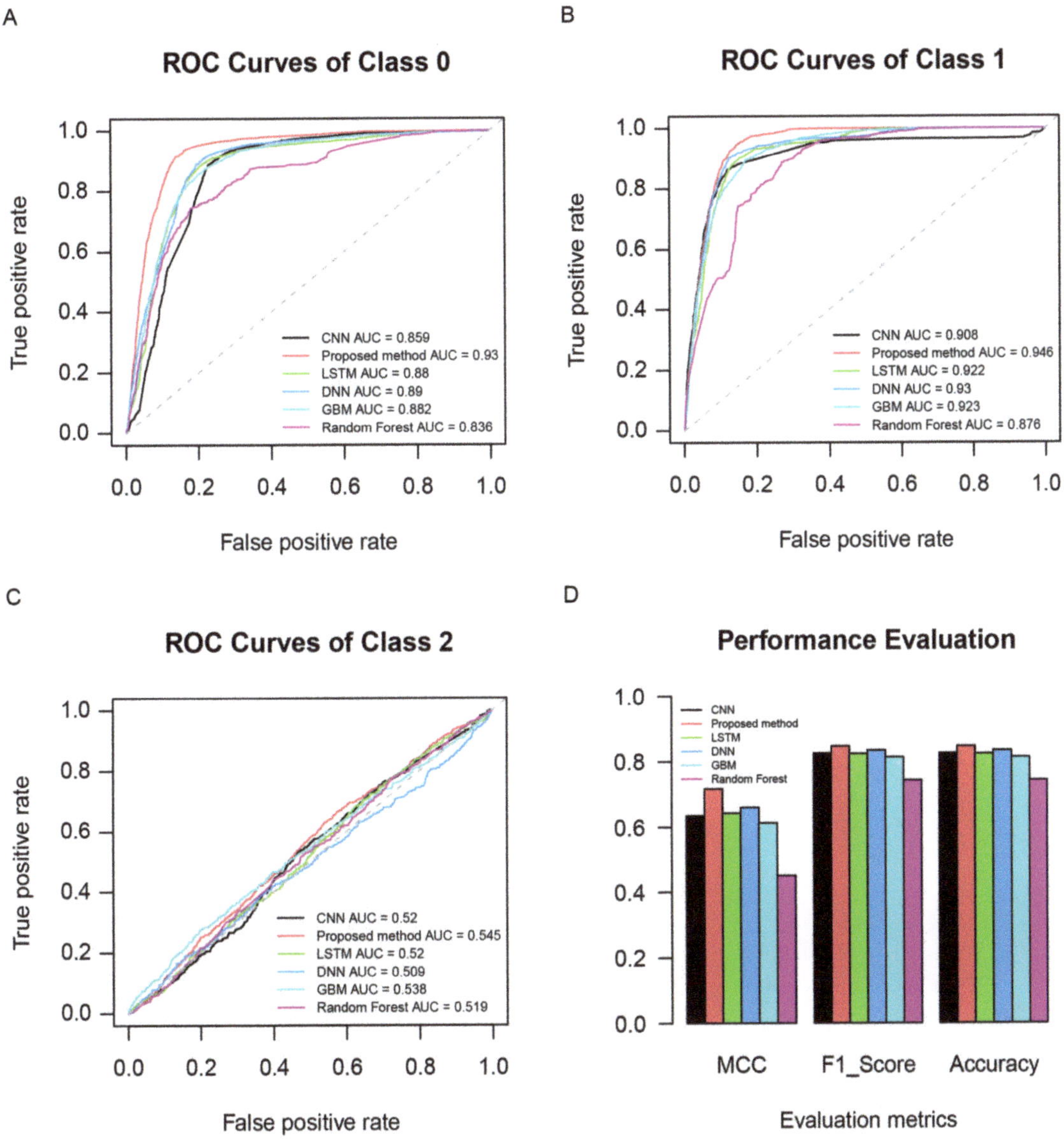

Figure 11. The ROC curves comparing the proposed method, gradient boosting machine (GBM), deep neural network (DNN), random forest (RF), convolutional neural network (CNN), and long short-term memory (LSTM) models for (**A**) class 0, (**B**) class 1, and (**C**) class 2. (**D**) The MCC, F1 score, and accuracy bar plot curves.

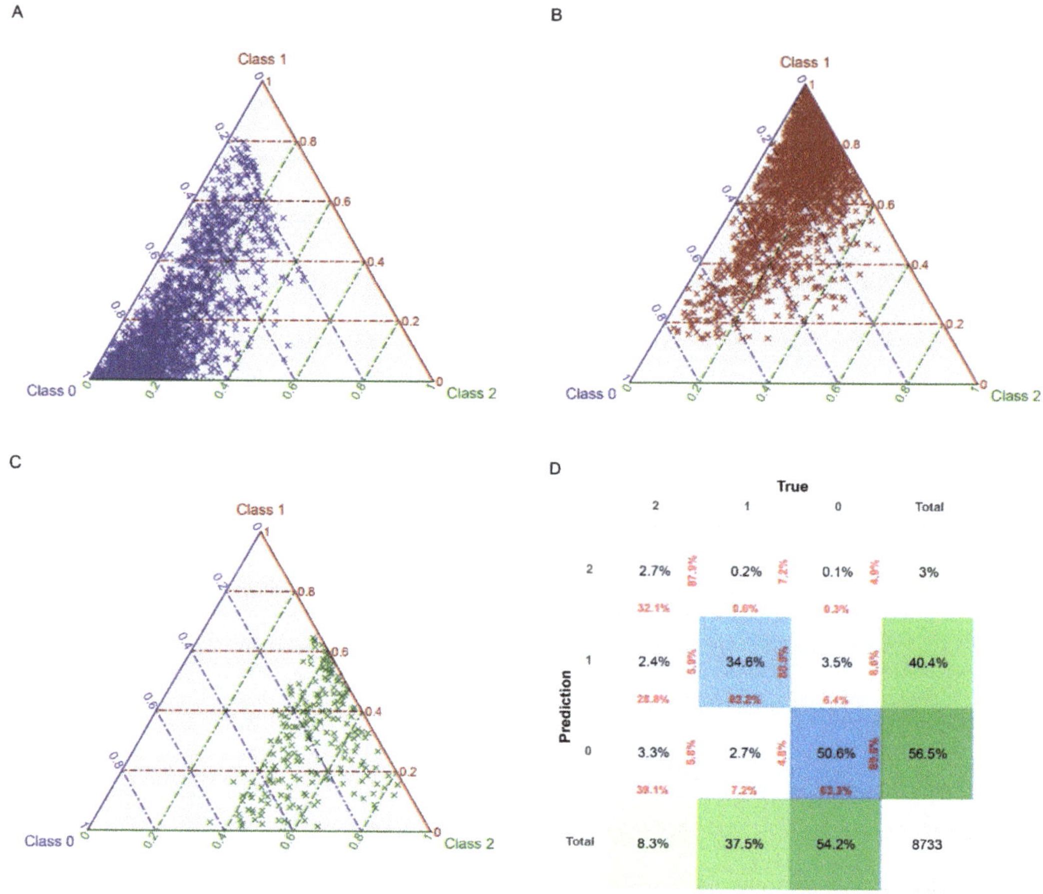

Figure 12. Ternary probability diagrams illustrating the findings of SafeNet on the test data for (**A**) class 0, (**B**) class 1, and (**C**) class 2. Class 0 is represented by blue crosses, class 1 by red crosses, and class 2 by green crosses; the color of the cross indicates the actual class, and the distance projected from each cross to the class axis represents the probability of that class in the model prediction. (**D**) Matrix of confusion illustrating the distribution of estimated and actual values. Each tile's center contains the normalized count (overall percentage) in black text. Column percentages are shown at the bottom of each tile, while row percentages are displayed on the right (both in red text). The sum tiles on the plot's right and bottom (in shades of green) indicate the overall distribution of predictions and targets. Note that the color intensity is proportional to the number of counts.

5. Discussions

Several studies have investigated the physical mechanisms of ionospheric pre-earthquake perturbations [47–51]. Perhaps the most widely accepted hypotheses regarding these mechanisms were presented by Pulinets, et al. [52] and Kuo, Lee, and Huba [27], who proposed complex lithosphere–atmosphere–ionosphere coupling as the physical basis for the generation of short-term earthquake precursors. Specifically, the lower crust and upper mantle generate gases such as radon (Rn) at lithostatic pressure during the buildup to an earthquake, and this gas could form large-scale domains in the rock. When reaching a certain vertical extent, these gas domains could become hydrostatically unstable and force their way upward through the lithosphere. Rapid lithospheric degassing can be expected to trigger several atmospheric processes near the Earth's surface, leading to changes in air conductivity and, therefore, changes to the near-ground atmospheric electric field. It is known that, as a part of the global electric circuit, the ionosphere immediately reacts to changes in near-ground electric properties, and an electric field induced within the ionosphere can cause ion drift and irregularities in electron concentrations. To better understand the mechanisms of ionospheric pre-earthquake perturbations, Freund [53] performed a series of experiments on a loaded rock and found that stress activation of p-hole charge carriers in the Earth's crust led to regional positive ground potential. In these scenarios, ionospheric perturbations are expected.

Crustal strain measurements often fail to detect any unusual changes before earthquakes [54], and the measurement of Rn gas content is among the most reported earthquake precursors [55,56]. Moreover, continuous monitoring of soil gas radon and water radon concentrations along with the Amritsar (Punjab, India) seismic zone correlation showed that the amplitude of radon gas anomalies was positively correlated with earthquake magnitude [57]. In addition, recent studies revealed a significant decrease in radon concentration within continuous measurements of radon concentration in the atmosphere before the 2018 earthquake in northern Osaka, Japan [58] and peculiar changes in radon concentration in the atmosphere two months before the 1995 Kobe earthquake in Japan [59]; Fu et al. [60] studied the radon gas anomalies in northern and northeastern Taiwan before the earthquake and observed a significant increase in soil radon concentration from a few days to a few weeks before the earthquake. Finally, it is reasonable that the amount of emitted radon could depend on the rupture length of the fault, that is, the magnitude of the seismic event. This study considers Rn gas emissions as the most reasonable result to be initially triggered by an earthquake. Because Rn decay can emit alpha particles, we propose a new hypothesis to explain the physical mechanisms of earthquake-induced ionospheric perturbations, as schematically depicted in Figure 13. In this hypothesis, various crustal movements in the pre-earthquake stage lead to rock fragmentation, melting, mineral dissolution, or phase change, and daughter isotopes of some radioactive parent isotopes retained in certain minerals or rocks are released in large quantities. In this situation, the Rn gas content in the area near the epicenter is abnormal before the earthquake, and furthermore the Rn gas decays quickly (half-life is about 3.8 days). During Rn decay, many alpha particles are released. The energy of an alpha particle is 5.2 MeV, and the ionization energy required by an atmospheric molecule is 32 eV. Therefore, one alpha particle is sufficient to generate 150,000 pairs of positive and negative ions, thereby creating an excess of positive airborne ions near the Earth's surface.

Because pre-earthquake field ionization occurs over relatively wide areas, we suggest that the positively charged air bubble expands owing to its internal electrostatic repulsion. Furthermore, as the prevailing electric field has an acceleration effect on positive airborne ions relative to the Earth's surface, the only direction of the airborne ions is upward. Therefore, many airborne ions would rise rapidly and shorten the vertical potential difference between the Earth's surface and the ionosphere. In response, the ionospheric plasma would be expected to polarize, causing the electrons located at the bottom of the ionosphere to be pulled downward. Thus, the physical properties of the ionosphere respond to changes in the vertical distribution of electrons and ions in the ionospheric plasma. In this way, it will result in a vertical electric field, E. The electric field pushes upward current flow from the atmosphere into the ionosphere. The injected current could result in the ionosphere being subjected to an enforced electric field. The perpendicular component of the electric field causes plasma $E \times B$ motion, which results in fluctuations in the ionosphere's density [27,61].

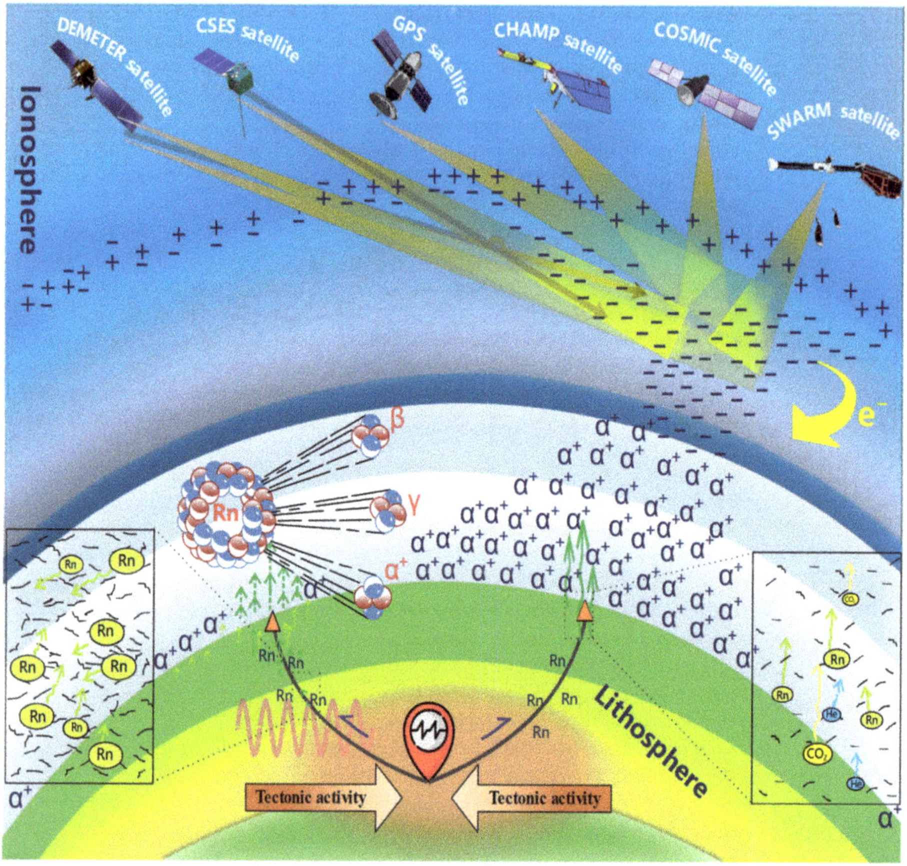

Figure 13. The proposed physical mechanisms of ionospheric perturbations induced by earthquakes. A large area of rock is broken and torn before the earthquake, after which a channel is opened to continuously release radon gas to generate radioactive decay. A gas bubble, laden with positive airborne ions generated during the process of radon decay at the ground-to-air interface, expands upward through the atmosphere, carrying up the Earth's ground potential and eliciting a polarization response in the ionosphere. This leads to a redistribution of the electrons at the lower edge of the ionosphere and thus modifies its physical properties. Thus, the satellite will receive anomalous signals from the ionosphere.

Moreover, the ionospheric anomalies before and after the earthquake could be positive or negative, and the probability of positive and negative perturbations is almost the same [62]. Simulations of anomalous electric fields show that if the anomalous electric field is westward, then the density enhancement occurs at the equator and the electron density decreases upward at the polarities [62]. If there is an eastward anomalous electric field, the positive and negative anomaly positions will be opposite. According to our knowledge, the anomalous electric field will be significantly different from one earthquake or different times of the same earthquake, so both positive and negative anomalies could be observed before and after the earthquake. In addition, Yao, et al. [63] also found that both positive and negative anomalies could occur before the earthquake by analyzing the GIM TEC of all Ms 7.0+ earthquakes in 2010. In addition, Zhao et al. studied the Wenchuan earthquake and also found that some of the GPS station data near the epicenter showed positive anomalies and some showed negative anomalies [64].

In the development of earthquake mechanisms at this stage, more experimental and actual evidence of radon gas generation in pressurized rocks is needed because it can provide a reasonable explanation for the uncertain relationship between pre-seismic electromagnetic anomalies and actual seismic events. In addition, in this endeavor, we suggest setting up ground-based observations of DC electric fields or using VLF noise anomalies to monitor the electrical activity of the atmosphere and lithosphere.

6. Conclusions

This study proposed the SafeNet deep learning framework for pre-earthquake ionospheric perturbation identification. SafeNet was trained and tested using 9017 independent earthquakes of magnitude 4.8 and above that occurred from April 2014 to April 2020, and the corresponding plasma and magnetic field data from the Swarm A satellite for about 6 years. The results indicated that electromagnetic pre-earthquake data within a circular region centered on the epicenter and with a radius given by the Dobrovolsky formula, with a model input window size of 70 consecutive points and nighttime sequence data, yielded the best performance in discriminating electromagnetic pre-earthquake perturbations, with an F1 score of 0.846 and an MCC value of 0.717. The study also concluded that the larger the magnitude of the earthquake, the better the performance of the SafeNet model in identifying possible pre-earthquake ionospheric anomalies. The results also suggest that the SafeNet model achieves a good performance for probable pre-seismic perturbation identification on the unbalanced dataset. In addition, based on constraints from this study, we proposed a new hypothesis on the physical mechanisms of earthquake-induced ionospheric perturbations.

In order to have a better understanding of the lithosphere, atmosphere, and ionosphere coupling mechanisms, to study pre-earthquake anomalies using electromagnetic satellites, the analysis of the spatial-temporal correlation of multi-sphere and multi-parameters by a remote sensing technique before earthquakes occur has become a hot research topic in recent years. However, it is difficult to match the different parameters and data of each sphere in both time and space. The existing multi-parameter earthquake studies are mainly focused on specific earthquakes or specific remote sensing parameters, have not sufficiently considered the other remote-sensing parameters of other spheres, and have not formed a complete chain of multi-parameter correlation analyses. The deep learning technology can overpass such limitations, combining the remote sensing parameters of multiple spheres in time and space and carrying out the analysis based on a consistent spatial-temporal framework, which could provide global earthquake cases and effectively explain the earthquake coupling mechanism models, and also expand the current tools for earthquake monitoring, providing new perspectives for earthquake prediction.

Supplementary Materials: The following are available online at https://www.mdpi.com/article/10.3390/rs13245033/s1, Text S1. Swarm Alpha nighttime tracks before the M7.9 Sumatra 2 March 2016 earthquake. Text S2. Swarm Alpha nighttime tracks before the M7.8 Ecuador 16 April 2016 earthquake. Text S3. Performance metrics. Text S4. The SafeNet model structure. Figure S25. The flowchart of the proposed deep learning framework. Table S1. Search space of parameters for the SafeNet model.

Author Contributions: Conceptualization, X.Z. and X.S.; data curation, P.X.; investigation, D.M., A.D.S., and X.Z.; methodology, P.X. and A.D.S.; project administration, A.D.S. and X.S.; software, P.X.; visualization, P.X. and D.M.; writing—original draft, P.X.; writing—review and editing, D.M., A.D.S., X.Z., and X.S. All authors have read and agreed to the published version of the manuscript.

Funding: This work is supported by the Special Fund of the Institute of Earthquake Forecasting, China Earthquake Administration (Grant No. 2021IEF0706, 2020IEF0510, 2021IEF0708) and funded by the National Key R&D Program of China under Grant No. 2018YFC1503505; Pianeta Dinamico-Working Earth (Italian Ministry of University and Research) and Limadou-Science+ (Italian Space Agency) Projects; National Natural Science Foundation of China (Grant number: 41974084); the China Postdoctoral Science Foundation (Grant number: 2021M691190).

Institutional Review Board Statement: Not applicable.

Informed Consent Statement: Not applicable.

Data Availability Statement: The Swarm data can be downloaded from the ESA Swarm FTP and HTTP Server swarm-diss.eo.esa.int. The Kp indices used in this paper were provided by the World Data Center for Geomagnetism, Kyoto (http://wdc.kugi.kyoto-u.ac.jp/wdc/Sec3.html).

Conflicts of Interest: The authors declare no conflict of interest.

References

1. Moore, G.W. Magnetic Disturbances preceding the 1964 Alaska Earthquake. *Nature* **1964**, *203*, 508–509. [CrossRef]
2. Davies, K.; Baker, D.M. Ionospheric effects observed around the time of the Alaskan earthquake of March 28, 1964. *J. Geophys. Res.* **1965**, *70*, 2251–2253. [CrossRef]
3. Parrot, M.; Li, M. Demeter results related to seismic activity. *URSI Radio Sci. Bull.* **2015**, *2015*, 18–25.
4. De Santis, A.; Balasis, G.; Pavón-Carrasco, F.J.; Cianchini, G.; Mandea, M. Potential earthquake precursory pattern from space: The 2015 Nepal event as seen by magnetic Swarm satellites. *Earth Planet. Sci. Lett.* **2017**, *461*, 119–126. [CrossRef]
5. Marchetti, D.; De Santis, A.; D'Arcangelo, S.; Poggio, F.; Piscini, A.; Campuzano, S.A.; De Carvalho, W.V.J.O. Pre-earthquake chain processes detected from ground to satellite altitude in preparation of the 2016–2017 seismic sequence in Central Italy. *Remote Sens. Environ.* **2019**, *229*, 93–99. [CrossRef]
6. Marchetti, D.; De Santis, A.; Campuzano, S.A.; Soldani, M.; Piscini, A.; Sabbagh, D.; Cianchini, G.; Perrone, L.; Orlando, M. Swarm Satellite Magnetic Field Data Analysis Prior to 2019 Mw = 7.1 Ridgecrest (California, USA) Earthquake. *Geosciences* **2020**, *10*, 502. [CrossRef]
7. Zhu, K.; Fan, M.; He, X.; Marchetti, D.; Li, K.; Yu, Z.; Chi, C.; Sun, H.; Cheng, Y. Analysis of Swarm Satellite Magnetic Field Data Before the 2016 Ecuador (Mw = 7.8) Earthquake Based on Non-negative Matrix Factorization. *Front. Earth Sci.* **2021**, *9*, 1976. [CrossRef]
8. Christodoulou, V.; Bi, Y.; Wilkie, G. A tool for Swarm satellite data analysis and anomaly detection. *PLoS ONE* **2019**, *14*, e0212098. [CrossRef] [PubMed]
9. Akhoondzadeh, M.; De Santis, A.; Marchetti, D.; Piscini, A.; Jin, S. Anomalous seismo-LAI variations potentially associated with the 2017 Mw = 7.3 Sarpol-e Zahab (Iran) earthquake from Swarm satellites, GPS-TEC and climatological data. *Adv. Space Res.* **2019**, *64*, 143–158. [CrossRef]
10. Marchetti, D.; Akhoondzadeh, M. Analysis of Swarm satellites data showing seismo-ionospheric anomalies around the time of the strong Mexico (M_w = 8.2) earthquake of 08 September 2017. *Adv. Space Res.* **2018**, *62*, 614–623. [CrossRef]
11. Zhu, K.; Li, K.; Fan, M.; Chi, C.; Yu, Z. Precursor Analysis Associated With the Ecuador Earthquake Using Swarm A and C Satellite Magnetic Data Based on PCA. *IEEE Access* **2019**, *7*, 93927–93936. [CrossRef]
12. Akhoondzadeh, M.; De Santis, A.; Marchetti, D.; Piscini, A.; Cianchini, G. Multi precursors analysis associated with the powerful Ecuador (MW = 7.8) earthquake of 16 April 2016 using Swarm satellites data in conjunction with other multi-platform satellite and ground data. *Adv. Space Res.* **2018**, *61*, 248–263. [CrossRef]
13. Parrot, M. Statistical analysis of automatically detected ion density variations recorded by DEMETER and their relation to seismic activity. *Ann. Geophys.* **2012**, *55*. [CrossRef]
14. Yan, R.; Parrot, M.; Pinçon, J.-L. Statistical Study on Variations of the Ionospheric Ion Density Observed by DEMETER and Related to Seismic Activities. *J. Geophys. Res. Space Phys.* **2017**, *122*, 12,421–12,429. [CrossRef]
15. Li, M.; Shen, X.; Parrot, M.; Zhang, X.; Zhang, Y.; Yu, C.; Yan, R.; Liu, D.; Lu, H.; Guo, F.; et al. Primary Joint Statistical Seismic Influence on Ionospheric Parameters Recorded by the CSES and DEMETER Satellites. *J. Geophys. Res. Space Phys.* **2020**, *125*. [CrossRef]
16. De Santis, A.; Marchetti, D.; Pavon-Carrasco, F.J.; Cianchini, G.; Perrone, L.; Abbattista, C.; Alfonsi, L.; Amoruso, L.; Campuzano, S.A.; Carbone, M.; et al. Precursory worldwide signatures of earthquake occurrences on Swarm satellite data. *Sci. Rep.* **2019**, *9*, 20287. [CrossRef] [PubMed]
17. Marchetti, D.; De Santis, A.; Jin, S.; Campuzano, S.A.; Cianchini, G.; Piscini, A. Co-Seismic Magnetic Field Perturbations Detected by Swarm Three-Satellite Constellation. *Remote Sens.* **2020**, *12*, 1166. [CrossRef]
18. Bergen, K.J.; Johnson, P.A.; de Hoop, M.V.; Beroza, G.C. Machine learning for data-driven discovery in solid Earth geoscience. *Science* **2019**, *363*, eaau0323. [CrossRef] [PubMed]
19. Rouet-Leduc, B.; Hulbert, C.; Johnson, P.A. Continuous chatter of the Cascadia subduction zone revealed by machine learning. *Nat. Geosci.* **2018**, *12*, 75–79. [CrossRef]
20. Ross, Z.E.; Trugman, D.T.; Hauksson, E.; Shearer, P.M. Searching for hidden earthquakes in Southern California. *Science* **2019**, *364*, 767–771. [CrossRef]
21. Xiong, P.; Long, C.; Zhou, H.; Battiston, R.; Zhang, X.; Shen, X. Identification of Electromagnetic Pre-Earthquake Perturbations from the DEMETER Data by Machine Learning. *Remote Sens.* **2020**, *12*, 3643. [CrossRef]
22. Xiong, P.; Tong, L.; Zhang, K.; Shen, X.; Battiston, R.; Ouzounov, D.; Iuppa, R.; Crookes, D.; Long, C.; Zhou, H. Towards advancing the earthquake forecasting by machine learning of satellite data. *Sci. Total Environ.* **2021**, *771*, 145256. [CrossRef] [PubMed]
23. Xiong, P.; Long, C.; Zhou, H.; Battiston, R.; De Santis, A.; Ouzounov, D.; Zhang, X.; Shen, X. Pre-Earthquake Ionospheric Perturbation Identification Using CSES Data via Transfer Learning. *Front. Environ. Sci.* **2021**, *9*, 9255. [CrossRef]

24. Olsen, N.; Friis-Christensen, E.; Floberghagen, R.; Alken, P.; Beggan, C.D.; Chulliat, A.; Doornbos, E.; da Encarnação, J.T.; Hamilton, B.; Hulot, G.; et al. The Swarm Satellite Constellation Application and Research Facility (SCARF) and Swarm data products. *Earth Planets Space* **2013**, *65*, 1189–1200. [CrossRef]
25. Friis-Christensen, E.; Lühr, H.; Hulot, G. Swarm: A constellation to study the Earth's magnetic field. *Earth Planets Space* **2006**, *58*, 351–358. [CrossRef]
26. Pinheiro, K.J.; Jackson, A.; Finlay, C.C. Measurements and uncertainties of the occurrence time of the **1969**, *1978*, 1991, and 1999 geomagnetic jerks. *Geochem. Geophys. Geosyst.* **2011**, *12*. [CrossRef]
27. Kuo, C.L.; Lee, L.C.; Huba, J.D. An improved coupling model for the lithosphere-atmosphere-ionosphere system. *J. Geophys. Res. Space Phys.* **2014**, *119*, 3189–3205. [CrossRef]
28. De Santis, A.; Marchetti, D.; Spogli, L.; Cianchini, G.; Pavón-Carrasco, F.J.; Franceschi, G.D.; Di Giovambattista, R.; Perrone, L.; Qamili, E.; Cesaroni, C.; et al. Magnetic Field and Electron Density Data Analysis from Swarm Satellites Searching for Ionospheric Effects by Great Earthquakes: 12 Case Studies from 2014 to 2016. *Atmosphere* **2019**, *10*, 371. [CrossRef]
29. Dobrovolsky, I.; Zubkov, S.; Miachkin, V. Estimation of the size of earthquake preparation zones. *Pure Appl. Geophys.* **1979**, *117*, 1025–1044. [CrossRef]
30. Spogli, L.; Sabbagh, D.; Regi, M.; Cesaroni, C.; Perrone, L.; Alfonsi, L.; Di Mauro, D.; Lepidi, S.; Campuzano, S.A.; Marchetti, D.; et al. Ionospheric Response Over Brazil to the August 2018 Geomagnetic Storm as Probed by CSES-01 and Swarm Satellites and by Local Ground-Based Observations. *J. Geophys. Res. Space Phys.* **2021**, *126*. [CrossRef]
31. Kingma, D.P.; Ba, J. Adam: A method for stochastic optimization. *arXiv* **2014**, arXiv:1412.6980.
32. Abadi, M.; Barham, P.; Chen, J.; Chen, Z.; Davis, A.; Dean, J.; Devin, M.; Ghemawat, S.; Irving, G.; Isard, M. Tensorflow: A system for large-scale machine learning. In Proceedings of the 12th USENIX symposium on operating systems design and implementation (OSDI 16), Savannah, GA, USA, 2–4 November 2016; pp. 265–283.
33. Oh, K.-S.; Jung, K. GPU implementation of neural networks. *Pattern Recognit.* **2004**, *37*, 1311–1314. [CrossRef]
34. Snoek, J.; Larochelle, H.; Adams, R.P. Practical bayesian optimization of machine learning algorithms. In Proceedings of the Advances in Neural Information Processing Systems, Lake Tahoe, CA, USA, 3–8 December 2012; pp. 2951–2959.
35. Bergstra, J.; Yamins, D.; Cox, D.D. Making a science of model search: Hyperparameter optimization in hundreds of dimensions for vision architectures. In Proceedings of the 30th International Conference on International Conference on Machine Learning, Atlanta, GA, USA, 16–21 June 2013; Volume 28, pp. I-115–I-123.
36. Japkowicz, N.; Stephen, S. The class imbalance problem: A systematic study. *Intell. Data Anal.* **2002**, *6*, 429–449. [CrossRef]
37. Matthews, B.W. Comparison of the predicted and observed secondary structure of T4 phage lysozyme. *Biochim. Biophys. Acta* **1975**, *405*, 442–451. [CrossRef]
38. Friedman, J.H. Greedy Function Approximation: A Gradient Boosting Machine. *Ann. Stat.* **2001**, *29*, 1189–1232. [CrossRef]
39. LeCun, Y.; Bengio, Y.; Hinton, G. Deep learning. *Nature* **2015**, *521*, 436–444. [CrossRef] [PubMed]
40. Geurts, P.; Ernst, D.; Wehenkel, L. Extremely randomized trees. *Mach. Learn.* **2006**, *63*, 3–42. [CrossRef]
41. Krizhevsky, A.; Sutskever, I.; Hinton, G.E. Imagenet classification with deep convolutional neural networks. In Proceedings of the Advances in Neural Information Processing Systems, Lake Tahoe, CA, USA, 3–8 December 2012; pp. 1097–1105.
42. Hochreiter, S.; Schmidhuber, J. Long short-term memory. *Neural Comput.* **1997**, *9*, 1735–1780. [CrossRef]
43. Němec, F.; Santolík, O.; Parrot, M.; Berthelier, J.J. Spacecraft observations of electromagnetic perturbations connected with seismic activity. *Geophys. Res. Lett.* **2008**, *35*. [CrossRef]
44. Němec, F.; Santolík, O.; Parrot, M. Decrease of intensity of ELF/VLF waves observed in the upper ionosphere close to earthquakes: A statistical study. *J. Geophys. Res. Space Phys.* **2009**, *114*. [CrossRef]
45. Píša, D.; Němec, F.; Parrot, M.; Santolík, O. Attenuation of electromagnetic waves at the frequency ~1.7 kHz in the upper ionosphere observed by the DEMETER satellite in the vicinity of earthquakes. *Ann. Geophys.* **2012**, *55*, 157–163. [CrossRef]
46. Píša, D.; Němec, F.; Santolík, O.; Parrot, M.; Rycroft, M. Additional attenuation of natural VLF electromagnetic waves observed by the DEMETER spacecraft resulting from preseismic activity. *J. Geophys. Res. Space Phys.* **2013**, *118*, 5286–5295. [CrossRef]
47. Pulinets, S.; Ouzounov, D. Lithosphere–Atmosphere–Ionosphere Coupling (LAIC) model—An unified concept for earthquake precursors validation. *J. Asian Earth Sci.* **2011**, *41*, 371–382. [CrossRef]
48. Ouzounov, D.; Pulinets, S.; Liu, J.-Y.; Hattori, K.; Han, P. Multiparameter Assessment of Pre-Earthquake Atmospheric Signals. *Pre-Earthq. Process.* **2018**, 339–359. [CrossRef]
49. Freund, F.T.; Heraud, J.A.; Centa, V.A.; Scoville, J. Mechanism of unipolar electromagnetic pulses emitted from the hypocenters of impending earthquakes. *Eur. Phys. J. Spec. Top.* **2021**, *230*, 47–65. [CrossRef]
50. Wu, L.-X.; Qin, K.; Liu, S.-J. GEOSS-Based Thermal Parameters Analysis for Earthquake Anomaly Recognition. *Proc. IEEE* **2012**, *100*, 2891–2907. [CrossRef]
51. Hayakawa, M.; Kasahara, Y.; Nakamura, T.; Muto, F.; Horie, T.; Maekawa, S.; Hobara, Y.; Rozhnoi, A.A.; Solovieva, M.; Molchanov, O.A. A statistical study on the correlation between lower ionospheric perturbations as seen by subionospheric VLF/LF propagation and earthquakes. *J. Geophys. Res. Space Phys.* **2010**, *115*. [CrossRef]
52. Pulinets, S.A.; Ouzounov, D.P.; Karelin, A.V.; Davidenko, D.V. Physical bases of the generation of short-term earthquake precursors: A complex model of ionization-induced geophysical processes in the lithosphere-atmosphere-ionosphere-magnetosphere system. *Geomagn. Aeron.* **2015**, *55*, 521–538. [CrossRef]

53. Freund, F.T. Pre-earthquake signals—Part I: Deviatoric stresses turn rocks into a source of electric currents. *Nat. Hazards Earth Syst. Sci.* **2007**, *7*, 535–541. [CrossRef]
54. Soter, S. Macroscopic seismic anomalies and submarine pockmarks in the Corinth–Patras rift, Greece. *Tectonophysics* **1999**, *308*, 275–290. [CrossRef]
55. Riggio, A.; Santulin, M. Earthquake forecasting: A review of radon as seismic precursor. *Boll. Di Geofis. Teor. Ed Appl.* **2015**, *56*.
56. Gold, T.; Soter, S. Fluid ascent through the solid lithosphere and its relation to earthquakes. *Pure Appl. Geophys.* **1985**, *122*, 492–530. [CrossRef]
57. Kumar, A.; Walia, V.; Singh, S.; Bajwa, B.S.; Dhar, S.; Yang, T.F. Earthquake precursory studies at Amritsar Punjab, India using radon measurement techniques. *Int. J. Phys. Sci.* **2013**, *7*, 5669–5677.
58. Muto, J.; Yasuoka, Y.; Miura, N.; Iwata, D.; Nagahama, H.; Hirano, M.; Ohmomo, Y.; Mukai, T. Preseismic atmospheric radon anomaly associated with 2018 Northern Osaka earthquake. *Sci. Rep.* **2021**, *11*, 7451. [CrossRef] [PubMed]
59. Omori, Y.; Nagahama, H.; Yasuoka, Y.; Muto, J. Radon degassing triggered by tidal loading before an earthquake. *Sci. Rep.* **2021**, *11*, 4092. [CrossRef]
60. Fu, C.-C.; Lee, L.-C.; Yang, T.F.; Lin, C.-H.; Chen, C.-H.; Walia, V.; Liu, T.-K.; Ouzounov, D.; Giuliani, G.; Lai, T.-H.; et al. Gamma Ray and Radon Anomalies in Northern Taiwan as a Possible Preearthquake Indicator around the Plate Boundary. *Geofluids* **2019**, *2019*, 1–14. [CrossRef]
61. Kuo, C.L.; Huba, J.D.; Joyce, G.; Lee, L.C. Ionosphere plasma bubbles and density variations induced by pre-earthquake rock currents and associated surface charges. *J. Geophys. Res. Space Phys.* **2011**, *116*. [CrossRef]
62. Liu, J.; Wan, W.; Zhou, C.; Zhang, X.; Liu, Y.; Shen, X. A study of the ionospheric disturbances associated with strong earthquakes using the empirical orthogonal function analysis. *J. Asian. Earth Sci.* **2019**, *171*, 225–232. [CrossRef]
63. Yao, Y.B.; Chen, P.; Zhang, S.; Chen, J.J.; Yan, F.; Peng, W.F. Analysis of pre-earthquake ionospheric anomalies before the global M = 7.0+ earthquakes in 2010. *Nat. Hazards Earth Syst. Sci.* **2012**, *12*, 575–585. [CrossRef]
64. Zhao, B.; Wang, M.; Yu, T.; Xu, G.; Wan, W.; Liu, L. Ionospheric total electron content variations prior to the 2008 Wenchuan Earthquake. *Int. J. Remote Sens.* **2010**, *31*, 3545–3557. [CrossRef]

remote sensing

MDPI

Article

Statistical Study of Ionospheric Equivalent Slab Thickness at Guam Magnetic Equatorial Location

Yuqiang Zhang [1], Zhensen Wu [1,*], Jian Feng [2], Tong Xu [2], Zhongxin Deng [2], Ming Ou [2], Wen Xiong [2] and Weimin Zhen [2]

1 School of Physics and Optoelectronic Engineering, Xidian University, Xi'an 710126, China; yqzhang_3@stu.xidian.edu.cn
2 China Research Institute of Radiowave Propagation (CRIRP), Qingdao 266107, China; fengj@crirp.ac.cn (J.F.); xut@crirp.ac.cn (T.X.); dengzx@crirp.ac.cn (Z.D.); oum@crirp.ac.cn (M.O.); xiongw@crirp.ac.cn (W.X.); zhenwm@crirp.ac.cn (W.Z.)
* Correspondence: wuzhs@mail.xidian.edu.cn

Abstract: The ionospheric equivalent slab thickness (τ) is defined as the ratio of the total electron content (TEC) to the F2-layer peak electron density (*Nm*F2), and it is a significant parameter representative of the ionosphere. In this paper, a comprehensive statistical analysis of the diurnal, seasonal, solar, and magnetic activity variations in the τ at Guam (144.86°E, 13.62°N, 5.54°N dip lat), which is located near the magnetic equator, is presented using the GPS-TEC and ionosonde *Nm*F2 data during the years 2012–2017. It is found that, for geomagnetically quiet days, the τ reaches its maximum value in the noontime, and the peak value in winter and at the equinox are larger than that in summer. Moreover, there is a post-sunset peak observed in the winter and equinox, and the τ during the post-midnight period is smallest in equinox. The mainly diurnal and seasonal variation of τ can be explained within the framework of relative variation of TEC and *Nm*F2 during different seasonal local time. The dependence of τ on the solar activity shows positive correlation during the daytime, and the opposite situation applies for the nighttime. Specifically, the disturbance index (DI), which can visually assess the relationship between instantaneous τ values and the median, is introduced in the paper to quantitatively describe the overall pattern of the geomagnetic storm effect on the τ variation. The results show that the geomagnetic storm seems to have positive effect on the τ during most of the storm-time period at Guam. An example, on the 1 June 2013, is also presented to analyze the physical mechanism. During the positive storms, the penetration electric field, along with storm time equator-ward neutral wind, tends to increase upward drift and uplift F region, causing the large increase in TEC, accompanied by a relatively small increase in *Nm*F2. On the other hand, an enhanced equatorward wind tends to push more plasma, at low latitudes, into the topside ionosphere in the equatorial region, resulting in the TEC not undergoing severe depletion, as with *Nm*F2, during the negative storms. The results would complement the analysis of τ behavior during quiet and disturbed conditions at equatorial latitudes in East Asia.

Keywords: τ; geomagnetic equator; magnetic storm

Citation: Zhang, Y.; Wu, Z.; Feng, J.; Xu, T.; Deng, Z.; Ou, M.; Xiong, W.; Zhen, W. Statistical Study of Ionospheric Equivalent Slab Thickness at Guam Magnetic Equatorial Location. *Remote Sens.* **2021**, *13*, 5175. https://doi.org/10.3390/rs13245175

Academic Editor: Fabio Giannattasio

Received: 23 November 2021
Accepted: 16 December 2021
Published: 20 December 2021

1. Introduction

The ionospheric equivalent slab thickness (EST, usually named τ) is defined as the ratio of the total electron content (TEC) to the F2-layer peak electron density (*Nm*F2) and it thus represents an imaginary equivalent depth of the ionosphere, with a constant uniform density of the F2 peak. Strictly speaking, when TEC is obtained by means of GNSS (Global Navigation Satellite System) signals, the τ value will contain a plasmaspheric component in addition to the 'pure' ionospheric component [1]. In the view of satellite-to-ground radio communications, τ is a useful parameter in that it includes information on both the topside and bottomside ionosphere. The τ is, thus, very helpful in understanding the nature of variations of the upper atmosphere and is, therefore, capable of addressing many useful ionospheric parameters, such as Chapman scale height H, topside 'half-peak density height' h0.5top, and temperature changes in the ionosphere/thermosphere systems [2–7]. In addition, according to the definition of τ, it connects VTEC and *Nm*F2, belonging to different fields of remote sensing, which can be used to deduce *Nm*F2 from the TEC, and vice versa [8–13].

The ionospheric slab thickness has been a subject of study since the 1960s [8–33]. They found that the diurnal, seasonal, and solar activity variations of the τ show significant dependence on the location of the observing station. The greatest variability in τ is observed during periods of geomagnetic storms, and the influence of geomagnetic activity on the τ shows different patterns as the observing station and solar activity vary. Modelling studies have also been carried out to provide global τ [8–13,26,29].

The climatology of τ near the magnetic equator has been mentioned few times in previous studies. Odeyemi et al. [24,25] investigated the corresponding morphologies of ionospheric profile and peak parameters over an equatorial station in west Africa, Ilorin (8.50°N, 4.68°E), during a low solar activity year 2010, and they found that the post-sunset increase in τ is more prominent during solstice than equinoxes, of which the December solstice perceives a relatively higher magnitude than the June solstice in 2010. Àlàgbé [30] studied the correlation between the τ and B0 at the Burkina Faso station (12.4°N, 1.5°W, dip 5.9°N), and they found that the maximum of τ at the station appeared in the daytime both high and low years of solar activity. The first peak occurred at between 05LT–06LT in the low solar activity year, and the daytime peak appears at about 11LT; the peak at sunrise in the high solar activity year is not apparent, while the daytime peak occurs at 12LT–13LT. The value of τ in high solar activity year is generally higher than that of low solar activity year. Duarte-Silva et al. [31] studied the τ at Palmas and São Jose′ dos Campos during one year of extremely low solar activity (from March 2009 to February 2010), and these two places are located at the inner edge of the anomaly region and the southern crest of the anomaly, respectively. They found that the τ at Palmas begins to increase gradually until its maximum value is reached between 13LT–15LT, and the maximum average peak values of τ, of about 477 km, were observed over Palmas during daytime (08LT–16LT) throughout the December solstice months. In addition, the study of τ behavior during the storm time at equatorial latitudes is relatively less. Therefore, long-term monitoring and analyzing observations at equatorial latitudes are needed to give a more comprehensive study of the climatology of τ and geomagnetic storm effect on it at this latitude.

Guam station (144.86°E, 13.62°N, 5.54°N dip lat) is located at the equatorial north latitudes, where it lies between western Pacific Ocean and east of the Eurasian continent, and the surrounding area is dominated by marine environment. The ionospheric data are relatively sparse at this region, and there are few specific studies investigating the τ in this region. In order to strengthen the understanding of the variation of ionospheric thickness in East Asia near the magnetic equator, this paper uses the GPS-TEC and *fo*F2 data at Guam station, between 2012 and 2017, to statistically analyze τ dependence with season, local time, and solar activity. Specifically, we investigate the effect of geomagnetic storms on the τ variation in this station. This paper is organized as follows: Section 2 briefly describes the data and methodology employed in this study. In Section 3, τ variation is presented, with an emphasis on its variations during geomagnetic storms. Section 4 discusses the derived results and use one example to analyze the causes of storm enhancement of τ. Finally, conclusions of our study are found in Section 5.

2. Data and Methods of Analysis

The GPS TEC data in this paper are derived from UNAVCO (University NAVSTAR Consortium) database (http://www.unavco.org/ accessed on 23 November 2021). After downloading the receiver-independent exchange data (RINEX format) from UNAVCO, the GPS-TEC program developed by Seemala (https://Seemala.blogspot.com accessed on 23 November 2021) is applied to derive TEC values and elevation angle cutoff of 30° is adopted in order to eliminate the multi-path effects. In addition, the ionospheric shell height, used to retrieve TEC, is 350 km. Uwamahoro et al. [34] explained, in detail, how the GPS-TEC software works. Moreover, the GPS-TEC software used in this study has been compared with other techniques, such as the one presented in Ref. [35], alongside the European Geostationary Navigation Overlay System (EGNOS) algorithm, which was taken as a reference in Ref. [36]. Generally, both software is consistent with EGNOS algorithm, but the Gopi software was found to be closer to EGNOS in low-latitudes. It is worth noting that the same GPS-TEC software used in the current work has extensively been used to derive TEC in ionosphere and in τ climatology studies that involved TEC computation [25,37–43]. Moreover, we used the regional kriging interpolation method to convert the STEC obtained by software to VTEC, rather than directly use the average values of VTEC data calculated by the software.

Simultaneous *fo*F2 data are obtained from the GIRO (Global Ionospheric Radio Observation) database (http://umlcar.uml.edu/DIDbase/ accessed on 23 November 2021) [44]. In the data processing, we filtered out the ionosonde data, which had a confidence score (CS) of less than 80 (maximum 100). The *Nm*F2 values are computed from the critical frequency of the F2 layer (*fo*F2) scaled from the recorded automatically ionograms, where $NmF2 = 1.24 \times 10^{10}\ (foF2)^2$. Then, the TEC and *Nm*F2 data, with a 15 min resolution, are used to compute the equivalent τ by using the following relation:

$$\tau = \mathrm{TEC}/NmF2 = \mathrm{TEC}/(1.24 \times 10^{10}\ (foF2)^2) \tag{1}$$

where TEC is measured in TEC units (10^{16} electrons/m^2) and *fo*F2 is the critical frequency of the F2-layer, with *fo*F2 given in MHz and τ in meters. For simplicity, we transform it in kilometers in the following.

Since (dayside) ionospheric density is produced, mainly, by solar EUV radiation (which is inferred by F10.7), we use F10.7 index to study the effect of solar activity on τ, and they are downloaded from NOAA (ftp://ftp.swpc.noaa.gov/pub/indices/old_indices/ accessed on 23 November 2021). Figure 1 shows the overall pattern of the solar F10.7 index during the period 2010–2017. It can be seen from Figure 1 that the F10.7 in 2014 was the largest among these 8 years, with an average value of 145.9 SFU (Solar Flux Unit). The F10.7 in 2010, 2016, and 2017 was small, with the average values being 80 SFU, 88.7 SFU, and 77.3 SFU, respectively. Due to the very high rate of data availability in 2016 (94.6%), this paper selects 2016 as the year of low solar activity and 2014 as the year of high solar activity to study the influence of solar activity on τ.

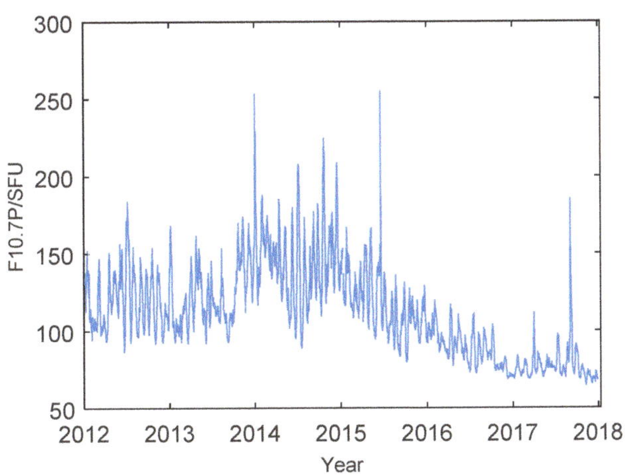

Figure 1. F10.7 variations during the years 2012–2017.

Previous studies have found that geomagnetic storms have effect on the τ variation [15,27,32,33]. In order to obtain the τ variation during geomagnetic quiet condition, we selected all the geomagnetic storms satisfying $Dst_{min} < -30$ during the period 2010–2017, according to the Dst (http://wdc.kugi.kyoto-u.ac.jp/ accessed on 23 November 2021) index, and the τ data, during the selected geomagnetic storm, are excluded from the overall τ dataset. Then, the τ data during the geomagnetic quiet condition are divided into 288 grids, according to January–December and 01–23LT. Recently, Pignalberi et al. [22] used similar methodology to analyze τ at mid latitudes. The seasons are defined as $\pm$45 days around vernal and autumnal equinoxes, June solstice, and December solstice.

The geomagnetic index Dst is used in this paper to select geomagnetic storm with minimum Dst < -50. To quantitatively describe the overall pattern of the geomagnetic storm effect on the τ variation, we define the disturbance index (DI):

$$DI = \tau_s/\tau_m - 1 \tag{2}$$

where τ_s and τ_m are storm-time and monthly median τ, respectively. Figure 2 shows an example during 2013 in Guam. Figure 2a illustrates the Dst variation, and the arrow 'main phase onset (MPO)' marks universal time of the magnetic main phase onset at 18:00 UT on 6 June. Figure 2b shows the variation of measured τ (blue lines) during 6–8 June and corresponding τ median (red) in June. Figure 2c presents the calculated DI index according to the measured τ value and its Median value. In addition, DI index in (storm time 1) S1 time period is used to study the characteristics of DI index during geomagnetic storm, while DI index in time periods (quiet time 1) Q1 and (quiet time 2) Q2 are used to study the one during geomagnetic quiet condition.

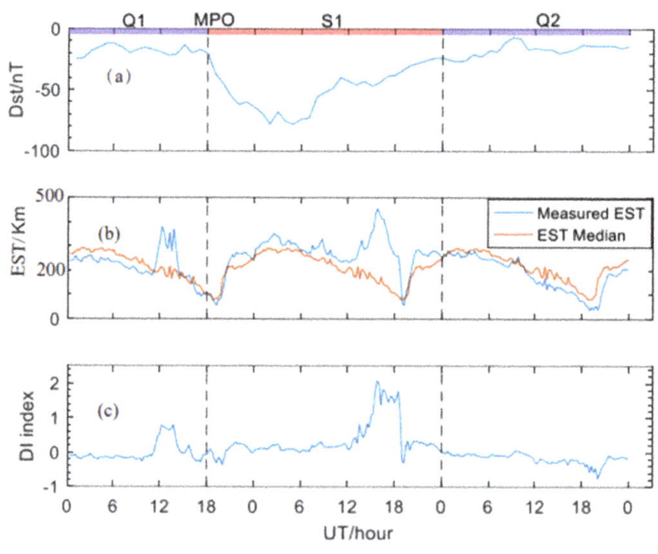

Figure 2. An example for the calculation of DI index during the period 6–8 June 2013.

3. Results

According to the grid division of January–December and 0LT–23LT, Figure 3 gives the distribution of mean and standard deviation of τ on this grid. It can be seen from the figure that the first peak of τ appears at 11LT–12LT in all seasons, and it is the maximum value throughout the day. Moreover, there is a second peak that appears at 18LT–22LT in winter and equinox. During the nighttime, τ is significantly smaller than that in the daytime. From the perspective of seasonal variations, the τ in winter and equinox are generally greater than that in summer. In addition, the annual maximum appeared in

October 12LT, reaching 437.6km, and the annual minimum appeared in April 5LT, reaching 105.2km. Moreover, the τ shows great variability during the night-time, especially during the pre-sunrise and post-sunset period, which is consistent with previous studies [7,22].

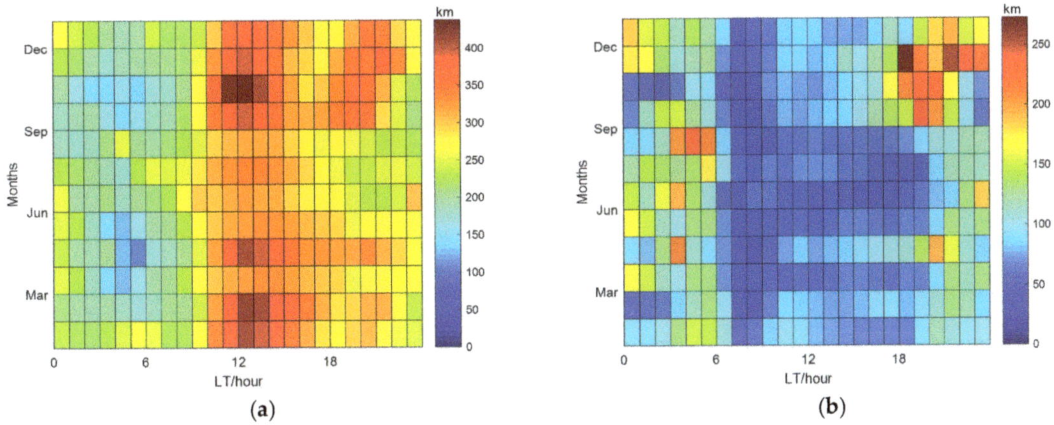

Figure 3. Map of mean (**a**) and standard deviation (**b**) of equivalent slab thickness (τ) under geomagnetically quiet condition, according to January–December and 0LT–23LT division in 2012–2017.

Figure 4 compared the distribution of mean and standard deviation of τ in the year of high solar activity and low solar activity, based on the same grid division. It can be seen from the figure that the EST in high/low activity year show some common features. That is, the first peak of τ appears from 11LT to 12LT in all season. The τ in 0LT–10LT is smaller than that in other periods, both during the years of high or low solar activity. The results also demonstrate that the τ during the daytime in winter and equinox is larger than the τ in summer. As for the differences between τ during the high and low solar activity year, it is worth noting that the τ in the 10LT–22LT time period of the high solar activity year seems greater than the average value of all years, while for other time periods, the τ is smaller. The opposite situation applies for the change in low solar activity year. In addition, nighttime and solar terminator hours are also characterized by the highest dispersion, in the same way as Figure 3b. Specifically speaking, τ shows more variability during the post-sunset period in the high solar activity year, while τ shows more variability during the pre-sunrise period in the low solar activity year.

Among previous studies of τ variation near geomagnetic equator, Àlàgbé. [30] found the maximum of τ at Burkina Faso (12.4°N, 1.5°W, dip 5.9°N) appeared during the day, regardless of the solar activity of the year. The daytime peak appears at about 11LT in the low solar activity year; it appears at 12–13LT in the high solar activity year. Duarte-Silva et al. [31] found that the τ at Palmas station (10.12°S, 48.21°W, 7.73°S dip) in the low solar activity have the following rules: the τ begins to increase gradually after sunrise, and its maximum value appeared between 13LT to 15LT. Moreover, annual maximum of τ reached 550 km in summer (that is, December) at the station. The τ pattern in this paper is similar to these findings, but it is not completely consistent. The ionospheric longitudinal difference may contribute to the differences between them [45–47].

Figure 5 presents the diurnal variations of mean and standard deviation of τ in summer, winter, and equinox at a geomagnetically quiet condition. It can be seen from Figure 5 that the first peak of τ in winter is significantly higher than that in summer, and the equinox τ is almost the same as the winter τ during this period (range 2 (R2) interval). There is another peak of τ in post sunset period in winter and equinox, and the summer τ is also significantly smaller than that in winter and equinox during this period (range 3 (R3) interval). Moreover, the τ during the post-midnight period is the smallest in equinox, especially during the period of 03LT–06LT (range 1 (R1) interval). In previous studies, the sunrise peak of τ is a hot topic [21,26,33]. However, Figure 5 shows that, starting from 04LT in winter and summer, and 05LT in equinox, the τ gradually increases, and there is no peak. There is only a small peak appears at 06LT in winter. In addition, it can also be seen from the figure that the nighttime is characterized by the high dispersion, especially during the pre-sunrise and post-sunset period. During the pre-sunrise period, τ has highest variability in the summer, and τ has highest variability in the equinox during the post-sunset period.

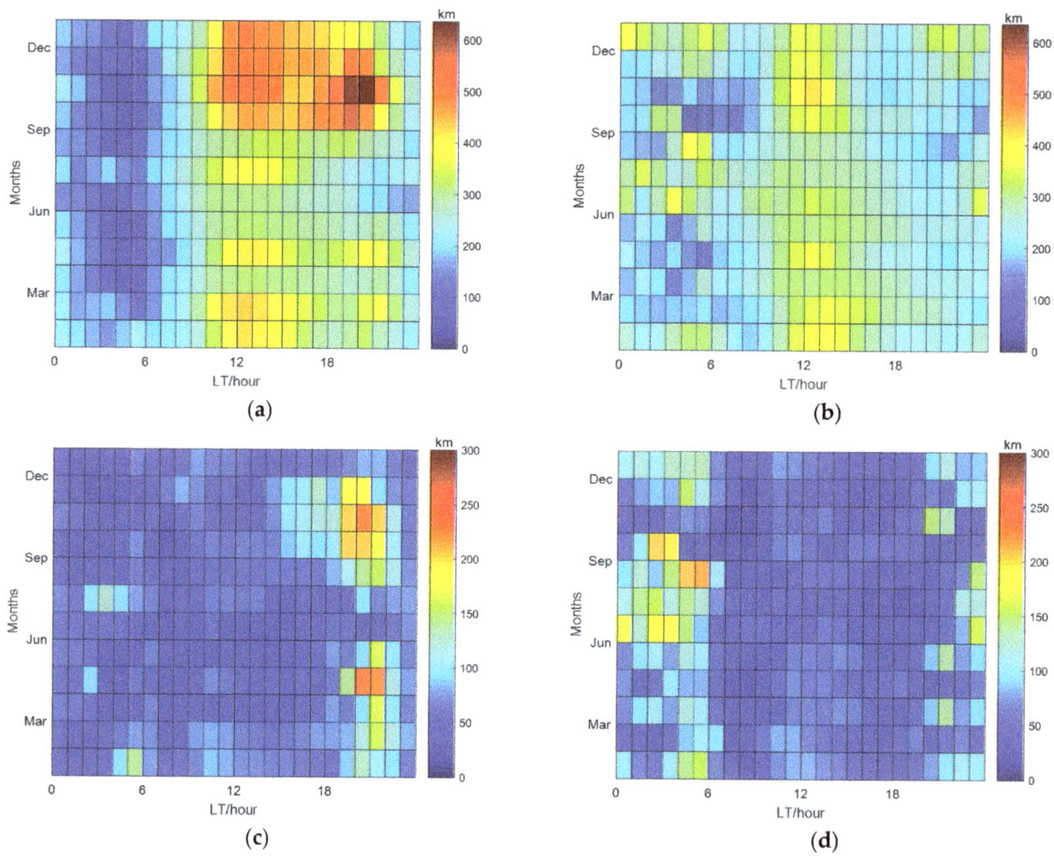

Figure 4. Map of mean and standard deviation of equivalent slab thickness (τ) under geomagnetically quiet condition. (**a**) Mean of τ in high solar activity year 2014; (**b**) Mean of τ in low solar activity year 2016; (**c**) Standard deviation of τ in high solar activity year 2014; (**d**) Standard deviation of τ in low solar activity year 2016.

Taking into account the rapid changes during the sunrise and sunset period, mean daytime (08LT–16LT) and night-time (20LT–04LT) values of τ, for magnetically quiet days during the solar maximum phase 2014 and the solar minimum phase 2016, for three seasons (summer, winter, and equinox) are calculated (Table 1) to investigate the dependence of τ on solar activity. It can be seen from the table that, regardless of the high and low years of solar activity, the τ during the day is greater than the τ at night, and the seasonal characteristics of daytime τ (the τ in winter and equinox are greater than the summer τ) are more obvious in the years of high solar activity. The most important finding is that the mean τ during the day, in the high solar activity, is greater than that in the low solar activity, whereas the opposite situation applies for the τ at night.

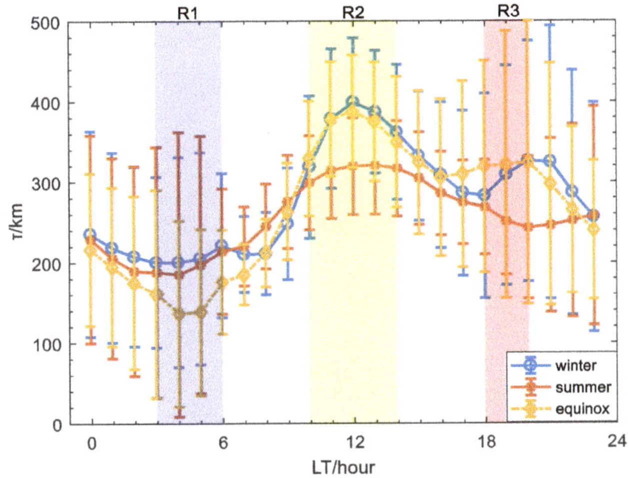

Figure 5. Diurnal variations of mean and standard deviation of equivalent slab thickness (τ) in winter, summer, and equinox at geomagnetically quiet condition.

Table 1. Mean daytime (08:00–16:00LT) and night-time (20:00–04:00LT) values of ionospheric equivalent slab thickness (τ) for magnetically quiet days during the high solar activity year 2014 and the low solar activity year 2016.

Year	Seasons	Mean Daytime Values of τ (km)	Mean Nighttime Values of τ (km)
2014	Winter	388.9	232.1
	Summer	303.9	188.9
	Equinox	364.6	250.7
2016	Winter	299.3	255.5
	Summer	288.6	238.5
	Equinox	301.2	268.4

The ionosphere is not the same every day, even under undisturbed geomagnetic and solar conditions. In order to compare storm-time τ variation with quiet time τ variations. Figure 6 displays the mean DI index during quiet-time. In this case, the index is

$$DI = \tau_q/\tau_s - 1$$

where τ_q and τ_m are quiet-time and monthly median τ, respectively. According to if the DI is positive or negative, the daily variation of mean and standard deviation of DI index are shown in two panels, respectively. The blue curves represent the positive DI variation, and the red curves represent the negative DI variation. It is important to note that thick lines correspond to the average values of positive and negative DIs for the entire period, and vertical lines correspond to STD. As it shown, the τ is most stable in the daytime (08LT–19LT), and the DI index ranges from −0.12 to 0.14 during this period. On the contrary, the τ is more variable during the period of night-time to sunrise period (22LT–07LT) with DI index more/less than 0.19/−0.15. The τ is most variable during the sunrise period, with maximum/minimum having reached 0.46/−0.3. It is consistent with the above results that the great variability of τ at sunrise, indicating that the τ values fluctuate sharply around sunrise. Previous observations and model simulations also illustrate that the day-to-day variability reaches the maximum at dawn, and they suggested that the day-to-day variability of neutral winds in the E region (≤~130 km) is the primary driver of the day-to-day variability of dawn ionosphere at the geomagnetic equator [48–51].

For the post sunset period (19LT–22LT), the DI increases/decreases from 0.12/−0.11 to 0.23/−0.15, whereas it decreases/increases from 0.21/−0.2 to 0.11/−0.11 during the post sunrise period (07LT–08LT).

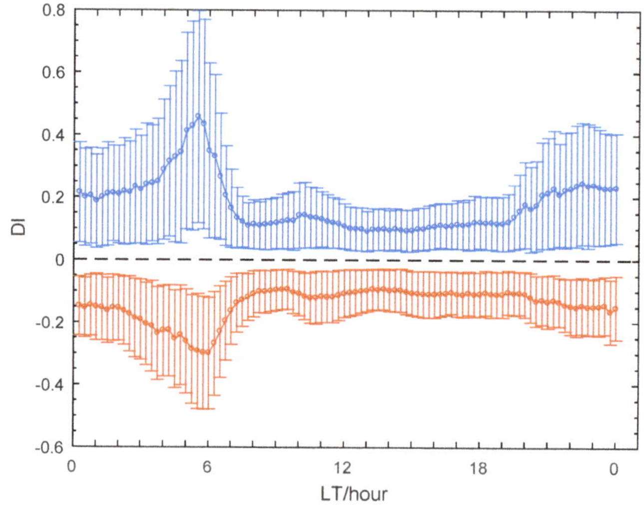

Figure 6. Diurnal variation of the mean and standard deviation (STD) of DI index at geomagnetically quiet conditions.

Figure 7 presents the storm-time DI variation. The storm-time DI pattern is basically consistent with that of quiet-time but with more variability. During the daytime (08LT–19LT), the τ has smaller variability, and the DI indices are between −0.2 to 0.27. By contrast, the τ is more variable during the nighttime to sunrise period (23LT–07LT), with DI indices more/less than 0.287/0.168. In addition, the τ is also most variable during the sunrise period (05LT–07LT), and maximum/minimum DI index are −0.45 and 0.88, respectively. For the post-sunset period (19LT–21LT), the DI increases/decreases from 0.22/−0.15 to 0.44/−0.17, and it decreases/increases from 0.26/−0.19 to 0.15/−0.12 during the sunrise period (07LT–08LT).

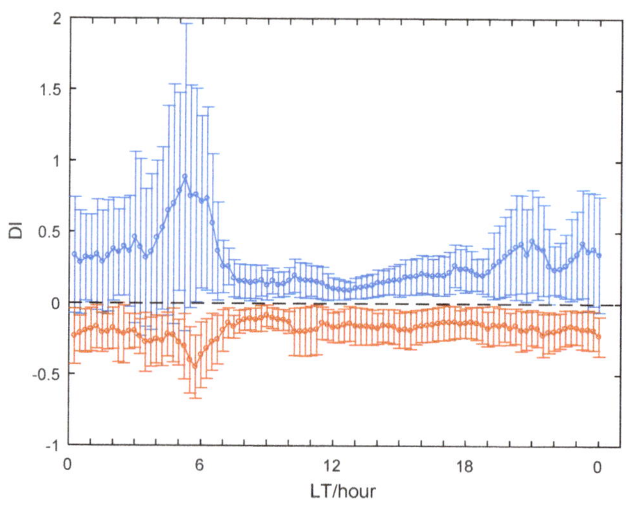

Figure 7. Same as Figure 6 but for geomagnetic storm periods.

To better illustrate the geomagnetic effect on the τ, Figure 8a compares the storm-time and quiet-time DI variations. One can find that the storm-time DI is more perturbed than quiet-time DI. Most of the storm-time positive DI are larger than quiet-time positive DI values, and this phenomenon is more apparent during night-time, especially during sunrise period. On the other hand, the storm-time negative DI are a little smaller than quiet-time negative DI values, except the pre-dawn period (4:30LT–6:30LT) with large difference. In order to illustrate the difference between them, we calculated the DI difference based on the sign of the DI index, and the result is shown in Figure 8b. As can be seen from Figure 8b, the positive DI difference during the storm is below 0.106 from 08LT to 19LT during the day. It is relatively large at sunrise (04LT–07LT) and sunset (19LT–21LT), reaching 0.45 and 0.24, respectively. The negative DI difference is less than −0.08 in all time periods, except for the period of sunrise, which reached −0.15 between 05LT–06LT. Therefore, we preliminarily draw the conclusion that geomagnetic storms have positive effects on the τ during most of the storm period, except they have negative effects on the τ during the sunrise period (4:30LT–6:30LT) in some cases.

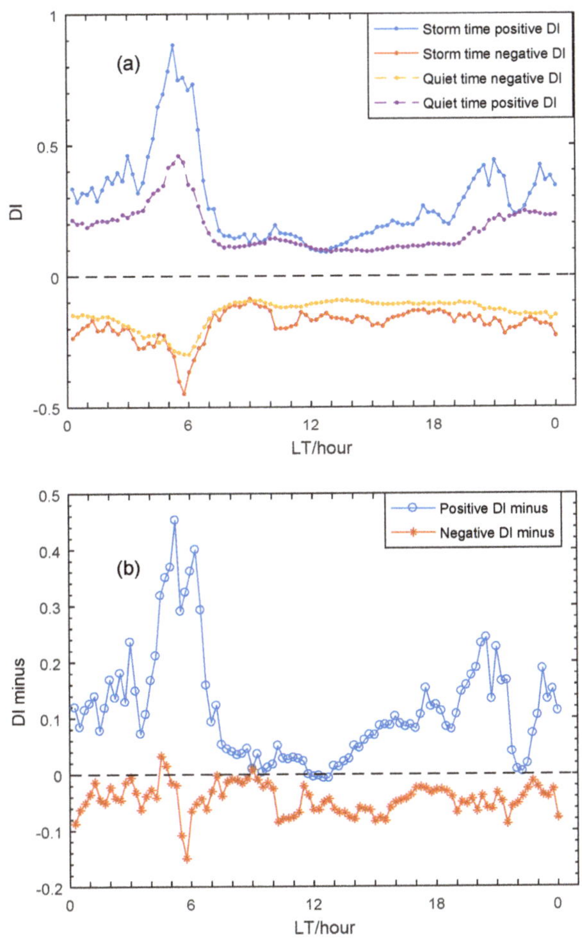

Figure 8. (**a**) Comparison of diurnal DI index for geomagnetically quiet condition and geomagnetic storm condition (**b**) The minus between the DI index during geomagnetic storm periods and DI index at geomagnetically quiet time, according to the sign of DI.

4. Discussion

The Guam station is located near the magnetic equator. The most important factor affecting local ionospheric state is the fountain effect and the ionospheric equatorial anomaly (EIA) caused by it. At the same time, the geographic latitude of Guam Station is 13.43°N, and the trans-equator neutral wind can affect its ionospheric state significantly. Strong geomagnetic storms lead to positive or negative ionospheric storms, depending on season, local time, and the universal time (UT) arrival time of storms [52–55]. Different scales of neutral atmosphere waves, such as gravity waves, thermospheric tides, and planetary waves, driven by lower atmosphere processes, can also influence the ionosphere directly through plasma transport, or indirectly through electrodynamic processes [56,57]. Therefore, the τ in this paper displays a complicated pattern. Considering that geomagnetic activity is the main contributor to the relative ionospheric variability, we first discussed the τ under geomagnetic quiet conditions in Guam.

In order to explain the diurnal variation of τ, Figure 9 gives the quiet-time variation of TEC and NmF2 in Guam. In addition, it can be easily determined that the local time variation of τ depends on the variation in TEC and NmF2, obtained from Equation (1), as the following:

$$\frac{d\tau}{dt} = \frac{1}{NmF2}\frac{dTEC}{dt} - \frac{TEC}{NmF2^2}\frac{dNmF2}{dt} \tag{3}$$

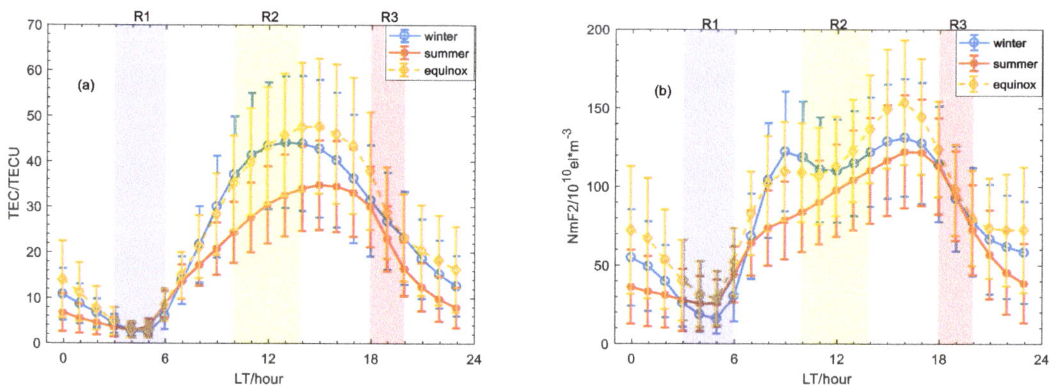

Figure 9. (**a**) Diurnal variations of mean and standard deviation of total electron content (TEC) in Guam during winter, summer, and equinox, respectively. (**b**) Diurnal variations of mean and standard deviation of F2-layer peak electron density (NmF2) in Guam during winter, summer, and equinox, respectively.

To better understand the possible formation mechanisms of the peak of τ, the temporal variation in the ionospheric parameters, TEC and NmF2, are given in Figure 10. According to Equation (3), the increase in the ratio variation of τ can be divided into following categories: (1) TEC increase and NmF2 decrease; (2) TEC and NmF2 increase, but the ratio of TEC increase is larger than that of NmF2; (3) TEC and NmF2 decreases but the ratio of NmF2 decrease is larger than that of TEC. The opposite situation applies for the decrease in τ. As shown in Figure 9, TEC and NmF2 consistently increase with the solar zenith angle, which increases after sunrise. Meanwhile, new electrons produced by the photoionization, in the magnetic equatorial ionosphere, began to be transported upward and diffused downward due to the fountain effect. Considering that the photoionization plays a much more important role in controlling the overall variability in the integrated quantity (i.e., TEC) than a localized quantity (i.e., NmF2) [58], the increases of ratio in TEC are larger than that in NmF2, as presented by Figure 10, and it causes the increases in τ during the forenoon.

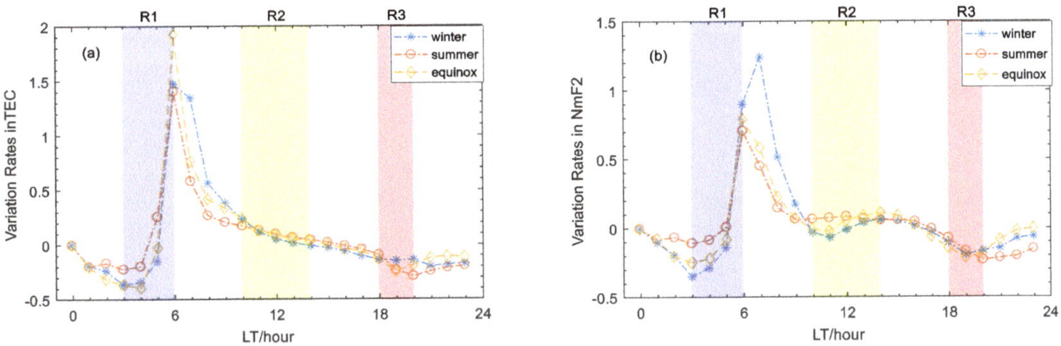

Figure 10. (**a**) Variation rates in TEC for winter, summer, and equinox, respectively. (**b**) Variation rates in NmF2 for winter, summer, and equinox, respectively.

Figure 5 shows that the τ in Guam reaches its maximum value at noontime, consistent with previous results for equatorial latitudes [24,25,30,31]. Moreover, Figure 5 also shows that the first peak of τ, in winter and equinox, are larger than in summer and this phenomenon is a new finding of this study. It is observed from Figure 9b that *Nm*F2 in winter and equinox show 'noon bite-out' during the period of first τ peak, whereas TEC do not show this phenomenon during the full year (R2 interval). Apparently, these variations, together, result in seasonal minimum of the first peak in summer, according to Equation (1). Noon bite-out in the low latitude to low equator has long been known, and previous studies illustrate that this feature is mainly caused by the upward $\mathbf{E} \times \mathbf{B}$ drifts in the F region during daytime and noon bite-out was absent in the daytime TEC variation [59,60]. Lee [61] used TEC, *Nm*F2, and *hm*F2 data, observed by the Jicamarca Digisonde and SAMI2, to model the noon bite-out under geomagnetic quiet condition and found that the reason why the noontime bite-out in TEC is absent is, principally, the upward $\mathbf{E} \times \mathbf{B}$ drifts because the drifts mainly account for the absent noontime bite-out in bottomside TEC and topside TEC. As for the absence of noon bite-out in summer, it might be caused by the neutral winds in Guam. Chen et al. [62] also suggested that the reason is seasonally dependent trans-equator plasma transport, induced by neutral winds. Since the meridional winds are equatorward during the summer daytime in Guam, plasma outflow, induced by trans-equator transport, results in significantly electron density depletion before the fountain effect induced bite-out. As a result, the pre-noon *Nm*F2 peak is restrained. Thus, noontime bite-out variation pattern does not appear in summer in Guam.

As the solar zenith angel decreases and vertical drift velocity takes a downward trend, the electron density continues to decrease after noontime. This indicates a reduction in the *Nm*F2 and TEC after midday, as illustrated in Figure 9. As described above, the photoionization plays a much more important role in controlling the overall variability in the TEC than *Nm*F2. The variation of *Nm*F2 is, therefore, less sensitive to the solar zenith angle compared to TEC. As shown in Figure 10, the ratio in decrease in TEC is larger than that in *Nm*F2, which causes the decrease in τ in the afternoon.

Figure 5 depicts that there is another peak of τ that appears around sunset in winter and equinox. As shown in Figure 9, TEC and *Nm*F2 continue to decrease during this period, due to the solar radiation decreases (R3 interval). It is the fact that the ratio of decreasing in TEC is smaller than that in *Nm*F2 results in the increase in τ, based on Equation (3) and Figure 10. It is generally accepted that pre-reversal enhancement (PRE), which refers to the strong upward $\mathbf{E} \times \mathbf{B}$ drift during the post-sunset period, appears from the equator to the low latitude region. Moreover, the electrodynamic process, such as $\mathbf{E} \times \mathbf{B}$, has stronger and more direct control only in the F-region height and density. Ionosphere generally exists at heights between 60–1000 km, whereas the plasmasphere could reach tens of thousands kilometers. Therefore, the plasmasphere receives more solar radiation than ionosphere after sunset, and TEC thus decreases slower than *Nm*F2. Based on these scenarios, there is an increase in τ during sunset period. In addition, night-time enhancement in electron density, at low latitudes, has been studied by several groups of workers and the post sunset enhancement is known to be caused by the PRE. It is necessary to note the night-time enhancement in electron density is said to occur if its value is greater than the exponentially decaying background value. Su et al. [63] reported that the frequency of occurrence of the post-sunset TEC enhancement has two maxima in equinox, and these equinoctial maxima are separated by a broad summer minimum at the eastern low-latitude station Taiwan (11.5°geomagnetic latitude). Therefore, the post-sunset τ does not illustrate an increase in summer, whereas they show another peak in winter and equinox.

Agrawal et al. [64] has shown that pre reversal increase in the vertical $\mathbf{E} \times \mathbf{B}$ drift, the primary source for the electron density enhancement, is strongest at equinox in both the eastern and western longitudes, and the night-time enhancement in *Nm*F2 and TEC is, therefore, greatest in equinox, as displayed by Figure 9. However, there is no solar radiation after sunset, and F region only constitutes a part of space environment when evaluating TEC. Therefore, the TEC enhancement is not apparent as *Nm*F2 enhancement. It causes the equinox/solstice ratio in *Nm*F2 to grow larger than that in TEC during the post-midnight to sunrise period, as shown in Figure 9. Figure 9 also shows that the post-midnight enhancement is most obvious in summer, consistent with previous studies. As the recombination process takes place steadily, one can find that the TEC and *Nm*F2 continues to decreases during the post-midnight period, respectively. In addition, the downward transport of plasma will slow the decrease in *Nm*F2 in the nighttime. Therefore, the decrease in τ to the ratio of decrease in TEC is larger than that in *Nm*F2, as shown in Figure 10. Moreover, the τ in equinox is smallest, which is caused by the aforementioned largest equinoctial *Nm*F2 and the faster decrease in TEC in equinox than in solstice, as shown in Figures 9 and 10, respectively.

In previous studies, the sunrise peak of τ has been a hot topic [21,22,26,33]. These studies indicate that the pre sunrise peak is a widely observed feature from low to high latitudes, and τ even reaches a maximum at sunrise for specific seasons. However, Figure 5 shows that, starting from 04LT in winter and summer, and 05LT in equinox, the τ gradually increases, and τ gets very low values (around 100 km) during the pre-sunrise period. For equatorial latitudes, the similar phenomenon was also seen for other longitudes, as follows. Àlàgbé [30] studied the correlation between the τ (TF2 in his paper) and B0 at the Burkina Faso station (12.4°N, 1.5°W, dip 5.9°N), and they found that the τ at Burkina Faso station (12.4°N, 1.5°W, dip 5.9°N) seemed to also have a very low value (less than 100 km) at pre-sunrise period. Duarte-Silva et al. [31] have also found that the Palmas station (10.12°S, 48.21°W, 7.73°S dip lat), located in the inner edge of the anomaly region, also seems to get very low values around the pre-sunrise period. Odeyemi et al. [24,25] have studied the τ over Africa Ilorin (8.50°N, 4.68°E), and they found that the τ also seems to get very low values on some months during the low solar activity period. Therefore, the very low values at pre-sunrise hours might not be a special phenomenon in Guam because it could also be seen at other longitudinal equatorial latitude station. Moreover, the TEC should give a more pronounced contribution than *Nm*F2 during the pre-sunrise period. Therefore, the τ begins to increase during the pre-sunrise period, as can be seen in Figure 5, but it does not reach a peak. To further understand the morphology of τ during the pre-sunrise period, more observations and modelling are needed to provide a statistical picture and study the physical mechanism of this phenomenon in the future.

Among previous studies on τ, the dependence of τ on solar activity has shown different correlations in different latitudes [16,19]. Àlàgbé [30] shows that the day-time τ on the magnetic equator, in the high solar activity, is significantly greater than that in the low solar activity. Jayachandran et al. [21] also found that τ at noon is positively correlated with solar activity when studying τ in low and middle latitudes. The results shown in Table 1 confirm their conclusions. As for the τ decreases with solar activity at night, it should be due to the fact the H+-O+ transition height in the low solar activity is lower than that in the high solar activity at night. It is known that the τ is sensitive to the variations of H+/O+ ratio at the F2 peak or equivalent to the transition level at which [O+] = [H+]. Large downward fluxes of H+ can decrease the O+ to H+ transition levels, thereby increasing the topside content and, hence, the slab thickness.

Geomagnetic activity is another key factor affecting the perturbations of the τ. Previous studies have shown that the influence of geomagnetic activity on the τ depends on the location of the observing station and solar activity. For the mid latitude, the τ seems to systematically enhance during periods of geomagnetic disturbances [15,32,65]. However, magnetic activity does not appear to have significant influence on the τ variations at low latitude [22,66]. Until now, there is no specific statistical study analyzing the geomagnetic storms effect on the τ at the near equator latitudes (best to our knowledge). Therefore, we made a statistical study of the geomagnetic storms effect on the τ in Guam, based on the aforementioned DI index during the geomagnetic storm period from the year 2010 to 2017. Comparing with the geomagnetic quiet time DI index shown in Figure 6, the results in Figure 8 suggest that the geomagnetic storm seems to have a positive effect on the τ during most of the storm period near the equatorial station Guam.

During geomagnetic storms, the main process affecting τ (TEC/*Nm*F2) in Guam includes: (1) penetration electric field driven by the solar wind ranging from high latitude to equator; (2) the equator-ward neutral wind resulting from particle precipitation and Joule heating at high latitude, sometimes accompanied with travelling atmosphere disturbance (TAD); (3) the disturbance dynamo electric fields produced by the globally altered thermospheric winds during magnetic storms; (4) composition changes, driven by storm time, altered neutral winds. These coupled drivers together with background thermosphere-ionosphere generate ionospheric storms and control their strengths in complicated ways. In the following, we will start from these aspects and use an example on the 1 June 2013 in Guam to analyze why the τ tends to increase when the positive or negative ionospheric storm occurs.

Figure 11 shows the geophysical conditions during 1–2 June 2013. As IMF Bz turns southward at 0:00UT, Dst began to decrease sharply, reaching a minimum of −124 nT at 8:00UT, and then, the geomagnetic storm entered the recovery phase. The DI indices of τ, TEC and *Nm*F2 of 1 June are given in Figure 12. As shown in the figure, with the start of the main phase, caused by southward reversal of IMF Bz, a TEC positive storm occurred during the main phase, but the increase in *Nm*F2 during this period was very small, resulting in a τ positive storm during this period (T1 interval in Figure 12).

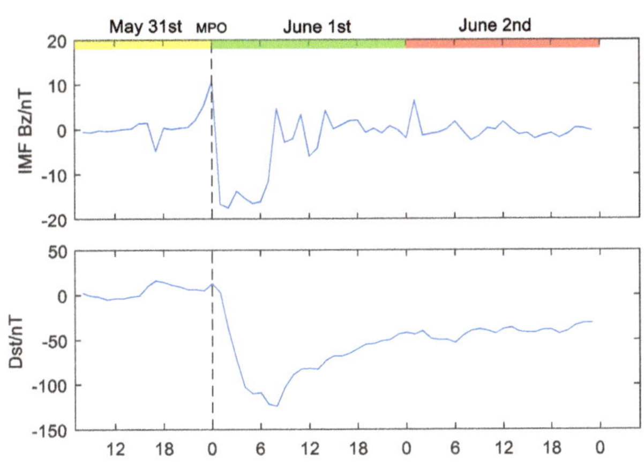

Figure 11. The geophysical conditions during 1–2 June 2013.

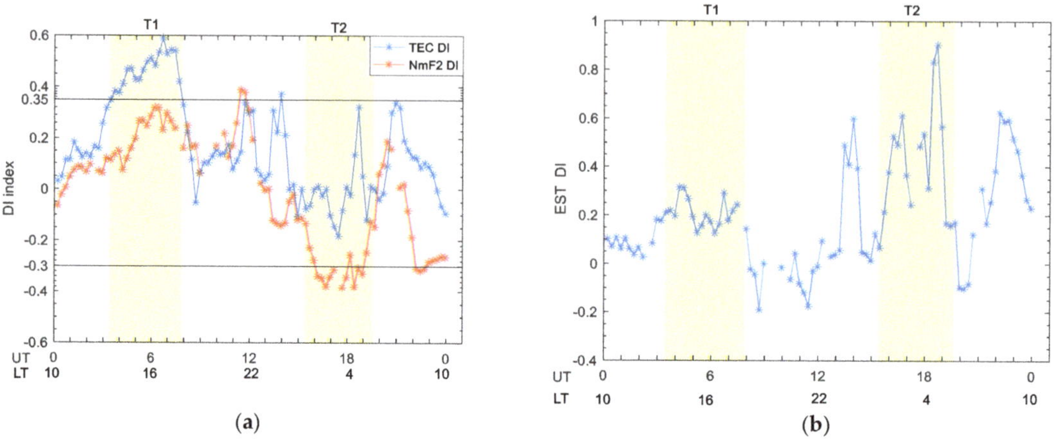

Figure 12. (**a**) The DI index of TEC and *Nm*F2 on 1 June 2013. (**b**) The DI index of τ on 1 June 2013.

It is generally accepted that the penetration electric field and equatorward neutral wind should be the primary drivers of the positive disturbance at this latitude [43,58]. From the fact that TEC positive storm occurred during the main phase of the magnetic storm, caused by the southward reversal of IMF BZ in Figure 12, we infer that there is a penetrating electric field during this period. The daytime penetrating electric field tends to move the plasma from the equatorial region into higher latitudes through the fountain effect. Meanwhile, the increased upward vertical drifts also transport local plasma in the bottomside ionosphere or the F2 layer to higher altitudes where the chemical recombination rate is low, so plasma accumulates in the topside ionosphere. On the other hand, the plasma, transported upward from the bottomside ionosphere, would be compensated by daytime ionization. As a result, a large increase in TEC is accompanied by a rather small increase in *fo*F2 and τ tends to enhance during the positive ionospheric storm, as shown in T1 interval. In order to confirm above theory, Figure 13 shows the DI index of ionospheric bottomside TEC (BTEC) and topside TEC (TTEC) relative to their monthly median value on 1 June. Among them, BTEC is obtained by integrating the electron density below the ionospheric peak height *hm*F2, while TTEC is obtained by subtracting BTEC from the TEC. Figure 13 shows that the ionospheric positive storm in this time period is mainly caused by the TTEC, while the BTEC gives little contribution. From the perspective of plasma scale height, which has a close relationship with τ, the increased upward drifts changed the shape of the topside ionosphere and increased the topside effective plasma scale height dh/d(ln(Ne)) significantly, so the whole τ is thus increased. Moreover, the enhanced equatorward wind during storm time should also play an important role in producing changes in electron densities, as it can raise the F2 region to a higher altitude to inhibit field-aligned ambipolar diffusion, thus causing changes in the shape of the ionospheric density profile [67,68].

As the geomagnetic storm entered the recovery phase, *Nm*F2 experienced a negative storm, but the disturbance of TEC during this period was small, resulting in a positive disturbance of τ (T2 interval in Figure 12). It is well known that the disturbance dynamo electric fields produced by the globally altered thermospheric winds during magnetic storms and composition changes brought by equatorward neutral winds are the main drivers of negative storms in the region we are interested in [54]. However, TEC and *Nm*F2 might have contrasting behavior during the recovery phase of the geomagnetic storm, as shown in Figure 12. Since the topside ionosphere, at equatorial latitudes, connected with middle and low latitudes of the region near the F2 peak at the flux tube, the storm time enhanced equatorward wind tends to push more plasma at low latitudes from the region near the F2 peak into the topside ionosphere in the equatorial region [69]. Consequently, the topside TEC in the equatorial region can undergo an obvious enhancement, and it causes τ to increase, combining with the negative disturbance of *Nm*F2. Figure 13 confirms this theory, BTEC experienced a large negative disturbance, but TTEC did not, and even was in a positive disturbance state for most of the time during T2 interval. Therefore, the positive disturbance of TTEC, caused by equatorial neutral wind during a storm, makes TEC decrease little when *Nm*F2 has a negative storm occur, resulting in an increase in τ.

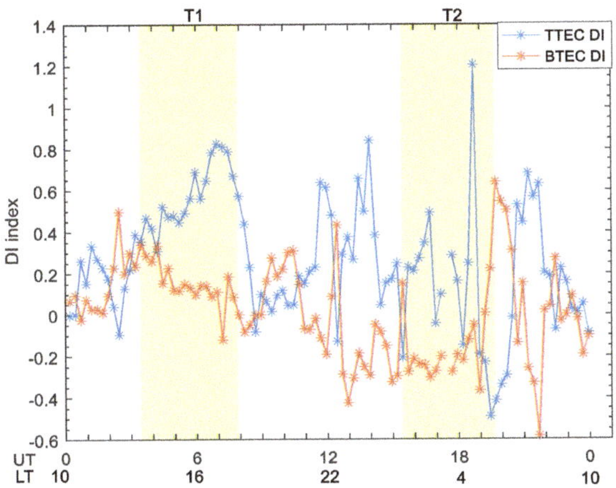

Figure 13. The DI index of TTEC and BTEC on 1 June 2013.

5. Conclusions

Based on the TEC and *Nm*F2 data from the years 2012–2017, this paper statistically analyzed the τ at equatorial latitude Guam. The results show great diurnal, seasonal, solar, and geomagnetic activity variation. A brief review of observations made by other researchers has also been presented, and we obtained the following results, which confirm similar studies:

1. The peak of τ appeared at noon, consistent with previous studies on equatorial latitudes.
2. There is a post-sunset peak in τ observed during the winter and equinox, and it means that *Nm*F2 is decreasing faster than TEC, which can be associated with the higher post-sunset TEC enhancement occurrence in equinox/winter, proving previous conclusions. In addition, the τ continues to decrease after post-midnight and the equinoctial one is smallest due to the largest nighttime *Nm*F2 enhancement in equinox and the faster decrease in TEC in equinox than that in solstice.
3. The dependence of τ on the solar activity are different for daytime and nighttime: the daytime τ seems to increase with solar activity, as TEC is more sensitive to the solar activity than *Nm*F2, whereas the nighttime one decreases with solar activity at night, and it should be due to the fact the H+-O+ transition height in the low solar activity is lower than that in the high solar activity at night.
4. The τ has more variability during nighttime than daytime, during both the geomagnetically quiet and disturbed conditions, and the greatest variability of τ appeared at sunrise.

Moreover, we obtained some new results, which provide interesting insights into τ of this region:

1. The τ at noon is larger in winter and equinox than in summer. It is probably due to the absence of *Nm*F2 noontime bite-out in summer at this region.
2. There is no pre sunrise peak in τ and τ get low values during the pre-sunrise period at Guam. However, previous studies indicate that pre sunrise peak is a widely observed feature, from low to high latitude, and τ even reaches maximum at sunrise for specific seasons. The contradiction is probably due to Guam being located at equatorial latitude, as the low values in the pre sunrise period could also be seen at

other longitudinal equatorial latitude station. In addition, longitudinal difference might also contribute to the difference.

3. The geomagnetic storm seems to have a positive effect on the τ during most of the storm period in Guam, except at sunrise period, when the τ attains large variability, even at the geomagnetically quiet condition. This study also provides a new physical explanation for the observed effect of geomagnetic storm on τ in Guam. During the positive storms, the penetration electric field along with storm time equator-ward neutral wind tends to increase upward drift and uplift F region, causing the large increase in TEC accompanied by relatively small increase in *Nm*F2. On the other hand, an enhanced equatorward wind tends to push more plasma, at low latitudes, into the topside ionosphere in the equatorial region, resulting in the TEC, which does not undergo severe depletion as *Nm*F2 does during the negative storms. Therefore, the geomagnetic storm seems to enhance τ both during the positive and negative ionospheric storms.

Author Contributions: Conceptualization, Y.Z., Z.W. and T.X.; methodology, Z.D.; investigation, J.F. and Y.Z.; validation, Y.Z.; formal analysis, Y.Z. and W.X.; resources, M.O.; visualization, Z.D. and W.Z.; funding acquisition, M.O. All authors have read and agreed to the published version of the manuscript.

Funding: This work was supported by the National Key R&D Program of China(Grant No. 2018YFF 0103700), Natural Science Basic Research Program of Shaanxi (Program No. 2020JQ-331).

Institutional Review Board Statement: Not applicable.

Informed Consent Statement: Not applicable.

Data Availability Statement: The use of ionosonde data provided by GIRO (Global Ionospheric Radio Observation) database (http://spase.info/SMWG/Observatory/GIRO accessed on 23 November 2021). The use of GPS data provided by UNAVCO database (University NAVSTAR Consortium) (http://www.unavco.org/ accessed on 23 November 2021).The use of Dst data provided by WDC (http://wdc.kugi.kyoto-u.ac.jp/ accessed on 23 November 2021). The use of F10.7 data provided by NOAA (ftp://ftp.swpc.noaa.gov/pub/indices/old_indices/ accessed on 23 November 2021).

Acknowledgments: We acknowledge the use of ionosonde data provided by GIRO (Global Ionospheric Radio Observation) database (http://umlcar.uml.edu/DIDbase accessed on 23 November 2021), GPS data provided by UNAVCO database (University NAVSTAR Consortium) (http://www.unavco.org/ accessed on 23 November 2021), Dst data provided by WDC (http://wdc.kugi.kyoto-u.ac.jp/ accessed on 23 November 2021) and F10.7 data provided by NOAA (ftp://ftp.swpc.noaa.gov/pub/indices/old_indices/ accessed on 23 November 2021).

Conflicts of Interest: The authors declare no conflict of interest.

References

1. Breed, A.M.; Goodwin, G.L.; Vandenberg, A.-M.; Essex, E.A.; Lynn, K.J.W.; Silby, J.H. Ionospheric total electron content and ionospheric slab thickness determined in Australia. *Radio Sci.* **1997**, *62*, 1635–1643. [CrossRef]
2. Wright, J.W. A model of the F-region above hmaxF2. *J. Geophys. Res.* **1960**, *65*, 185–191. [CrossRef]
3. Titheridge, J.E. The ionospheric slab thickness of the mid-latitude ionosphere. *Planet Space Sci.* **1973**, *21*, 1775–1793. [CrossRef]
4. Jakowski, N.; Putz, E.; Spalla, P. Ionospheric storm characteristics deduced from satellite radio beacon observations at three European stations. *Ann. Geophys.* **1990**, *8*, 343–352.
5. Jakowski, N.; Mielich, J.; Hoque, M.M.; Danielides, M. Equivalent ionospheric slab thickness at the mid-latitude ionosphere during solar cycle 23. In Proceedings of the 38th COSPAR Scientific Assembly, Bremen, Germany, 18–25 July 2010.
6. Jakowski, N.; Hoque, M.M.; Mielich, J.; Hall, C. Equivalent ionospheric slab thickness of the ionosphere over Europe as an indicator of long-term temperature changes in the thermosphere. *J. Atmos. Terr. Phys.* **2017**, *163*, 92–101.
7. Huang, Z.; Yuan, H. Climatology of the ionospheric slab thickness along the longitude of 120° E in China and its adjacent region during the solar minimum years of 2007–2009. *Ann. Geophys.* **2015**, *33*, 1311–1319. [CrossRef]
8. Maltseva, O.A.; Mozhaeva, N.S.; Nikitenko, T.V. Comparison of model and experimental ionospheric parameters at high latitudes. *Adv. Space Res.* **2013**, *51*, 599–609. [CrossRef]
9. Maltseva, O.A.; Mozhaeva, N.S.; Nikitenko, T.V. Validation of the Neustrelitz Global Model according to the low latitude ionosphere. *Adv. Space Res.* **2014**, *54*, 463–472. [CrossRef]

10. Maltseva, O.A.; Mozhaeva, N.S. The Use of the Total Electron Content Measured by Navigation Satellites to Estimate Ionospheric Conditions. *Int. J. Nav. Obs.* **2016**, *2016*, 7016208. [CrossRef]
11. Froń, A.; Galkin, I.; Krankowski, A.; Bilitza, D.; Hernández-Pajares, M.; Reinisch, B.; Li, Z.; Kotulak, K.; Zakharenkova, I.; Cherniak, I.; et al. Towards Cooperative Global Mapping of the Ionosphere: Fusion Feasibility for IGS and IRI with Global Climate VTEC Maps. *Remote Sens.* **2020**, *12*, 3531. [CrossRef]
12. Gerzen, T.; Jakowski, N.; Wilken, V.; Hoque, M.M. Reconstruction of F2 layer peak electron density based on operational vertical total electron content maps. *Ann. Geophys.* **2013**, *31*, 1241–1249. [CrossRef]
13. Krankowski, A.; Shagimuratov, I.I.; Baran, L.W. Mapping of foF2 over Europe based on GPS-derived TEC data. *Adv. Space Res.* **2007**, *39*, 651–660. [CrossRef]
14. Bhonsle, R.V.; Da Rosa, A.V.; Garriott, O.K. Measurement of Total Electron Content and the Equivalent ionospheric slab thickness of the Mid latitude Ionosphere. *Radio Sci.* **1965**, *69*, 929–937.
15. Kersley, L.; Hajeb-Hosseinieh, H. Dependence of ionospheric slab thickness on geomagnetic activity. *J. Atmos. Terr. Phys.* **1976**, *38*, 1357–1360. [CrossRef]
16. Jin, S.; Cho, J.-H.; Park, J.-U. Ionospheric slab thickness and its seasonal variations observed by GPS. *J. Atmos. Sol. Terr. Phys.* **2007**, *69*, 1864–1870. [CrossRef]
17. Huang, Y.N. Some results of ionospheric slab thickness observations at Lunping. *J. Geophys. Res.* **1983**, *88*, 5517–5522. [CrossRef]
18. Davies, K.; Liu, X.M. ionospheric slab thickness in middle and low latitudes. *Radio Sci.* **1991**, *26*, 997–1005. [CrossRef]
19. Minakoshi, H.; Nishimuta, I. Ionospheric electron content and equivalent ionospheric slab thickness at lower mid-latitudes in the Japanese zon. In *Proc. Beacon Satellite Symposium (IBSS)*; University of Wales: Wales, UK, 1994; Volume 144.
20. Gulyaeva, T.L.; Jayachandran, B.; Krishnankutty, T.N. Latitudinal variation of slab thickness. *Adv. Space Res.* **2004**, *33*, 862–865. [CrossRef]
21. Jayachandran, B.; Krishnankutty, T.; Gulyaeva, T. Climatology of ionospheric slab thickness. *Ann. Geophys.* **2004**, *22*, 25–33. [CrossRef]
22. Pignalberi, A.; Nava, B.; Pietrella, M.; Cesaroni, C.; Pezzopane, M. Midlatitude climatology of the ionospheric equivalent slab thickness over two solar cycles. *J. Geod.* **2021**, *95*, 124. [CrossRef]
23. Guo, P.; Xu, X.; Zhang, G.X. Analysis of the ionospheric equivalent ionospheric slab thickness based on ground-based GPS/TEC and GPS/COSMIC RO. *J. Atmos. Sol. Terr. Phys.* **2011**, *73*, 839–846. [CrossRef]
24. Odeyemi, O.O.; Adeniyi, J.O.; Oladipo, O.A.; Olawepo, A.O.; Adimu, I.A.; Oyeyemi, E.O. Ionospheric slab thickness investigation on slab-thickness and B0 over an equatorial station in Africa and comparison with IRI model. *J. Atmos. Sol. Terr. Phys.* **2018**, *179*, 293–306. [CrossRef]
25. Odeyemi, O.O.; Adeniyi, J.O.; Oyeyemi, E.O.; Panda, S.K.; Jamjareegulgarn, P.J.; Olugbon, B.; Oluwadare, E.J.; Akala, A.O.; Olawepo, A.J.; Adewale, A.A. Morphologies of ionospheric-equivalent slab thickness and scale height over equatorial latitude in Africa. *Adv. Space Res.* **2021**, in press. [CrossRef]
26. Huang, H.; Liu, L.; Chen, Y.; Le, H.; Wan, W. A global picture of ionospheric slab thickness derived from GIM TEC and COSMIC radio occultation observations. *J. Geophys. Res. Space Phys.* **2016**, *121*, 867–880. [CrossRef]
27. Fox, M.W.; Mendillo, M.; Klobuchar, J.A. Ionospheric equivalent ionospheric slab thickness and its modeling applications. *Radio Sci.* **1991**, *26*, 429–438. [CrossRef]
28. Spalla, P.; Ciraolo, L. TEC and foF2 comparison. *Ann. Geophys.* **1994**, *37*, 929–938. [CrossRef]
29. Jakowski, N.; Hoque, M.M. Global equivalent ionospheric slab thickness model of the Earth's ionosphere. *J. Space Weather Space Clim.* **2021**, *11*, 10. [CrossRef]
30. Àlàgbé, G.A. Quiet- and storm-time correlation of F2-layer ionospheric slab thickness and B0 at an equatorial station. *Int. J. Res. Rev. Appl. Sci.* **2012**, *13*, 133–138.
31. Duarte-Silva, M.H.; Muella, M.T.A.H.; Silva, L.C.C.; de Abreu, A.J.; Fagundes, P.R. Ionospheric slab thickness at the equatorial anomaly region after the deep solar minimum of cycle 23/24. *Adv. Space Res.* **2015**, *56*, 1961–1972. [CrossRef]
32. Gulyaeva, T.; Stanislawska, I. Night-day imprints of ionospheric slab thickness during geomagnetic storm. *J. Atmos. Sol. Terr. Phys.* **2005**, *67*, 1307–1314. [CrossRef]
33. Stankov, S.M.; Warnant, R. Ionospheric slab thickness—Analysis, modelling and monitoring. *Adv. Space Res.* **2009**, *44*, 1295–1303. [CrossRef]
34. Uwamahoro, J.C.; Giday, N.M.; Habarulema, J.B.; Katamzi-Joseph, Z.T.; Seemala, G.K. Reconstruction of storm-time total electron content using ionospheric tomography and artificial neural networks: A comparative study over the African region. *Radio Sci.* **2018**, *53*, 1328–1345. [CrossRef]
35. Ciraolo, L.; Azpilicueta, F.; Brunini, C.; Meza, A.; Radicella, S. Calibration errors on experimental slant total electron content (TEC) determined with GPS. *J. Geodesy.* **2007**, *81*, 111–120. [CrossRef]
36. Abe, O.; Villamide, X.O.; Paparini, C.; Radicella, S.; Nava, B.; Rodríguez Bouza, M. Performance evaluation of gnss-tec ionospheric slab thicknessimation techniques at the grid point in middle and low latitudes during different geomagnetic conditions. *J. Geodesy* **2017**, *91*, 409–417. [CrossRef]
37. Adewale, A.; Oyeyemi, E.; Adeniyi, J.; Adeloye, A.; Oladipo, O. Comparison of total electron content predicted using the IRI-2007 model with GPS observations over Lagos, Nigeria. *Indian J. Radio Space Phys.* **2011**, *40*, 21–25.

38. Seemala, G.; Valladares, C. Statistics of total electron content depletions observed over the South American continent for the year 2008. *Radio Sci.* **2011**, *46*, RS5019. [CrossRef]
39. Olwendo, O.; Baki, P.; Mito, C.; Doherty, P. Characterization of ionospheric GPS Total Electron content (GPS TEC) in low latitude zone over the Kenyan region during a very low solar activity phase. *J. Atmos. Sol. Terr. Phys.* **2012**, *84*, 52–61. [CrossRef]
40. Akala, A.; Seemala, G.; Doherty, P.; Valladares, C.; Carrano, C.; Espinoza, J.; Oluyo, S. Comparison of equatorial GPS-TEC observations over an African station and an American station during the minimum and ascending phases of solar cycle 24. *Ann. Geophys.* **2013**, *31*, 2085–2096. [CrossRef]
41. Matamba, T.M.; Habarulema, J.B.; McKinnell, L.-A. Statistical analysis of the ionospheric response during geomagnetic storm conditions over South Africa using ionosonde and GPS data. *Space Weather.* **2015**, *13*, 536–547. [CrossRef]
42. Akala, A.O.; Adewusi, E.O. Quiet-time and storm-time variations of the African equatorial and low latitude ionosphere during 2009–2015. *Adv. Space Res.* **2020**, *66*, 1441–1459. [CrossRef]
43. De Dieu Nibigira, J.; Sivavaraprasad, G.; Ratnam, D.V. Performance analysis of IRI-2016 model TEC predictions over Northern and Southern Hemispheric IGS stations during descending phase of solar cycle 24. *Acta Geophys.* **2021**, *69*, 1509–1527. [CrossRef]
44. Reinisch, B.W.; Galkin, T.A. Global ionospheric radio observatory (GIRO). *Earth Planets Space* **2011**, *63*, 377–381. [CrossRef]
45. England, S.L.; Immel, T.J.; Huba, J.D.; Hagan, M.E.; Maute, A.; DeMajistre, R. Modeling of multiple effects of atmospheric tides on the ionosphere: An examination of possible coupling mechanisms responsible for the longitudinal structure of the equatorial ionosphere. *J. Geophys. Res. Space Res.* **2010**, *124*, A05308. [CrossRef]
46. Chen, Y.; Liu, L.; Le, H.; Wan, W.; Zhang, H. Equatorial ionization anomaly in the low-latitude topside ionosphere: Local time evolution and longitudinal difference. *J. Geophys. Res. Space Res.* **2016**, *121*, 7166–7182. [CrossRef]
47. Lühr, H.; Häusler, K.; Stolle, C. Longitudinal variation of F region electron density and thermospheric zonal wind caused by atmospheric tides. *Geophys. Res. Lett.* **2007**, *34*, L16102. [CrossRef]
48. Aggson, T.L.; Herrero, F.A.; Johnson, J.A.; Pfaff, R.F.; Laakso, H.; Maynard, N.C.; Moses, J.J. Satellite observations of zonal electric fields near sunrise in the equatorial ionosphere. *J. Atmos. Sol. Terr. Phys.* **1995**, *57*, 19–24. [CrossRef]
49. Kelley, M.C.; Rodrigues, F.S.; Pfaff, R.F.; Klenzing, J. Observations of the generation of eastward equatorial electric fields near dawn. *Ann. Geophys.* **2014**, *32*, 1169–1175. [CrossRef]
50. Zhang, R.; Liu, L.; Chen, Y.; Le, H. The dawn enhancement of the equatorial ionospheric vertical plasma drift. *J. Geophy. Res.: Space Phys.* **2015**, *120*, 688–697. [CrossRef]
51. Zhou, X.; Liu, H.-L.; Lu, X.; Zhang, R.; Maute, A.; Wu, H.; Yue, X.; Wan, W. Quiet-time day-to-day variability of equatorial vertical E × B drift from atmosphere perturbations at dawn. *J. Geophys. Res. Space Phys.* **2020**, *124*, e2020JA027824. [CrossRef]
52. Balan, N.; Rao, P.B. Dependence of the ionospheric response on the local time of sudden commencement and the intensity of the geomagnetic storms. *J. Atmos. Terr. Phys.* **1990**, *52*, 269–275. [CrossRef]
53. Kashcheyev, A.; Migoya-Orue, Y.; Amory-Mazaudier, C.; Fleury, R.; Nava, B.; Alazo-Cuartas, K.; Radicella, S.M. Multivariable comprehensive analysis of two great geomagnetic storms of 2015. *J. Geophys. Res.: Space Phys.* **2018**, *123*, 5000–5018. [CrossRef]
54. Buonsanto, M.J. Ionospheric storms—A review. *Space Sci. Rev.* **1999**, *88*, 563–601. [CrossRef]
55. Mendillo, M. Storms in the ionosphere: Patterns and processes for total electron content. *Rev. Geophys.* **2006**, *44*, 335–360. [CrossRef]
56. Immel, T.J.; Sagawa, S.L.; England, S.B.; Henderson, M.E.; Hagan, S.B.; Mende, H.U.; Frey, C.; Swenson, M.; Paxton, L.J. Control of equatorial ionospheric morphology by atmospheric tides. *Geophys. Res. Lett.* **2006**, *33*, L15108. [CrossRef]
57. Rishbeth, H.; Mendillo, M. Patterns of F2-layer variability. *J. Atmos. Sol. Terr. Phys.* **2001**, *63*, 1661–1680. [CrossRef]
58. Fang, T.-W.; Fuller-Rowell, T.; Yudin, V.; Matsuo, T.; Viereck, R. Quantifying the sources of ionosphere day-to-day variability. *J. Geophys. Res. Space Phys.* **2018**, *123*, 9682–9696. [CrossRef]
59. Rajaram, G.; Rastogi, R.G. Equatorial Electron Densities—Seasonal and Solar Cycle Changes. *J. Atmos. Terr. Phys.* **1977**, *39*, 1175–1182. [CrossRef]
60. Radicella, S.M.; Adeniyi, J.O. Equatorial ionospheric electron density below the F2 peak. *Radio Sci.* **1999**, *34*, 1153–1163. [CrossRef]
61. Lee, C.C. Examination of the absence of noontime bite-out in equatorial total electron content. *J. Geophys. Res. Space Res.* **2012**, *117*, A09303. [CrossRef]
62. Chen, Y.; Liu, L.; Le, H.; Zhang, H. Equatorial North-South Difference of Noontime Electron Density Bite-out in the F2-layer. *J. Geophys. Res. Space Res.* **2020**, *125*, e2020JA028124. [CrossRef]
63. Su, Y.Z.; Bailey, G.J.; Balan, N. Night-time enhancements in TEC at equatorial anomaly latitudes. *J. Atmos. Terr. Phys.* **1993**, *56*, 1619–1628. [CrossRef]
64. Agrawal, A.; Maski, K.; Vijay, S.K.; MaShra, S.D. An Analytical Study of Nighttime Enhancement in foF2 during 23rd& 24th Solar Cycle at Low Latitude Station. *Int. J. Eng. Trends Tech.* **2018**, *61*, 25–30.
65. Fox, M.W.; Mendillo, M.; Spalla, P. The variation of ionospheric slab thickness during geomagnetic storms. In Proceedings of the International Beacon Satellite Symposium (IBSS), Tucuman, Argentina, 27–30 March 1990.
66. Bhuyan, P.K.; Lakha, S.; Tyagi, T.R. Equivalent ionospheric slab thickness of the ionosphere over 26°N through the ascending half of a solar cycle. *Ann. Geophys.* **1986**, *4*, 131–136.
67. Balan, N.; Shiokawa, K.; Otsuka, Y.; Kikuchi, T.; Vijaya Lekshmi, D.; Kawamura, S.; Yamamoto, M.; Bailey, G.J. A physical mechanism of positive ionospheric storms at low latitudes and midlatitudes. *J. Geophys. Res. Space Phys.* **2010**, *115*, A02304. [CrossRef]

68. Lei, J.; Zhu, Q.; Wang, W.; Burns, A.G.; Zhao, B.; Luan, X.; Zhong, J.; Dou, X. Response of the topside and bottomside ionosphere at low and middle latitudes to the October 2003 superstorms. *J. Geophys. Res. Space Phys.* **2015**, *120*, 6974–6986. [CrossRef]
69. Lei, J.; Zhong, J.; Mao, T.; Hu, L.; Yu, T.; Luan, X.; Dou, X.; Sutton, E.; Yue, X.; Lin, J.; et al. Contrasting behavior of the F2 peak and the topside ionosphere in response to the 2 October 2013 geomagnetic storm. *J. Geophys. Res. Space Phys.* **2016**, *121*, 10549–10563. [CrossRef]

remote sensing

MDPI

Article

Measurement of Aspect Angle of Field-Aligned Plasma Irregularities in Mid-Latitude E Region Using VHF Atmospheric Radar Imaging and Interferometry Techniques

Jenn-Shyong Chen [1,*], Chien-Ya Wang [2] and Yen-Hsyang Chu [3]

[1] Center for General Education, China Medical University, Taichung 406040, Taiwan
[2] Department of Optoelectric Physics, Chinese Culture University, Taipei 11114, Taiwan; cywang@ulive.pccu.edu.tw
[3] Department of Space Science and Engineering, National Central University, Taoyuan 320317, Taiwan; yhchu@jupiter.ss.ncu.edu.tw
* Correspondence: james.chen@mail.cmu.edu.tw

Abstract: Multireceiver and multifrequency radar imaging were carried out with the 46.5 MHz MU radar in Japan (34.85°N and 136.10°E) to examine the aspect sensitivity of field-aligned plasma irregularities (FAIs) in the mid-latitude ionosphere E region. A radar beam was directed to geographic north and at 51° zenith angle, which was normal to the geomagnetic field line around 100 km height. Nineteen receivers and five carrier frequencies were used for radar imaging to retrieve the power distribution in the radar volume, and then the aspect angle along the geomagnetic field line was calculated according to the angular power distribution. Retrieval algorithms such as Fourier, Capon, and norm-constrained Capon (NC-Capon) were employed, in which the NC-Capon was applied to FAIs for the first time and found to be more suitable for the present study. The aspect angles estimated by the NC-Capon ranged between 0.1° and 0.4° mostly, and averaged around 0.2°, which were the same order to the previous measurements with radar interferometry (RI), made for equatorial electrojet irregularities and the lower mid-latitude sporadic E region. For comparison, RI-estimated aspect angles were also investigated and found to be close to that of NC-Capon, but distributed over a wider extent of angles.

Keywords: radar imaging; aspect angle; field-aligned plasma irregularities; mid-latitude E region; norm-constrained Capon; VHF radar

Citation: Chen, J.-S.; Wang, C.-Y.; Chu, Y.-H. Measurement of Aspect Angle of Field-Aligned Plasma Irregularities in Mid-Latitude E Region Using VHF Atmospheric Radar Imaging and Interferometry Techniques. *Remote Sens.* **2022**, *14*, 611. https://doi.org/10.3390/rs14030611

Academic Editor: Fabio Giannattasio

Received: 26 December 2021
Accepted: 17 January 2022
Published: 27 January 2022

1. Introduction

Aspect sensitivity of refractive irregularities in the atmosphere describes a dependence of radar echo intensity on angular direction. In horizontally structured atmosphere, the radar echoes are usually strongest near the zenith and drop off with off-zenith angle, which is evident in observations throughout the low and middle atmosphere with the radar at MF, HF, and VHF bands. For the field-aligned plasma irregularities (FAIs) in the ionosphere, however, the radar echoes are strongest at the beam direction normal to the geomagnetic field line, and decay very quickly with off-perpendicular angle along the geomagnetic field line. Aspect angle is a measurement of aspect sensitivity, which can be defined by the half width of half power, or the standard deviation of Gaussian fitting in the angular power distribution; the former is about $\sqrt{ln4}$ (≒1177) times the latter if the angular power distribution is characterized by a Gaussian function. The order of aspect angle in the lower atmosphere is within several degrees [1,2]. By contrast, it can be as small as several times 0.1° or less in FAI echoes, depending on latitude and height [3–6]. It has been addressed in [3–6] that the FAIs' aspect angle in the direction of geomagnetic field line relates partly with the values of electron-neutral collision frequency (ν_e), electron gyrofrequency (Ω_e), ion-neutral collision frequency (ν_i), ion gyrofrequency (Ω_i), the ratio ν_e/Ω_e, among others, and moreover, is also subjected to nonlinear coupling process of unstable waves in the plasma irregularities.

Because of highly localized FAI echoes in the radar volume, and much, much smaller aspect angle in the ionosphere than in the lower atmosphere, many of the approaches/techniques proposed for measurement of aspect angle in the lower and middle atmosphere, such as those used in [1,2,7–13] (i.e., beam swing, spaced antenna, spectral width, etc.), are difficult to apply to FAIs. A proper approach to measurement of FAIs' aspect angle is the radar interferometry (RI) technique that was first applied to equatorial ionosphere by Kudeki and Farley [3] and observed the aspect angles of electrojet irregularities ranging between 0.1° and 0.4° generally. A later study for equatorial spread F irregularities found an even smaller value of aspect angle, i.e., a few times 0.01° [4]. Additionally, RI was applied to the lower mid-latitude sporadic E region (~25.6°N and ~120.8°E geography; ~16.33°N and ~167.34°W in geomagnetic coordinate) but with a different process and definition [14], in which the aspect angles of layer-type and clump-type plasma irregularities were estimated, and were found to be the same order of magnitude as that of equatorial electrojet irregularities.

In this study, we attempted to apply the approach of coherent radar imaging (CRI) to estimate the aspect angle of mid-latitude E region FAIs. CRI has been employed for study of aspect sensitivity in the lower and middle atmosphere [2,15,16]. The technical requirement of CRI is the use of separate antennas as independent receiving channels to collect the radar echoes. The multiple-channel echoes received can be employed to retrieve the in-beam angular power distribution through a beamforming process. The algorithm of beamforming can be linear-based, e.g., the Fourier method, or adaptive, such as the Capon [15] and norm-constrained Capon methods [17–20], and others. CRI can be regarded as an advanced version of RI.

To meet the requirement of CRI, the Middle and Upper (MU) Atmosphere radar, a pulsed radar operated at 46.5 MHz and maintained by Kyoto University, Japan, was employed in this study. The MU radar is located at 34.85°N and 136.10°E in geographic coordinates, and 26.36°N and 153.78°W in geomagnetic coordinates. A total of 25 receiving channels can be assigned, and moreover, frequency hopping of five carrier frequencies between radar pulses is available. Such flexible capabilities of operation allow radar image processing not only in 1D range imaging and 2D angular imaging, but also 3D spatial imaging. To this end, the retrieval methods used consistently, such as the Fourier, Capon, and norm-constrained Capon methods, were examined to find their suitability for the present study. As far as we know, the norm-constrained Capon method has not yet been applied to full 3D radar imaging. In addition to radar imaging, several sets of subarrays aligned approximately with the direction of the geomagnetic field line are also available to test the RI technique in measurement of aspect angle. Therefore, comparison of measurements is possible to validate the usability of radar imaging and RI, or evaluate which kind of method is more suitable for estimating the aspect angle of FAIs with the MU radar configuration. To complete the study, the effect of signal-to-noise ratio (SNR) on the estimate of aspect angle was also examined.

The rest of this paper is organized as follows. In Section 2, measurement techniques are described, including experimental setup and data processing of radar imaging and RI. Section 3 provides the results and discussion, and Section 4 states the conclusions.

2. Measurement Techniques

Radar experimental setup, radar imaging, and radar interferometry are described in Sections 2.1–2.3, respectively.

2.1. Radar Experimental Setup

Figure 1 shows the antenna array configuration of the MU radar, in which each plus symbol denotes a single antenna with an independent transceiver. The diameter of the array is about 103 m and the array can be partitioned into 25 subarrays, as denoted by the letters from A1 to F5. The separation of adjacent subarrays is 19.55 m, and central carrier frequency is 46.5 MHz. A detailed description of the radar characteristics can be found through the website http://www.rish.kyoto-u.ac.jp/English/MU/index.html (accessed on 25 December 2021). Some operational parameters related to the present study are described below and listed in Table 1.

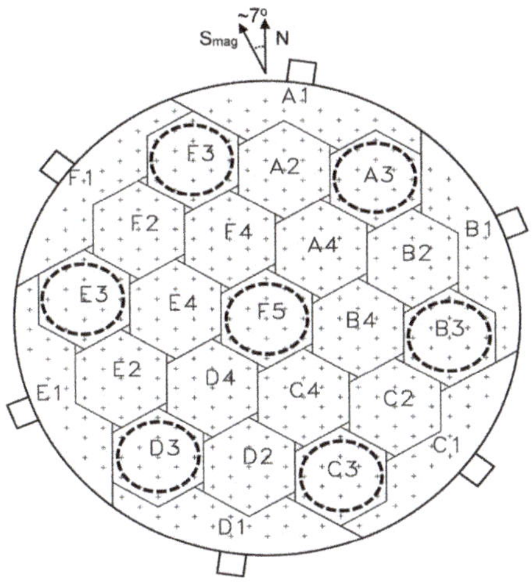

Figure 1. Configuration of MU radar array. Each plus symbol denotes a single antenna with an independent transceiver. The diameter of the array is about 103 m, and the subarrays denoted by the letters from A1 to F5 are individual receiving arrays. The 19 subarrays in the interior and the seven subarrays encircled are employed in radar imaging. The geomagnetic south pole (S_{mag}) is about 7° west of geographic north.

Table 1. Radar parameters.

Parameters	Values
Beam direction	Geographic north, 51° zenith
Interpulse period (IPP)	0.0015 s
Pulse length	4 μs
Sampling time step	4 μs
Number of sampling gates	128
Integration times	1
Pulse codes	no
Carrier frequencies	46.25, 46.375, 46.5, 46.625, and 46.75 MHz
Receiver channels	Full array and 19 subarrays A2–F5, exclusive of A1, B1, C1, D1, E1, F1
Sampling range (height)	120.0–196.8 km (~75.518–123.850 km)

Full array was used for transmission of radio waves, and the radar beam was steered to geographic north and at the zenith angle of 51°. As a result, the radar beam bearing was perpendicular to the geomagnetic field line around the height of 100 km. Moreover, five carrier frequencies were employed in sequential radar pulses that were operated repeatedly. In reception, the whole array and the 19 subarrays in the interior of the array were assigned to 20 receiving channels to collect the radar echoes from five different carrier frequencies. This multireceiver and multifrequency operation allows us to execute 1D range imaging, 2D angular imaging, and 3D spatial imaging. The imaging here means the retrieval of power distribution within the radar volume.

The sampling range interval was between 120.0 and 196.8 km. The corresponding sampling height interval was between about 75.518 and 123.850 km, which was estimated by projecting the sampling range at vertical direction and ignoring the curvature of the Earth surface. Sampling time was 0.0075 s for each carrier frequency returns (=5 × IPP = 5 × 0.0015 s). In calculation of covariance function for radar imaging, 256 raw data points were taken for ensemble average, resulting in an estimate of 1.92 s (=256 × sampling time = 256 × 0.0075 s).

2.2. Radar Imaging

The retrieval methods used for radar imaging in this study were Fourier, Capon, and norm-constrained Capon. We compared their effectiveness for deriving aspect angle in the 2D and 3D imaging processes. For 2D angular imaging, only multireceiver data are needed. For 3D spatial imaging, both multireceiver and multifrequency data are required. These methods of radar imaging were examined and employed in several studies in the research field of atmosphere [15–25]. A brief mathematical description of the radar imaging employed in this study is provided in Appendix A.

Two-dimensional angular imaging yields the power density (termed brightness distribution hereafter) as a function of angle for each sampling gate, and there could be more than one local maximum in the brightness distribution. Nevertheless, only the one with the maximum brightness level was used in the study. That is, the mean location of the brightness center encircled by the highest contour level was determined first, and then Gaussian-function fitting was applied to the brightness values around the estimated mean location of the brightness center to obtain a brightness width, i.e., the Gaussian-fitted standard deviation. The fitting was executed only in the direction along the geomagnetic field line, the aspect angle of which we surveyed. It was estimated that the geomagnetic south pole was about 7° west of geographic north in the field of view of the MU radar in the observation. Therefore, it was expected that the brightness distribution of FAI echoes is field-aligned and usually has a slant of about 7° in the 2D imaging surface viewed from the radar beam direction [25]. Notice that the imaging surface is the transverse plane of the radar beam, and the origin is at the radar beam bearing. Abscissa and ordinate of the imaging surface are zonal and elevation directions relative to the radar beam bearing, respectively (Figure 2b,c; explained later).

Since the aspect angle of FAIs could be as small as only 0.1~0.5°, the imaging was processed with an angular step of 0.05° in both zonal and elevation directions. To confirm the Gaussian-fitting result, we also used the peak-find process built in the computing software (i.e., MATLAB) to obtain the half-power peak width. The half-power peak width or the Gaussian-fitted standard deviation is an indication of the aspect angle of the FAIs, in which the former is about $\sqrt{In4}$ (≒1177) times the latter if the angular brightness distribution around the mean location of the brightness center, in the geomagnetic field-aligned direction, is characterized by a Gaussian function.

It should be noticed that the aspect angle estimated from 2D angular imaging is a mean value within the range interval of the sampling gate. If one expects to observe the spatial structure and estimate the aspect angles at different ranges in the radar volume, 3D spatial imaging is one of the solutions. In the 3D radar imaging executed in this study, the range step of imaging was 37.5 m, giving 17 range locations in the sampling gate of 600 m length, i.e., at the range locations of −300, −262.5, −225, ... , 300 m relative to the range center of the sampling gate. A smaller range step is certainly available, but it consumes more computation time and may not be helpful for improving the range resolution due to the finite number of carrier frequencies used in observation. For each range location, the Gaussian fitting and peak-find process can be performed for the 2D-angular brightness distribution to obtain two brightness widths in the geomagnetic field-aligned direction, which denote two possible values of aspect angles at that range location. The results from 19 subarrays and 7 subarrays (indicated by the dashed circles in Figure 1) are compared and shown later.

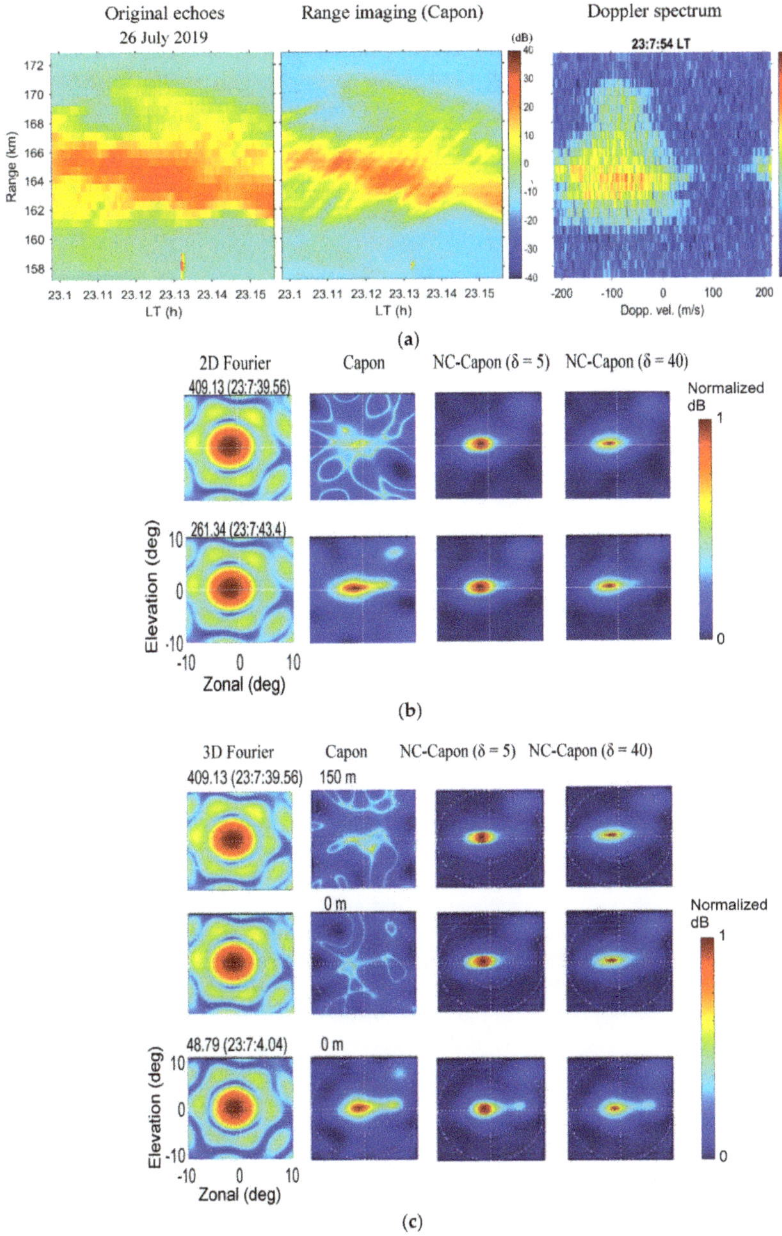

Figure 2. One-, two-, and three-dimensional radar imaging. (**a**) Left: radar echoes; middle: 1D range imaging; right: one example of range-Doppler spectrum; (**b**,**c**) are the brightness distribution obtained from 2D and 3D radar imaging with Fourier, Capon, and norm-constrained Capon methods, respectively. Two examples are shown. Brightness values in (**b**,**c**) are self-normalized and displayed in dB scale. SNR is given in the Fourier panel title, present in linear scale, and following the SNR value is the data time in the format hh:mm:ss LT, on 26 July 2019. The numbers in the Capon column of (**c**) are the range positions of the slices of brightness distribution in the sampling gate, which are relative to the central range of the sampling gate.

2.3. Radar Interferometry

Radar interferometry (RI) has been widely used in atmospheric studies such as mean angle of arrival of radar echoes [26], lightning detection [27], meteor radar wind [28], among others. Use of RI in measurement of the aspect angle of FAIs was first reported by Kudeki and Farley in 1989 [3] for the equatorial electrojet region. In their approach, normalized cross spectra were calculated for all receiver pairs and then a suitable fitting of a parabola to the absolute values (termed coherence) of the cross spectra was made to retrieve aspect angles at various spectral lines (i.e., Doppler shifts). Lu et al. [6] extended the study of [3] and provided the following normalized cross spectrum:

$$S_{mn}(kD_{mn}, w) = e^{jkD_{mn}\langle\theta\rangle_w} \exp\left[-\frac{1}{2}k^2 D_{mn}^2 \theta_{rms}^2(w)\right], \quad (1)$$

where k is wave number, w is Doppler shift, D_{mn} is the length of the baseline between receiver antennas m and n, $\langle\theta\rangle_w$ is the mean angular location of the irregularity pattern at the Doppler shift w, and $\theta_{rms}(w)$ denotes the angular spread value relative to $\langle\theta\rangle_w$, which can be regarded as the aspect angle at w. Gaussian function is then employed to fit the coherence of the cross spectra to derive the parameter θ_{rms}.

In this study, we did not examine the aspect angle at different Doppler shifts. Therefore, the cross-correlation function $S_{mn}(kD_{mn})$ was estimated in time domain. Unfortunately, there was no set of subarrays with the baseline parallel to the direction of the geomagnetic field line exactly. As shown in Figure 1, the three subarrays (A2, F5, D2) and the two sets of five subarrays (A2, F4, F5, D4, D2; A2, A4, F5, C4, D2) had the alignments of about 7.60° east of geographic north, and thus had about 14.85° difference from the geomagnetic southern pole. Another set of subarrays, F3, F4, F5, C4, and C3, was aligned in the direction of about 28.07° west of geographic north, and thus had a difference of about 20.82° from the geomagnetic south pole. These sets of subarrays were used for RI in this study.

3. Results

The aspect angles retrieved from radar imaging and RI are shown and discussed in this section. Moreover, the effect of SNR on the estimate of aspect angle is also examined.

3.1. Radar Imaging in 1D, 2D, and 3D

First, the benefits and shortages of the Fourier, Capon, and NC-Capon methods in 2D and 3D radar imaging were investigated, as shown in Figure 2. Figure 2a shows the radar echoes collected by the full array of the radar, where the original range-time intensity of FAIs is present in the left panel and 1D range-imaging result is displayed in the middle panel. In the range imaging, the radar echoes collected from the full array and five carrier frequencies were processed with the standard Capon method, and the range step was 5 m. It is clearly seen in the range-imaged brightness distribution that these FAI echoes were positive range rate, indicating a northern component of motion. This direction of motion can be revealed from the typical range–Doppler spectrum of the echoes provided in the right panel of Figure 2a. As shown, the Doppler velocities spread out over a wide range, but were negative mostly and centered on −100 m/s, indicating a motion of away from the radar site according to our definition of Doppler velocity direction in calculation. Since the tilted radar beam was directed to geographic north, apparently a northern component of motion existed in the FAIs structure. Certainly, the zonal motion might also contribute a part to the Doppler spectrum, but could not be as large as 100 m/s because the radar beam was directed to geographic north and had a very narrow beamwidth (~3.6°). Notice that some visible spectral power at the Doppler velocities of around 200 m/s and in the range interval between 164 and 166 km were aliases of some negative Doppler velocities that were smaller than the negative Nyquist velocity (about −215 m/s).

Figure 2b shows 2D angular brightness distributions of two examples retrieved from different methods. The value given in the title of the Fourier panel is SNR in linear scale, and following the SNR value is the data time on 26 July 2019 (hh:mm:ss LT). Obviously, the first example shows that the Fourier and Capon methods were not good for retrieving the brightness distribution (present in dB scale). The Fourier result had a coarser angular resolution and suffered from serious contamination of sidelobes arising from the weighting vector used in retrieval (refer to the Appendix A). The Capon result, on the other hand, provided an unreliable distribution, which is known to be related to the sensitivity of the method to small errors in the received signals. Nevertheless, the Capon method is still workable for many of the echoes, as shown in the example displayed in the second row of Figure 2b. By contrast, the NC-Capon method exported more stable and reliable outcomes, and moreover, a larger δ value yielded a higher angular resolution. A comparison of the results from different δ values is given later. Readers can also find the meaning of the δ value in Appendix A.

Figure 2c shows two examples of the angular brightness distributions in 3D imaging retrieved from different methods. The upper two rows provide two slices of brightness distribution at the range position of 150 m and 0 m in the sampling gate, and the bottom row presents the brightness distribution at the central position (0 m) of the sampling gate in another example. In general, the features resulting from different methods were similar to that in 2D imaging.

Figure 3 shows the scatter plots of brightness width (degree) vs. SNR (dB). The outcomes of Gaussian fitting and peak-find methods, 2D and 3D imaging, and Capon and NC-Capon methods are present, where the δ value in the NC-Capon method was 60. In this study, we define the half width of half power as representative of aspect angle, and therefore the brightness width obtained from Gaussian fitting was multiplied by $\sqrt{ln4}$. This correction was also made for the results of the RI technique shown later. The brightness centers with angular locations within 2° of the radar beam direction were adopted, and moreover, the scatter plots of 3D imaging included only the brightness widths from the slices of angular distributions at the range positions of −75, −37.5, 0, 37.5, and 75 m in the sampling gate. The reason for this selection of outcomes is shown later in Figure 4. Several findings from Figure 3 are listed below:

1. Brightness width was SNR-dependent; generally, a larger SNR produced a smaller brightness width.
2. The brightness widths retrieved from the Capon method (Figure 3a,c) were more divergent at higher SNR as compared with the result of NC-Capon (Figure 3b,d), regardless of 2D or 3D imaging, and in Gaussian fitting or peak-find processes. Moreover, a sudden drop of brightness width at large SNR can be seen in the Capon result. This drop, together with divergence of brightness widths at larger SNR, implied a failure in using the Capon method for these cases.
3. As revealed from the outcomes at higher SNR, Figure 3b,d shows that Gaussian-fitted brightness widths were about 0.1° larger than those of the peak-find method.

Based on the consequences of what is shown in Figures 2 and 3, the NC-Capon method is definitely more suitable than the standard Capon method for retrieval of brightness width and thereby aspect angle. In the rest of this paper, therefore, only the results of the NC-Capon and peak-find methods are present. Readers can find additional maps of brightness distribution via the web link provided in [29].

As mentioned, in the scatter plot of 3D imaging, only the outcomes from the five slices of angular distributions at the range positions of −75, −37.5, 0, 37.5, and 75 m in the sampling gate were included. This was based on the features shown in Figure 4, where the profiles of mean brightness widths within five sampling gates are shown for three δ values: 20, 40, and 60. The left and right panels demonstrate the results from the outcomes with SNR > 1 and >20, respectively. The horizontal bar denotes the width of two times the standard deviation of brightness width. It can be seen clearly in the left panel that in each sampling gate, the mean value of brightness width varied with the range and had a smaller value around the central range of the sampling gate (indicated by g_1 to g_5). If a higher threshold of SNR was used, i.e., SNR > 20 as given in the right panel, the dependence of brightness width on the range in the sampling gate existed still although the dependence was mitigated significantly. One cause of this characteristic could be the volume weighting effect. In this study, range-weighting and beam-weighting effects were not corrected in either 2D or 3D imaging. This is under the consideration that the echoes of the FAIs are highly localized, and therefore correction of the weighting effect could produce overcorrected brightness values around the edge of the radar volume. To avoid the influence of the weighting effect on estimate of brightness width, we used the outcomes with the brightness centers located around the center of the sampling gate, that is, the range interval within −75 and 75 m, and the angular interval between −2° and 2°.

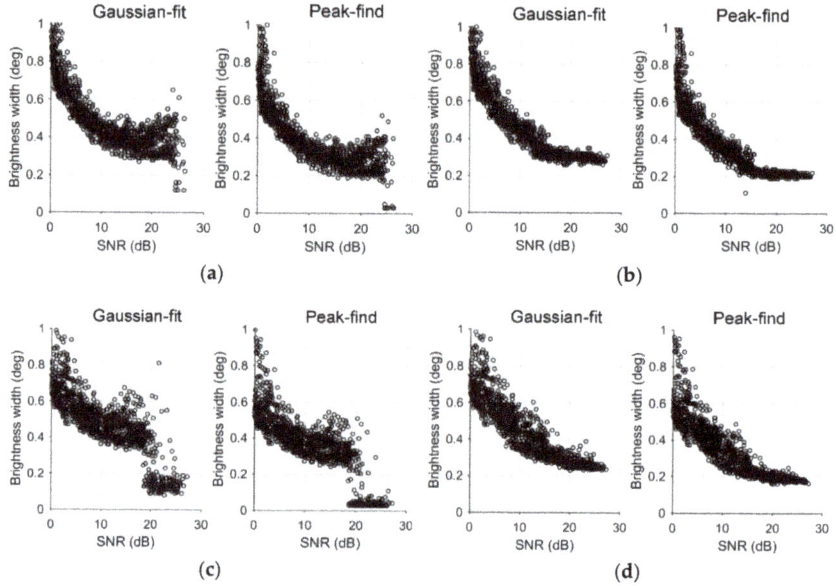

Figure 3. Scatter plots of brightness width (aspect angle) vs. SNR, retrieved from Capon and NC-Capon methods: (**a**,**b**) show 2D radar imaging results obtained from Capon and NC-Capon methods, respectively. The parameter of δ is 60 in NC-Capon method. (**c**,**d**) are similar to (**a**,**b**) but show 3D radar imaging results. Both outcomes of Gaussian-fit and peak-find methods are present.

Figure 4 shows that mean brightness width varied with the δ value used in the NC-Capon method. In view of this, a dependence of the outcome on δ value should be examined further to find a suitable value of δ for the following data analysis. Figure 5a shows the results from the 2D NC-Capon method, where the upper two panels are the scatter plots of brightness width vs. SNR resulting from the δ values of 10 and 30. Obviously, the brightness width at larger SNR decreased with increased δ value. To reduce the influence of SNR, therefore, the outcomes with SNR > 10 were used to estimate mean brightness width, and the results are shown in the lower panel of Figure 5a. As seen, the change of mean brightness width was tiny when the δ value was larger than 40, although the standard deviation of brightness widths increased slightly with δ value. Similar behavior can also be found in the 3D NC-Capon results, as shown in Figure 5b.

More issues of the NC-Capon method that deserve inspection are computation time and brightness value at different δ values. Figure 6 provides an example of 3D NC-Capon showing the variations of brightness sum and computation time with δ value, where the values on ordinate are self-normalized. Note that the brightness sum in panel (a) was the summation of brightness values in the radar volume just from a single estimate; the other estimates had similar scenarios (not shown). On the other hand, the computation time shown in panel (b) was recorded from fifty estimates. As seen, the self-normalized brightness sum decreased with increased δ value, and had tiny changes at the δ values larger than ~40. Meanwhile, the computation time also decreased quickly with increased δ value.

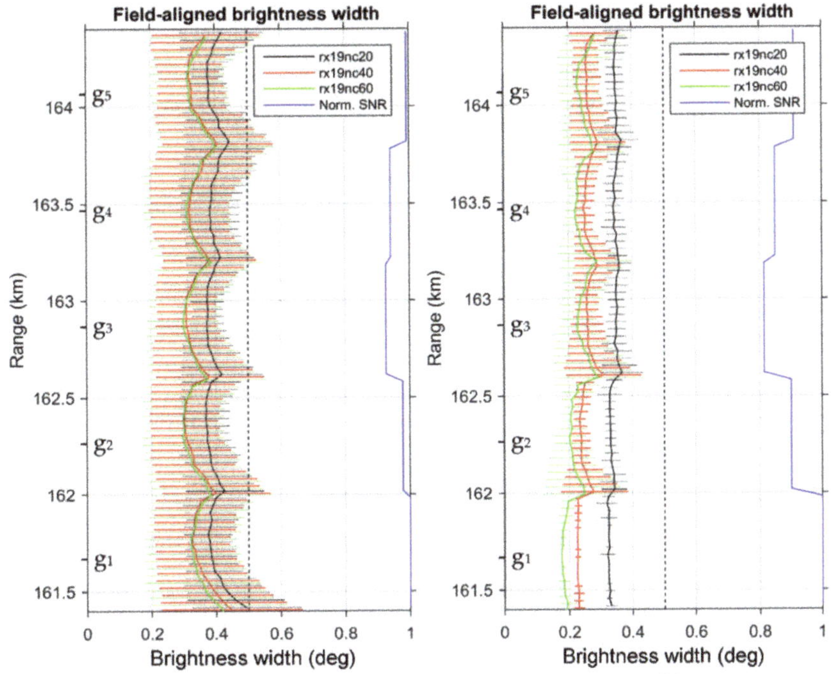

Figure 4. Profiles of brightness widths obtained from the NC-Capon method with three different δ values (20, 40, and 60). The brightness widths for statistical estimates are collected with the threshold of (**a**) SNR > 1 and (**b**) SNR > 20; g_1 to g_5 indicate the range center locations of five sampling gates.

Based on the investigation results shown in Figures 5 and 6, we can use the value larger than 40 for δ to save much computation time, but should not be too large to avoid magnifying the divergence of brightness widths. In view of this, the consequences of δ = 60 are presented in the remaining discussion of the 2D and 3D NC-Capon methods.

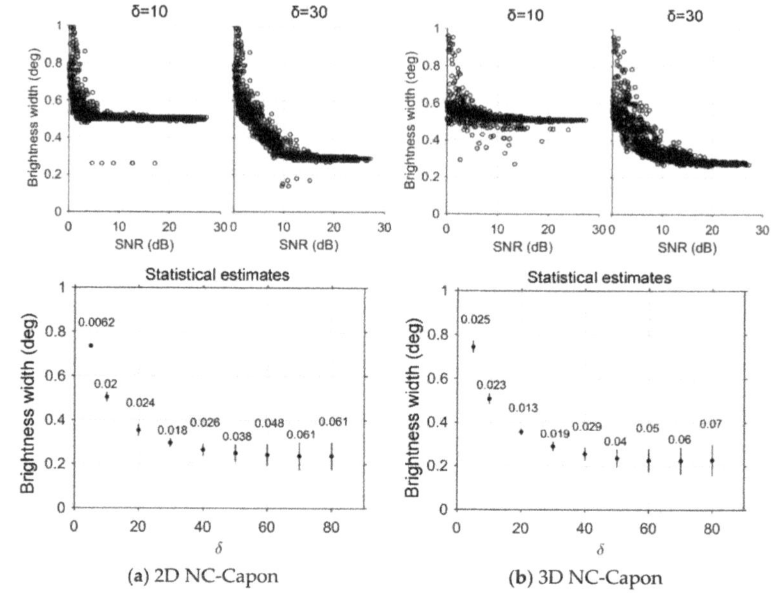

Figure 5. (a) Upper: brightness width (aspect angle) vs. SNR for 2D radar imaging, retrieved from the NC-Capon method with δ = 10 and 30, respectively. Bottom: brightness width vs. δ value; dots are mean values and short vertical segments denote two times the standard deviation of brightness widths. The brightness widths are selected from SNR > 10 and angular center location within 2°. The value of standard deviation is indicated above the segment. (b) The same as (a), but for 3D radar imaging, in which the restriction of range interval between −75 and 75 m is added to select the outcomes for statistical estimates.

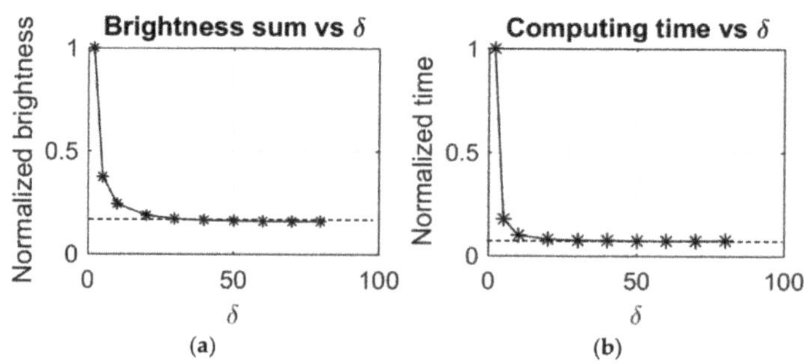

Figure 6. (a) Sum of brightness values vs. δ value, resulting from a single estimate. (b) Computation time of 50 estimates vs. δ value. Ordinate values are self-normalized. The computation is performed with NC-Capon method for 3D radar imaging.

3.2. SNR Effect

As seen in the scatter plots of brightness width vs. SNR shown in Figure 5, a lower SNR led to a larger brightness width. In many previous studies, finite correction of SNR effect for the coherence of the covariance function had been suggested to improve the consequences. Usually, this can be made through multiplying the coherence by $(1 + 1/SNR_1)^{1/2}(1 + 1/SNR_2)^{1/2}$, in which SNR_1 and SNR_2 are the respective SNRs of the two data sets used in the analysis. Such a correction will increase the coherence. However, a smaller SNR magnifies the coherence greatly, and could lead to a coherence larger than 1 sometimes, which is irrational and may cause the coherence matrix to be more singular, or lead to improper Gaussian fitting in the RI approach. In view of this, only the coherence with SNR > 1 was corrected, and moreover, the corrected coherence was assumed to be 0.99 when it was larger than 1.

Figure 7a,b provides the comparisons of original and SNR-corrected results of 2D and 3D imaging, respectively. The original scatter plots of 2D and 3D imaging show a significant decrease of brightness width with SNR for the echoes of SNR < ~15 dB. After SNR-correction, the dependence of brightness width on SNR was mitigated appreciably in the 2D imaging result, as shown in Figure 7a. By contrast, SNR-corrected coherence yielded less alleviation of SNR-dependent features in the 3D imaging result, as shown in Figure 7b.

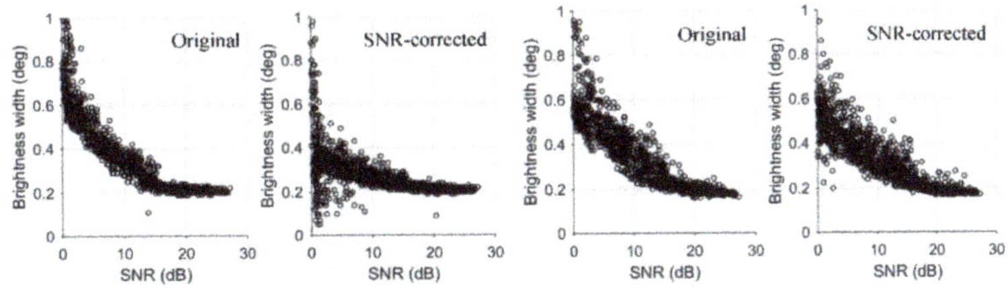

(a) 2D NC-Capon, δ = 60 (b) 3D NC-Capon, δ = 60

Figure 7. Scatter plots of brightness width vs. SNR. (**a**) Results of 2D radar imaging with norm-constrained Capon method, δ = 60. The left and right subpanels show the outcomes from the original and SNR-corrected coherence functions, respectively. (**b**) The same as (**a**) but for 3D radar imaging.

Based on the consequences shown in Figure 7, it is suggested that the outcomes with SNR > 10 dB or even >15 dB can be adopted to calculate the brightness width of the echoes. On the condition of SNR > 10 dB, most of the aspect angles, represented by brightness widths, spread from ~0.15° to ~0.4°. This order of magnitude is consistent with the RI measurements of aspect angles at the equator [3,6] and lower mid-latitude E region [5,14].

3.3. Radar Interferometry

As mentioned in Section 2, four sets of subarrays in the MU radar array, of which baselines are aligned approximately with the direction of the geomagnetic field line, are available for radar interferometry (RI). To compare with the results of radar imaging, however, expression (1) without the factor of Doppler shift was employed. Figure 8 exhibits the results, where the scatter plots of brightness width vs. SNR are shown. Several features can be found in Figure 8:

1. The scatter plots from the original coherence values show that brightness widths were SNR dependent, as seen in the left subpanels of panels (a)–(d). The scenario was similar to that of 2D and 3D imaging shown in Figure 7. Nevertheless, there were more outliers than in 2D and 3D imaging, i.e., more divergent in the distribution.
2. Correction of the SNR effect indeed reduced the brightness width significantly, as shown in the right subpanels of panels (a)–(d). Many brightness widths at lower SNR were now closer to the events at higher SNRs. However, brightness widths still spread over a wider interval of angle than in 2D and 3D imaging.
3. The outcomes of the two sets of subarrays shown in (b) and (c) were more concentrated than in (a) and (d), although the baselines of the subarrays of (b) and (c) were not exactly linear.

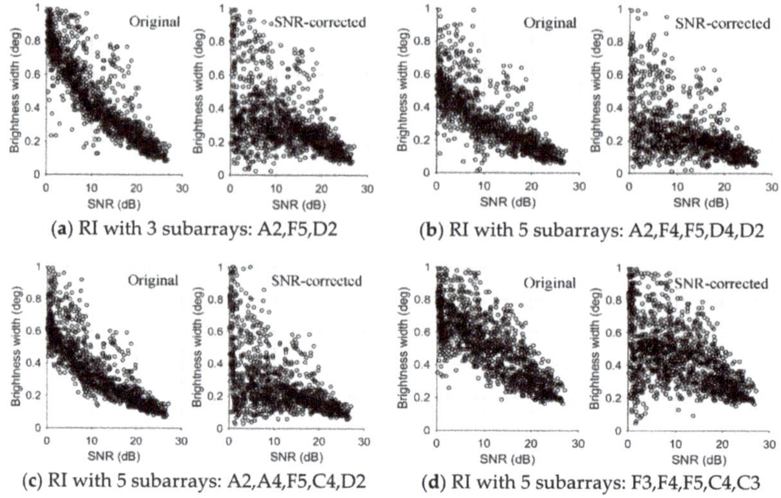

Figure 8. Scatter plots of brightness width vs. SNR resulting from the radar-interferometry approach proposed by [3]. Four sets of receiving subarrays are used, as indicated in the panels (**a**–**d**). Left and right subpanels of each panel show the outcomes from the original and SNR-corrected coherence functions, respectively.

For feature (3), it can be understood that (b) and (c) have more subarrays than (a), and the baselines of (b) and (c) are parallel to the direction of the geomagnetic field line more than that of (d). Accordingly, the subarrays in (b) and (c) are more suitable for retrieval of aspect angle of FAIs. As for the features (1) and (2), the aspect angles have more outliers and spread in a broader interval of angle as compared with 2D and 3D imaging results; one of the causes could be the certainty that the baselines of subarrays did not align with the geomagnetic field lines exactly, and another cause was the smaller length of baselines between subarrays. The maximum length of baseline was about 70 m in the subarrays used in Figure 8a–c, and about 80 m in Figure 8d, which were much smaller than that used in [3] and [6] (~310 m).

3.4. Statistical Comparison of Aspect Angles

Since different methods may result in different brightness widths (i.e., aspect angles), we compared their brightness widths, as shown in Figure 9. In panel (a), the results from 2D and 3D imaging with 19 subarrays and 7 subarrays (legend " … rx19" and " … rx7", respectively), and RI (legend "RI: … ") are shown. The 7 subarrays used here are indicated by the dashed circles in Figure 1. The two sets of subarrays for RI are those used in Figure 8c,d. Correction of the SNR effect has been made for coherence and the threshold of 10 dB was given for SNR to select the outcomes for statistical calculation. The δ value was 60 in the NC-Capon method. As a result, the profile of mean brightness width denoted by RI:A2A4-D2 in panel (a) had smaller values than the others below the range of 162 km, while that denoted by RI:F3F4-C3 had the largest values above the range of 162 km. In addition, both mean profiles of RI were more range-variant than other profiles. On the other hand, the standard deviations of RI aspect angles were mostly larger than the others above 162 km, as observed in the right panel of Figure 9a. In the study, we also checked the RI results from the set of subarrays E3, E4, F5, B4, and B3, of which baseline alignment turns aside from the direction of the geomagnetic field line more than that of the subarrays present here, and found that the resultant aspect angles spread over a much broader interval of angle (not shown). It is thus demonstrated that the aspect angles resulting from RI are highly dependent on the baseline alignment of the subarrays. In contrast to the RI results, the mean profiles of 2D imaging, denoted by "2Dd60rx19" and

"2Dd60rx7", had the smallest variation through the range, and moreover, the standard deviations of brightness widths of 2D imaging were generally the smallest among the results. There was not much difference between the mean profiles of "2Dd60rx19" and "2Dd60rx7", but the standard deviation marked by "2Dd60rx7" had values slightly larger than that of "2Dd60rx19", suggesting that a smaller number of subarrays may deteriorate the outcomes of 2D imaging. On the other hand, 3D imaging provided consequences approximately between 2D imaging and RI results. In general, the outcomes of 3D imaging resulting from 19 subarrays were closer to that of 2D imaging.

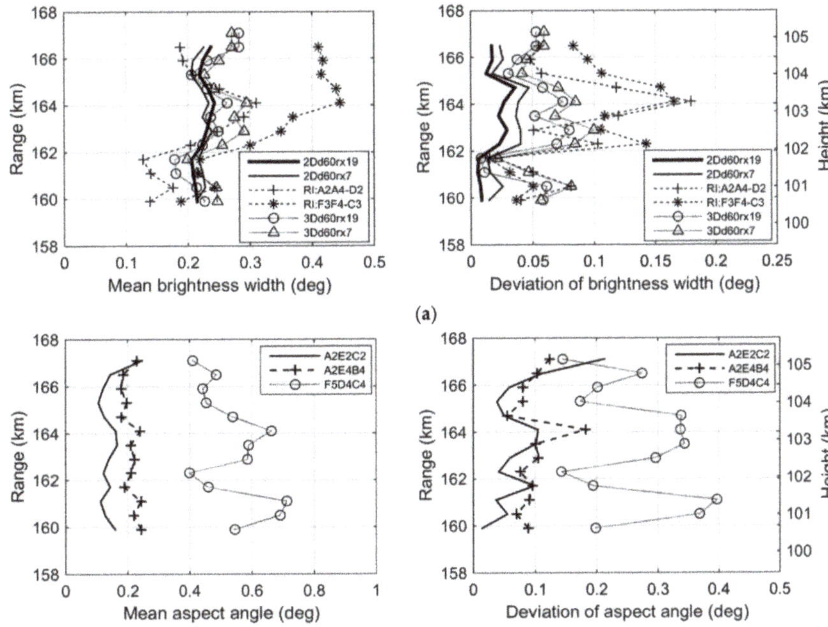

(a)

(b)

Figure 9. Profiles of mean and standard deviation of brightness widths estimated from various approaches. (**a**) Results from 2D and 3D radar imaging with the norm-constrained Capon method ($\delta = 60$), and from the radar interferometry (RI) used by Kudeki and Farley [3]. In radar imaging, the locations of echo centers within 2° and between −75 and 75 m in the sampling gate are adopted for statistical calculation. In RI, the outcomes shown in Figure 8c,d are used. Only the data with SNR larger than 10 are taken. (**b**) Results from the RI approach proposed by Wang et al. [14].

Another outcome of aspect angle estimated from the definition of Wang et al. [14] is provided in Figure 9b. The method of [14] was also based on RI but three non-coplanar receiving antennas/subarrays were used to derive a mean echo center in the radar volume. These locations of echo centers obtained were generally geomagnetic field-aligned. The aspect angle was then defined to be half the angle subtended by the field-aligned length of echo center locations. There are many associations of three receiving subarrays available in the MU radar array. Among them, the results from the three sets of subarrays—A2E2C2, A2E4B4, and F5D4C4—are shown, which are typical and represent larger, medium, and smaller triangle configurations of receiving subarrays, respectively. Note that the abscissa scale in panel (b) is twice that in panel (a). Obviously, the estimated aspect angles depend on the baseline length between receiving antennas, in which a larger triangle configuration yielded smaller and more concentrated values of aspect angles (solid curves). The smallest triangle configuration (F5D4C4) provided the largest mean and most divergent outcomes (circles). In quantitative analysis, the mean aspect angles of the medium triangle configuration (A2E4B4) were around 0.2°, which were closer to those of 2D radar imaging. The largest triangle configuration (A2E2C2), however, produced mean aspect angles less than 0.2° and close to 0.1° at some heights; these were the smallest values among all results. Nevertheless, the deviations of aspect angles yielded by the largest and medium triangle configurations mostly ranged between 0.05° and 0.2°, which were comparable to those in panel (a) but still lager than the results of 2D radar imaging (<0.05°).

Based on the comparison made in Figure 9, 2D radar imaging can yield the brightness width/aspect angle of FAIs varying with range more smoothly and converging on smaller interval. It should be known, however, that the outcome from 2D imaging is an average throughout the range of the sampling gate (600 m in the present study); 3D imaging, on the other hand, is likely to be capable of inspecting the variation of aspect angle or irregularity structure within the sampling gate. A problem for 3D imaging is that its computation time is much longer than that of 2D imaging, especially when using the NC-Capon method. A test shows that the computation time of 3D imaging can be about 150 times that of 2D imaging. The iteration number in the computation of the NC-Capon method is unpredictable because it depends on the characteristics of the individual data set. In addition, the more subarrays employed and more slice numbers given in the processing, the more time consumed in computation, although using more subarrays and slice numbers for radar imaging may export a higher resolution of brightness distribution.

4. Conclusions

In this study, radar imaging and radar interferometry (RI) techniques were applied to the measurement of the aspect angle of field-aligned plasma irregularities (FAIs) in the mid-latitude ionospheric E region. This was achieved by means of multireceiver and multifrequency capabilities built in the Middle and Upper (MU) Atmosphere radar (34.85°N and 136.10°E) of Japan. The brightness distribution retrieved from 2D and 3D radar imaging with three retrieval methods, i.e., Fourier, Capon, and norm-constrained Capon, were investigated. It was shown that the norm-constrained Capon method could produce more reliable brightness distribution for most circumstances and thereby yield a more trustworthy value of aspect angle.

By taking the radar data of SNR > 10 dB for 2D and 3D radar imaging, the estimated mean aspect angles ranged from 0.2° to 0.3°, and the standard deviations of aspect angles were less than 0.1° and could be as small as 0.01° at some heights for 2D radar imaging. These values are consistent with those measured in the equatorial electrojet (EEJ) region and in the lower mid-latitude E region with RI technique. With the MU radar, the RI results are also acceptable for certain receiving configurations, but the derived aspect angles spread over an angular range wider than that of radar imaging. This is attributed to a much shorter baseline used for RI in the MU radar array, as compared with the radar used for the EEJ region, and/or different approaches/definitions of aspect angle made for the lower mid-latitude E region.

In conclusion, 2D radar imaging with the norm-constrained Capon method is preferable to other methods, examined in this study, for measurement of aspect angle with the MU radar configuration. Three-dimensional radar imaging is also recommended for further studies of variation in aspect angle and irregularity structure in the radar volume. Nevertheless, 3D radar imaging with the norm-constrained Capon method consumes a much longer time in computation, which is not advantageous to a great amount of data analysis for the ionosphere. In the future, variations of aspect angle with time and altitude can be investigated with 2D radar imaging for the mid-latitude ionosphere. The results can assist in understanding the spatial and temporal characteristics of plasma irregularities in the ionosphere within the radar volume and at the temporal resolution of instrument integration time.

Author Contributions: Conceptualization, J.-S.C. and Y.-H.C.; methodology, J.-S.C., C.-Y.W., and Y.-H.C.; software, J.-S.C. and C.-Y.W.; validation, J.-S.C., C.-Y.W., and Y.-H.C.; formal analysis, J.-S.C. and C.-Y.W.; investigation, J.-S.C. and C.-Y.W.; data curation, J.-S.C. and C.-Y.W.; writing—original draft preparation, J.-S.C.; writing—review and editing, J.-S.C., C.-Y.W., and Y.-H.C.; visualization, J.-S.C. and C.-Y.W.; supervision, J.-S.C. and Y.-H.C.; project administration, J.-S.C. and Y.-H.C.; funding acquisition, J.-S.C. and Y.-H.C. All authors have read and agreed to the published version of the manuscript.

Funding: This research was funded by Ministry of Science and Technology, and China Medical University, Taiwan (ROC), under grants MOST110-2111-M-039-001 and CMU110-MF-69, respectively.

Institutional Review Board Statement: Not applicable.

Informed Consent Statement: Not applicable.

Data Availability Statement: Data sharing is not applicable to this article.

Acknowledgments: Experiments were conducted with the support of the MUR International Collaborative Research Program, Research Institute for Sustainable Humanosphere, Kyoto University, Japan.

Conflicts of Interest: The authors declare no conflict of interest.

Appendix A. Radar Imaging

The equations of radar imaging using multiple receivers and multiple frequencies are provided briefly below. Given the signals from N receivers and M carrier frequencies,

$$\mathbf{S}(t) = \begin{array}{l}[S_{11}(t)S_{12}(t)\ldots S_{1N}(t) \\ S_{21}(t)S_{22}(t)\ldots S_{2N}(t) \\ \vdots \\ S_{M1}(t)S_{M2}(t)\ldots S_{MN}(t)]^T\end{array}, \tag{A1}$$

where t is time, $S_{ij}(t)$ is the signal received by receiver j at carrier frequency i, and the superscript T denotes the transpose of matrix.

First, covariance function of $\mathbf{R}(t)$ is estimated:

$$\mathbf{R}(t) = \mathbf{S}(t)\,\mathbf{S}^H(t) = \begin{bmatrix} \mathbf{V}_{11}\mathbf{V}_{12}\cdots\mathbf{V}_{1M} \\ \mathbf{V}_{21}\mathbf{V}_{22}\cdots\mathbf{V}_{2M} \\ \vdots\,\vdots\,\ddots\,\vdots \\ \mathbf{V}_{M1}\mathbf{V}_{N2}\cdots\mathbf{V}_{MM} \end{bmatrix}, \tag{A2}$$

where the superscript H denotes the conjugate and transpose (Hermitian) operator, and $\mathbf{V}_{pq}$ is an $N \times N$ matrix composed of covariance functions of all receiver pairs at the carrier frequencies p and q. As a result, $\mathbf{R}(t)$ is an *MN-by-MN* matrix.

Second, the so-called brightness (power density), B, can be estimated with the following expression:

$$B(\boldsymbol{a}, r) = \frac{1}{W^2(\boldsymbol{a}, r)}\mathbf{w}^H\mathbf{R}\mathbf{w}, \tag{A3}$$

where r is range, and $\boldsymbol{a}$ = [sinθsinφ, sinθcosφ, cosθ], θ and φ are zenithal and azimuthal angles, respectively. $W(\boldsymbol{a}, r)$ is the spatial weighting function formed by radar beam and range weighting functions. The weighting vector $\mathbf{w}$ is given as

$$\mathbf{w} = \begin{array}{l}[e^{-j(2k_1r - k_1\boldsymbol{a}\bullet\boldsymbol{D}_1)}\; e^{-j(2k_1r - k_1\boldsymbol{a}\bullet\boldsymbol{D}_2)} \ldots e^{-j(2k_1r - k_1\boldsymbol{a}\bullet\boldsymbol{D}_N)} \\ e^{-j(2k_2r - k_2\boldsymbol{a}\bullet\boldsymbol{D}_1)}\; e^{-j(2k_2r - k_2\boldsymbol{a}\bullet\boldsymbol{D}_2)} \ldots e^{-j(2k_2r - k_2\boldsymbol{a}\bullet\boldsymbol{D}_N)} \\ \vdots\;\vdots\;\ddots\;\vdots \\ e^{-j(2k_Mr - k_M\boldsymbol{a}\bullet\boldsymbol{D}_1)}\; e^{-j(2k_Mr - k_M\boldsymbol{a}\bullet\boldsymbol{D}_2)} \ldots e^{-j(2k_Mr - k_M\boldsymbol{a}\bullet\boldsymbol{D}_N)}]^T\end{array}, \tag{A4}$$

where $\boldsymbol{D}_j$ is the location coordinates of receiver j, and k_i is the wave number of carrier frequency i. Expression (A3) is a linear beamforming, called the Fourier method in this study. The weighting pattern of (A4) contains appreciable sidelobes that will contaminate the imaging result of (A3); cases of 2D imaging are shown in Figure 2b,c. On the other hand, the estimator of the standard Capon method is:

$$B(\boldsymbol{a}, r) = \frac{1}{W^2(\boldsymbol{a}, r)} \frac{1}{\mathbf{e}^H \mathbf{R}^{-1} \mathbf{e}}, \tag{A5}$$

where the superscript −1 is the inverse of matrix, and vector **e** is identical to (A4). In obtaining (A5), a constrained condition, min(B = $\mathbf{w}^H\mathbf{R}\mathbf{w}$) subject to $\mathbf{e}^H\mathbf{w}$ = 1, is given to derive an optimal weighting vector:

$$\mathbf{w}_C = \frac{\mathbf{R}^{-1}\mathbf{e}}{\mathbf{e}^H \mathbf{R}^{-1} \mathbf{e}}. \tag{A6}$$

Substituting (A6) into (A3) yields (A5). The estimator (A5) suppresses the contamination of the sidelobes in the weighting pattern of (A4); however, it is sensitive to small errors in the received signals, which may cause the covariance matrix R to be singular or badly scaled, leading to an inaccurate estimate of R^{-1}.

To mitigate the sensitivity of (A5) to small errors in the received signals, the norm-constrained Capon method uses two constraints:

$$\min(B = \mathbf{w}^H\mathbf{R}\mathbf{w}) \text{ subject to } \mathbf{e}^H\mathbf{w} = N \times M, \tag{A7}$$

$$|\,\mathbf{w}^H\mathbf{w}\,| \leq \delta\,(N \times M), \tag{A8}$$

where δ is a user-defined value. Expression (A7) is basically the same constraint used in derivation of (A6) except for the multiplier $N \times M$. To meet the demand of norm constraint (A8), a diagonal-loading value σ is applied to (A6):

$$\mathbf{w}_{NC} = \frac{(\mathbf{R} + \sigma\mathbf{I})^{-1}\mathbf{e}}{\mathbf{e}^H (\mathbf{R} + \sigma\mathbf{I})^{-1}\mathbf{e}}\,(N \times M), \tag{A9}$$

where **I** is the identity matrix. A suitable $\mathbf{w}_{NC}$ can be obtained by giving a proper value of δ and then increasing the value of σ from 0 until (A9) satisfies the constraint (A8). The brightness is thus estimated with the following expression:

$$B(\mathbf{a}, r) = \mathbf{w}_{NC}^H (\mathbf{R} + \sigma\mathbf{I}) \mathbf{w}_{NC}. \tag{A10}$$

Note that selection of a proper δ value is another issue of the NC-Capon method; it could depend on the data and radar system characteristics. The effect of δ value on the outcomes of brightness width (aspect angle) were examined and shown in Figures 5 and 6.

In this study, the spatial weighting function $W(a, r)$ was ignored because of highly localized FAI echoes. Moreover, in use of these equations for 1D range imaging, only one receiver is needed and so the receiver-indexed terms and parameters are either omitted or reduced to unit value. For 2D angular imaging, only one carrier frequency is required and thus the frequency-indexed terms and parameters are either left out or replaced by one. Notice that the zenith orientation is toward the bearing of the radar beam because of a tilt beam used, and then the range goes along the radar beam direction and the angular surface is the transverse of the radar beam direction.

One more note is that in multiple frequency operation, the M carrier frequencies are used with sequential radar pulses and repeat in observation, therefore the time difference between the echoes from the first and M-th pulses is M times that of the interpulse period (IPP). If the target moves very fast, such as meteor heads, a significant change of location occurs during the time period of $M \times$ IPP so that range imaging of the target could deteriorate or fail.

References

1. Hocking, W.K.; Fukao, S.; Tsuda, T.; Yamamoto, M.; Sato, T.; Kato, S. Aspect sensitivity of stratospheric VHF radar wave scatterers, particularly above 15-km altitude. *Radio Sci.* **1990**, *25*, 613–627. [CrossRef]
2. Chen, J.-S.; Furumoto, J. Measurement of atmospheric aspect sensitivity using coherent radar imaging after mitigation of radar beam weighting effect. *J. Atmos. Ocean. Technol.* **2013**, *30*, 245–259. [CrossRef]
3. Kudeki, E.; Farley, D. Aspect sensitivity of equatorial electrojet irregularities and theoretical implications. *J. Geophys. Res.* **1989**, *94*, 426–434. [CrossRef]
4. Farley, D.T.; Hysell, D.L. Radar measurement of very small aspect angles in the equatorial ionosphere. *J. Geophys. Res.* **1996**, *101*, 5177–5184. [CrossRef]
5. Huang, C.-M.; Kudeki, E.S.; Frank, J.; Liu, C.-H.; Röttger, J. Brightness distribution of mid-latitude E region echoes detected at Chung-LI VHF radar. *J. Geophys. Res.* **1995**, *100*, 14703–14715. [CrossRef]
6. Lu, F.; Farley, D.T.; Swartz, W.E. Spread in aspect angles of equatorial E region irregularities. *J. Geophys. Res.* **2008**, *113*, A11309.
7. Briggs, B.H. Radar measurements of aspect sensitivity of atmospheric scatterers using spaced-antenna correlation techniques. *J. Atmos. Terr. Phys.* **1992**, *54*, 153–165. [CrossRef]
8. Hooper, D.; Thomas, L. Aspect sensitivity of VHF scatterers in the troposphere and stratosphere from comparison of powers in off-vertical beams. *J. Atmos. Terr. Phys.* **1995**, *57*, 655–663. [CrossRef]
9. Worthington, R.M.; Palmer, R.D.; Fukao, S. Complete maps of the aspect sensitivity of VHF atmospheric radar echoes. *Ann. Geophys.* **1999**, *17*, 1116–1119. [CrossRef]
10. Palmer, R.D.; Larsen, M.F.; Fukao, S.; Yamamoto, M. On the relationship between aspect sensitivity and spatial interferometric in-beam incident angles. *J. Atmos. Sol. Terr. Phys.* **1998**, *60*, 37–48. [CrossRef]
11. Tsuda, T.; VanZandt, T.E.; Saito, H. Zenith-angle dependence of VHF specular reflection echoes in the lower atmosphere. *J. Atmos. Sol. Terr. Phys.* **1997**, *59*, 761–775. [CrossRef]
12. Chu, Y.-H.; Chao, J.-K.; Liu, C.-H.; Röttger, J. Aspect sensitivity at tropospheric heights measured with vertically pointed beam of the Chung-Li VHF radar. *Radio Sci.* **1990**, *25*, 539–550. [CrossRef]
13. Zecha, M.; Bremer, J.; Latteck, R.; Singer, W.; Hoffmann, P. Properties of midlatitude mesosphere summer echoes after three seasons of VHF radar observations at 54°N. *J. Geophys. Res.* **2003**, *108*, 8439. [CrossRef]
14. Wang, C.-Y.; Chu, Y.-H.; Su, C.-L.; Kuong, R.-M.; Chen, H.-C.; Yang, K.-F. Statistical investigations of layer-type and clump-type plasma structures of 3-m field-aligned irregularities in nighttime sporadic E region made with Chung-Li VHF radar. *J. Geophys. Res.* **2011**, *116*, A12311. [CrossRef]
15. Palmer, R.D.; Gopalam, S.; Yu, T.-Y.; Fukao, S. Coherent radar imaging using Capon's method. *Radio Sci.* **1998**, *33*, 1585–1598. [CrossRef]
16. Chilson, P.B.; Yu, T.-Y.; Palmer, R.D.; Kirkwood, S. Aspect sensitivity measurements of polar mesosphere summer echoes using coherent radar imaging. *Ann. Geophys.* **2002**, *20*, 213–223. [CrossRef]
17. Chen, J.-S.; Wang, C.-Y.; Su, C.-L.; Chu, Y.-H. Meteor observations using radar imaging techniques and norm-constrained Capon method. *Planet. Space Sci.* **2020**, *184*, 104884. [CrossRef]
18. Hashiguchi, H.; Manjo, T.; Yamamoto, M. Development of Middle and Upper atmosphere radar real-time processing system with adaptive clutter rejection. *Radio Sci.* **2018**, *53*, 83–92. [CrossRef]
19. Nishimura, K.; Nakamura, T.; Sato, T.; Sato, K. Adaptive Beamforming Technique for Accurate Vertical Wind Measurements with Multichannel MST Radar. *J. Atmos. Ocean. Technol.* **2012**, *29*, 1769–1775. [CrossRef]
20. Kamio, K.; Nishimura, K.; Sato, T. Adaptive sidelobe control for clutter rejection of atmospheric radars. *Ann. Geophys.* **2004**, *22*, 4005–4012. [CrossRef]
21. Yu, T.-Y.; Palmer, R.D. Atmospheric radar imaging using multiple-receiver and multiple-frequency techniques. *Radio Sci.* **2001**, *36*, 1493–1503. [CrossRef]
22. Hassenpflug, G.; Yamamoto, M.; Luce, H.; Fukao, S. Description and demonstration of the new Middle and Upper atmosphere radar imaging system: 1-D, 2-D, and 3-D imaging of troposphere and stratosphere. *Radio Sci.* **2008**, *43*, RS2013. [CrossRef]
23. Yu, T.-Y.; Furumoto, J.; Yamamoto, M. Clutter suppression for high-resolution atmospheric observations using multiple receivers and multiple frequencies. *Radio Sci.* **2010**, *45*, 1–15. [CrossRef]
24. Chen, J.-S.; Furumoto, J.; Yamamoto, M. Three-dimensional radar imaging of atmospheric layer and turbulence structures using multiple receivers and multiple frequencies. *Ann. Geophys.* **2014**, *32*, 899–909. [CrossRef]
25. Chen, J.-S.; Wang, C.-Y.; Chu, Y.-H.; Su, C.-L.; Hashiguchi, H. 3-D radar imaging of E-region field-aligned plasma irregularities by using multireceiver and multifrequency techniques. *IEEE Trans. Geosci. Remote Sens.* **2018**, *56*, 5591–5599. [CrossRef]
26. Lin, F.-F.; Wang, C.-Y.; Su, C.-L.; Shiokawa, K.; Saito, S.; Chu, Y.-H. Coordinated observations of F region 3 m field-aligned plasma irregularities associated with medium-scale traveling ionospheric disturbances. *J. Geophys. Res. Space Phys.* **2016**, *121*, 3750–3766. [CrossRef]
27. Yin, W.; Jin, W.; Zhou, C.; Liu, Y.; Tang, Q.; Liu, M.; Chen, G.; Zhao, Z. Lightning detection and imaging based on VHF radar interferometry. *Remote Sens.* **2021**, *13*, 2065. [CrossRef]

28. Su, C.-L.; Chen, H.-C.; Chu, Y.-H.; Chung, M.-Z.; Kuong, R.-M.; Lin, T.-H.; Tzeng, K.-J.; Wang, C.-Y.; Wu, K.-H.; Yang, K.-F. Meteor radar wind over Chung-Li (24.9°N, 121°E), Taiwan, for the period 10–25 November 2012 which includes Leonid meteor shower: Comparison with empirical model and satellite measurements. *Radio Sci.* **2014**, *49*, 597–615. [CrossRef]
29. Supplementary Brightness Maps of Radar Imaging. Available online: https://drive.google.com/drive/folders/1u-2wSzr5jdeMzXG7F_ZC0I-8SziWtoJC?usp=sharing (accessed on 27 August 2021).

remote sensing

MDPI

Article

Magnetic Field and Electron Density Scaling Properties in the Equatorial Plasma Bubbles

Paola De Michelis [1,*], Giuseppe Consolini [2], Tommaso Alberti [2], Roberta Tozzi [1], Fabio Giannattasio [1], Igino Coco [1], Michael Pezzopane [1] and Alessio Pignalberi [1]

[1] Istituto Nazionale di Geofisica e Vulcanologia, Via di Vigna Murata 605, 00143 Roma, Italy; roberta.tozzi@ingv.it (R.T.); fabio.giannattasio@ingv.it (F.G.); igino.coco@ingv.it (I.C.); michael.pezzopane@ingv.it (M.P.); alessio.pignalberi@ingv.it (A.P.)

[2] INAF—Istituto di Astrofisica e Planetologia Spaziali, Via del Fosso del Cavaliere 100, 00133 Roma, Italy; giuseppe.consolini@inaf.it (G.C.); tommaso.alberti@inaf.it (T.A.)

* Correspondence: paola.demichelis@ingv.it

Abstract: The ionospheric plasma density irregularities are known to play a role in the propagation of electromagnetic signals and to be one of the most important sources of disturbance for the Global Navigation Satellite System, being responsible for degradation and, sometimes, interruptions of the signals received by the system. In the equatorial ionospheric F region, these plasma density irregularities, known as plasma bubbles, find the suitable conditions for their development during post-sunset hours. In recent years, important features of plasma bubbles such as their dependence on latitude, longitude, and solar and geomagnetic activities have been inferred indirectly using their magnetic signatures. Here, we study the scaling properties of both the electron density and the magnetic field inside the plasma bubbles using measurements on board the Swarm A satellite from 1 April 2014 to 31 January 2016. We show that the spectral features of plasma irregularities cannot be directly inferred from their magnetic signatures. A relation more complex than the linear one is necessary to properly describe the role played by the evolution of plasma bubbles with local time and by the development of turbulent phenomena.

Keywords: plasma turbulence; ionospheric irregularities

Citation: De Michelis, P.; Consolini, G.; Alberti, T.; Tozzi, R.; Giannattasio, F.; Coco, I.; Pezzopane, M.; Pignalberi, A. Magnetic Field and Electron Density Scaling Properties in the Equatorial Plasma Bubbles. *Remote Sens.* **2022**, *14*, 918. https://doi.org/10.3390/rs14040918

Academic Editor: Michael E. Gorbunov

Received: 8 January 2022
Accepted: 10 February 2022
Published: 14 February 2022

1. Introduction

One of the most interesting features of the equatorial ionospheric F region is the existence of plasma density irregularities, which find the suitable conditions for their development during post-sunset hours [1]. In the last few years, different terminologies have been used to term the equatorial ionospheric irregularities. Often, these are identified as equatorial spread-F, in virtue of the spread of the experimental trace either in frequency or in amplitude affecting the ionosonde observations, which is the marking of the existence of electron density irregularities in the reflecting layer (e.g., Booker and Wells [2]). With the advent of in situ satellite observations, they started using terms such as "depletions", "bite-outs," or "plasma holes", because the irregularities were recorded mostly at locations where the electron density is depleted with respect to the background ionosphere (e.g., McClure et al. [3], Dyson and Benson [4]). Then, terms such as "plumes" or "wedges" appeared to characterize the morphology of turbulent regions and the development of such irregularities, as observed by radars (e.g., Scannapieco and Ossakow [5], Woodman and La Hoz [6]). In the last few decades, to identify the generation process of these irregularities, the term "bubble" has taken hold more and more (e.g., Burke [7]). It is worth noting that recent studies (e.g., Kil et al. [8]) introduced also the expression "plasma depletion shell" to highlight the three-dimensional structure of bubbles. The formation of these irregularities is recognized to be driven by the Rayleigh–Taylor instability mechanism [6,9,10] that generates in the ionosphere a situation similar to that occurring when a heavy fluid flows over a lighter one. The absence of sunlight, which leads to a different rate of recombination in the ionospheric layers, causes a density gradient between the upper and lower ionospheric F region with a plasma density in the upper F region higher than in the lower one. Such steep vertical density gradients realize the conditions for the growth of the Rayleigh–Taylor instability. These plasma density irregularities expand vertically and successively move along the magnetic field lines on each side of the geomagnetic equator. Typically, they also drifted eastward from vertical polarization electric fields associated with zonal neutral winds in the F region.

The result is the formation of magnetic field-aligned plasma-depleted regions, which are usually known as equatorial plasma bubbles. They are unstable to plasma density perturbations and polarization electric fields so that secondary irregularities can be generated, which are mainly the result of cascading processes. Nowadays, many important characteristics of these irregularities are known. They have different scale sizes, from hundreds of kilometers down to a few decameters [11,12], are mainly observed within a narrow band of $\pm 20^{\circ}$ magnetic latitude [9], are observable in a wide range of altitudes [6,9] up to around 1000/1500 km [13–15], and their spatial and temporal distributions are strongly controlled by various geophysical parameters such as solar [16] and geomagnetic activity [17], season [16,18], local time, and longitude [8,19–21]. What drives the scientific community to investigate the plasma density irregularities is the knowledge of their role in the propagation of electromagnetic signals. They can be responsible for the disturbance of radio waves and are one of the most important sources of disturbance for the Global Navigation Satellite System (GNSS) signal. Indeed, in the worst case, plasma density irregularities can cause a loss of lock event, a condition for which a GNSS receiver can no longer track the signal sent by the satellite, with a consequent degradation of the positioning accuracy. In the last few years, the relevance of the GNSS in our society has substantially increased, as many critical infrastructures and economies are dependent on the positioning, navigation, and timing services of GNSS, so that our society is nowadays vulnerable to damages due to the malfunction of these systems. For this reason, one of the research priorities in the space weather community is to improve the knowledge of these plasma density irregularities, trying to better understand their features and generation mechanisms.

An interesting aspect of these irregularities, which requires more study, is their possible turbulent nature. Indeed, the equatorial ionosphere has proper physical conditions for the establishment of a fully developed turbulence [22,23]. Through numerical simulations, it has been possible to reproduce the ionospheric irregularities and study the corresponding dynamics under collisional and inertial regime flows [24,25]. It has been found that the plasma irregularities may be characterized by energy spectra with slopes ranging between $-5/3$ and -3 associated with the occurrence of energy and/or enstrophy cascades, respectively [26,27]. Over the years, these numerical models have been improved trying to describe the ionospheric two-dimensional plasma turbulence in a more realistic way. It has been found that after sunset, at apex heights between 600 and 1000 km, large-scale irregularities present a dynamics typical of inertial regime flow [28] and that although the foundations of the Kraichnan's theory could be undetermined in the topside equatorial ionospheric F region, the numerical simulation results seem to confirm the probable existence of large plasma density irregularities. These are characterized by velocity fluctuation spectra following power laws with spectral indices similar to those predicted for energy and enstrophy cascades in two-dimensional turbulent flows [23]. In the last twenty years, thanks essentially to the rapid increase in the performance of computers, it has been possible to improve the models describing the evolution of plasma depletions. Thus, new models capable of reproducing the three-dimensional turbulent structures of equatorial plasma bubbles with high spatial and temporal resolutions have been developed (see for example Yokoyama et al. [29], Yokoyama [30]). The results of such simulations confirm observations of plasma density data recorded, for example, on board DE-2 satellite. In this case, it has been found that plasma density is characterized by an energy spectrum with a slope around $-5/3$ for scale sizes larger than 1 km in all regions characterized by strong plasma density irregularities [28]. Nowadays, the numerical simulations have become useful tools both to study the plasma bubbles and their nonlinear evolution, and to understand some of their features, which cannot be fully understood from theoretical predictions. Therefore, it is extremely important to combine the numerical simulation results with real observations.

In order to detect and characterize plasma density irregularities, different methods, approaches, instruments, and data can be used such as, for example, ionosondes, airglow imager, coherent/incoherent scatter radars [31], as well as Global Positioning System (GPS) Total Electron Content (TEC) [32,33] in the topside ionosphere. The ground-based observations are mostly fixed in a single location and cannot give the spatial variations of plasma density irregularities in a large area. Conversely, observations from Low Earth Orbit (LEO) satellites such as for example DMSP (Defense Meteorological Satellite Program), ROCSAT-1 (Republic of China Satellite), C/NOFS (Communications/Navigation Outage Forecasting System), and COSMIC (Constellation Observing System for Meteorology, Ionosphere, and Climate) permitted climatology studies and provided the latitude-local time distributions of plasma density irregularities. The recently launched Global-scale Observation of Limb and Disk (GOLD) mission, which is located at a geo-stationary orbit, seems to be a promising mission to study bubbles, offering the opportunity to observe the Earth's complete disk continuously from the geostationary orbit [34,35]. A different method for the detection of ionospheric irregularities based on the diamagnetic effect was suggested a few years ago. Plasma density irregularities are indeed associated with magnetic field perturbations [36,37] that can be used as a proxy for plasma irregularities. There are different ways to describe the so-called diamagnetic effect but, perhaps, the easiest one is to consider that the sum of magnetic and plasma pressures has to be constant in a quasi-stationary state. This means that when the plasma pressure is reduced, the magnetic pressure and consequently the magnetic field strength must increase. The equatorial plasma bubbles identify regions of reduced plasma density; consequently, they have to be characterized by a magnetic field strength higher than the ambient one. That was confirmed using simultaneous observations of magnetic field and electron density recorded by the CHAllenging Minisatellite Payload (CHAMP) satellite. These measurements revealed the increase of a few nT in the main field intensity inside the plasma density depletions [38] Park et al. [39] quantitatively investigated the actual balance between plasma and magnetic pressures within the plasma bubbles, showing that a dominant part of the magnetic pressure change could be explained in term of plasma density change under specific assumptions. Based on the high correlation between the variations of plasma density and those associated with magnetic field strength, an extensive survey of magnetic signatures related to plasma density irregularities was used for statistical studies [15,40,41] devoted to the analysis of the global distribution of equatorial plasma irregularities in the topside ionosphere. These studies revealed features of the magnetic signatures that closely reflected those of the plasma bubbles previously obtained using different methods as GPS scintillation [19], in situ plasma density measurements [42,43], and radio wave propagation [44]. Furthermore, magnetic field measurements recorded on board the CHAMP satellite with different time resolution (50 Hz and 1 Hz) permitted addressing some properties of the plasma density irregularities related to their different spatial sizes. Evidently, these results do not reflect directly the features of plasma depletions but their associated magnetic signatures. However, the features obtained using the magnetic signatures of the plasma depletions are consistent with those obtained by directly using the electron density measurements. For this reason, the diamagnetic effect, as an indirect way of sampling plasma density structures, can be extremely useful in studying plasma irregularities.

While it is correct to assume that some plasma bubbles features can be inferred using the corresponding magnetic signatures, the assumption that plasma bubbles scaling properties can be inferred using the associated magnetic field scaling properties may be not necessarily correct as well. The purpose of this study is to understand if the hypothesis of a linear relation between the scaling properties of the electron density and of the magnetic field is valid. So, needless to say, the clarification of this point is of crucial importance.

2. Data Description

To investigate the scaling features of the electron density irregularities and of their magnetic signatures at low latitude, we use measurements on board the Swarm A satellite during its near-polar orbit around the Earth at an altitude of approximately 460 km [45]. Swarm A is one of the satellites constellation launched by ESA for Earth observation. Specifically, the ESA mission consists of three identical satellites named Alpha, Bravo, and Charlie (A, B, and C), which were launched on 22 November 2013 into a near-polar orbit. Swarm A and C fly side-by-side at an altitude of 462 km (initial altitude) and at an inclination angle of 87.35°, whereas Swarm B flies at a higher orbit of 511 km (initial altitude) and at an inclination angle of 87.75°.

We consider measurements collected from 1 April 2014 to 31 January 2016 at low and mid latitudes ($|Lat| < 40°$) under conditions of high solar activity. It is known indeed that the equatorial plasma bubbles occur mainly during periods characterized by high extreme ultraviolet solar flux [41]. In the selected time interval, the value of the solar radio flux index (F10.7) is equal to (100 ± 30) sfu. This is not an extremely high value for the solar activity, but the selected period, which covers a part of the descending phase of Solar Cycle 24, is the only period of moderate/high solar activity during which Swarm is flying. Electron density and magnetic field vector data at a rate of 1 Hz are selected from the ESA dissemination server (http://earth.esa.int/swarm accessed on 12 February 2022). To highlight the magnetic signatures associated with plasma bubbles, we remove from the magnetic field measurements the contribution due to the magnetic field of internal origin, as modeled by CHAOS-6 [46]. CHAOS-6 is the latest generation of the CHAOS series of global high-resolution geomagnetic field models, which spans a period between 1997 and 2016. It is derived primarily from magnetic satellite data (Swarm, CHAMP, Ørsted, and SAC-C), although ground-based activity indices and observatory monthly means are also used. It is able to model the dominating core field, the crustal field, and the magnetic fields from large-scale magnetospheric currents, but there is no explicit representation of the ionospheric field or fields due to magnetosphere—ionosphere coupling currents. One of the applications of this model is the removal from the observed magnetic field of the large contributions from the prime sources when the weaker effects such as ionospheric currents, plasma irregularities, or even ocean tidal motions want to be investigated. We use the CHAOS-6 package to compute the geomagnetic field of internal origin, which is successively subtracted to the magnetic field observed by Swarm. Lastly, we focus on the 18:00–24:00 local time (LT) sector and select periods characterized by low/moderate geomagnetic activity. To select these periods, we used the Kp index, which is designed to measure the solar particle radiation through its magnetic effects, and today, it is considered a proxy for the energy input from the solar wind to Earth. It is a global geomagnetic activity index based on 3-hour measurements from ground-based magnetometers around the world. The Kp index ranges from 0 to 9, where a value of 0 means that the geomagnetic activity is absent, and a value of 9 means that an extreme geomagnetic storming is occurring. Data are distributed by GFZ (Helmholtz Centre Potsdam) [47] and they are redistributed by various data centers and databases. To select periods characterized by low/moderate geomagnetic activity, we consider all those time intervals characterized by $Kp \leq 3$. According to previous results [40], the probability of observing plasma bubbles is maximum under these conditions.

To select the plasma bubbles events, we use the Swarm Level-2 Ionospheric Bubble Index (IBI) product, which was evaluated using plasma density and magnetic field data with a cadence of 1 Hz. A full description of the data file and explanations regarding the IBI data can be found in the L2-IBI product description file (https://earth.esa.int/eogateway/documents/20142/37627/Swarm-Level-2-IBI-product-description.pdf/3e9f6c3a-1ffc-ea53-0161-a18b63f90c6f, accessed on 13 February 2022). This product is composed of three parameters: Bubble Index, Bubble Probability, and Bubble Flag. The Bubble Index can assume three different values: 1, when the analyzed data are affected by a plasma bubble, 0 when there is not a plasma bubble, and −1 when conditions do not allow unequivocally identifying a plasma bubble in the data [48]. Furthermore, when IBI = 1, there is an additional flag, the Bubble Flag (BF), indicating the quality of plasma bubbles detection that is related to exceeding a certain probability threshold. When BF = 1, we are in the presence of a high correlation between plasma density depletion and magnetic field signature. Here, we chose the condition IBI = 1 and BF = 1 to identify the plasma bubbles. However, we note that this automatic detection algorithm cannot distinguish equatorial plasma bubbles from plasma blobs, which are localized plasma density enhancements. This means that the events identified by the L2-IBI processor can contain also blob events. However, the occurrence probability of equatorial plasma bubbles is significantly higher than that of blobs mainly during solar maximum years.

3. Method and Data Analysis

The scaling features of the electron density and magnetic field at different temporal scales are evaluated using the so-called generalized q^{th}-order structure function, namely S_q. For a signal $X(t)$, S_q is given by the following equation:

$$S_q(\tau) = \langle | X(t+\tau) - X(t) |^q \rangle, \tag{1}$$

where t is the time, τ is the time delay, and $\langle ... \rangle$ stands for a statistical average. When $S_q(\tau)$ has a power-law behavior as a function of τ:

$$S_q(\tau) = \tau^{\gamma(q)}, \tag{2}$$

the analyzed signal is characterized by a scale invariance, and $\gamma(q)$ is the so-called q^{th} order scaling exponent, which defines the scaling nature of the increments of the signal $X(t)$. A signal possesses a global scale-invariance if $\gamma(q) = Hq$; that is, it is a linear function of the order q. Conversely, a signal is defined as multifractal, that is, it has a local scale-invariance, if $\gamma(q)$ is a nonlinear convex function of q. Here, we investigate the second-order scaling exponent, $\gamma(2)$, which is related to the power spectral density (PSD) exponent, β (PSD$\simeq f^{-\beta}$), via the following relation:

$$\beta = \gamma(2) + 1. \tag{3}$$

The use of the structure function analysis, rather than the standard Fourier analysis, as a method to detect scaling features is motivated by the fact that the analysis of scaling/spectral features in the real time space, using signal increments, is less affected by the choice of a specific decomposition basis, as it is instead the case for the Fourier analysis. Furthermore, being geophysical signals, such as magnetic field and electron density fluctuations in the ionosphere, generally characterized by power-law spectra, i.e., $PSD(f) \simeq f^{-\beta}$, with spectral exponents β ranging between 1 and 3, one deals with nonstationary signals with stationary increments [49]. In this case, the Fourier analysis, which requires stationarity, might produce unreliable results especially when evaluating spectral properties over short time intervals. Conversely, the structure function analysis works with signal increments that are expected to be nearly stationary, and thus, the corresponding results over short time windows/intervals are more reliable. Furthermore, the structure function analysis is a companion method that is widely used in the framework of turbulence-related studies.

The measurements collected by instruments on board a satellite require some precautions to be correctly used. The satellite crosses regions characterized by different physical processes, which change in space and time. Consequently, the collected measurements, such as in our case electron density and magnetic field strength, present properties with a local character. The scaling properties of these real-time series also acquire a local character, and it is necessary to apply methods of analysis capable of evaluating their local scaling features describing the different regions explored by the satellite. The structure function approach allows doing this. Among the different methods to evaluate the structure functions, here, we apply the local detrended structure function analysis (DSFA) method discussed in De Michelis et al. [50,51] and already successfully applied in a series of previous works showing results that are well in agreement with the present literature on the scaling and/or spectral features of turbulent fluctuations in the ionosphere. In detail, we apply the analysis on a moving window of $N = 301$ points using a time series obtained from the original one after removing its linear trend, if present. This gives us the opportunity to remove from the initial signal all the possible large-scale variations that can affect a correct estimation of the scaling features. The choice of a moving window of 301 points is the result of an accurate series of test over a wide set of simulated multifractional Brownian motions [52–54], which are characterized by a local dependence (nonstationarity) of the scaling/spectral features. Indeed, this number of points is the right compromise to get an estimation of local scaling exponents sufficiently precise (with a typical error $\leq$ 7%), being 1 order of magnitude larger than the longest investigated timescale ($\tau = 30$ s). In each moving window, we evaluate the second-order structure functions associated with the electron density and magnetic field strength time series, respectively. These are evaluated for time delay values (τ) ranging from 1 to 40 s, and the estimated second-order scaling exponents are associated with the position of the satellite at the center of the fixed window. We assume that the results obtained in the time domain are valid in the space domain too (see e.g., [55,56] and references therein). Thus, considering that the Swarm A orbital velocity is around 8 km/s, the results obtained in the time domain between 1 and 30 s are valid in the spatial domain between $\sim$8 and $\sim$ 250 km, i.e., $\delta r \simeq V_s \tau$ where V_s is the satellite velocity. We remark that the previous conversion between the timescale and the corresponding spatial scale is based on the assumption of the Taylor's hypothesis according to which instantaneous spatial averages coincide with temporal averages calculated from the recorded signal. Given this hypothesis as valid, which can be applied for frozen turbulence, we assume that the temporal evolution of turbulent structures is longer than that required for the satellite to cross them [22,55–57]. Hence, by evaluating the scaling properties of the magnetic field strength and electronic density associated with the plasma bubbles, we can investigate the properties of these plasma density irregularities from both a magnetic and plasma point of view in a range of scales between some kilometers and a few hundred kilometers.

The structure function analysis is initially done on the entire time series of around $5 \cdot 10^7$ points (22 months with a time resolution of 1 Hz) and only later, the scaling exponent values of the electron density (Ne) and magnetic field strength ($|\mathbf{B}|$) associated with the occurrence of plasma bubbles are selected.

4. Results

Figure 1 reports a map of the values of the second-order structure function exponent, $\gamma(2)$, which was obtained analyzing N_e and $|\mathbf{B}|$ inside the plasma bubbles. These two different parameters are characterized by similar values of $\gamma(2)$, supporting the previous findings by Lühr et al. [15]. However, to look for any difference, we start by plotting in Figure 2 the corresponding histograms. The second-order structure function exponents range from 0.4 to 2 with the highest probability occurring at (1.0 ± 0.2) in both cases.

Taking into account the relation between the second-order scaling exponent and the Fourier power spectral density exponent ($\beta = \gamma(2) + 1$), we notice that the spectral features obtained in the case of electron density ($\beta \simeq 2$) are in agreement with previous results. For example, Livingston et al. [58] analyzing the spectra of electron density data recorded from the Atmospheric Explorer-E satellite (AE-E) in the interval 20:00–02:00 LT found a distribution of spectral slopes in the range between 1 and 3 centered around 1.9, which is a mean value that agrees with our results and is close to those of Dyson et al. [59], who used data recorded by the OGO satellite, Basu et al. [60], who analyzed AE-E measurements, and Le et al. [61], who analyzed ROCSAT-1 burst mode measurements at 1024 Hz sampling. Concerning the spectral scaling exponents obtained for the magnetic field strength, there are not many previous analyses with which to compare them. Lühr et al. [15] used magnetic field data recorded by CHAMP to investigate the spectral properties of the signal and found values of the spectral indices ranging between 1.4 and 2.6 with a peak around 1.9. Again, these results agree with ours. Although the mean values of the two distributions reported in Figure 2 are equal, their shapes are different, in which the $\gamma(2)$ distribution for N_e is skewed. Indeed, the skewness λ_3 values of $\gamma(2)$ distributions of N_e and $|\mathbf{B}|$ are $\lambda_3^{N_e} \simeq 0.45$ and $\lambda_3^{|\mathbf{B}|} \simeq 0.08$, respectively.

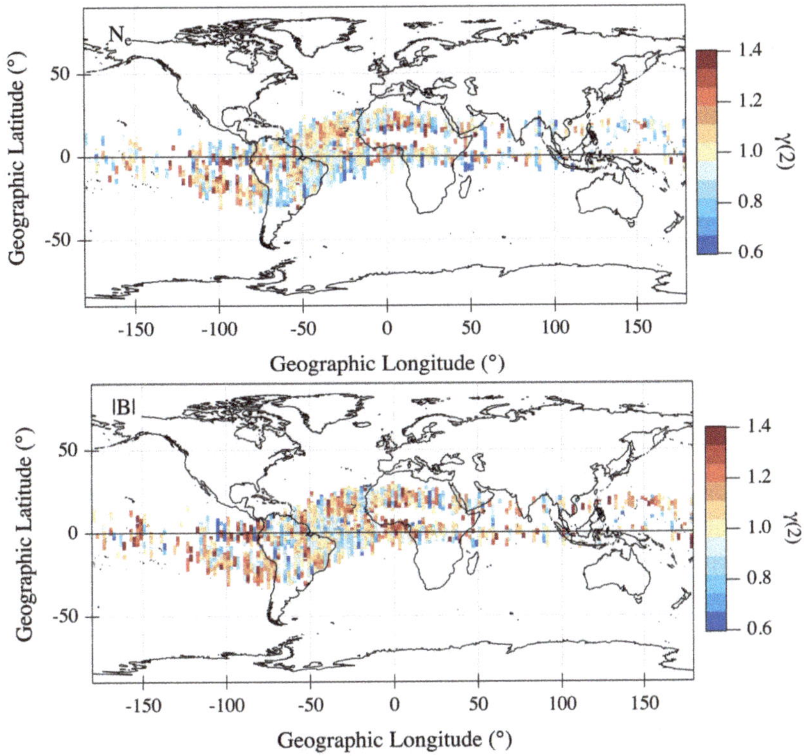

Figure 1. Values of the second-order scaling exponent, $\gamma(2)$, associated with electron density (**top**) and magnetic field strength (**bottom**) inside the plasma bubbles in the geographic latitude–longitude plane. Data are selected according to the following conditions: 18:00 $\leq$ LT$\leq$ 24:00, $\pm 40°$ geographic latitude, $Kp \leq 3$.

Figure 3 reports the values of the second-order structure function exponent obtained analyzing N_e and $|\mathbf{B}|$ inside the plasma bubbles according to QD magnetic latitude. For N_e, the highest occurrence of $\gamma(2)$ is found for $\gamma(2) < 1$ at a latitudinal distance from the equator of around 10° in magnetic latitude in both hemispheres, while in the case of $|\mathbf{B}|$, the highest occurrence of $\gamma(2)$ is found for a value around 1 occurring always in the same regions. Thus, the peak of the highest occurrence of plasma bubbles is associated with a range of $\gamma(2)$ values that is smaller for N_e than $|\mathbf{B}|$.

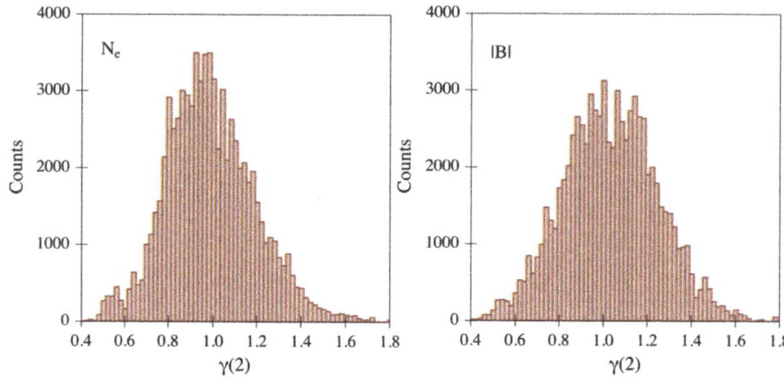

Figure 2. Histograms of the second-order scaling exponent values associated with electron density (**left**) and magnetic field strength (**right**) inside the plasma bubbles, respectively. Data are selected according to the following conditions: 18:00 $\leq$ LT $\leq$ 24:00, $\pm 40^\circ$ geographic latitude, $Kp \leq 3$.

To go into the observed differences between the two distributions, we repeat our analysis considering two-hour time intervals from 18:00 LT to 24:00 LT. Figure 4 reports the distribution of $\gamma(2)$ values relative to N_e and $|\mathbf{B}|$ in the three different selected time intervals. A double peak distribution of $\gamma(2)$ clearly appears analyzing the scaling properties of $|\mathbf{B}|$ in the range 20:00 $\leq$ LT $<$ 22:00. The $\gamma(2)$ distribution has one peak in the next local time interval (22:00 $\leq$ LT $\leq$ 24:00) and a rather flat shape in the previous one (18:00 $\leq$ LT $<$ 20:00). In the case of $\gamma(2)$ values associated with N_e, the data distribution has only one peak in the second and third local time intervals while a distribution with a rather flat shape appears in the first local time interval. The emergence of a bimodal distribution for the magnetic field $\gamma(2)$ exponent could be an indication of a non-unique relation between the scaling features of electron density and magnetic field fluctuations in plasma bubbles. Furthermore, as clearly shown in several works (see, e.g., [62]) (and references therein), in the second LT interval here considered, plasma bubbles are still evolving at Swarm altitude, so that the bimodality could be a reflection of the nonstationarity of the plasma bubble structures.

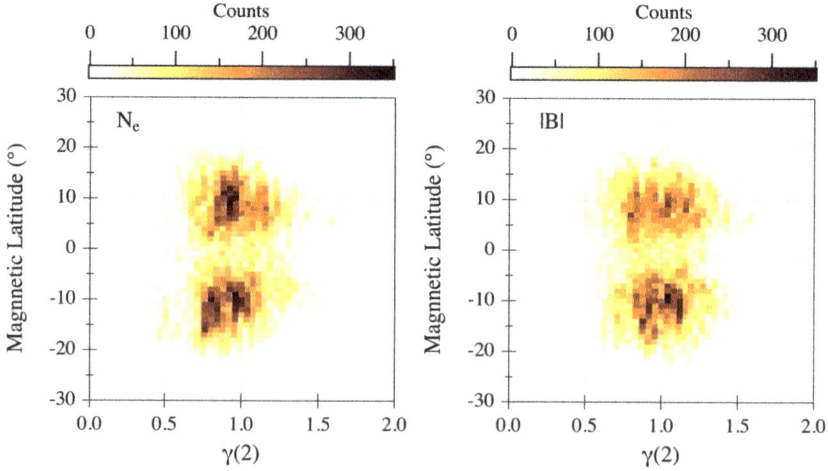

Figure 3. Dependence on QD magnetic latitude of the second−order scaling exponent values associated with N_e (**left**) and $|\mathbf{B}|$ (**right**) inside the plasma bubbles.

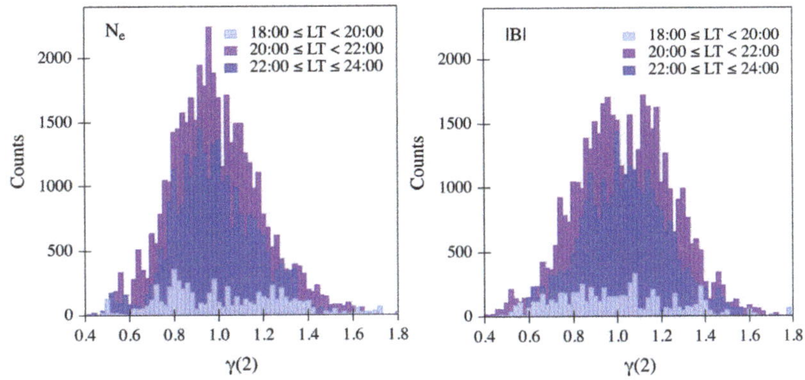

Figure 4. Histograms of the second-order scaling exponent values associated with electron density (**left**) and magnetic field strength (**right**) inside the plasma bubbles, respectively. Data are selected according to the following conditions: 18:00 $\leq$ LT $<$ 20:00 (light blue), 20:00 $\leq$ LT $<$ 22:00 LT (blue) and 22:00 $\leq$ LT $\leq$ 24:00 (purple). In all cases, the geographic latitude is $\pm 40°$ and $Kp \leq 3$.

To get more insights on the bimodal character of the $\gamma(2)$ distribution that appears, analyzing the scaling properties of $|\mathbf{B}|$ inside the plasma bubbles between 20:00 LT and 22:00 LT, we check how the $\gamma(2)$ values of the two peaks are distributed in latitude and longitude. Taking into account that the two different populations are characterized by values of $\gamma(2) < 1$ and $\gamma(2) > 1$, respectively, Figure 5 shows their distributions in the latitude–longitude plane. The two geographic distributions seem similar, although the values $\gamma(2) < 1$ seem to prefer the areas closest to the equator. This feature is well highlighted on the bottom of Figure 5, where the $\gamma(2)$ values are sorted according to magnetic latitude. Here, the red line clearly shows how the plasma bubbles around the magnetic equator ($\pm 4°$) are often characterized by a magnetic field strength with a second-order scaling exponent $\gamma(2) < 1$. The values of the scaling exponents of the magnetic field strength seem to depend on magnetic latitude at least during the hours following their formation. To better investigate this point, we report in the left panels of Figure 6 the joint probability distributions of $\gamma(2)$ values relative to N_e ($\gamma_{N_e}(2)$) and $|\mathbf{B}|$ ($\gamma_{|\mathbf{B}|}(2)$) along with the mean values of $\gamma_{|\mathbf{B}|}(2)$ exponent (white circles) for fixed $\gamma_{N_e}(2)$ at the three different selected two-hour time intervals. There is a large spreading of the observed distributions with most of the values concentrated around $[\gamma_{N_e}(2), \gamma_{|\mathbf{B}|}(2)] = [1, 1]$, which indicates that the relation between electron density and magnetic field scaling exponents can be assumed to be linear only in a zero-order approximation. However, looking at the trend of the mean values of $\gamma_{|\mathbf{B}|}(2)$ fixed $\gamma_{N_e}(2)$, we can realize how the relation between the two scaling indices significantly departs from a linear trend for $\gamma_{N_e}(2) < 0.9$ at LT $\geq 20:00$. This result suggests that a more complex relation might exist between plasma density and magnetic field variations. Furthermore, the departure from a linear relation increases with the midnight approaching. We can conjecture that the observed departure from linearity occurs when the fluctuations of the electron density show scaling features similar to those expected for convective turbulence, i.e., $\beta \sim 5/3$ [63]. In the right panels of Figure 6, we report the conditioned probably distribution density functions of $\gamma_{|\mathbf{B}|}(2)$ values for fixed $\gamma_{N_e}(2) = 1$. The observed distributions are multimodal, thus supporting the hypothesis that a linear relation between magnetic field intensity and plasma density spectral features has to be considered valid only on average being a very crude approximation. The multimodal structure of the observed distribution could also be the counterpart for different physical processes at the origin of the observed fluctuations, or it could be due to the role played by other physical quantities, such as for example the plasma temperature fluctuations, that could modify the relation between electron density and magnetic field second-order scaling exponents. Moreover, a very interesting result stands in the clear bimodal character of the distribution of the magnetic field $\gamma(2)$ exponent in the first LT sector. Indeed, there is not a one-to-one correspondence between the magnetic field $\gamma(2)$ exponent and that of the electron density fluctuations, i.e., $\gamma_{|\mathbf{B}|}(2) < \gamma_{N_e}(2) \equiv 1$. A possible explanation could be that, as shown in Gruzinov et al. [63], we are still in the early stage of a rising bubble so that the spectral features are still not stationary.

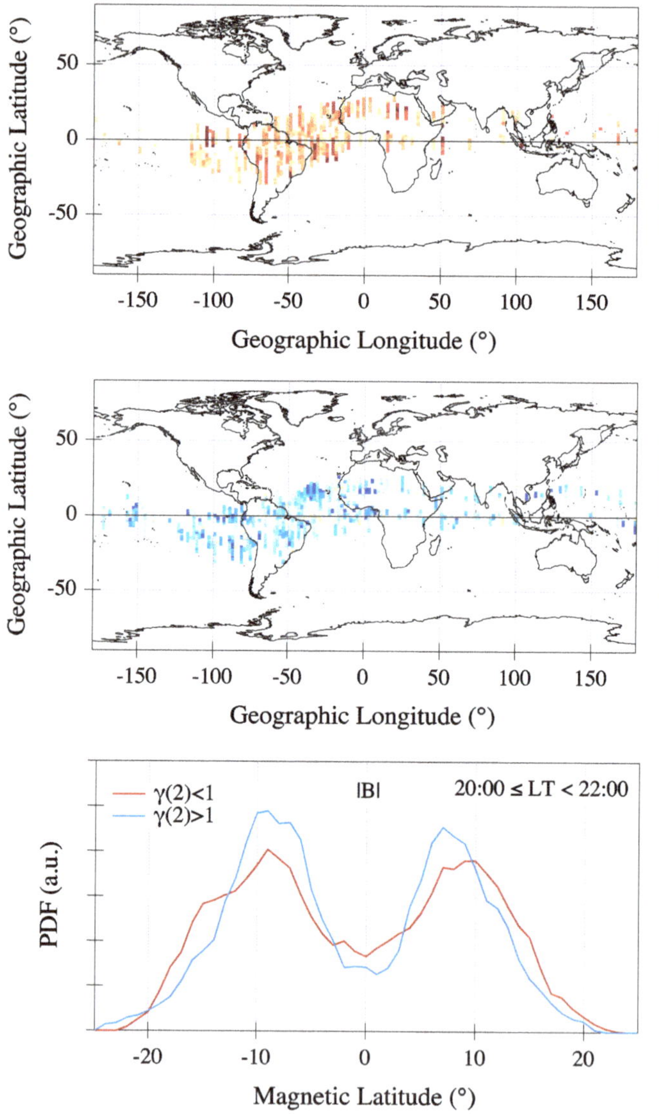

Figure 5. Values of the second-order scaling exponent associated with magnetic field strength inside the plasma bubbles in the geographic latitude–longitude plane during the time interval 20:00 $\leq$ LT $<$ 22:00 LT for $\gamma(2) < 1$ and $\gamma(2) > 1$, respectively. On the bottom, the $\gamma(2) < 1$ and $\gamma(2) > 1$ values are sorted according to magnetic latitude.

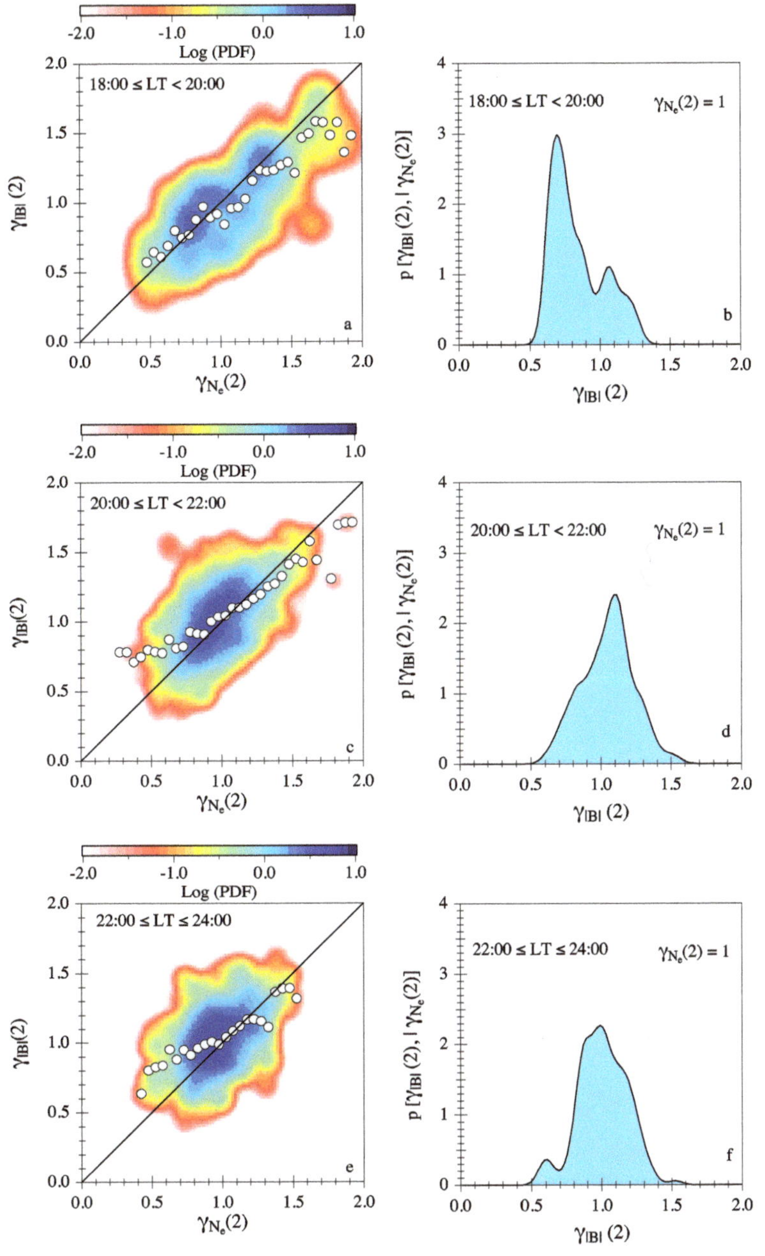

Figure 6. On the left, the joint probability density functions of second-order scaling exponents for N_e and for $|\mathbf{B}|$ in the three different selected two-hour time intervals. White circles refer to the mean values of the magnetic field $\gamma_{|\mathbf{B}|}(2)$ exponent at fixed electron density $\gamma_{N_e}(2)$ exponent values. The black line is the bisector line for which $\gamma_{|\mathbf{B}|}(2) = \gamma_{N_e}(2)$. On the right panels, the conditioned probability distribution density functions of $\gamma_{|\mathbf{B}|}(2)$ values for fixed $\gamma_{N_e}(2) = 1$.

We conclude our analysis reporting in Figure 7 the probability density functions of the location of the plasma bubbles as a function of both the magnetic latitude and LT. These distributions suggest the tendency of plasma bubbles to move away from the equator as the time goes by. This could indicate that the plasma bubbles observed near the midnight sector have not been generated in that LT sector but could have been generated near sunset, have moved at high altitudes, and successively have started to descend in altitude, moving along the magnetic field lines. Indeed, the flux tube intersecting the 460 km Swarm A orbit at $\sim 10^{\circ}$ magnetic latitude has an apex height around 600 km at the equator where the plasma bubbles are generated. This is a crucial point in the interpretation of the LT evolution of the scaling features of magnetic field and density fluctuations inside the plasma bubbles. It makes you understand how, moving away from the sunset, Swarm A crosses older plasma bubbles at locations where the turbulence had time to fully develop.

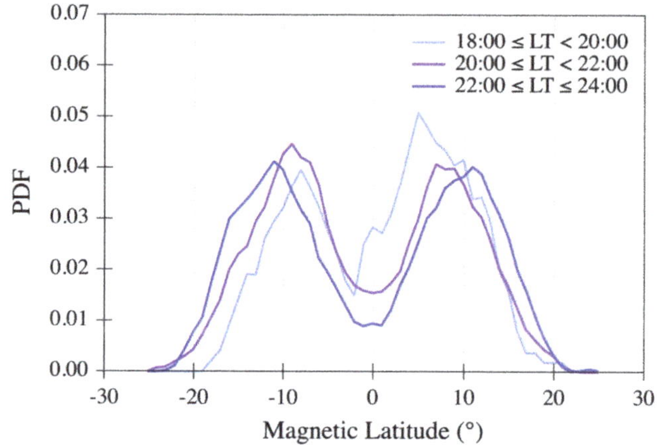

Figure 7. Dependence on magnetic latitude and local time of location of the plasma bubbles.

5. Discussion

Our findings seem to suggest that a non-trivial linear relation exists between the spectral features of the plasma density and magnetic field in the plasma irregularities which depends on the local time and latitude. Actually, the fact that the magnetic field and electron density scaling properties are not simply linearly related is somehow expected. Usually, when we study the plasma bubbles dynamics, we start with a few assumptions.

First, following the derivation of the diamagnetic current in Alken et al. [64], i.e., considering the standard magnetohydrodynamic equations for a plasma in a steady-state configuration, that is, $\partial_t \to 0$, and neglecting gravity, we can write that the plasma pressure force (∇p) is counter-balanced by the Lorentz force ($\mathbf{j} \times \mathbf{B}$), i.e.,

$$\nabla p = \mathbf{j} \times \mathbf{B}, \tag{4}$$

where $\mathbf{B}$ is the ambient magnetic field. Combining Equation (4) with the Maxwell's equation describing the Ampère's circuital law in absence of displacement currents

$$\nabla \times \mathbf{B} = \mu_0 \mathbf{j}, \tag{5}$$

where μ_0 is the permeability of free space, we can write

$$\nabla p = \frac{1}{\mu_0} (\nabla \times \mathbf{B}) \times \mathbf{B}. \tag{6}$$

By using the vector calculus identity

$$(\nabla \times \mathbf{B}) \times \mathbf{B} = (\mathbf{B} \cdot \nabla)\mathbf{B} - \nabla\left(B^2/2\right), \tag{7}$$

we can rewrite Equation (4) as

$$\nabla(p + \frac{B^2}{2\mu_0}) = \frac{1}{\mu_0}(\mathbf{B} \cdot \nabla)\mathbf{B}. \tag{8}$$

The term on the left of Equation (8) is the sum of plasma pressure force and magnetic pressure one, while on the right, the term represents magnetic tension due to the curvature of the field lines. Taking into account that the plasma structures are much smaller than the geomagnetic field curvature radius, it is possible to assume a linear field line geometry. In this case, we can neglect the magnetic curvature term, $(\mathbf{B} \cdot \nabla)\mathbf{B} \sim 0$, in the momentum balance so that a change in plasma pressure is, in first approximation, compensated by a change in the magnetic pressure (second assumption). Thus, according to Lühr et al. [37] and taking into account that $p = k_B N_e(T_e + T_i)$, we can write that across the irregularity walls, the following relation holds:

$$\Delta\left(\frac{B^2}{2\mu_0}\right) + k_B \Delta[N_e(T_e + T_i)] = 0 \tag{9}$$

where Δ denotes a finite scale difference (i.e., the gradient in Equation (4) computed at the bubble intensity scales), k_B is the Boltzmann constant, N_e is the electron density, and T_e and T_i are electron and ion temperatures, respectively.

Third, considering that the strength of the ambient magnetic field is around four orders of magnitude greater than the magnetic field generated by the equatorial plasma bubbles, Equation (9) can be linearized as:

$$\Delta B \approx -\frac{\mu_0 k_B}{B}\Delta[N_e(T_e + T_i)]. \tag{10}$$

Fourth, assuming a constant plasma temperature, i.e., assuming a local thermodynamic equilibrium $T_e + T_i = T_{eq}$, according to the latter expression, a high correlation must exist between ΔB and ΔN_e across plasma bubble walls, and plasma density irregularities can be detected reliably through magnetic field data, being valid the following relation

$$\Delta B \approx C_0 \, \Delta N_e \tag{11}$$

where C_0 is a constant. Thus, the scaling properties of the electron density and magnetic field data should be equal, leading to a universal character of the statistical properties of fluctuations as

$$\Delta B \approx C_0 \, \Delta N_e \sim \tau^{\gamma(1)}, \tag{12}$$

that can be generalized as

$$\langle|\Delta B|^q\rangle \approx \langle|\Delta N_e|^q\rangle \sim \tau^{\gamma(q)}. \tag{13}$$

Conversely, our analysis find a discrepancy between the theoretical prediction by Lühr et al. [37] and the experimental results.

As observed by Alken et al. [64], studying the global flow of the gravity-driven and diamagnetic currents, the diamagnetic effect formula proposed by Lühr et al. [37] is strictly valid only when the plasma has reached equilibrium and the ambient field has no curvature. The Earth's magnetic field has a large dipole moment and the assumption to ignore the magnetic field line curvature could be one of the possible reasons responsible for the observed discrepancies. Furthermore, it is however true that the plasma bubbles are unstable systems that evolve rapidly in time. This means that the assumption of a plasma in equilibrium that, in first approximation, may be only valid in the early hours after sunset, is no longer true in the following hours when the turbulence inside plasma bubbles is fully developed [51]. This point is supported by the discrepancies from a linear relation between magnetic field and plasma density scaling exponents for increasing LT, as shown in Figure 6, and by the geographic location of plasma bubbles in LT, as shown in Figure 7. We remark that Equation (10), being a linear approximation of Equation (4), is expected to be valid in the first phases of the plasma bubbles rising, i.e., for LT near the dusk–night sector, as highlighted in Figure 6. There is also an additional issue related to the assumption of a constant plasma (electron and ion) temperature that has to be considered. This might be true in the first instants of plasma bubbles formation, i.e., when plasma bubbles rise just after the sunset and the instability growth is still quasi-linear. Conversely, when these structures start evolving, the emerging of turbulent fluctuations can strongly affect both plasma density and temperature fluctuations which could passively respond to turbulent fluctuations. This means that the assumption of negligible temperature fluctuations could be no longer valid for hours approaching midnight. This is a crucial point that will be investigated in a successive work.

Last but not least, we note that Equation (10) neglects the role of other terms that might affect the evolution of plasma bubbles, such as for instance the emergence of $\mathbf{E} \times \mathbf{B}$ convective turbulence and/or Hall MHD compressive turbulence, which might be at the basis of some discrepancies between density and magnetic field intensity fluctuations. In particular, we observe that some of the features of the observed fluctuations of the magnetic field, such as the spectral slope near k^{-2} are in good agreement with the emergence of 2D $\mathbf{E} \times \mathbf{B}$ turbulence of electron density, which is indeed isomorphic to the viscous convection of an ordinary fluid driven by temperature gradients [63]. These extra elements require the investigation of the dynamics of the magnetic field components with a further detail than that discussed here in the case of the magnetic field intensity.

6. Summary and Conclusions

In order to understand whether it is possible to study the dynamical features of plasma density irregularities by using independently either the magnetic field or the electron density measurements, this study focused on determining the relationship between the spectral features of electron density and magnetic field strength inside plasma bubbles. In the past, many important features of the equatorial plasma bubbles have been obtained indirectly from the analysis of their magnetic signatures, using the so-called diamagnetic effect. For example, it has been possible to investigate their dependence on latitude, longitude [8,19–21], and solar and geomagnetic activity [16,17] by using magnetic field data recorded by LEO satellites. However, the study of the dynamical features of these plasma density irregularities by using only the magnetic field might not be correct, since it is necessary to assume that the scaling properties of the electron density and magnetic field fluctuations are equal.

In order to address this point, we studied the scaling properties of both the electron density and the magnetic field inside plasma bubbles using measurements on board the Swarm A satellite from 1 April 2014 to 31 January 2016. These were obtained using the so-called generalized q^{th}-order structure function analysis evaluated by applying the local detrended structure function analysis (DSFA) method [50,51]. The considered method, along with the time resolution of the electron density and magnetic field data, permitted investigating spatial scales ranging from some kilometers to a few hundred kilometers. At this range of spatial scales, our findings suggest that it is not correct to hypothesize that the spectral features of the plasma irregularities can be inferred from their magnetic properties. For example, this type of assumption was done in the past by Lühr et al. [15], who using 50 Hz magnetic field measurements from CHAMP obtained the spectral properties of the magnetic data and considered them also valid to investigate the spectral features of the electron density irregularities.

Our findings suggest that a more complex relation may exist between the spectral features of the electron density and magnetic field in the plasma irregularities, which depends on the local time and latitude. Indeed, the results of this research support the idea that there are some relevant discrepancies between the spectral features of electron density fluctuations and magnetic field intensity ones. The discrepancies depend on the evolution of plasma bubbles with local time and on its turbulent nature. In the early stage of plasma bubble formation, i.e., when plasma bubbles rise just after the sunset and the instability growth is still quasi-linear, we find a quasi-linear relationship on average between the scaling properties of electron density and those of magnetic field. Conversely, when these structures develop, the evolution of turbulent fluctuations can strongly affect also other physical quantities such as the plasma temperature fluctuations, which could modify the spectral features. This means that the assumption of negligible temperature fluctuations could no longer be valid for hours approaching midnight. In other words, the electron temperature fluctuations could play an important role when turbulence is fully developed.

A better comprehension of the plasma bubbles dynamics and of the turbulence processes that characterize their time evolution may benefit from the use of very high-resolution vector magnetic field and plasma density measurements such as those that should be available from the future NanoMagSat mission [65]. This future satellite, which should fly at an altitude of 500 km and an inclination angle of 60°, should allow a quick full local time coverage and for the first time joint measurements of electron density and magnetic signals at very high frequencies. The mission should give the opportunity to investigate plasma variability at very short spatial scales ranging from a few meters to some kilometers.

Author Contributions: Conceptualization, P.D.M.; methodology, P.D.M. and G.C.; investigation, P.D.M., G.C. and T.A.; data curation, R.T., F.G., M.P., A.P. and I.C.; writing—original draft preparation, P.D.M.; writing—review and editing, all. All authors have read and agreed to the published version of the manuscript.

Funding: This research received financial support from Progetto INGV Pianeta Dinamico funded by MIUR (Fondo finalizzato al rilancio degli investimenti delle amministrazioni centrali dello Stato e allo sviluppo del Paese, legge 145/2018)-Task A1-2021 under Grant codice CUP D53J19000170001.

Institutional Review Board Statement: Not applicable.

Informed Consent Statement: Not applicable.

Data Availability Statement: Publicly available data were analyzed in this study. ESA-Swarm data are available from the ESA dissemination server (http://earth.esa.int/swarm, accessed on 13 February 2022). OMNI data are available at NASA CDAWeb site (https://cdaweb.gsfc.nasa.gov/index.html/, accessed on 13 February 2022). Elaborated data are available on request from the corresponding author.

Acknowledgments: The results presented rely on data collected by ESA-Swarm mission. We thank the European Space Agency that supports the Swarm mission. Swarm data can be accessed at http://earth.esa.int/swarm (accessed on 12 February 2022). The authors kindly acknowledge V. Papitashvili and J. King at the National Space Science Data Center of the Goddard Space Flight Center for permission to use the OMNI data and the NASA CDAWeb team for making these data available. This work was supported in part by Progetto INGV Pianeta Dinamico funded by MIUR (Fondo finalizzato al rilancio degli investimenti delle amministrazioni centrali dello Stato e allo sviluppo del Paese, legge 145/2018)-Task A1-2021 under Grant codice CUP D53J19000170001

Conflicts of Interest: The authors declare no conflict of interest.

Abbreviations

The following abbreviations are used in this manuscript:

2D	Two-dimensional
AE-E	Atmospheric Explorer-E
BF	Bubble Flag
CHAMP	CHAllenging Minisatellite Payload
C/NOFS	Communications/Navigation Outage Forecasting System
COSMIC	Constellation Observing System for Meteorology, Ionosphere, and Climate
DMSP	Defense Meteorological Satellite Program
F10.7	Solar radio flux index

GOLD	Global-scale Observation of Limb and Disk
GPS	Global Positioning System
IBI	Ionospheric Bubble Index
LEO	Low Earth Orbit
LT	Local Time
MHD	Magnetohydrodynamic
OGO	Orbiting Geophysical Observatory
PSD	Power Spectral Density
QD	Quasi Dipole
ROCSAT-1	Republic of China Satellite
TEC	Total Electron Content

References

1. Kil, H.; Heelis, R.A. Global distribution of density irregularities in the equatorial ionosphere. *J. Geophys. Res. (Space Phys.)* **1998**, *103*, 407–418. [CrossRef]
2. Booker, H.G.; Wells, H.W. Scattering of radio waves by the F-region of the ionosphere. *Terr. Magn. Atmos. Electr. (J. Geophys. Res.)* **1938**, *43*, 249. [CrossRef]
3. McClure, J.P.; Hanson, W.B.; Hoffman, J.H. Plasma bubbles and irregularities in the equatorial ionosphere. *J. Geophys. Res. (Space Phys.)* **1977**, *82*, 2650. [CrossRef]
4. Dyson, P.L.; Benson, R.F. Topside sounder observations of equatorial bubbles. *Geophys. Res. Lett.* **1978**, *5*, 795–798. [CrossRef]
5. Scannapieco, A.J.; Ossakow, S.L. Nonlinear equatorial spread F. *Geophys. Res. Lett.* **1976**, *3*, 451–454. [CrossRef]
6. Woodman, R.F.; La Hoz, C. Radar observations of F region equatorial irregularities. *J. Geophys. Res. (Space Phys.)* **1976**, *81*, 5447–5466. [CrossRef]
7. Burke, W.I. Plasma bubbles near the dawn terminator in the topside ionosphere. *Planet. Space Sci.* **1979**, *27*, 1187–1193. [CrossRef]
8. Kil, H.; Heelis, R.A.; Paxton, L.J.; Oh, S.J. Formation of a plasma depletion shell in the equatorial ionosphere. *J. Geophys. Res. (Space Phys.)* **2009**, *114*, A11302. [CrossRef]
9. Kelley, M.C. *The Earth's Ionosphere: Plasma Physics and Electrodynamics*, 2nd ed.; Academic Press: Burlington, MA, USA, 2009.
10. Schunk, R.; Nagy, A. *Ionospheres: Physics, Plasma Physics, and Chemistry*; Cambridge University Press: Cambridge, UK, 2009. [CrossRef]
11. Hysell, D.L.; Seyler, C.E. A renormalization group approach to estimation of anomalous diffusion in the unstable equatorial F region. *J. Geophys. Res. (Space Phys.)* **1998**, *103*, 26731–26738. [CrossRef]
12. Tsunoda, R.T.; Livingston, R.C.; McClure, J.P.; Hanson, W.B. Equatorial plasma bubbles: Vertically elongated wedges from the bottomside F layer. *J. Geophys. Res. (Space Phys.)* **1982**, *87*, 9171–9180. [CrossRef]
13. Anderson, D.N.; Mendillo, M. Ionospheric conditions affecting the evolution of equatorial plasma depletions. *Geophys. Res. Lett.* **1983**, *10*, 541–544. [CrossRef]
14. Cherniak, I.; Zakharenkova, I.; Sokolovsky, S. Multi-Instrumental Observation of Storm-Induced Ionospheric Plasma Bubbles at Equatorial and Middle Latitudes. *J. Geophys. Res. (Space Phys.)* **2019**, *124*, 1491–1508. [CrossRef]
15. Lühr, H.; Xiong, C.; Park, J.; Rauberg, J. Systematic study of intermediate-scale structures of equatorial plasma irregularities in the ionosphere based on CHAMP observations. *Front. Phys.* **2014**, *2*, 15. [CrossRef]
16. Smith, J.; Heelis, R.A. Equatorial plasma bubbles: Variations of occurrence and spatial scale in local time, longitude, season, and solar activity. *J. Geophys. Res. (Space Phys.)* **2017**, *122*, 5743–5755. [CrossRef]
17. Gurram, P.; Kakad, B.; Bhattacharyya, A.; Pant, T.K. Evolution of Freshly Generated Equatorial Spread F (F-ESF) Irregularities on Quiet and Disturbed Days. *J. Geophys. Res. (Space Phys.)* **2018**, *123*, 7710–7725. [CrossRef]
18. Hei, M.A.; Heelis, R.A.; McClure, J.P. Seasonal and longitudinal variation of large-scale topside equatorial plasma depletions. *J. Geophys. Res. (Space Phys.)* **2005**, *110*, A12315. [CrossRef]
19. Tsunoda, R.T. Control of the seasonal and longitudinal occurrence of equatorial scintillations by the longitudinal gradient in integrated E region pedersen conductivity. *J. Geophys. Res. (Space Phys.)* **1985**, *90*, 447–456. [CrossRef]
20. Aa, E.; Zou, S.; Liu, S. Statistical Analysis of Equatorial Plasma Irregularities Retrieved From Swarm 2013-2019 Observations. *J. Geophys. Res. (Space Phys.)* **2020**, *125*, e27022. [CrossRef]
21. Chou, M.Y.; Wu, Q.; Pedatella, N.M.; Cherniak, I.; Schreiner, W.S.; Braun, J. Climatology of the Equatorial Plasma Bubbles Captured by FORMOSAT-3/COSMIC. *J. Geophys. Res. (Space Phys.)* **2020**, *125*, e27680. [CrossRef]
22. Kintner, P.M.; Seyler, C.E. The status of observations and theory of high latitude ionospheric and magnetospheric plasma turbulence. *Space Sci. Rev.* **1985**, *41*, 91–129. [CrossRef]
23. Hysell, D.L.; Shume, E.B. Electrostatic plasma turbulence in the topside equatorial F region ionosphere. *J. Geophys. Res. (Space Phys.)* **2002**, *107*, 1269. [CrossRef]
24. Hassam, A.B.; Hall, W.; Huba, J.D.; Keskinen, M.J. Spectral characteristics of interchange turbulence. *J. Geophys. Res. (Space Phys.)* **1986**, *91*, 13513–13522. [CrossRef]
25. Zargham, S.; Seyler, C.E. Collisional and inertial dynamics of the ionospheric interchange instability. *J. Geophys. Res. (Space Phys.)* **1989**, *94*, 9009–9027. [CrossRef]

26. Kraichnan, R.H. Inertial Ranges in Two-Dimensional Turbulence. *Phys. Fluids* **1967**, *10*, 1417–1423. [CrossRef]
27. Kraichnan, R.H.; Montgomery, D. Two-dimensional turbulence. *Rep. Prog. Phys.* **1980**, *43*, 547–619. [CrossRef]
28. McDaniel, R.D.; Hysell, D.L. Models and DE II observations of inertial-regime irregularities in equatorial spread F. *J. Geophys. Res. (Space Phys.)* **1997**, *102*, 22233–22246. [CrossRef]
29. Yokoyama, T.; Shinagawa, H.; Jin, H. Nonlinear growth, bifurcation, and pinching of equatorial plasma bubble simulated by three-dimensional high-resolution bubble model. *J. Geophys. Res. (Space Phys.)* **2014**, *119*, 10474–10482. [CrossRef]
30. Yokoyama, T. A review on the numerical simulation of equatorial plasma bubbles toward scintillation evaluation and forecasting. *Prog. Earth Planet. Sci.* **2017**, *4*, 37. [CrossRef]
31. Jin, H.; Zou, S.; Chen, G.; Yan, C.; Zhang, S.; Yang, G. Formation and Evolution of Low-Latitude F Region Field-Aligned Irregularities During the 7-8 September 2017 Storm: Hainan Coherent Scatter Phased Array Radar and Digisonde Observations. *Space Weather* **2018**, *16*, 648–659. [CrossRef]
32. Akala, A.O.; Awoyele, A.; Doherty, P.H. Statistics of GNSS amplitude scintillation occurrences over Dakar, Senegal, at varying elevation angles during the maximum phase of solar cycle 24. *Space Weather* **2016**, *14*, 233–246. [CrossRef]
33. Zakharenkova, I.; Astafyeva, E. Topside ionospheric irregularities as seen from multisatellite observations. *J. Geophys. Res. (Space Phys.)* **2015**, *120*, 807–824. [CrossRef]
34. Karan, D.K.; Daniell, R.E.; England, S.L.; Martinis, C.R.; Eastes, R.W.; Burns, A.G.; McClintock, W.E. First Zonal Drift Velocity Measurement of Equatorial Plasma Bubbles (EPBs) From a Geostationary Orbit Using GOLD Data. *J. Geophys. Res. (Space Phys.)* **2020**, *125*, e28173. [CrossRef]
35. Cai, X.; Burns, A.G.; Wang, W.; Coster, A.; Qian, L.; Liu, J.; Solomon, S.C.; Eastes, R.W.; Daniell, R.E.; McClintock, W.E. Comparison of GOLD Nighttime Measurements With Total Electron Content: Preliminary Results. *J. Geophys. Res. (Space Phys.)* **2020**, *125*, e27767. [CrossRef]
36. Aggson, T.L.; Burke, W.J.; Maynard, N.C.; Hanson, W.B.; Anderson, P.C.; Slavin, J.A.; Hoegy, W.R.; Saba, J.L. Equatorial bubbles updrafting at supersonic speeds. *J. Geophys. Res. (Space Phys.)* **1992**, *97*, 8581–8590. [CrossRef]
37. Lühr, H.; Rother, M.; Maus, S.; Mai, W.; Cooke, D. The diamagnetic effect of the equatorial Appleton anomaly: Its characteristics and impact on geomagnetic field modeling. *Geophys. Res. Lett.* **2003**, *30*, 1906. [CrossRef]
38. Lühr, H.; Maus, S.; Rother, M.; Cooke, D. First in-situ observation of night-time F region currents with the CHAMP satellite. *Geophys. Res. Lett.* **2002**, *29*, 1489. [CrossRef]
39. Park, J.; Stolle, C.; Lühr, H.; Rother, M.; Su, S.Y.; Min, K.W.; Lee, J.J. Magnetic signatures and conjugate features of low-latitude plasma blobs as observed by the CHAMP satellite. *J. Geophys. Res. (Space Phys.)* **2008**, *113*, A09313. [CrossRef]
40. Stolle, C.; Lühr, H.; Rother, M.; Balasis, G. Magnetic signatures of equatorial spread F as observed by the CHAMP satellite. *J. Geophys. Res. (Space Phys.)* **2006**, *111*, A02304. [CrossRef]
41. Xiong, C.; Lühr, H.; Ma, S.Y.; Stolle, C.; Fejer, B.G. Features of highly structured equatorial plasma irregularities deduced from CHAMP observations. *Ann. Geophys.* **2012**, *30*, 1259–1269. [CrossRef]
42. Park, J.; Min, K.W.; Kim, V.P.; Kil, H.; Lee, J.J.; Kim, H.J.; Lee, E.; Lee, D.Y. Global distribution of equatorial plasma bubbles in the premidnight sector during solar maximum as observed by KOMPSAT-1 and Defense Meteorological Satellite Program F15. *J. Geophys. Res. (Space Phys.)* **2005**, *110*, A07308. [CrossRef]
43. Huang, C.Y.; Burke, W.J.; Machuzak, J.S.; Gentile, L.C.; Sultan, P.J. DMSP observations of equatorial plasma bubbles in the topside ionosphere near solar maximum. *J. Geophys. Res. (Space Phys.)* **2001**, *106*, 8131–8142. [CrossRef]
44. Whalen, J.A. Dependence of equatorial bubbles and bottomside spread F on season, magnetic activity, and E $\times$ B drift velocity during solar maximum. *J. Geophys. Res. (Space Phys.)* **2002**, *107*, 1024. [CrossRef]
45. Friis-Christensen, E.; Lühr, H.; Knudsen, D.; Haagmans, R. Swarm An Earth Observation Mission investigating Geospace. *Adv. Space Res.* **2008**, *41*, 210–216. [CrossRef]
46. Finlay, C.C.; Olsen, N.; Kotsiaros, S.; Gillet, N.; Tøffner-Clausen, L. Recent geomagnetic secular variation from Swarm and ground observatories as estimated in the CHAOS-6 geomagnetic field model. *Earth Planets Space* **2016**, *68*, 112. [CrossRef]
47. Matza, J.; Stolle, C.; Yamazaki, Y.; Bronkalla, O.; Morschhauser, A. The Geomagnetic Kp Index and Derived Indices of Geomagnetic Activity. *Space Weather* **2021**, *19*, e2020SW002641. [CrossRef]
48. Park, J.; Noja, M.; Stolle, C.; Lühr, H. The Ionospheric Bubble Index deduced from magnetic field and plasma observations on board Swarm. *Earth Planets Space* **2013**, *65*, 1333–1344. [CrossRef]
49. Davis, A.; Marshak, A.; Wiscombe, W.; Cahalan, R. Multifractal characterizations of nonstationarity and intermittency in geophysical fields: Observed, retrieved, or simulated. *J. Geophys. Res. (Space Phys.)* **1994**, *99*, 8055–8072. [CrossRef]
50. De Michelis, P.; Consolini, G.; Tozzi, R. Magnetic field fluctuation features at Swarm's altitude: A fractal approach. *Geophys. Res. Lett.* **2015**, *42*, 3100–3105. [CrossRef]
51. De Michelis, P.; Consolini, G.; Tozzi, R.; Pignalberi, A.; Pezzopane, M.; Coco, I.; Giannattasio, F.; Marcucci, M.F. Ionospheric Turbulence and the Equatorial Plasma Density Irregularities: Scaling Features and RODI. *Remote Sens.* **2021**, *13*, 759. [CrossRef]
52. Peltier, R.F.; Lévy-Véhel, J. *Multifractional Brownian Motion: Definition and Preliminary Results;* Technical Report; INRIA: Rocquencourt, France, 1995.
53. Benassi, A.; Jaffard, S.; Roux, D. Elliptic Gaussian random processes. *Rev. Mat. Iberoam.* **1997**, *13*, 19–90. [CrossRef]
54. Consolini, G.; De Marco, R.; De Michelis, P. Intermittency and multifractional Brownian character of geomagnetic time series. *Nonlinear Process. Geophys.* **2013**, *20*, 455–466. [CrossRef]

55. Basu, S.; MacKenzie, E.; Basu, S.; Coley, W.R.; Sharber, J.R.; Hoegy, W.R. Plasma structuring by the gradient drift instability at high latitudes and comparison with velocity shear driven processes. *J. Geophys. Res. (Space Phys.)* **1990**, *95*, 7799–7818. [CrossRef]
56. Consolini, G.; De Michelis, P.; Alberti, T.; Giannattasio, F.; Coco, I.; Tozzi, R.; Chang, T.T.S. On the Multifractal Features of Low-Frequency Magnetic Field Fluctuations in the Field-Aligned Current Ionospheric Polar Regions: Swarm Observations. *J. Geophys. Res. (Space Phys.)* **2020**, *125*, e27429. [CrossRef]
57. Consolini, G.; Quattrociocchi, V.; D'Angelo, G.; Alberti, T.; Piersanti, M.; Marcucci, M.F.; De Michelis, P. Electric Field Multifractal Features in the High-Latitude Ionosphere: CSES-01 Observations. *Atmosphere* **2021**, *12*, 646. [CrossRef]
58. Livingston, R.C.; Rino, C.L.; McClure, J.P.; Hanson, W.B. Spectral characteristics of medium-scale equatorial F region irregularities. *J. Geophys. Res. (Space Phys.)* **1981**, *86*, 2421–2428. [CrossRef]
59. Dyson, P.L.; McClure, J.P.; Hanson, W.B. In situ measurements of the spectral characteristics of F region ionospheric irregularities. *J. Geophys. Res. (Space Phys.)* **1974**, *79*, 1497. [CrossRef]
60. Basu, S.; McClure, J.P.; Basu, S.; Hanson, W.B.; Aarons, J. Coordinated study of equatorial scintillation and in situ and radar observation of nighttime F region irregularities. *J. Geophys. Res. (Space Phys.)* **1980**, *85*, 5119–5130. [CrossRef]
61. Le, G.; Huang, C.S.; Pfaff, R.F.; Su, S.Y.; Yeh, H.C.; Heelis, R.A.; Rich, F.J.; Hairston, M. Plasma density enhancements associated with equatorial spread F: ROCSAT-1 and DMSP observations. *J. Geophys. Res. (Space Phys.)* **2003**, *108*, 1318. [CrossRef]
62. Tsunoda, R.T. Observations of Equatorial Spread F: A Working Hypothesis. *Ionos. Dyn. Appl.* **2021**, *3*, 201–280. [CrossRef]
63. Gruzinov, A.V.; Kukharkin, N.; Sudan, R.N. Two-Dimensional Convective Turbulence. *Phys. Rev. Lett.* **1996**, *76*, 1260–1263. [CrossRef]
64. Alken, P.; Maus, S.; Richmond, A.D.; Maute, A. The ionospheric gravity and diamagnetic current systems. *J. Geophys. Res. (Space Phys.)* **2011**, *116*, A12316. [CrossRef]
65. Hulot, G.; Leger, J.M.; C., L.; Deconinck, F.; Coïsson, P.; Vigneron, P.; Alken, P.; Chulliat, A.; Finlay, C.C.; Grayver, A.; et al. NanoMagSat, a 16U nanosatellite constellation high-precision magnetic project to initiate permanent low-cost monitoring of the Earth's magnetic field and ionospheric environment. In Proceedings of the EGU21-The EGU General Assembly 2021, Online, 19–30 April 2021. [CrossRef]

remote sensing

MDPI

Article

Pressure-Gradient Current at High Latitude from Swarm Measurements

Giulia Lovati [1], Paola De Michelis [2,*], Giuseppe Consolini [3] and Francesco Berrilli [4]

1 Dipartimento di Fisica, Università di Roma Sapienza, 00185 Rome, Italy; giulia.lovati@uniroma1.it
2 Istituto Nazionale di Geofisica e Vulcanologia, 00143 Rome, Italy
3 INAF—Istituto di Astrofisica e Planetologia Spaziali, 00133 Rome, Italy; giuseppe.consolini@inaf.it
4 Dipartimento di Fisica, Università di Roma Tor Vergata, 00133 Rome, Italy; francesco.berrilli@roma2.infn.it
* Correspondence: paola.demichelis@ingv.it

Abstract: The pressure-gradient current is among the weaker ionospheric current systems arising from plasma pressure variations. It is also called diamagnetic current because it produces a magnetic field which is oriented oppositely to the ambient magnetic field, causing its reduction. The magnetic reduction can be revealed in measurements made by low-Earth orbiting satellites flying close to ionospheric plasma regions where rapid changes in density occur. Using geomagnetic field, plasma density and electron temperature measurements recorded on board ESA Swarm A satellite from April 2014 to March 2018, we reconstruct the flow patterns of the pressure-gradient current at high-latitude ionosphere in both hemispheres, and investigate their dependence on magnetic local time, geomagnetic activity, season and solar forcing drivers. Although being small in amplitude these currents appear to be a ubiquitous phenomenon at ionospheric high latitudes characterized by well defined flow patterns, which can cause artifacts in the main field models. Our findings can be used to correct magnetic field measurements for diamagnetic current effect, to improve modern magnetic field models, as well as to understand the impact of ionospheric irregularities on ionospheric dynamics at small-scale sizes of a few tens of kilometers.

Keywords: ionosphere F region; high latitude; pressure-gradient current; diamagnetic current; swarm measurements

Citation: Lovati, G.; De Michelis, P.; Consolini, G.; Berrilli, F. Pressure-Gradient Current at High Latitude from Swarm Measurements. *Remote Sens.* **2022**, *14*, 1428. https://doi.org/10.3390/rs14061428

Academic Editor: Michael E. Gorbunov

Received: 27 January 2022
Accepted: 9 March 2022
Published: 15 March 2022

1. Introduction

The Earth's magnetosphere is characterized by different plasma regions, each one with particular density and temperature distributions [1]. Different currents flow in the magnetosphere characterized by patterns and intensities changing in time and space. Some of these currents are created by plasma space inhomogeneity and driven by plasma pressure gradients. The neutral sheet current, the magnetopause current and the ring current are examples of pressure-gradient currents, which flow in the Earth's magnetosphere [2].

This phenomenon also occurs in the ionosphere where the pressure-gradient current is among the weaker ionospheric current systems arising from plasma pressure variations. Indeed, due to the coupling between geomagnetic field and plasma pressure gradient, electrons and ions drift in opposite directions generating an electric current whose intensity is of the order of a few nA/m^2. This current is also called diamagnetic because it produces a magnetic field, which is oriented oppositely to the ambient magnetic field, causing its reduction. The magnetic reduction associated with the pressure-gradient current can be revealed in measurements made by low-Earth orbiting (LEO) satellites flying close to ionospheric plasma regions where rapid changes in density occur. Anyway, identifying diamagnetic current by using its magnetic signature is not easy due to the weak intensity of generated magnetic perturbation, that is about 10,000 time smaller than the ambient geomagnetic field. This is why the studies investigating this current are quite recent, since high-accuracy satellite magnetic field measurements are available. Due to its origin, the diamagnetic current can be revealed at both low and high latitudes, and more generally in all those regions where the plasma pressure gradients are greatest.

Several studies focus on low latitude, in the equatorial belt, where irregular plasma density structures of various scale sizes, from centimeters up to hundreds of kilometers, can be observed after sunset and where exists a region with increased plasma density, known as Appleton or equatorial ionization anomaly. About twenty years ago, Lühr and collaborators [3] were the first to reveal the presence of these currents in the ionosphere. By using the high-resolution magnetic field data in combination with plasma density observations recorded by CHAMP (CHAllenging Minisatellite Payload) satellite [4], they revealed the magnetic signatures of plasma density variations at low latitudes and reconstructed the strength and geographic location of diamagnetic field caused by the Appleton anomaly. They estimated the strength of this field up to 5 nT at a variety of local times, showed how the diamagnetic effect could severely affect the correct estimation of the ionospheric current distributions from space, and how it could be capable of producing artifacts in the new generation main field models [5].

In the following years some other papers dealt with this subject. Interesting is the study carried out by Alken et al. [6] where the global gravity and diamagnetic current systems in the ionospheric topside F2 region were modeled by using the Thermosphere Ionosphere Electrodynamics General Circulation Model (TIEGCM) [7] with empirical density, wind and temperature inputs. The results obtained from this model permitted to reconstruct the diamagnetic current structure at low latitude, which was found to be prominent near the equatorial ionization anomaly, and to calculate the associated magnetic field. Although the obtained results showed some discrepancies with what found using the diamagnetic effect formula proposed by Lühr et al. [3], the study confirmed the need to pay attention to all those ionospheric current systems that, although smaller than others well-known ionospheric current systems, could play a role in the ionospheric dynamics and help in the development of more accurate magnetic models.

Successively, by considering 10 years of CHAMP measurements (2000–2010) and two years of Swarm data (late 2013–2015), Alken [8] reconstructed the gravity and diamagnetic currents flowing at low latitude in the topside F2 region. He found that the strength of these currents depended on season, local time and solar activity. These currents were strongest during spring and fall and their magnetic signatures ranged from 1 to 7 nT, depending on solar activity level. Another important feature of these currents was their presence until midnight. For this reason, the assumption made by many magnetic models according to which the ionospheric currents were negligible during the nighttime and geomagnetically quiet periods, was not correct.

Recently, awareness of the existence of a large amount of plasma density irregularities at high latitude, especially at polar latitudes, has led to the hypothesis that diamagnetic currents had to exist also in these regions, where they could play a role in the ionospheric dynamics. Here, the pressure-gradients are mainly related to the convection pattern and to electron precipitation, that characterizes both the polar cap and the auroral oval [9]. In particular, soft electron precipitation caused by the direct entry of magnetosheath electrons in the cusp region and from electrons of ionospheric origin that are energized at intermediate altitudes by wave-particle interactions with dispersive Alfvén waves, are supposed to generate density anomalies with elevated electron temperature in the F region ionosphere [10]. At the present time, few studies have been published on the diamagnetic currents at high latitudes: we know only two papers where these currents are investigated analysing their magnetic signatures. The first paper is by Park et al. [9] and the second one by Laundal et al. [11]. Using magnetic data from CHAMP satellite and a combination of plasma density measurements from both the Digital Ion Drift Meter and the Planar Langmuir Probe on board CHAMP, Park et al. [9] presented a climatological study of the diamagnetic signatures at high latitude due to ionospheric plasma irregularities, finding that ionospheric irregularities did not have a uniform zonal spatial distribution, but exhibited a maximum at pre-midnight ionosphere and dayside cusp. The geographic location of the irregularities also depended on geomagnetic activity and season. In this paper, the authors do not reconstruct the diamagnetic currents but use their magnetic signatures in order to investigate the occurrence rate of ionospheric irregularities. Successively, analyzing two years of magnetic and plasma density measurements from Swarm constellation, Laundal et al. [11] showed that the magnetic field variations recorded at satellite altitude were often associated with plasma pressure variations in the polar cap. They found that superposed on the magnetic perturbations due to the auroral electrojet currents was an irregular pattern, anticorrelated with plasma density measurements, that was the signature of the diamagnetic current. They suggested that this current could dominate the disturbances in the magnetic field strength at small-scale size of a few tens of kilometers.

Here, using geomagnetic field, plasma density and electron temperature measurements on board Swarm constellation, we reconstruct the flow patterns of the diamagnetic current at high-latitude ionosphere in both hemispheres, and investigate their dependence on geomagnetic activity, season and solar forcing drivers.

2. Data and Method

We study the pressure-gradient current at ionospheric high latitude using measurements on board Swarm constellation [12]. This constellation consists of three identical satellites launched into a near polar orbit at the end of 2013. Two of these satellites (Swarm A and Swarm C) fly side-by-side (1.4° separation in longitude) at an altitude around 460 km (initial altitude), whereas the third satellite (Swarm B) flies at higher orbit, around 510 km (initial altitude). These satellites are equally equipped and simultaneously acquire magnetic field, plasma density and electron temperature measurements, thanks to the presence in each satellite of a Vector Field Magnetometer (VFM) and a Electric Field Instrument (EFI) [12,13].

In this study, we use magnetic field components, electron density N_e and electron temperature T_e measurements acquired by Swarm A [14], at a rate of 1 Hz, during a period of four years from 1 April 2014 to 28 February 2018. Data are selected from the ESA dissemination server (http://earth.esa.int/swarm accessed on 14 March 2022). Starting from April 2014, the selected time interval excludes the period where Swarm constellation was stabilizing its orbit and makes possible to assume that Swarm A is approximately at the same altitude for all the period under analysis. The choice of February 2018 is due to the lack of the Auroral Electrojet index data beyond this period.

Since the ionospheric currents are mainly organized by Earth's magnetic field, the study is done using a magnetic coordinate system, the non-orthogonal Quasi-Dipole (QD) system of coordinates [15–17]. We select data with a quasi-dipole magnetic latitude $\lambda_{QD} \geq |50°|$ in order to focus on mid- and high-latitude ionospheric topside F2 region, and we use the Magnetic Local Time (MLT) [18] instead of universal time (UT) in order to consider the position respect to the Sun. To report our data on the λ_{QD}-MLT plane, we realize maps of our parameters with a $1° \times 1°$ binning (1° longitude corresponds to 4 min in MLT). Time series belonging to each bin are initially filtered removing values beyond 3σ from the mean value in order to take off possible spikes. The value representative of each bin corresponds to the average value of the filtered time series.

We use the Auroral Electrojet AE index to study the dependence of the pressure-gradient currents on the geomagnetic activity. This index is a good proxy of the global geomagnetic activity at high latitude, well describing the auroral zone magnetic activity due to the increase of the ionospheric currents flowing below and within the auroral oval. The AE time series has been downloaded from the Space Physics Data Facility of the NASA Goddard Space Flight Center (OMNI data set http://cdaweb.gsfs.nasa.gov/) where it is available only until 28 February 2018.

To analyse the dependence on the solar activity of the flow patterns of the diamagnetic current at high latitude we consider two different proxies: the Mg II core-to-wing ratio and the F10.7 index. The Mg II index refers to a broad absorption feature with narrow emission peaks in the core. This emission doublet, near 280 nm, originates in the Sun's chromosphere, while the line wings part originates in the photosphere, showing much less variability. Therefore, the ratio of line core intensity to wing intensity provides a good estimate of solar variability, avoiding degradation effects [19]. For the selected period Mg II is provided by GOME-2 (Global Ozone Monitoring Experiment) mission instruments [20]. F10.7 index, instead, is the radio emission at 10.7 cm, that originates from regions of intense magnetic field, characterized by structures like plages, networks, and sunspots [21]. It is recorded by the Dominion Radio Astrophysical Observatory (Canada) with three flux determinations for each day [22]. The time series containing these two proxies have been downloaded from the Space Physics Data Facility of the NASA Goddard Space Flight Center (OMNI data set http://cdaweb.gsfs.nasa.gov/ accessed on 14 March 2022) in the case of F10.7, and from UVSAT Bremen University (GOME-2A, http://www.iup.uni.bremen.de/UVSAT/Datasets/mgii accessed on 14 March 2022) in the case of MgII.

To reconstruct the flow patterns of the pressure-gradient current at high latitude we can observe that for time scales longer than about 1 min the ionospheric electrodynamics is approximately in a steady-state with divergence-free current density ($\vec{J}$), and time-invariant electric field ($\vec{E}$). From the generalized Ohm's law, the current density $\vec{J}$ can be expressed as the sum of different contributes due to the neutral wind ($\vec{J}_w$), the global electrostatic electric field ($\sigma\vec{E}$, with σ equal to the conductivity tensor), gravitational force ($\vec{J}_g$), and pressure-gradient force ($\vec{J}_P$). At large spatial scale the electric force and the neutral wind dominate the pressure-gradient and the gravitational forces. Thus, the first two forces are responsible for the main ionospheric currents while the other two generate currents that can usually be neglected. However, these currents can become important on smaller spatial scale, and can have non-negligible effects on low Earth orbit satellite because of their magnetic perturbations. By assuming that it is possible to ignore the effects of neutral collisions, the current density due to pressure-gradient force can be expressed as in Equation (1) [15,23].

$$\vec{J}_P = -\frac{\nabla P \times \vec{B}}{B^2} \tag{1}$$

where $P = N_e k_B (T_i + T_e)$ is the plasma pressure, k_B is the Boltzmann constant, T_e is the electron temperature, T_i is the ion temperature, and $\vec{B}$ is the ambient magnetic field, which is produced by all sources within and outside the solid Earth up to the so-called magnetopause. The plasma pressure-gradient current, which is mainly in the zonal direction due to the strong vertical gradient of the plasma density, is naturally responsible of variations in the magnetic field strength. In first approximation, the magnetic field variations associated with this current can be evaluated by considering the balance between the plasma and magnetic pressures. According to Lühr et al. [3], in a quasi-stationary plasma where the structures are much smaller than the geomagnetic field curvature radius, the signature in the magnetic variation strength due to plasma pressure gradient is equal to:

$$\Delta B = -\frac{\mu_0 k_B}{B} \Delta(N_e(T_e + T_i)) \tag{2}$$

where μ_0 is the susceptibility of free space, and k_B is the Boltzmann constant. Here, the main assumption is that the measured magnetic fluctuations are spatial on the time scale for the satellite to traverse them. This interpretation is valid when the wave frequency in the plasma rest frame is much less than the Doppler frequency due to spacecraft motion across the horizontal spatial structure. Assuming a local thermodynamic equilibrium, the magnetic field perturbations are anticorrelated with variations in plasma density: a plasma density depletion is associated with a magnetic field enhancement while a reduction in the magnetic field is related to an increase in plasma density. The current density driven by pressure gradient flows around the density enhancement in a direction that produces a magnetic field capable of reducing the ambient magnetic field.

To reconstruct the flow patterns of the plasma pressure-gradient current at Swarm altitude, it is necessary to estimate the plasma pressure. This requires to know the temperature and electron density distributions within the region of interest, being $P = N_e k_B(T_e + T_i)$, according to the equation of state. Swarm satellite measures only the electron density and electron temperature along its orbit but, at its altitude, the temperature of oxygen ions (O^+), which can be assumed as the main ionic species present, is of the same order of magnitude as the electron temperature and we make the assumption that $T_e \simeq T_i$ [11,24]. Thus, in first approximation, we assume that plasma pressure is equal to $P = 2N_e k_B T_e$. From P values it is possible to compute the pressure gradient (∇P) on the horizontal plane. In order to approximate the geometry of the system, we consider the gradient in spherical coordinates:

$$\nabla P = \frac{\partial P}{\partial r}\vec{r} + \frac{1}{r}\frac{\partial P}{\partial \theta}\vec{\theta} + \frac{1}{r\sin\theta}\frac{\partial P}{\partial \phi}\vec{\phi} \tag{3}$$

where θ is the magnetic co-latitude, and ϕ is the longitude, which corresponds to MLT. Since Swarm A flies approximately at a constant altitude, we have $r \simeq const$ and in addition we are not able to evaluate the gradient along this direction. Therefore, we estimate the pressure gradient in the plane perpendicular to the radial direction:

$$\nabla P_{\perp} = \frac{1}{r}\frac{\partial P}{\partial \theta}\vec{\theta} + \frac{1}{r\sin\theta}\frac{\partial P}{\partial \phi}\vec{\phi} \tag{4}$$

However, considering Equation (1), we remark that the neglected $\frac{\partial P}{\partial r}$ term provides a small contribution to the plasma pressure-gradient current being the magnetic field nearly parallel to it. Indeed, at high latitudes the magnetic field is approximately aligned with the radial direction and, compared to this term, the other two components result negligible. Therefore, in this approximation, the lack of information on $\frac{\partial P}{\partial r}$ does not affect significantly our analysis.

In order to estimate $\frac{\partial P}{\partial \theta}$ and $\frac{\partial P}{\partial \phi}$, we consider the matrix of pressure binned values, having a number of rows equal to QD-latitude bins and a number of column equal to MLT bins. Having selected latitudes from $|50°|$ upwards and a binning of $1° \times 1°$, it results in a 40×360 matrix. The gradient is evaluated through second order accurate central differences [25]. Once obtained $\frac{\partial P}{\partial \theta}$ and $\frac{\partial P}{\partial \phi}$, we complete the gradient estimation in spherical coordinate using Equation (4) considering r as fixed at the altitude of Swarm A, and θ equal to the mean QD-latitude of each bin under consideration. Finally, the current density due to pressure-gradient force is estimated according to Equation (1).

3. Results

3.1. Variation of Pressure-Gradient Current with Geomagnetic Activity Level

The analysis of the current density spatial distribution due to pressure-gradient force at high latitude in the Northern and Southern hemispheres starts investigating the dependence on the geomagnetic activity level. Using the AE index we select two different levels of geomagnetic activity. Quiet periods correspond to values of $AE \leq 80$ nT while disturbed ones to $AE \geq 120$ nT [26,27]. The two selected activity levels are chosen in order to clearly separate active and quiet periods according to the distribution function of AE index values, which is bimodal with a crossover around 100 nT. For this type of analysis we consider the whole dataset, that is, the electron density, electron temperature, and magnetic field recorded by Swarm A from 1 April 2014 to 28 February 2018.

Figures 1–3 report a polar view of the large-scale spatial distribution of electron density (N_e), electron temperature (T_e), and plasma pressure (P) in the Northern and Southern hemispheres during quiet and disturbed periods, respectively. Data are mapped into grid binned at $1° \times 1°$ in QD latitude and MLT coordinates. The two selected geomagnetic activity levels correspond to different spatial distributions of the plasma parameters in the two hemispheres. Focusing on plasma pressure, we notice that its spatial distribution shows a clear dependence on magnetic latitude and magnetic local time (see Figure 3).

It is characterized by a marked dayside/nigthside asymmetry, which mainly reflects the dayside/nigthside asymmetry of the electron density and temperature spatial distributions. Indeed, as reported in Figure 1, the electron density is generally almost twice in the dayside compared to the nightside due to the dependence of the electron density on the upper atmosphere ultraviolet ionization.

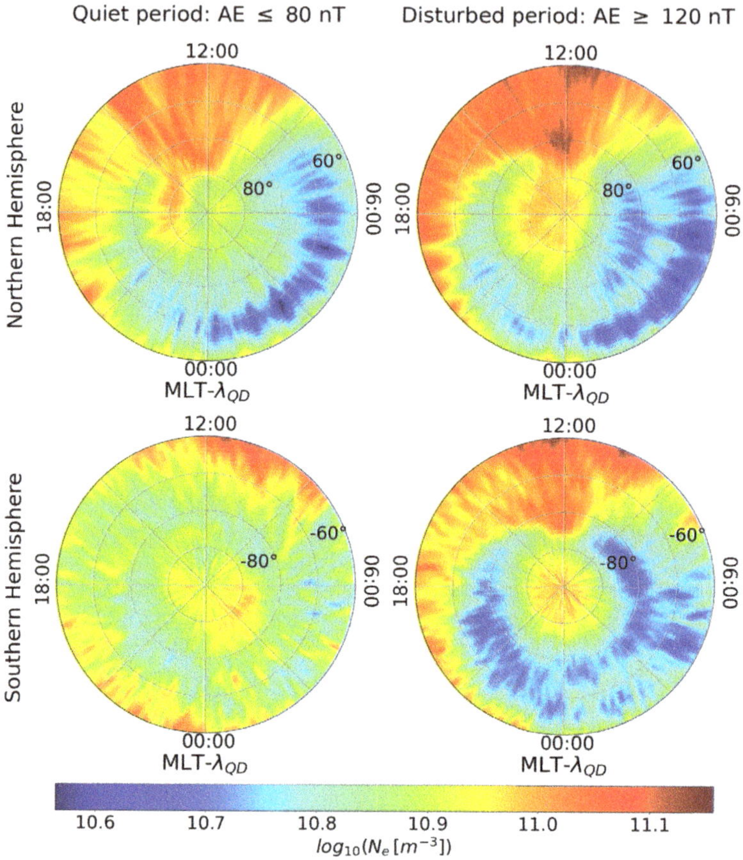

Figure 1. Polar view of the average spatial distribution of electron density (N_e) for quiet (**left** column) and disturbed (**right** column) geomagnetic activity conditions, for the Northern and Southern hemispheres, in QD-latitude ($\lambda_{QD} > 50°$ and $\lambda_{QD} < -50°$ for the Northern and Southern hemisphere, respectively) and MLT reference system. Maps are obtained using data recorded by Swarm A from 1 April 2014 to 28 February 2018. The concentric dashed circles represent contours of magnetic latitude separated by 10°.

Moreover, a marked dayside/nigthside asymmetry also characterizes the spatial distribution of the electron temperature (see Figure 2) which is, even in this case, due to the increase of ultraviolet photoionization. However, the spatial distribution of the electron temperature also depends on the particle precipitation caused by magnetosphere-ionosphere coupling. Indeed, soft particle precipitation heats the F region ionosphere and is a likely source of the observed correlation of the anomalous density with elevated particle temperature. Thus, the magnetosphere-ionosphere coupling is responsible of the temperature enhancement both in the cusp region around noon [28], where particles of solar origin are directly injected into the ionosphere, and at auroral and sub-auroral latitudes where particles arrive directly from the geomagnetic tail regions and plasmasphere during disturbed periods (see Figure 2).

Not all the features of the electron density and electron temperature are recognizable in the plasma pressure spatial distribution. For example, the so-called main ionospheric trough (MIT), which corresponds to a prominent plasma depletion at subauroral regions during the night hours in a latitudinally limited band between 60° and 70° (see Figure 1), is not clearly recognizable in the pressure data. It is due to the fact that there is an electron temperature enhancement owing to the joint action of particle precipitation and decreased collisional cooling in correspondence with the main ionospheric trough [29,30]. In the nightside, at auroral and sub-auroral latitudes this process is dominant and becomes more and more important with the increase of the geomagnetic activity level (see Figure 2). Due to the anticorrelation between the electron density and electron temperature, the plasma pressure does not show a particular behaviour in this region. Conversely, Figure 3 reveals a plasma pressure depletion at low latitudes, between 50° and 65°, in the nightside, between 21:00 and 06:00 MLT, whose position moves progressively to a lower latitude as the level of geomagnetic activity increases.

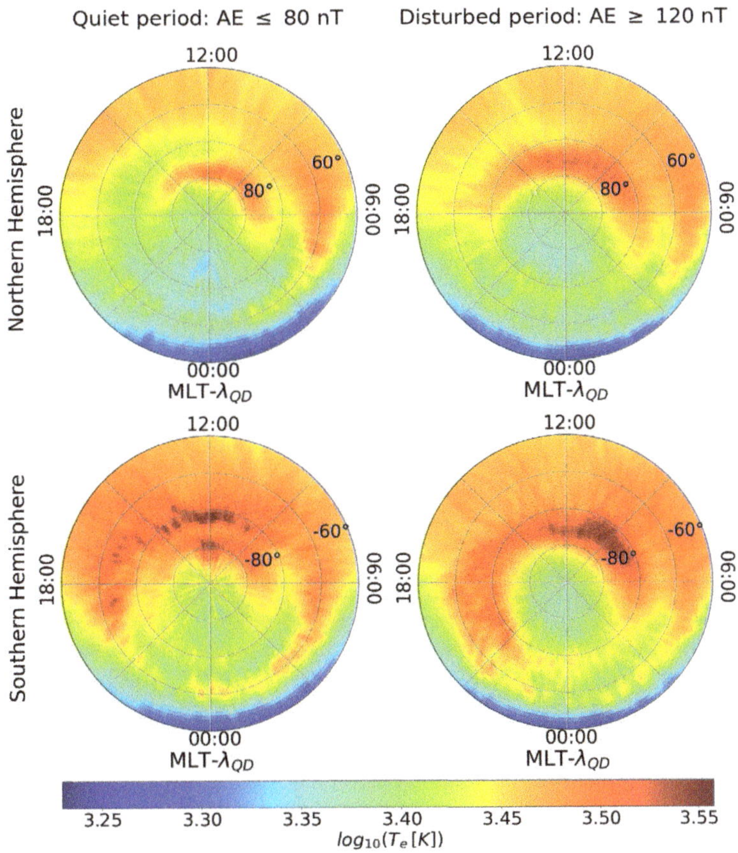

Figure 2. Polar view of the average spatial distribution of electron temperature (T_e) for quiet (**left** column) and disturbed (**right** column) geomagnetic activity conditions, for the Northern and Southern hemispheres, in QD-latitude ($\lambda_{QD} > 50°$ and $\lambda_{QD} < -50°$ for the Northern and Southern hemisphere, respectively) and MLT reference system. Maps are obtained using data recorded by Swarm A from 1 April 2014 to 28 February 2018. The concentric dashed circles represent contours of magnetic latitude separated by 10°.

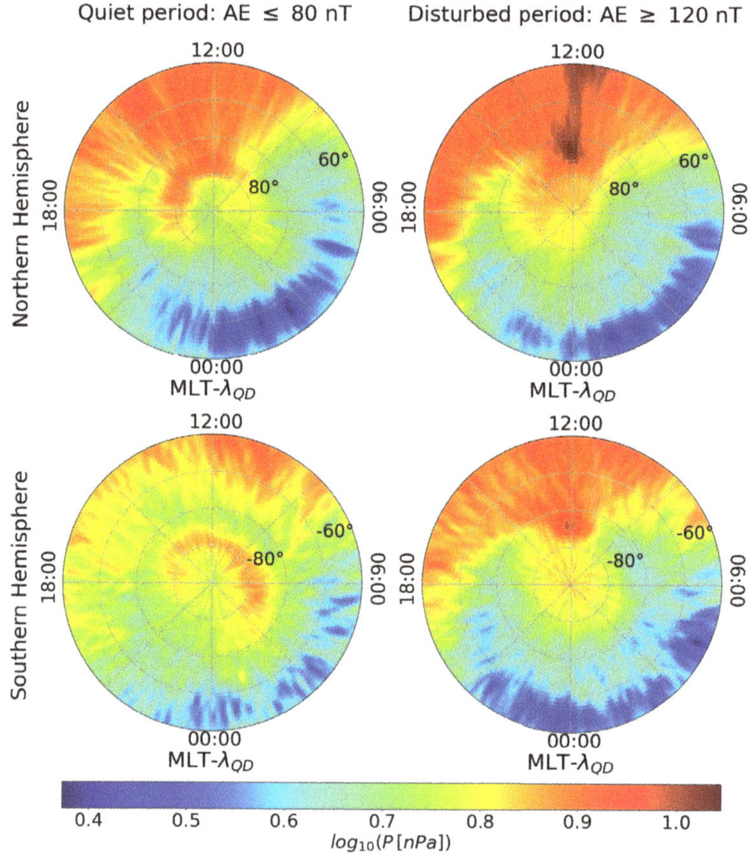

Figure 3. Polar view of the average spatial distribution of plasma pressure for quiet (**left** column) and disturbed (**right** column) geomagnetic activity conditions, for the Northern and Southern hemispheres, in QD-latitude ($\lambda_{QD} > 50°$ and $\lambda_{QD} < -50°$ for the Northern and Southern hemisphere, respectively) and MLT reference system. Maps are obtained using data recorded by Swarm A from 1 April 2014 to 28 February 2018. The concentric dashed circles represent contours of magnetic latitude separated by 10°.

Another interesting feature of the plasma pressure is its marked increase in the cusp region around noon, which becomes more marked in the disturbed periods in both hemispheres. This increase is associated with the observed electron temperature and density enhancements.

Lastly, Figure 3 shows two other remarkable features, which mainly reflect N_e spatial distribution features. There is an enhancement of plasma pressure from noon across the polar cap to the nightside during disturbed conditions in both hemispheres. This plasma pressure enhancement corresponds to the characteristic tongue of ionization (TOI) of N_e [31,32], which is a large-scale feature of the F-region polar ionosphere. Furthermore, there is a plasma pressure increases around dusk in disturbed conditions reflecting the well-known storm-enhanced density (SED) plumes, which are prominent ionospheric electron density increases at the dayside mid and high latitudes [33].

Figure 4 reports the flow patterns of the plasma pressure-gradient current (black arrows) at high latitudes in the Northern and Southern hemispheres during the two different geomagnetic activity levels. Only for graphical reasons, the black arrows are mapped into grids binned at 2° × 4° in QD latitude and MLT coordinates, where 4° magnetic longitude corresponds to 16 min. Below the flow patterns of the reconstructed current there is the corresponding plasma pressure map, which permits us to better visualize the regions where the current develops. As expected, the current tends to flow around the plasma pressure enhancements and is stronger in region where the plasma pressure changes rapidly. Some well defined patterns can be recognized in the plasma pressure-gradient current regardless of geomagnetic activity level.

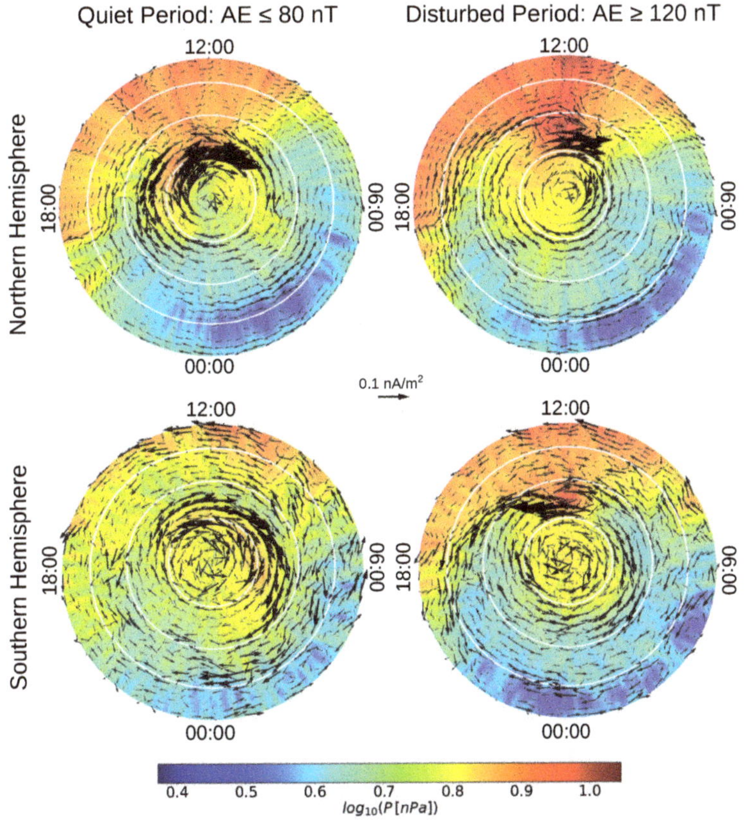

Figure 4. Flow patterns of pressure-gradient current (black arrows) superimposed on the plasma pressure spatial distribution for quiet (**left** column) and disturbed (**right** column) geomagnetic activity conditions, for the Northern and Southern hemispheres, in QD-latitude ($\lambda_{QD} > |50°|$) and MLT reference system. Maps are obtained using data recorded by Swarm A from 1 April 2014 to 28 February 2018. For graphical reasons, the current's vector field is mapped into grids binned at 2° × 4°, where 4° magnetic longitude corresponds to 16 min. The concentric white circles are plotted in 10° intervals, corresponding to QD-latitudes of |80°|, |70°| and |60°| starting from the centre, respectively.

A vortex exists around the cleft region, whose position shifts equatorward as magnetic activity increases, following the natural shift of center of the precipitation region. Around the cleft region the current flows counter-clockwise in the Northern hemisphere and clockwise in the Southern one. This current flows in a direction which reduces the ambient geomagnetic field. At high latitude the magnetic field is nearly vertical and the horizontal pressure-gradient current produces a magnetic field which is, in first approximation, parallel to the main field but with opposite direction. Thus, the current produces a reduction of the magnetic field inside the plasma region around which the current flows. The existence of this current structure is indirectly confirmed by Park et al. [9] analysing the diamagnetic signatures of high-latitude ionospheric irregularities. In both hemispheres they found depletions in magnetic field strength of about 1 nT concentrated in the cusp region that they associated with the presence of local plasma irregularities with scale sizes below some hundred kilometers. Anyway, this region can be also characterized by large anomalies in the electron density correlated with intense small-scale magnetic fluctuations, which are associated with incident Alvén waves producing a local heating. This means that in this region an opposite scenario to that implicitly assumed may be possible: the observed pressure gradients may be also produced by magnetic disturbances [10].

At high latitudes ($\lambda_{QD} > |75°|$) the current flows counter-clockwise in the Northern hemisphere and in opposite direction in the Southern one. The spatial distribution of the current seems to identify, especially during disturbed periods, another region characterized by an enhancement of plasma density, which is due to the propagation of plasma from the cusp to the polar cap. This region, that is visible in both hemispheres, is known to be characterized by the presence of plasma instability and the formation of ionospheric irregularities known as polar cap patches, where the density is at least twice that of the background [34]. However, the reconstructed flow patterns of the pressure-gradient currents capture plasma pressure variations which are not necessary due to the presence of plasma density irregularities as for example polar cap patches and auroral blobs. At lower latitudes in the Northern and Southern hemispheres the flow pattern of the pressure-gradient current identifies another region where the ambient magnetic field can be reduced by this current. In first approximation, this region corresponds to the auroral zone from 15:00 MLT to 09:00 MLT, passing by 24:00 MLT. Thus, the pressure-gradient currents characterize also the auroral zone and equatorward of the auroral zone on the nigthside.

In general, the flow patterns of the pressure-gradient current are similar in both hemispheres and what one finds is their shift towards lower latitudes with the increase of the geomagnetic activity.

Figure 5 reports the distributions of the pressure-gradient current values for disturbed and quiet geomagnetic activity conditions for the Northern and Southern hemispheres. The comparison among the four distributions shows that the pressure-gradient current intensity is slightly higher in the Southern hemisphere than in the Northern one, while there is no great difference between quiet to disturbed periods. At high latitudes the mean values of these currents are quite low, around 1 order of magnitude less than the same diamagnetic currents observed at low latitudes [35]. Our findings are in agreement with previous studies. Analyzing two years of Swarm measurements Laundal et al. [11] estimated the small variations in the magnetic field strength due to pressure-gradient currents at polar latitudes by evaluating the magnetic field intensity variations anticorrelated with plasma density variations. As result of their statistical analysis, they showed that the polar caps, the auroral zone, and equatorward of the auroral zone on the nightside were the regions where the magnetic field variations were well explained by plasma pressure gradients. Their findings are slightly different from the results by Park et al. [9], where the highest occurrence rates between the magnetic field and the electron density variations are found in the auroral oval. In both papers the flow pattern of the pressure-gradient current is not calculated, nor its intensity. The existence of this current is invoked to justify the variations of magnetic field correlated with those of plasma density and climatological analyses of their correspondence are proposed. This means that, to make a comparison with our results, it is necessary to consider that the pressure-gradient currents flow around the plasma pressure enhancement regions, which are often characterized by the occurrence of ionospheric irregularities and are the regions identified in the previous papers. It is also important to notice that both Laundal et al. [11] and Park et al. [9] assume that the variations in plasma pressure are dominated by electron density, under the hypothesis of a local thermodynamic equilibrium, but looking at Figure 2 we notice that also the temperature plays a role in the plasma pressure, mainly in the auroral and sub-auroral latitudes.

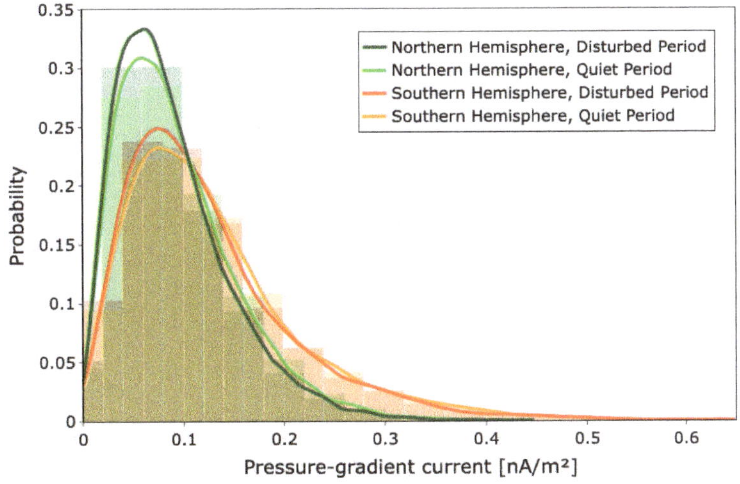

Figure 5. Distributions of the pressure-gradient current values for disturbed and quiet geomagnetic activity conditions for the Northern and Southern hemispheres. Solid line represents the continuous probability density curve and it is plotted above the more transparent relative histogram.

To evaluate the pressure-gradient current we have used the magnetic field measurements recorded by Swarm A. It contains all the different contributes coming from its sources both internal (core and crust) and external (magnetospheric and ionospheric currents) to the Earth. In order to investigate the possible effect on the pressure-gradient current flow patterns of remote currents, in particular of those current systems that flow about 300 km below Swarm A and are expected to produce a smooth magnetic signature on the measured magnetic field [36], we evaluate the pressure-gradient current considering in this last case the magnetic field strength associated with internal sources. The internal part of the magnetic field is estimated using the CHAOS model [37]. Thus, the pressure-gradient current is evaluated and compared with that obtained using the ambient magnetic field. Figure 6 reports the percentage difference between the current evaluated in the two ways. This procedure should leave only the contribute of external sources to the pressure-gradient current. The effect on the pressure-gradient currents at Swarm altitude of the remote current systems is very low: it does not exceed 0.3%. However, it is not uniform, depending on the geographic location, magnetic local time, and geomagnetic activity level. Our findings show that the features of the reconstructed pressure-gradient currents at Swarm altitude depend only partially on geomagnetic field variations due to external disturbances, mainly they depend on the main field structures and on the features of pressure gradients.

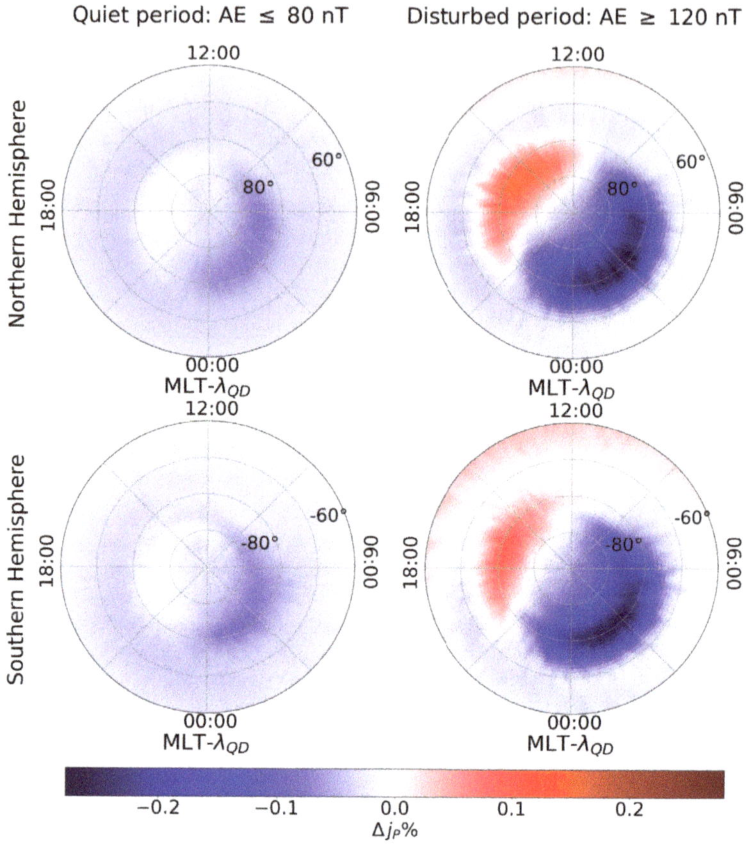

Figure 6. Difference in percentage between the pressure-gradient current evaluated using the magnetic field measured by Swarm A and that evaluated using the magnetic field of internal origin, in the case of quiet (**left** column) and disturbed (**right** column) geomagnetic activity conditions, for the Northern and Southern hemispheres, in QD-latitude ($\lambda_{QD} > 50°$ and $\lambda_{QD} < -50°$ for the Northern and Southern hemisphere, respectively) and MLT reference system.

3.2. *Variation of Pressure-Gradient Current with Season*

In order to investigate the seasonal dependence of the pressure-gradient current, we consider the period between 5 May and 5 August of each year to identify the Northern local summer and the Southern local winter, and the period between 5 November and 5 February of each year to identify the Northern local winter and Southern local summer. Thus, by portioning the entire dataset according to summer and winter solstices, we can study the dependence of the pressure-gradient current on solar illumination.

Figure 7 reports the obtained plasma pressure distribution according to the selected seasons for both hemispheres, together with the resulting flow patterns of the pressure-gradient current. The current is described using black arrows whose lengths are proportional to the intensity. Differently from previous figures, we use different scales for the plasma pressure in the two seasons. Indeed, the plasma pressure is almost three times higher in summer than in winter. Looking at Figure 7 we notice that from winter to summer, there is an increase of plasma pressure in the dayside ionosphere at all MLTs, especially between the noon and the dusk at middle latitudes. Plasma pressure increases even in the post-nightside sector at low latitudes in the Southern hemisphere, and at high latitudes in the polar cap. Lastly, an increase is observable in the polar cusp during summer in both hemispheres.

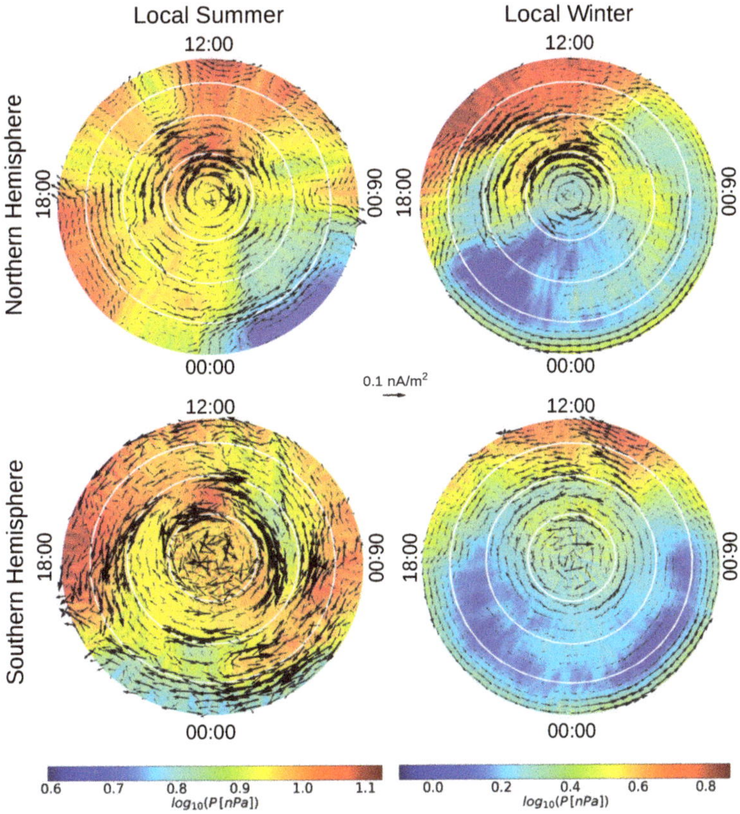

Figure 7. Flow patterns of pressure-gradient current (black arrows) superimposed on the plasma pressure distribution during summer (**left** column) and winter (**right** column) periods, in the Northern and Southern hemispheres. Maps are in QD-latitude ($\lambda_{QD} > |50°|$) and MLT reference system. The binning window is $1°\lambda_{QD} \times 4'$ MLT for the pressure map, while for the current's vector field the window is $2°\lambda_{QD} \times 16'$ MLT for graphical reasons. The concentric white circles are plotted in 10° interval, corresponding to QD-latitudes of |80°|, |70°| and |60°| starting from the centre.

The plasma pressure distributions reported in Figure 7 show that summer-winter asymmetry is more pronounced in the Southern hemisphere than in the Northern one. This is due to the different summer-winter asymmetry of N_e spatial distribution between the two hemispheres, which is a combination of two different effects, the first produced by the solar radiation and the second by the F region annual anomaly [38,39]. Indeed, N_e values are higher during summer than during winter, as a result of a larger solar radiation but, at the same time, they are higher during December solstice than during June one. Since in the Southern hemisphere the increase in N_e values happens during summer season, it amplifies the difference with respect to local winter. Conversely, the F region annual anomaly has the effect of reducing the asymmetry between local seasons in the Northern hemisphere.

The dependence of the plasma pressure on the seasons also involves a dependence of the pressure-gradient current on the solar illumination. The flow patterns of the currents and their strengths change from winter to summer. The pressure-gradient currents are visible especially in summer where the structures, which have previously identified by studying the dependence on the geomagnetic activity levels, can be easily recognizable. Even in this case, we can observe three regions characterized by the presence of these currents, one around cleft region, another at high latitudes around the magnetic poles, and the third confined at auroral latitudes. These regions should identify the zones characterized by the occurrence of ionospheric irregularities.

As regards the current intensity in the two hemispheres during the two selected seasons, we report the distributions of the pressure-gradient current values in Figure 8. We notice that the current intensity is higher during summer than winter, when more plasma is produced by sunlight and plasma pressure gradients are greater. The current intensity is greater in the Southern hemisphere than Northern one.

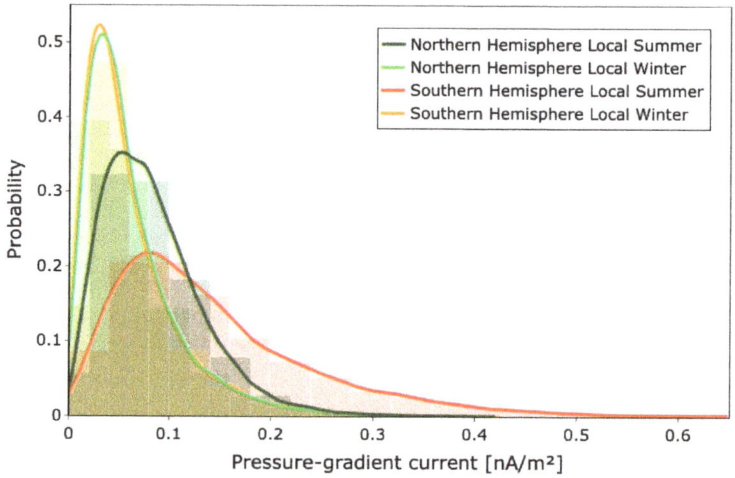

Figure 8. Distributions of the pressure-gradient current values for local summer and local winter conditions in the Northern and Southern hemispheres. Solid line represents the continuous probability density curve and is plotted above the more transparent relative histogram.

3.3. *Variation of Pressure-Gradient Current with Solar Activity*

To evaluate the level of solar activity during the selected period of Swarm observations we use, as previously described, the daily solar flux at 10.7 cm (F10.7) and the Mg II core-to-wing ratio as proxies. The values of F10.7 and Mg II from 1 April 2014 to 31 December 2017 are reported in the lower panel of Figure 9. Here, the vertical dotted lines indicate the transition from one year to the next, the horizontal lines identify the yearly mean values, and the shaded parts cover the values between the yearly mean values $\pm$ one standard deviation year by year for each index. The analysed period corresponds to the decreasing phase of the solar cycle, and is characterized by a decrease in the F10.7 and Mg II yearly mean values.

For what concerns the pressure-gradient current intensity, the most remarkable features emerging from Figure 9 are a decrease in maximum and minimum values of the current intensity with the decrease of solar activity as well as a decrease of the intensity around the cusp. Moreover, regardless of the solar activity level, the pressure-gradient current is substantially more intense in the Southern hemisphere than in the Northern one at almost all magnetic latitudes and MLTs. In the Southern hemisphere, besides the three main regions already identified, another region of enhanced current is visible. It is at a latitude between 70° and 80° in the nightside. This structure is well visible in 2014 and 2015 when the solar activity is higher.

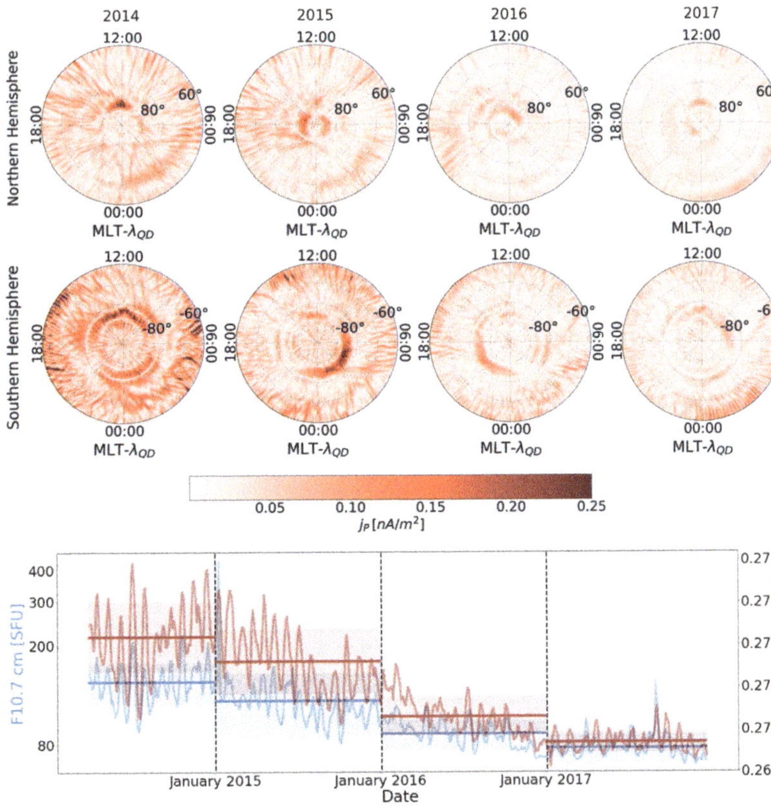

Figure 9. From top to bottom: Polar view of the average spatial distribution of the pressure-gradient current intensities, year by year, for the Northern and Southern hemisphere, respectively; F10.7 (blue) and the Mg II (red) indices during the 4 years of Swarm observations. The vertical dotted lines indicate the transition from one year to the next, the horizontal lines identify the yearly mean values, while the shaded parts cover the values between the yearly mean values ± one standard deviation year by year for each index.

4. Discussion and Conclusions

The aim of the present research was to examine the flow patterns of the pressure-gradient currents and their strengths at high-latitude ionospheric topside F2 region in both hemispheres, and investigate their dependence on magnetic local time, geomagnetic activity, season and solar forcing drivers. All this has been achieved thanks to the geomagnetic field, electron density and electron temperature variations observed by Swarm A satellite from 2014 to 2018. The flow patterns of the pressure-gradient currents have been estimated under the hypothesis that it is possible to ignore the effects of neutral collisions and consider the electron and ion temperature equal. Our study is the first to show polar flow patterns of the pressure-gradient currents and their strengths at Swarm altitude. From the observations the following conclusions can be drawn:

- During geomagnetically disturbed periods ($AE \geq 120$ nT) the plasma pressure gradients are particularly large around cleft region, where the electron density is changing rapidly. The pressure-gradient current flows around this plasma pressure enhancement region in both hemispheres. The existence of this flow pattern agrees with an increased probability of finding magnetic field variations correlated with plasma density ones around cleft region [9]. Anyway, we remark that additional contributions to the observed magnetic field variations and plasma pressure gradients can be due to incident Alfvén wave [10];

- Regardless of the level of geomagnetic activity, at high latitudes ($\lambda_{QD} > |75^{\circ}|$) the flow patterns of the pressure-gradient current identify another region characterized by large plasma pressure gradients, the polar cap. This region, that is observable in both hemispheres, is known to be characterized by the presence of plasma instabilities and the formation of ionospheric irregularities [34]. Additionally, in this region previous studies [11] found a high occurrence rate of magnetic field variations well explained by plasma density variations;
- At lower latitudes in both hemispheres the flow patterns of the pressure-gradient current identify another region where the plasma pressure is changing. In first approximation, it corresponds to the auroral oval and equatorward of the auroral oval on the nigthside. These flow patterns move to lower geomagnetic latitude with increasing geomagnetic activity;
- The pressure-gradient current mean intensity is quite low, around 1 order of magnitude less than the same current observed at low latitudes. In addition, the mean value found in our analysis is lower than that obtained by Laundal et al. [11] at high latitude. The reasons are probably due to the different method used to estimate the currents and to the different size of the window used to evaluate the pressure gradients. We use a window larger than the one used by Laundal et al. [11] and for this reason our pressure gradients are sharper and the pressure-gradient current intensities are smaller.
- Pressure-gradient current shows a clear dependence on solar illumination, and its intensity is influenced by F region annual anomaly. This is probably the reason why the asymmetry summer/winter is more marked in the Southern hemisphere than in the Northern one. Using the diamagnetic effect, Park et al. [9] and Laundal et al. [11] investigated the dependence on season of the ionospheric irregularity occurrences at high latitude. In some ways, our findings are in agreement with those reported by Laundal et al. [11], who found higher occurrence rates of magnetic field variations well explained by plasma pressure in summer than winter. However, the region characterized by a higher probability to find a correspondence between magnetic field and electron density variations is mainly confined in the polar cap in Laundal et al. [11], while the geographic location of our currents is wider. Nevertheless, a more precise comparison is not possible since there is not a distinction between the two hemispheres in Laundal et al. [11]. Conversely, the findings of our study do not seem to support the previous research by Park et al. [9], who reported higher occurrence rates of plasma density irregularities in winter than summer. This is probably due to a different data selection, Park et al. [9] studied the seasonal dependence considering measurements relative to geomagnetically active periods ($AE \geq 200$ nT). We study the seasonal dependence regardless of geomagnetic activity level and in our selection the percentage of data, that satisfies the AE threshold fixed by Park et al. [9], is approximately of 28%;
- Regardless of geomagnetic and solar activity, the pressure-gradient current intensity is always slightly greater in the Southern hemisphere than in the Northern one;
- The pressure-gradient current intensity decreases with the solar activity level.

The findings of this study confirm that the strength of the pressure-gradient currents is weak in the high-latitude ionospheric topside F2 region, at Swarm A altitude ($\simeq$460 km), but these currents appear preferentially at the same geographic locations regardless of geomagnetic activity, season and solar activity. This means that the magnetic field variations associated with these flow patterns can induce disturbances on the ionospheric magnetic field measurements and cause artifacts in main field models.

Author Contributions: Conceptualization, G.L., P.D.M. and G.C.; methodology, software and formal analysis, G.L.; investigation, G.L., P.D.M. and G.C.; writing original draft preparation, G.L. and P.D.M.; writing review and editing, all Authors; supervision, P.D.M. and F.B. All authors have read and agreed to the published version of the manuscript.

Funding: This research received no external funding.

Data Availability Statement: Publicly available dataset were analyzed in this study. Elaborated data are available on request from the corresponding author.

Conflicts of Interest: The authors declare no conflict of interest.

References

1. Cravens, T., Dessler, A., Houghton, J.; Rycroft, M. *Physics of Solar System Plasmas*; Cambridge University Press: Cambridge, UK, 1997.
2. Ganushkina, N.; Liemohn, M.; Dubyagin, S. Current systems in the Earth's magnetosphere. *Rev. Geophys.* **2018**, *56*, 309-332. [CrossRef]
3. Lühr, H.; Rother, M.; Maus, S.; Mai, W.; Cooke, D. The diamagnetic effect of the equatorial Appleton anomaly: Its characteristics and impact on geomagnetic field modeling. *Geophys. Res. Lett.* **2003**, *30*. [CrossRef]
4. Reigber, C.; Schwintzer, P.; Neumayer, K.; Barthelmes, F.; König, R.; Förste, C.; Balmino, G.; Biancale, R.; Lemoine, J.; Loyer, S.; et al. The CHAMP-only Earth gravity field model EIGEN-2. *Adv. Space Res.* **2003**, *31*, 1883–1888. [CrossRef]
5. Lühr, H.; Maus, S. Solar cycle dependence of quiet-time magnetospheric currents and a model of their near-Earth magnetic fields. *Earth Planets Space* **2010**, *62*, 843–848. [CrossRef]
6. Alken, P.; Maus, S.; Richmond, A.; Maute, A. The ionospheric gravity and diamagnetic current systems. *J. Geophys. Res. Space Phys.* **2011**, *116*, A12316. [CrossRef]
7. Richmond, A.; Ridley, E.; Roble, R. A thermosphere/ionosphere general circulation model with coupled electrodynamics. *Geophys. Res. Lett.* **1992**, *19*, 601–604. [CrossRef]
8. Alken, P. Observations and modeling of the ionospheric gravity and diamagnetic current systems from CHAMP and Swarm measurements. *J. Geophys. Res. Space Phys.* **2016**, *121*, 589–601. [CrossRef]
9. Park, J.; Ehrlich, R.; Lühr, H.; Ritter, P. Plasma irregularities in the high-latitude ionospheric F region and their diamagnetic signatures as observed by CHAMP. *J. Geophys. Res. Space Phys.* **2012**, *117*, A10322. [CrossRef]
10. Lotko, W.; Zhang, B. Alfvénic heating in the cusp ionosphere-thermosphere. *J. Geophys. Res. Space Phys.* **2018**, *123*, 10–368. [CrossRef]
11. Laundal, K.; Hatch, S.; Moretto, T. Magnetic effects of plasma pressure gradients in the upper F region. *Geophys. Res. Lett.* **2019**, *46*, 2355–2363. [CrossRef]
12. Friis-Christensen, E.; Lühr, H.; Hulot, G. Swarm: A constellation to study the Earth's magnetic field. *Earth Planets Space* **2006**, *58*, 351–358. [CrossRef]
13. Knudsen, D.; Burchill, J.; Buchert, S.; Eriksson, A.; Gill, R.; Wahlund, J.; Åhlén, L.; Smith, M.; Moffat, B. Thermal ion imagers and Langmuir probes in the Swarm electric field instruments. *J. Geophys. Res. Space Phys.* **2017**, *122*, 2655–2673. [CrossRef]
14. Lomidze, L.; Knudsen, D.; Burchill, J.; Kouznetsov, A.; Buchert, S. Calibration and validation of Swarm plasma densities and electron temperatures using ground-based radars and satellite radio occultation measurements. *Radio Sci.* **2018**, *53*, 15–36. [CrossRef]
15. Richmond, A. D. Ionospheric electrodynamics using magnetic apex coordinates. *J. Geomagn. Geoelectr.* **1995**, *47*, 191–212. [CrossRef]
16. Emmert, J.; Richmond, A.; Drob, D. A computationally compact representation of Magnetic-Apex and Quasi-Dipole coordinates with smooth base vectors. *J. Geophys. Res. Space Phys.* **2010**, *115*, A08322. [CrossRef]
17. Laundal, K.; Richmond, A. Magnetic coordinate systems. *Space Sci. Rev.* **2017**, *206*, 27–59. [CrossRef]
18. Baker, K.; Wing, S. A new magnetic coordinate system for conjugate studies at high latitudes. *J. Geophys. Res. Space Phys.* **1989**, *94*, 9139–9143. [CrossRef]
19. Peeters, P.; Simon, P.; White, O.; De Toma, G.; Rottman, G.; Woods, T.; Knapp, B. Mg II Core-to-Wing Solar Index from High Resolution GOME Data. 1997. Available online: https://earth.esa.int/workshops/ers97/papers/peeters3/index-2.html (accessed on 13 March 2022).
20. Snow, M.; Machol, J.; Viereck, R.; Woods, T.; Weber, M.; Woodraska, D.; Elliott, J. A revised magnesium II core-to-wing ratio from SORCE SOLSTICE. *Earth Space Sci.* **2019**, *6*, 2106–2114. [CrossRef]
21. Foukal, P. Extension of the F10.7 index to 1905 using Mt. Wilson Ca K spectroheliograms. *Geophys. Res. Lett.* **1998**, *25*, 2909–2912. [CrossRef]
22. Tapping, K. The 10.7 cm solar radio flux (F10.7). *Space Weather* **2013**, *11*, 394–406. [CrossRef]
23. Richmond, A.; Maute, A. Ionospheric electrodynamics modeling. In *Modeling the Ionosphere-Thermosphere System*; American Geophysical Union, Washington, DC, USA, 2014; pp. 57–71.
24. Kamide, Y.; Chian, A. *Handbook of the Solar-Terrestrial Environment*; Springer: Berlin/Heidelberg, Germany, 2007.
25. Quarteroni, A.; Sacco, R.; Saleri, F. *Numerical Mathematics*; Springer-Verlag New York, Inc.: New York, NY, USA, 2010; Volume 58.
26. Consolini, G. Self-organized criticality: A new paradigm for the magnetotail dynamics. *Fractals* **2002**, *10*, 275–283. [CrossRef]
27. Alberti, T.; Giannattasio, F.; De Michelis, P.; Consolini, G. Linear versus nonlinear methods for detecting magnetospheric and ionospheric current systems patterns. *Earth Space Sci.* **2020**, *7*, e2019EA000559. [CrossRef]

28. Milan, S.; Provan, G.; Hubert, B. Magnetic flux transport in the Dungey cycle: A survey of dayside and nightside reconnection rates. *J. Geophys. Res. Space Phys.* **2007**, *112*. [CrossRef]
29. Wang, W.; Burns, A.; Killeen, T. A numerical study of the response of ionospheric electron temperature to geomagnetic activity. *J. Geophys. Res. Space Phys.* **2006**, *111*. [CrossRef]
30. Prölss, G. Subauroral electron temperature enhancement in the nighttime ionosphere. *Ann. Geophys.* **2006**, *24*, 1871–1885. [CrossRef]
31. Knudsen, W. Magnetospheric convection and the high-latitude F 2 ionosphere. *J. Geophys. Res.* **1974**, *79*, 1046–1055. [CrossRef]
32. Foster, J.; Coster, A.; Erickson, P.; Holt, J.; Lind, F.; Rideout, W.; McCready, M.; Van Eyken, A.; Barnes, R.; Greenwald, R.; et al. Multiradar observations of the polar tongue of ionization. *J. Geophys. Res. Space Phys.* **2005**, *110*. [CrossRef]
33. Foster, J. Storm time plasma transport at middle and high latitudes. *J. Geophys. Res. Space Phys.* **1993**, *98*, 1675–1689. [CrossRef]
34. Spicher, A.; Clausen, L.; Miloch, W.; Lofstad, V.; Jin, Y.; Moen, J. Interhemispheric study of polar cap patch occurrence based on Swarm in situ data. *J. Geophys. Res. Space Phys.* **2017**, *122*, 3837–3851. [CrossRef]
35. Alken, P.; Maute, A.; Richmond, A. The *F*-Region Gravity and Pressure Gradient Current Systems: A Review. *Space Sci. Rev.* **2017**, *206*, 451–469. [CrossRef]
36. Olsen, N. A new tool for determining ionospheric currents from magnetic satellite data. *Geophys. Res. Lett.* **1996**, *23*, 3635–3638. [CrossRef]
37. Finlay, C.; Olsen, N.; Tøffner-Clausen, L. DTU candidate field models for IGRF-12 and the CHAOS-5 geomagnetic field model. *Earth Planets Space* **2015**, *67*, 114. [CrossRef]
38. Torr, D.; Torr, M.; Richards, P. Causes of the F region winter anomaly. *Geophys. Res. Lett.* **1980**, *7*, 301–304. [CrossRef]
39. Rishbeth, H.; Müller-Wodarg, I. Why is there more ionosphere in January than in July? The annual asymmetry in the F2-layer. *Ann. Geophys.* **2006**, *24*, 3293–3311. [CrossRef]

remote sensing

Article

On Turbulent Features of E × B Plasma Motion in the Auroral Topside Ionosphere: Some Results from CSES-01 Satellite

Giuseppe Consolini [1,*], Virgilio Quattrociocchi [1,2], Simone Benella [1], Paola De Michelis [3], Tommaso Alberti [1], Mirko Piersanti [2] and Maria Federica Marcucci [1]

[1] INAF-Istituto di Astrofisica e Planetologia Spaziali, Via del Fosso del Cavaliere 100, 00133 Roma, Italy; virgilio.quattrociocchi@inaf.it (V.Q.); simone.benella@inaf.it (S.B.); tommaso.alberti@inaf.it (T.A.); federica.marcucci@inaf.it (M.F.M.)

[2] Dipartimento Scienze Fisiche e Chimiche, Universitá degli Studi dell'Aquila, Via Vetoio, 67100 L'Aquila, Italy; mirko.piersanti@univaq.it

[3] Istituto Nazionale di Geofisica e Vulcanologia, Via di Vigna Murata 605, 00143 Roma, Italy; paola.demichelis@ingv.it

* Correspondence: giuseppe.consolini@inaf.it

Abstract: The recent Chinese Seismo-Electromagnetic Satellite (CSES-01) provides a good opportunity to investigate some features of plasma properties and its motion in the topside ionosphere. Using simultaneous measurements from the electric field detector and the magnetometers onboard CSES-01, we investigate some properties of the plasma **E** × **B** drift velocity for a case study during a crossing of the Southern auroral region in the topside ionosphere. In detail, we analyze the spectral and scaling features of the plasma drift velocity and provide evidence of the turbulent character of the **E** × **B** drift. Our results provide an evidence of the occurrence of 2D **E** × **B** intermittent convective turbulence for the plasma motion in the topside ionospheric F2 auroral region at scales from tens of meters to tens of kilometers. The intermittent character of the observed turbulence suggests that the macro-scale intermittent structure is isomorphic with a quasi-1D fractal structure, as happens, for example, in the case of a filamentary or thin-tube-like structure. Furthermore, in the analyzed range of scales we found that both magnetohydrodynamic and kinetic processes may affect the plasma dynamics at spatial scales below 2 km. The results are discussed and compared with previous results reported in the literature.

Keywords: plasma turbulence; auroral ionosphere; **E** × **B** plasma motion

Citation: Consolini, G.; Quattrociocchi, V.; Benella, S.; De Michelis, P.; Alberti, T.; Piersanti, M.; Marcucci, M.F. On Turbulent Features of E × B Plasma Motion in the Auroral Topside Ionosphere: Some Results from CSES-01 Satellite. *Remote Sens.* **2022**, *14*, 1936. https://doi.org/10.3390/rs14081936

Academic Editor: Michael E. Gorbunov

Received: 12 February 2022
Accepted: 9 April 2022
Published: 17 April 2022

1. Introduction

In the geospace environment, such as the Earth's ionosphere and magnetosphere, plasma dynamics can display multiscale features and chaos. As it occurs in several space and astrophysical plasmas, the chaotic nature of the plasma dynamics in the geospace environment is due to the occurrence of turbulent processes. This happens, for instance, at high-latitude ionospheric regions, where plasma dynamics was shown to be characterized by turbulence as revealed, for example, by the irregular and chaotic character of the electric and magnetic field fluctuations in the ULF (Ultra Low Frequency) and ELF (Extra Low Frequency) spectral ranges [1–3]. In addition, previous studies of the ionospheric electric, magnetic and electron density fluctuations have shown that the fluctuations are characterized by power-law spectral densities, scaling features and non-Gaussian statistics of spatial and temporal increments at small/short scales (see, e.g., [4–9] and references therein). In detail, in the auroral regions, it has been found by Kintner et al. [10] that the origin of electric field broadband spectra and the features of its fluctuations are probably caused by the occurrence of intermittent turbulence [10,11], due to the sporadic fast interactions between localized coherent plasma structures [4,12]. In any case, most of the observed fluctuations of fields and plasma parameters in the high-latitude ionospheric regions are characterized by very complex multiscale features that might affect the plasma dynamics [4,13,14].

The high variability of the electric and magnetic field fluctuations can affect the plasma dynamics through the term $\mathbf{E} \times \mathbf{B}$, which can be strongly influenced by the multiscale and turbulent character of the field fluctuations. The plasma $\mathbf{E} \times \mathbf{B}$ drift has been widely investigated in the ionospheric equatorial region, where the variations of the vertical plasma velocity drift play a relevant role in generating ionospheric irregularities [15–18]. Conversely, it is believed less relevant in the auroral region, especially at the typical altitudes where auroral emission occurs [19]. This is because at this altitude the presence of neutrals can interfere with the $\mathbf{E} \times \mathbf{B}$ drift disrupting the coherent plasma motion perpendicular to the magnetic field direction due to the ion-neutral collisions. This mechanism is at the origin of the well-known Hall current [19]. However, at higher altitudes where the neutrals' density is lower, the $\mathbf{E} \times \mathbf{B}$ drift may play a relevant role in the dynamics of the plasma also in the auroral regions, and can be relevant to the formation of plasma irregularities.

Recently, the Chinese Seismo-Electromagnetic Satellite (CSES-01) [20,21], launched on February 2018 and equipped with a large set of instruments, including an electric field detector (EFD) [22,23], a flux gate (HPM) and a search-coil (SCM) [24] magnetometer, has had the opportunity to explore the Southern auroral oval region during a time interval characterized by the occurrence of a magnetospheric substorm. During the crossing of the southern auroral region, the satellite has measured simultaneously the magnetic and electric field fluctuations [25]. CSES-01, which has been originally designed to study possible correlations between seismic events and iono/magnetospheric perturbations, flies in the topside F2 ionosphere at $\sim$500 km of altitude on a Sun-synchronous orbit. Thus, it can also provide the opportunity to explore the plasma dynamics at an altitude higher than that at which the typical auroral phenomena occur. Generally CSES-01 is operative at geographic latitude $|Lat| < 65^\circ$, and thus the observation of high-latitude phenomena is sporadic and limited to a few periods in which the Earth's dipole tilt angle is higher than $|15^\circ|$, corresponding to solstices. On 11 August 2018 from 21:41 UT to 21:45 UT, CSES-01 was in operational mode during a crossing of the Southern polar F2 topside ionosphere. Its measurements allowed us [25] to investigate some interesting features of the electric field fluctuations in the auroral region and to establish the very complex structure of electric field fluctuations probably due to the occurrence of intermittent turbulence [4,12].

This study sits out to investigate the spectral and scaling features of the plasma $\mathbf{E} \times \mathbf{B}$ drift velocity using magnetic and electric field measurements from the HPM, SCM and EFD experiments onboard CSES-01 during the crossing of the Southern auroral region occurred on 11 August 2018. In detail, this case study seeks to examine the multiscale character of the plasma drift velocity $\mathbf{E} \times \mathbf{B}$.

2. Data Description

We consider magnetic and electric field data recorded by HPM, SCM and EFD instruments [22,24] onboard CSES-01 during a crossing of the Southern polar ionosphere on 11 August 2018 from 21:42:30 UT to 21:45:00 UT. During this time interval CSES-01 partially crosses the Southern auroral oval region in its ascending orbit near 02:00 MLT when auroral precipitation is going on. This time interval is, indeed, characterized by a moderate/high auroral substorm activity, as indicated by the value of the auroral electrojet index, AE > 500 nT [25]. This period has already been investigated in previous papers [25,26], to which the reader may refer for more information.

We use magnetic field data with a resolution of 1 sample/s from HPM and with a resolution of 10 k samples/s from SCM and electric field measurements with a resolution of 5000 samples/s. Magnetic field data from HPM and SCM are joined and successively reduced at the same resolution of EFD, i.e., 5000 samples/s, in order to compute the plasma $\mathbf{E} \times \mathbf{B}$ drift velocity, i.e.,

$$\mathbf{v}_D = \frac{\mathbf{E} \times \mathbf{B}}{B^2}, \tag{1}$$

where **E** and **B** are the electric and magnetic field, respectively. Data are in the geographic coordinate system, $[x, y, z]$, where the X-component is pointing to the geographic North, the Y-component to East and the Z-component to Earth's center. Furthermore, the electric field components are corrected subtracting the term $\mathbf{E}_s = \mathbf{v}_s \times \mathbf{B}$, where $\mathbf{v}_s$ is the satellite velocity, which is due to the satellite motion in the Earth's magnetic field. Lastly, the velocity field is downsampled to the resolution of 250 samples/s by averaging the original 5000 samples/s data on consecutive boxes of 20 points. Thus, considering the satellite velocity, which is of the order of $v_s \simeq 8$ km/s, and the chosen resolution of 250 samples/s, we are capable of exploring velocity fluctuations up to scales of the order of ~30 m. This length scale is well below the ion inertial length scale, which is of the order of $2 \div 3$ km for O^+ and $10 \div 20$ m for e^-. This means that we are capable of exploring also the sub-ionic scales.

Figure 1 shows the electric and the magnetic field data collected by CSES-01 and used in our analysis. The values of the electric field along the three different directions are similar, ranging between -0.3 V/m and 0.15 V/m. Conversely, the magnetic field is stronger along the z-axis as expected at high latitude being the magnetic field nearly vertical.

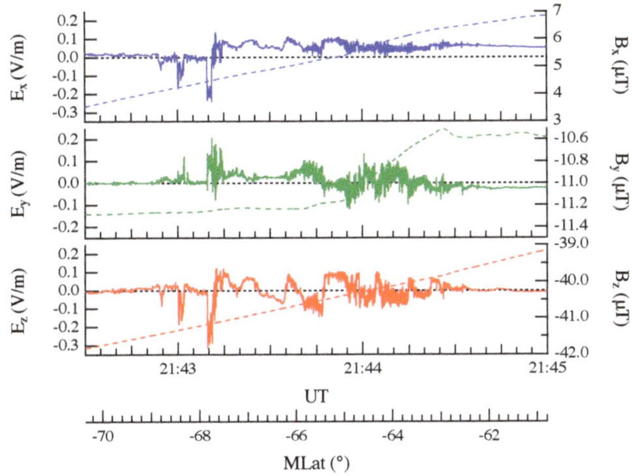

Figure 1. Electric (solid lines on the left vertical axes) and magnetic (dashed lines on the right vertical axes) field measurements collected by CSES-01 EFD and HPM instruments for the time interval under consideration. The three components are in the geographic coordinate system.

Figure 2 reports the plasma drift velocity, $\mathbf{v}_D$, computed using Equation (1). Due to the quasi-dipole magnetic field configuration, the velocity field of the plasma in the crossed polar region is mainly horizontal in the $[x, y]$-plane. Although the mean flow direction is along the Y component, the plasma drift velocity shows large fluctuations along the x and y axes, as confirmed by the values of the root mean square along the three axes $\mathbf{v}_D^{RMS} \equiv [0.83, 1.34, 0.35]$ km/s.

Anyway, the geographic coordinate system is not the optimal reference system for our analysis. Usually, the analysis of the fluctuation field is done in the parallel and perpendicular directions to the magnetic field. For this reason, we change coordinate system and compute the parallel and perpendicular components to the local magnetic field direction of the plasma $\mathbf{E} \times \mathbf{B}$ drift velocity. The component of the drift velocity, which is parallel ($v_\parallel$) to the magnetic field direction, is practically zero being $| v_\parallel / v_\perp | \ll 10^{-6}$ (we remark that it is expected to be zero by construction). The two perpendicular components, which describe the horizontal plasma motion, are reported in Figure 3. They are chosen so that $v_\perp^{(1)}$ and $v_\perp^{(2)}$ are almost along the direction of North (x) and East ($-y$) axis, respectively. In detail, for each point we consider the local magnetic field direction (i.e., the geomagnetic field direction that is essentially oriented in the z-direction) and define two unitary vectors in the plane perpendicular to geomagnetic field direction, with components mainly along x- and y-directions. This defines the local reference frame. Successively, we compute the components of the $\mathbf{E} \times \mathbf{B}$ along these two perpendicular unitary vectors.

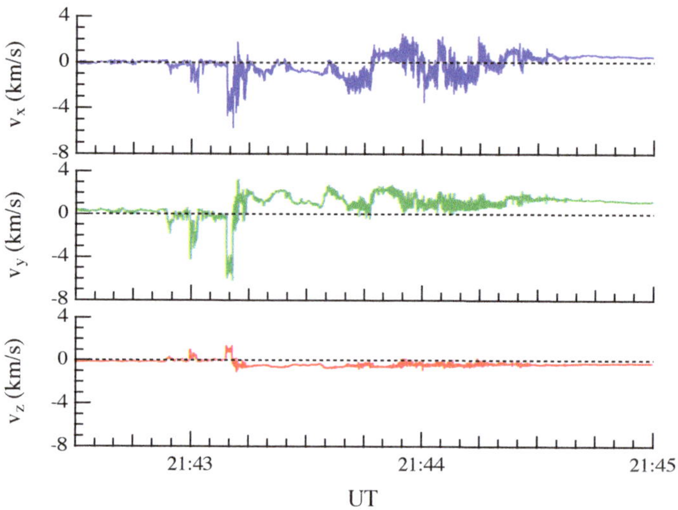

Figure 2. Drift velocity ($\mathbf{v}_D = (\mathbf{E} \times \mathbf{B})/B^2$) in the geographic coordinate system.

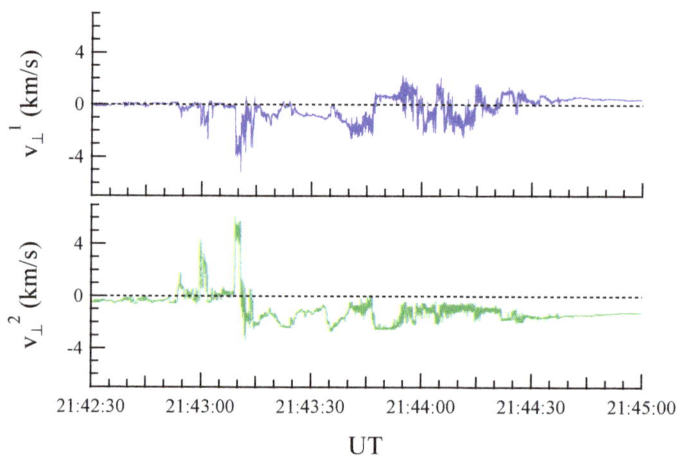

Figure 3. The two components of drift velocity ($\mathbf{v}_D = (\mathbf{E} \times \mathbf{B})/B^2$) perpendicular to the local magnetic field direction.

The effective dimension d^* of the drift velocity fluctuation field can be measured by applying the covariance analysis, which allows us to estimate the eigenvalues of the covariance matrix of the drift velocity. Indeed, knowing the eigenvalues of covariance matrix, we can compute the effective dimension, d^*, of the fluctuating field as

$$d^* = \frac{1}{\lambda_{max}} \sum_i \lambda_i \tag{2}$$

where λ_i are the eigenvalues and λ_{max} is the maximum eigenvalue. In our case, we get an effective dimension $d^* \simeq 1.35$. This quantity provides an indication of the dimension of the phase-space spanned by the field fluctuations. In our case, being $1 < d^* < 2$, we expect that the drift velocity fluctuation field can be confined in a 2D space. Consequently, we could assume to deal with a 2D plasma motion. Anyway, we remark that the effective dimension d^* is not to be confused with the fractal dimension associated with the plasma dynamics under the drift velocity.

Before proceeding in the analysis, we check the consistency of the obtained average velocity field with the macroscopic overall convection pattern, that can be reconstructed from measurements of the Super Dual Auroral Radar Network (SuperDARN) radars operating in the Southern polar hemisphere. SuperDARN is an international scientific radar network consisting of high frequency (HF) coherent scatter radars that is operated under international cooperation. The SuperDARN data are mainly used to monitor the dynamics of the ionosphere and upper atmosphere in the high- and mid-latitude regions through the production of global plasma convection maps. However, these data can also be used to study many others geospace phenomena including field aligned currents, geomagnetic storms, magnetospheric substorms, magnetic reconnection, and interhemispheric plasma convection asymmetries (see, e.g., [27–29] and references therein). Figure 4 reports a comparison between the ionospheric convection map as reconstructed from SuperDARN observations relative to the time interval 21:44-21:46 UT, corresponding to the central part CSES-01 track and the convection drift velocity obtained by $\mathbf{E} \times \mathbf{B}$ from CSES-01 measurements.

Global maps of ionospheric convection can be derived from the SuperDARN data using the technique developed by Ruohoniemi and Baker [30] and Shepherd and Ruohoniemi [31]. In this technique, the SuperDARN line-of-sight velocities are fitted to an expansion of the high-latitude electrostatic potential in spherical harmonic functions. In order to constraint the solution in regions where no SuperDARN data are available, additional data from statistical models are used. The convection map shown in Figure 4 has been computed using the statistical model by Thomas and Shepherd [32] for adding data where SuperDARN data are missing. This statistical model (named TS18) has been derived for the Southern and Northern hemispheres using measurements from all the mid-latitude, high-latitude, and polar radars available in the years from 2010 to 2016. The TS18 is capable of producing climatological patterns of ionospheric convection organized by solar wind, interplanetary magnetic field (IMF), and dipole tilt angle conditions. The TS18 model inputs used for the computation of the map in Figure 4 correspond the average values of IMF clock angle (223°), solar wind electric field ($E_{sw} = 2.2$ mV/m), and dipole tilt angle ($-17.7°$) for the time interval of interest (11 August 2018 from 21:42:30 UT to 21:45:00 UT). In Figure 4, the colored vectors represent the fitted SuperDARN velocity vectors. The convection map has been generated by using the Radar Software Toolkit 4.2 [33]. Lastly, the green arrows are designed to facilitate the reader in locating the direction of the plasma convection motion. On the bottom of the same figure, we report a zoom of the plasma convection pattern, which permits us to identify the CSES-01 trajectory during the analyzed period. The trajectory is the blue line, while the color lines refer to the reconstructed plasma $\mathbf{E} \times \mathbf{B}$ drift velocity ($\mathbf{v}_D$). The length of the lines is proportional to the intensity of the drift velocity while the color (from red to light violet) indicates the time. An overall agreement between the observed convection pattern (green arrow) and the reconstructed $\mathbf{E} \times \mathbf{B}$ drift velocity (colored lines) is found. This confirms the correct reconstruction of the convective plasma $\mathbf{E} \times \mathbf{B}$ drift velocity by using CSES-01 measurements. The only small discrepancy is observed in correspondence with the higher latitude part of CSES-01 trajectory, where there is a plasma flow reversal. This can be due to a localized vortex structure near the lower latitude boundary of the convection cells not resolved in the SuperDARN convection map.

In Figure 5, we report the motion of the velocity vector tip in the plane perpendicular to the magnetic field direction. The motion of the vector tip shows a certain degree of chaoticity, which may be an indication of the occurrence of turbulence in plasma motion.

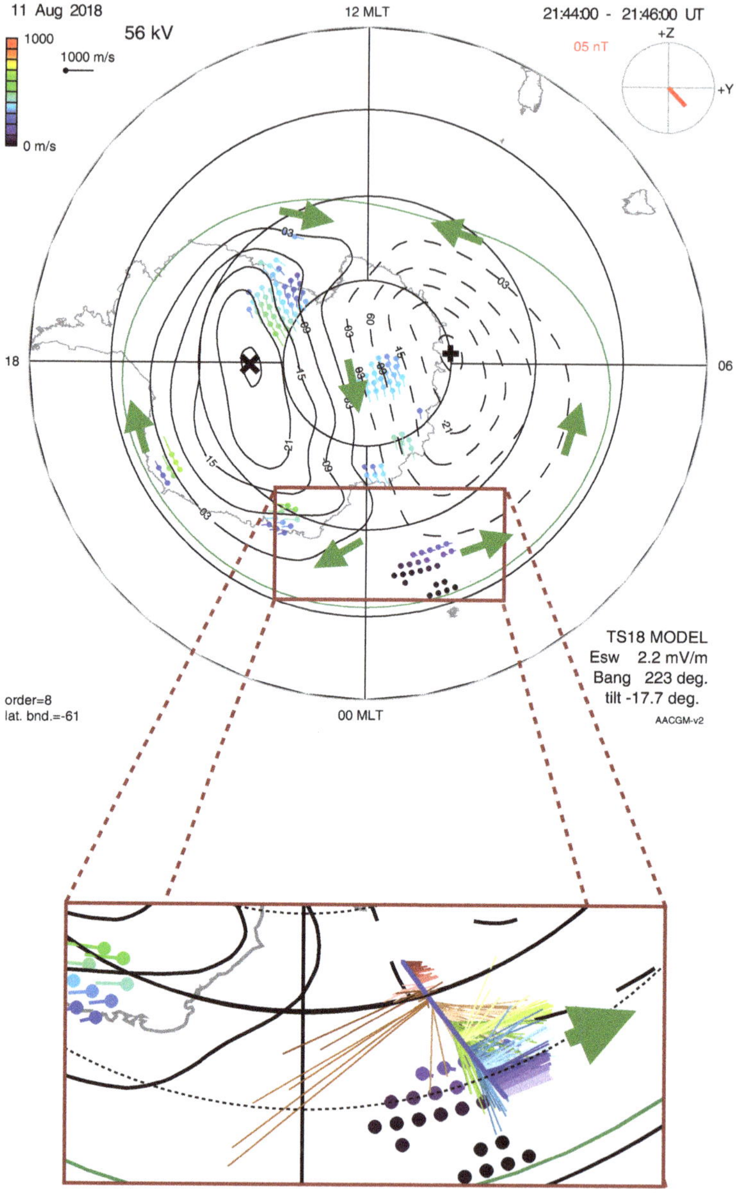

Figure 4. On the **top**: the instantaneous convection cells as reconstructed from SuperDARN observations in Antarctica. Green arrows show the overall plasma convection. On the **bottom**: a zoom of the region of the CSES-01 trajectory. Colored lines refer to the velocity vector field in the XY-plane. Colors (from red to light violet) indicate the time.

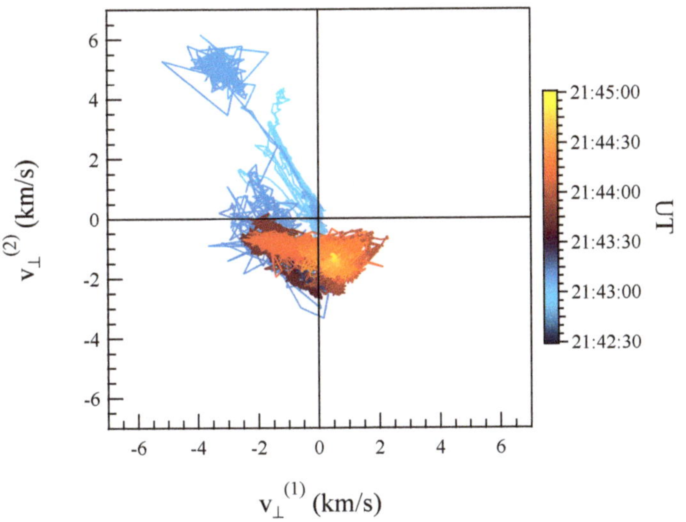

Figure 5. Motion of the velocity vector tip in the plane perpendicular to the magnetic field. The color is associated with the universal time UT (see the color bar).

3. Methods

The analysis of the spectral and scaling features of the plasma $\mathbf{E} \times \mathbf{B}$ drift is done using the standard Fourier power spectral analysis and the structure function analysis applied in fluid and MHD turbulence studies. While the Fourier analysis is a standard method to detect the presence or not of specific modes in the fluctuation field, the structure function analysis, as clearly discussed in Frisch [34], is one of the most powerful methods to investigate scaling features for fully developed turbulence. This approach is, indeed, a powerful method to detect the occurrence anomalous scaling features, i.e., what is generally called *intermittency* in fluid and MHD turbulence.

In the case of fully developed turbulence, a qth-order structure function, $S^{(q)}(\delta r)$, is defined as the qth-order moment of the longitudinal velocity increments, δv_l, at the spatial scale δr, i.e.,

$$S^{(q)}(\delta r) = \langle (v_l(r+\delta r) - v_l(r))^q \rangle \equiv \langle (\delta v_l)^q \rangle. \quad (3)$$

Looking at the definition of Equation (3) for a fixed spatial scale δr, the qth-order structure function represents the moment of order q of the distribution of the velocity increments at that scale. In the case of a turbulent flow, this quantity is expected to scale in the inertial range according to the following expression,

$$S^{(q)}(\delta r) \sim \delta r^{\zeta(q)}, \quad (4)$$

where the scaling exponents $\zeta(q)$, as stated by the usual Kolmogorov K41 theory of turbulence, are expected to be $\zeta(q) = q/3$ [34]. In particular, following the K41 theory and according to the Navier–Stokes equation for an inviscid fluid (i.e., in the limit of an infinite Reynolds number) the third order structure function ($S^{(3)}(\delta r)$) is expected to scale with $\zeta(3) = 1$. From the theory, it is possible to write

$$S^{(3)}(\delta r) = -\frac{4}{5}\epsilon\delta r. \quad (5)$$

where ϵ is the energy transfer rate along the cascade in the inertial range. Conversely, the $\zeta(q) = q/3$ dependence of the structure function scaling exponents is essentially a conjecture that implies that a global scale-invariance should exist. However, the observed scaling exponents can show departures from the linear dependence due to intermittency effects. The observed dependence of scaling exponents on the moment order is well described by a convex function of q, so that the global scale invariance is missing. By the way, in the case of intermittency effects, Equation (5) is also expected to hold.

Sometimes, the previous definition of the structure function is generalized considering the absolute value of the velocity differences, δv_l, i.e.,

$$S^{(q)}(\delta r) = \langle |\delta v_l(\delta r)|^q \rangle, \tag{6}$$

which allows both a better analysis of the dependence of scaling exponents on moment order (see, e.g., [35,36]) and the investigation of non-integer q. We remark that, although structure function analysis is generally defined for increments along the flow direction, it has also been generalized to the case of transverse directions, as well as, for time increments, assuming Taylor's hypothesis to be valid [37,38].

A simple way to investigate the emergence of intermittency or the validity of the existence of a global scale invariance is to compute the relative scaling of the qth-order structure function on the pth-order one, which is expected to scale as

$$S^{(q)}(\delta r) = \left[S^{(p)}(\delta r)\right]^{\gamma_p(q)}, \tag{7}$$

where $\gamma_p(q) = q/p$ in the case of a global scale invariance as the one predicted by K41 theory. We can note how $\gamma(q) \equiv \zeta(q)$ if Equation (5) holds and $p = 3$ [35,39]. This method has been introduced by Benzi et al. [35] and it is named as *extended self-similarity* (ESS), which is also valid in the case of low Reynolds number turbulence.

Another interesting quantity is the so-called *generalized kurtosis* $\Gamma_q(\delta r)$ [37], which is defined as

$$\Gamma_q(\delta r) \equiv \frac{S^{(q)}(\delta r)}{\left[S^{(2)}(\delta r)\right]^{q/2}}. \tag{8}$$

It is expected to be constant in the inertial range when global scale invariance holds. Conversely, departures from a constant value are generally referred as the evidence for the occurrence of *intermittency*. We remark that the emergence of intermittency manifests also in the lack of a scale-invariant shape of the probability density functions (PDFs) of the velocity increments $\delta v(\delta r)$ with the scale δr.

We apply the described analysis to the drift velocity field.

4. Results

We start our study of the properties of the plasma $\mathbf{E} \times \mathbf{B}$ drift velocity by investigating its spectral features. Figure 6 reports the trace, $S(f)$, of the power spectral density (PSD) of the components of the drift velocity field, which are perpendicular to the local magnetic field direction. The trace of PSD is defined as the sum of the PSDs of the two perpendicular components. The observed shape of PSD is quite consistent with the spectra expected in the case of two dimensional (2D) $\mathbf{E} \times \mathbf{B}$ convective turbulence in a quasi-steady state [40] or two-dimensional magnetohydrodynamic (MHD) turbulence in absence of Alfvén effect [41]. Indeed, for example, numerical simulations by Gruzinov et al. [40] revealed that when ionospheric plasma density irregularities break up into fingers leading to the formation of stable shear flows, the omnidirectional energy spectrum scaling exponent approaches to $\alpha \simeq -2$ after reaching a quasi-steady state rather than $\alpha \simeq -5/3$ as expected from the K41 theory prediction in the case of isotropic turbulence. In both cases [40,41], the spectral scaling exponents are steeper than the K41 theory prediction, as also occurs in our case being $\alpha \simeq -1.7$. Consistent with the literature, the observed PSD of the components of the plasma drift velocity field seems to support the idea of the occurrence of a strong 2D turbulence in a low$-\beta$ plasma medium [42], where β, which is the ratio of plasma to magnetic pressure, is a parameter classifying plasma conditions being usually $\beta \ll 1$ for the Earth's ionospheric plasma.

Let us now move to the analysis of the scaling features by the generalized structure functions, $S^{(q)}(\tau)$,

$$S^{(q)}(\tau) = \langle |v_i(t+\tau) - v_i(t)|^q \rangle, \tag{9}$$

where τ is the timescale, i.e., the time delay used to compute the velocity increments. The analysis is done in the plane perpendicular to the geomagnetic field local direction by considering separately the two components and then averaging the obtained structure functions. The averaging procedure is justified by the fact that the observed scaling features in the two separate perpendicular directions are isotropic, i.e., we do not observe significant differences. The timescale can be associated with a spatial scale δr by considering the satellite velocity v_s, i.e., $\delta r = \tau v_s$. Here, we investigate the timescale in the range from 4 ms to 3 s, approximately corresponding to a range of spatial scale $\delta r \in [0.03, 23.50]$ km. Clearly, this assumption, which is analogous to the Taylor's hypothesis, is strictly valid if the transit time is faster than the evolution time of the velocity field at the investigated time scale, an assumption that is nearly valid in the low-frequency range, that is, where we can assume that the temporal fluctuations are principally due to Doppler-shifted and stationary spatial variations (see, e.g., [11,25,43] and references therein). We will return on this point in Section 5.

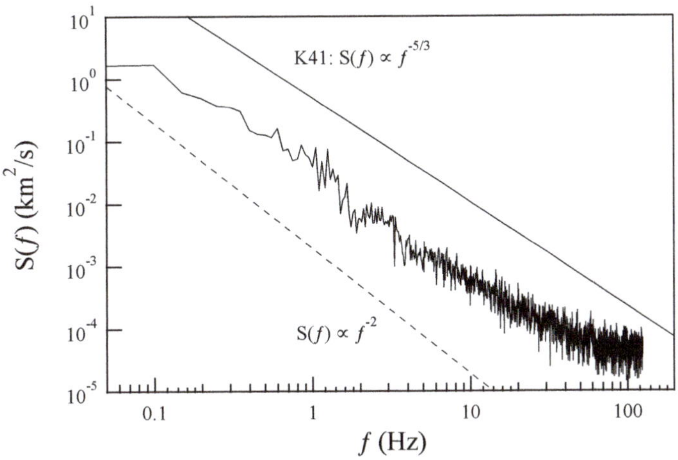

Figure 6. The trace, $S(f)$, of the PSD of the perpendicular components of the $\mathbf{E} \times \mathbf{B}$ drift velocity. The two power laws are that expected for K41 theory ($\sim f^{-5/3}$) and that observed for 2D $\mathbf{E} \times \mathbf{B}$ convective turbulence ($\sim f^{-2}$) in a quasi-steady state [40].

In the case of scale-invariant signals, the generalized q-order structure functions, $S^{(q)}(\tau)$, are expected to depend on the scale τ according to a power law, i.e.,

$$S^{(q)}(\tau) \simeq \tau^{\gamma(q)}, \tag{10}$$

where $\gamma(q)$ are the scaling exponents, which for a signal characterized by a global scale invariance are expected to depend linearly on the moment order q, i.e., $\gamma(q) = \alpha q$. In particular, the K41 theory of homogeneous turbulence predicts $\gamma(q) = q/3$.

Figure 7 shows the generalized structure functions, $S^{(q)}(\tau)$, of the velocity field components perpendicular to the magnetic field direction. Good scaling regimes are observed. In particular, the scaling of the structure functions is observed over near two orders of magnitude from $\tau \sim 0.04$ s to $\tau \sim 4$ s. Furthermore, the third-order structure function, $S^{(3)}(\tau)$, displays a range of time scales where a reasonable agreement with the expected linear scaling predicted by the K41 fluid turbulence theory can be found. This is very clear by plotting the compensated structure functions, $\tau^{-1}S^{(q)}(\tau)$ for $q = 2$, 3 and 4 as shown in Figure 8. This result suggests that to compute the scaling exponents of the structure functions, we can apply the ESS method introduced by Benzi et al. [35], which is based on the investigation of the relative scaling of the structure functions as reported in Equation (7).

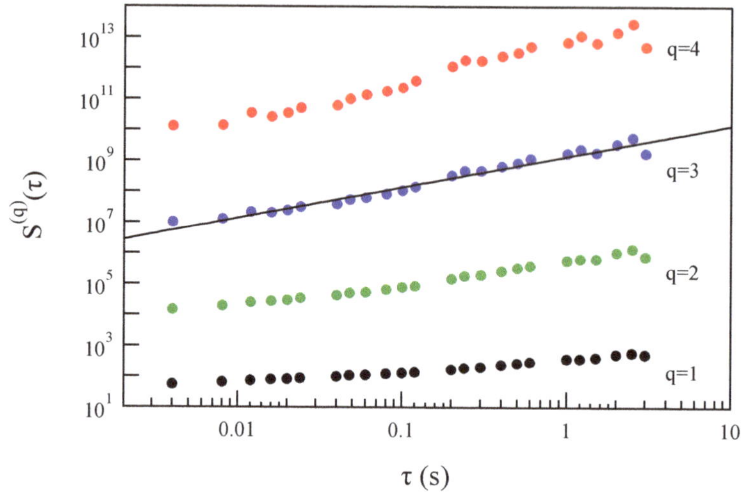

Figure 7. The generalized structure functions, $S^{(q)}(\tau)$, of the velocity field components perpendicular to the magnetic field direction. The black line refers to a linear scaling, i.e., $S^{(3)}(\tau) \simeq \tau$.

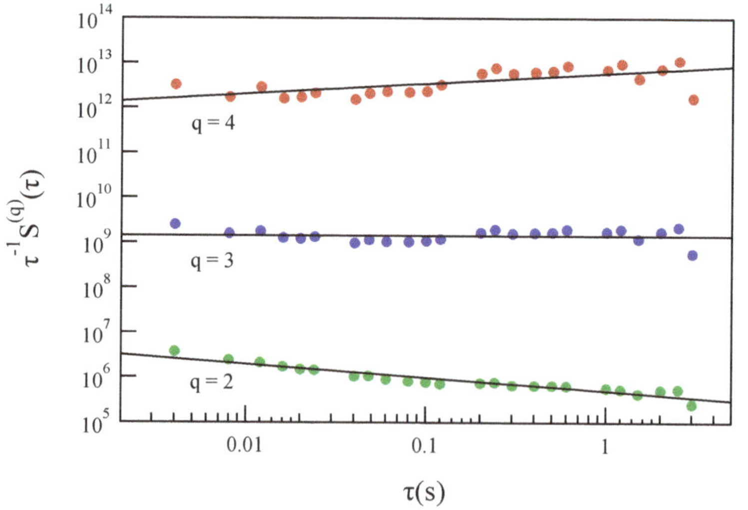

Figure 8. The compensated generalized structure functions, $\tau^{-1}S^{(q)}(\tau)$, of the velocity field components perpendicular to the magnetic field direction for $q = 2$, 3 and 4. The black lines refer to power law fits.

Figure 9 shows relative scaling of the qth-order structure functions, $S^{(q)}(\tau)$, on the 3rd-order one for a selected number of moment orders q. For all the moment orders, the qth-order structure functions show a power-law dependence on the corresponding 3rd-order one in agreement with Equation (7). In other terms, ESS is a property of the observed velocity fluctuations perpendicular to the magnetic field direction.

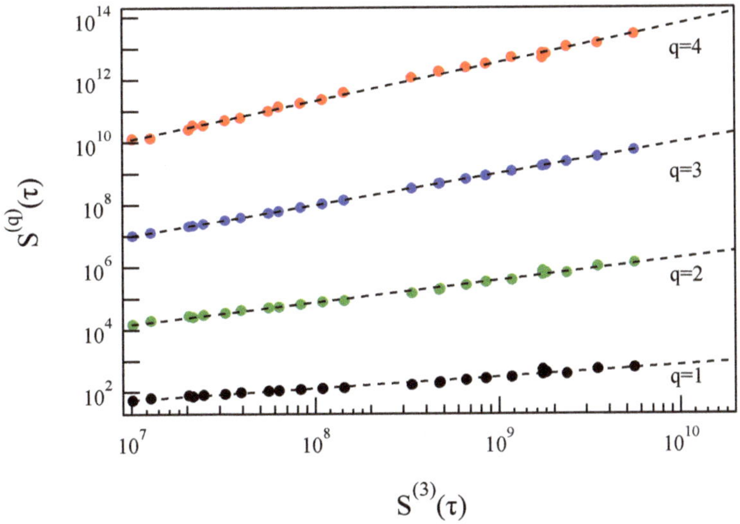

Figure 9. The relative scaling of qth-order structure functions, $S^{(q)}(\tau)$, of the velocity field components perpendicular to the magnetic field direction versus the corresponding 3rd-order one. The black dashed lines are power-law fits.

The relative scaling exponents $\gamma(q)$ are reported as a function of the moment order q in Figure 10. A clear departure on the linear scaling is observed, being the dependence of the scaling exponents on moment order of a convex function. The departure from a linear scaling does not support the occurrence of a global scale-invariance, suggesting instead the emergence of anomalous scaling features, i.e., *intermittency*.

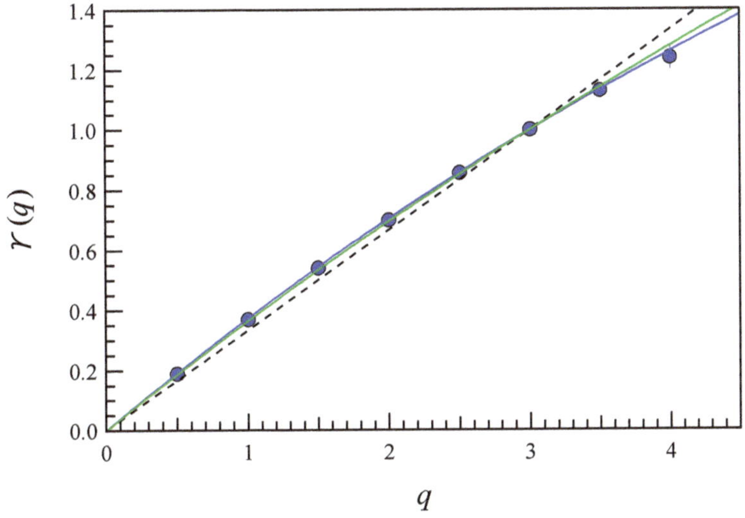

Figure 10. The relative scaling exponents $\gamma(q)$ as a function of the moment order q. The dashed line is the expected trend of $\gamma(q)$ for the K41 theory of turbulence ($\gamma(q) = q/3$ being $\zeta(3) = 1$). The blue solid line is a nonlinear best fit done using the Meneveau and Sreenivasan *P-model* [44]. The solid green line is the $\gamma(q)$ trend for the She-Leveque model of Equation (12) [45].

To characterize and quantify the degree of intermittency, that is how the scaling exponents $\gamma(q)$ deviate from the Kolmogorov prediction, some intermittency models were proposed, which attempt to explain in particular the anomalous scaling exponents. A model capable of explaining the anomalous scaling exponents is the multifractal model by Meneveau and Sreenivasan [44], known as the *P-model*. It is a simple model able to describe the energy-cascading process in the inertial range whose scheme is based on the generalized two-scale Cantor set with equal scales, but unequal weights. Thus, we compare the observed trend of the $\gamma(q)$ exponents with the one predicted by *P-model*. In detail, we fit the trend of $\gamma(q)$ as a function of moment order q using the following expression,

$$\gamma(q) = 1 - \log_2\left(p^{\frac{q}{3}} + (1-p)^{\frac{q}{3}}\right), \tag{11}$$

where p is the weight of the measure repartition in the multiplicative process of the two-scale Cantor set. The *P-model* is an excellent approximation of the observed trend of $\gamma(q)$ and we get $p = [0.25 \pm 0.01]$. Being the value of p significantly different from one-half ($p \neq 0.5$) we may surely assert that the drift velocity fluctuation field has a multifractal structure and that the observed energy repartition at the different scales is an anomalous multiplicative process as the one described by the *P-model*.

In the framework of fluid turbulence, other models have been proposed to explain the convex trend of $\gamma(q)$. Among these, the She-Leveque (SH) model [45] relates the anomalous scaling with the dimension of the dissipative structure in turbulence. In detail, in the case of fluid turbulence, the SH model predicts for $\gamma(q)$ the following behavior:

$$\gamma(q) = \frac{q}{9} + 2\left[1 - \left(\frac{2}{3}\right)^{\frac{q}{3}}\right]. \tag{12}$$

The previous expression, which is specialized for the case of 3D fluid turbulence, was generalized later by Politano and Pouquet [46] (see also ref. Biskamp and Müller [47]). The generalized expression of SH model assumes the following form:

$$\gamma(q) = (1-x)\frac{q}{3} + C_0\left[1 - \left(1 - \frac{x}{C_0}\right)^{\frac{q}{3}}\right], \tag{13}$$

where C_0 is the co-dimension of the intermittent structures and x is the scaling exponent of the dynamic timescale associated with the most intermittent structure (see [46,47]). Thus, in the case of 3D fluid turbulence, one gets $C_0 = 2$ and $x = 2/3$.

The SH-model (see Equation (13)) can be also written as,

$$\gamma(q) = \left(1 - \frac{C_0}{3}\right)\frac{q}{3} + C_0\left(1 - \beta^{\frac{q}{3}}\right) \tag{14}$$

where $\beta = 1 - x/C_0$ is a parameter related to the degree of intermittency. Indeed, for $\beta \to 1$, there is no intermittency, while if $\beta \leq 1$, intermittency is present. In our case, we get $\beta \simeq 2/3$.

Table 1 shows the values for the first 4 scaling exponents in comparison with Ruiz-Chavarria et al. [48] results from fluid turbulence, with Benzi et al. [49] for 3D convective turbulence, Biskamp and Schwarz [50] simulations on 2D MHD turbulence (in this case data refer to Elsässer variables), and SH 3D fluid model. The observed values are very well in agreement in spite of the different physical scenarios, i.e., $\mathbf{E} \times \mathbf{B}$ convective 2D plasma motion, 3D fluid turbulence [48], 2D MHD simulations [50] and theoretical models [45]. A very good agreement is found with exponents from fluid and convective turbulence. In particular, the good agreement with the SH model [45] suggests that the co-dimension of the intermittent structures is $C_0 \simeq 2$.

Table 1. Observed scaling exponents $\gamma(q)$ and comparison with results from models and literature.

q	This Work	From Ref. [48]	From Ref. [49]	From Ref. [50]	She-Leveque [45]
1	$[0.37 \pm 0.01]$	$[0.37 \pm 0.01]$	0.36	0.42	0.364
2	$[0.70 \pm 0.01]$	$[0.70 \pm 0.01]$	0.70	0.75	0.696
3	1	$[1.00 \pm 0.01]$	1	1	1
4	$[1.24 \pm 0.02]$	$[1.28 \pm 0.02]$	1.27	1.18	1.28

Data from Ref. [50] refer to scaling exponents of Elsässer variables.

Figure 11 reports the generalized kurtosis $\Gamma_4(\tau)$ as a function of the timescale τ. Γ_4 is greater than 3 for all the considered timescales and exhibits an increasing trend for decreasing values of τ. These features provide a strong indication of the non-Gaussian character of the velocity increments (being $\Gamma_4 > 3$) that is related to the occurrence of intermittency. Another interesting feature of the trend of Γ_4 is the presence of a local maximum at $\tau \simeq 0.2$ s. This timescale is near the one corresponding to the O^+ ion inertial length, η, assuming a density $n_{O^+} \in [1,\ 2] \cdot 10^5$ cm^{-3} (this range of values for the oxygen ion density n_{O^+} is estimated from electron density n_{e^-} [51] assuming quasi-neutrality and that oxygen ions are the more relevant ions at CSES-01 altitude). Indeed, if we consider the Taylor's hypothesis valid, we have $\tau_\eta^{O^+} = \eta/v_s \simeq 0.3$ s being $v_s \sim 7.8$ km/s the satellite speed.

We compute the probability density functions (PDFs) of the velocity increments in the time scale interval from 4 ms to ~0.5 s. Figure 12 shows the evolution of the standard deviation normalized PDF with the timescale. PDF collapsing does not occur since the shape of the PDFs evolves across different timescale τ, supporting the emergence of intermittency in the velocity field.

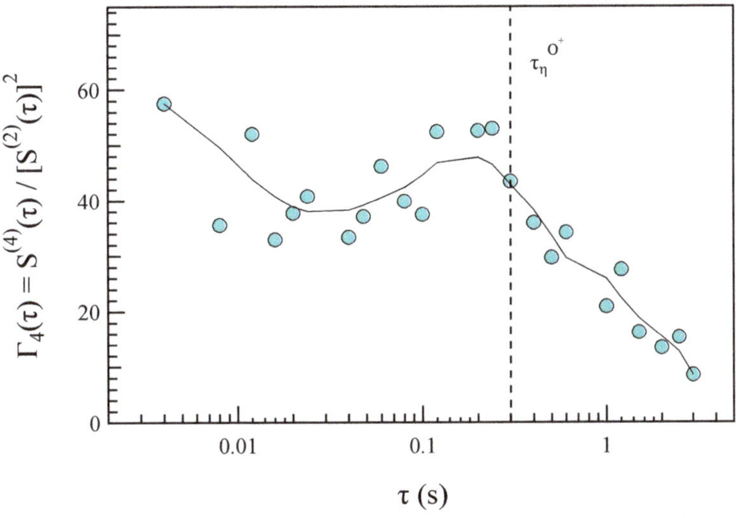

Figure 11. The generalized kurtosis $\Gamma_4(\tau)$. The solid curve is a guide for eye. The vertical dashed line indicates the expected timescale $\tau_\eta^{O^+} \simeq 0.3$ s corresponding to the O^+ inertial length, η, assuming a density in the range $[1,2] \times 10^5$ cm^{-3}.

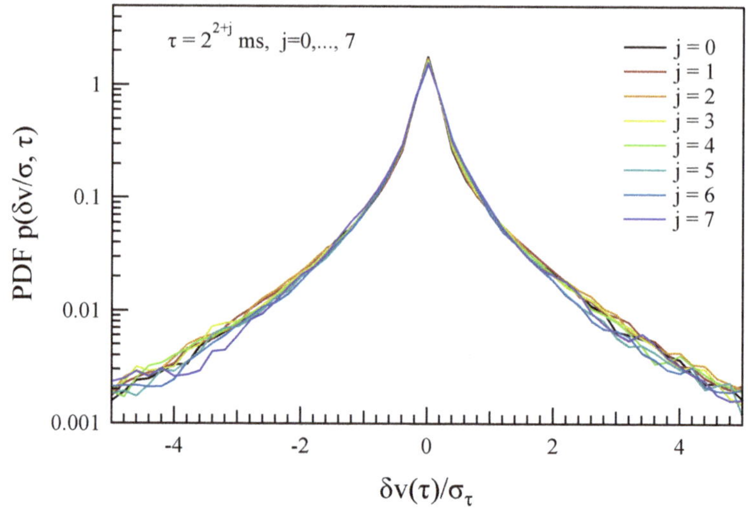

Figure 12. The PDFs of the velocity increments in the time scale interval from 4 ms to ∼0.5 s. Data are rescaled by the corresponding standard deviation.

We remark that the PDFs are not Gaussian being, indeed, characterized by a leptokurtic shape, which can be well fitted by the following function:

$$p(x) = \frac{N_0}{\left[1 + \frac{1}{\kappa}\left(\frac{x}{x_0}\right)^2\right]^{\kappa}} \exp\left(-\left|\frac{x}{x_c}\right|\right), \tag{15}$$

where $x = \delta v(\tau)/\sigma_\tau$, x_0 and x_c are characteristic scales, N_0 is a normalization factor and κ is the exponent governing the tail behavior. In the case of the smallest-scale PDF at $\tau = 4$ ms we get $x_0 = [0.130 \pm 0.003]$, $x_c = [2.06 \pm 0.03]$ and $\kappa = [0.6 \pm 0.1]$.

To better characterize the evolution of the PDFs as a function of the timescale, we estimate the Kullback–Leibler (KL) distance between the PDFs. In detail, we compute the following quantity:

$$KL(\tau \mid \tau_0) = \int_{-5}^{+5} p(x,\tau) \log_2 \frac{p(x,\tau)}{p(x,\tau_0)} dx \tag{16}$$

where $x = \delta v/\sigma$ and τ and τ_0 are the timescales of the two considered PDFs. Here, we set $\tau_0 = 4$ ms, i.e., the minimum available timescale.

To quantify the significance of the measured KL-distance we compute a critical threshold via Monte Carlo simulation. In detail, we generate a series of 1000 random samples x_i containing the same number of values of the actual time series and following the PDF $p(x, \tau_0)$ of Equation (15). Successively, we compute the reciprocal KL-distance between the distributions of this set of 1000 random samples and evaluate the cumulative distribution of the values of KL-distance. The value corresponding to the 95% level of the cumulative distribution of the KL-distance is set as the critical threshold to discriminate the significance of the measured KL-distance. This threshold value is $KL^* = 1 \times 10^{-3}$.

Figure 13 shows a clear increasing of KL-distance with the time scale well above the 95% significance threshold, which is the evidence for the absence of PDF-collapsing. In other words, the PDFs evolve with the timescale of increments so that there is not an invariant shape of the PDFs. This is another counterpart of the absence of a global scale-invariance of increments at different timescales, i.e., of the emergence of intermittency. Furthermore, we observe a short intermediate range of timescales from 0.1 s to 0.6 s where there is a plateau in KL distance. This plateau is well in agreement with the range of timescales $\tau_{\eta}^{O^+}$ associated with the O^+ inertial length η.

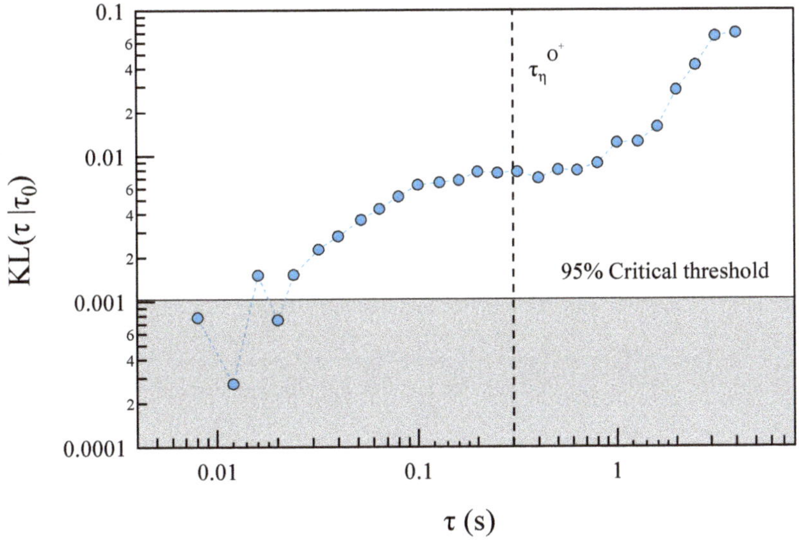

Figure 13. The Kullback–Leibler (KL) distance between the PDFs. The vertical dashed line is in correspondence of the timescale $\tau_{\eta}^{O^+} \sim 0.35$ s associated with the O^+ inertial length η for a density of $\sim 10^5$ cm^{-3}. The solid horizontal line indicate the 95% critical threshold, $KL^* = 1 \times 10^{-3}$, below which the KL-distance between PDFs is not significant (grey region).

5. Discussion

In this study we provided a first analysis of the scaling features of plasma $\mathbf{E} \times \mathbf{B}$ drift velocity in the auroral regions using data collected by the CSES-01 for a case study during a crossing of the Southern polar ionosphere. In detail, our study investigates the drift velocity fluctuations in a range of timescales $\tau \in [0.004, 3]$ s, approximately corresponding to a range of spatial scale $\delta r \in [0.03, 23.50]$ km, assuming that Taylor's hypothesis is valid.

The reconstructed 3D convective velocity field is anisotropic, the vertical velocity being very small. This is due to the geomagnetic field configuration in the polar regions, which is mainly vertical. We found a general good consistency of the reconstructed convective velocity field and the overall large scale plasma convection as observed by SuperDARN observations.

Regarding the spectral features, we found evidence of the anisotropic character of the fluctuation field and of the existence of a large interval of frequencies (wave-numbers via Taylor's hypothesis) where the spectral features follow a power law. The observed spectral exponent of the PSDs in the plane perpendicular to the main geomagnetic field is well in agreement with that expected in the case of 2D $\mathbf{E} \times \mathbf{B}$ convective turbulence ($|\alpha| \lesssim 2$) in a quasi-steady state [40]. Furthermore, the observed spectral features are also in agreement with the occurrence of 2D MHD turbulence and in strong low-β 2D plasma turbulence. Indeed, low-β plasma, due to the strong mean magnetic field the plasma motion, is mainly confined in the perpendicular plane, so that the turbulence is essentially 2D. The presence of a strong mean field prevents the bending of magnetic field lines in the mean field direction [50]. Similar spectral features have been found by Kelley and Kintner [52] and Cerisier et al. [53] in the high-latitude ionosphere for the electric field fluctuations, which are strongly correlated to the $\mathbf{E} \times \mathbf{B}$ velocity in that region, and in inertial range of strong turbulence in low-β plasmas [42].

We remark that the previous considerations on the spectral properties are based on the assumption of the validity of the Taylor's hypothesis, which is assumed to be true in the investigated frequency range. Indeed, according to Tchen et al. [42], the conversion of Eulerian observations into Lagrangian ones using the Taylor's hypothesis is strictly valid only if the flow velocity is very high, or in the case of frozen turbulence, i.e., in the lack of a nonlinear dispersion relation. In other words, the assumption of a linear relationship between f and k is restricted to those cases in which the plasma drift velocity is very high or, if this is not the case, for large wave numbers, as it occurs in the inertial range.

From a general point of view, a difference between 3D and 2D MHD turbulence is the occurrence of an inverse magnetic helicity cascade, which is capable of generating self-organization, i.e., large-scale coherent structures [41]. This scenario seems to be very well in agreement with the occurrence of large scale plasma motion, which manifests in the formation of the two macroscopic convection cells observed in the polar ionosphere (see Figure 4 top panel). The convective motion is clearly forced by the magnetic reconnection phenomenon at the magnetopause, which enhances the plasma convection in the magnetosphere–ionosphere system. In other words, it seems that the nature of the observed turbulence could be isomorphic to a forced convective turbulence.

The analysis of the scaling features of the velocity field increments evidences the occurrence of anomalous scaling properties, supporting the intermittent character of the velocity fluctuation field. The emergence of intermittency is also supported by generalized kurtosis Γ_4, which shows an increasing trend for decreasing timescale τ, and by the absence of collapsing of the PDFs of the velocity increments at different timescales as shown in Figure 12 and by the evolution of KL-distance with the timescales, as shown in Figure 13.

In spite of the different character of the observed turbulence with respect to the ordinary 3D fluid turbulence, the observed anomalous scaling properties are consistent with what is found in the case of the 3D fluid turbulence [48] and theoretical derivations [44,45]. This is a very interesting result that requires a deeper theoretical investigation, but that in any case seems to support the similarity of 2D and 3D plasma turbulence [41], although we have to remark that in the case of 2D MHD turbulence, a more intermittent character is expected with respect to 3D MHD turbulence [50]. Furthermore, the comparison with SH model suggests that the co-dimension of the intermittent structures is $C_0 \simeq 2$ and $\beta \simeq 2/3$.

Being $C_0 \simeq 2$, in a first approximation considering the case of a 3D system, the fractal dimension of the intermittent structures is $d_F = D - C_0 = 1$ (where $D = 3$ is the system dimension), which suggests that the *macro-scale* structure is isomorphic with a quasi-1D fractal structure, i.e., a flux tube or a line. This is quite well in agreement with what can be observed in Figure 4 when the satellite enters in the convection region, where the structure of the $\mathbf{E} \times \mathbf{B}$ drift velocity resembles a quasi-cylindrical configuration, i.e., a structure of co-dimension $C_0 = 2$ and a dimension 1.

Another interesting result from our analysis stands in the behavior of the generalized kurtosis Γ_4 and the KL-distance as a function of timescale τ. As expected from the occurrence of intermittency and anomalous scaling properties we observe an increase of the generalized kurtosis Γ_4 for decreasing timescales and a departure of the shape of the PDFs from the one at the smallest investigated timescale $\tau_0 = 4$ ms. However, around the typical timescale associated with the oxygen inertial length we observe a maximum in the generalized kurtosis Γ_4 and a plateau in the increasing trend of the KL-distance. This result suggests that in the investigated range of scales, the turbulence associated with $\mathbf{E} \times \mathbf{B}$ convective motion could be more complex than the standard single fluid 2D MHD turbulence. In other words, the multi-ion character of ionospheric plasma could manifest in contiguous multiple regimes of different physical processes, both MHD and kinetic, in the range of timescales from $\tau = 5$ ms to $\tau = 1$ s [13,14].

6. Summary and Conclusions

In this work, we have investigated the scaling features of the $\mathbf{E} \times \mathbf{B}$ drift motion in the topside F2 polar ionosphere for a case study relative to a crossing of the Southern auroral region by the Chinese Seismo-Electromagnetic satellite (CSES-01) during a geomagnetically disturbed period.

We have found a clear evidence for the occurrence of a turbulent character of the $\mathbf{E} \times \mathbf{B}$ drift velocity in the analyzed region. The observed turbulence seems to be isomorphic to the 2D $\mathbf{E} \times \mathbf{B}$ convective turbulence [40]. Furthermore, the drift velocity fluctuations display an intermittent character as evidenced by the scaling analysis and the behavior of the generalized kurtosis $\Gamma_4(\tau)$. The observed intermittency features of the $\mathbf{E} \times \mathbf{B}$ drift velocity suggests the hypothesis that intermittency can be related to a macro-scale structure, which is consistent with a quasi-1D fractal structure (thin-flux-like or filamentary structures).

Another interesting result is the emergence of the occurrence of both MHD and kinetic processes in the range of investigated timescales, which can affect the dynamics of the plasma motion at the smallest spatial scales, i.e., below 2 km. This is clearly shown by the trend of the generalized kurtosis Γ_4 and the Kulback–Leibler distance, which show some deviations from the general trend at a scale consistent with the O^+ inertial length. Our findings support the view of very complex dynamics and the idea that it is necessary to consider the multi-species character of the ionospheric plasma medium in order to properly understand and correctly model the small scale ionospheric dynamics, not being sufficient the use of a single fluid theory as is the MHD theory [13,14].

Clearly, further research is required to disentangle and characterize the features of the MHD and kinetic processes involved at the smallest spatial scales. A better comprehension of the observed turbulence processes may benefit from the use of very high-resolution measurements such as those available from the future NanoMagSat mission [54].

Author Contributions: Conceptualization, G.C. and V.Q.; methodology, G.C. and T.A.; investigation, V.Q., G.C., S.B., T.A. and P.D.M.; data curation, M.P. and M.F.M.; writing—original draft preparation, V.Q., G.C. and P.D.M.; writing—review and editing, all. All authors have read and agreed to the published version of the manuscript.

Funding: This research is supported by the contract ASI LIMADOU Scienza+ n. 2020-31-HH.0 and Italian PNRA under contract PNRA18 00289-A Space weather in Polar Ionosphere: the Role of Turbulence

Institutional Review Board Statement: Not applicable.

Informed Consent Statement: Not applicable.

Data Availability Statement: This work made use of the data from the CSES mission (http://www.leos.ac.cn/, accessed on 1 October 2020), a project funded by China National Space Administration and China Earthquake Administration in collaboration with Italian Space Agency and Istituto Nazionale di Fisica Nucleare.

Acknowledgments: This work is in the framework of the CSES-LIMADOU Collaboration (http://cses.roma2.infn.it). We acknowledge the Italian Space Agency (ASI) and the Italian National Project for Anctartic Research (PNRA) for supporting this work in the framework of contract ASI "LIMADOU Scienza+" n° 2020-31-HH.0 and PNRA18 00289-A "Space weather in Polar Ionosphere: the Role of Turbulence". V. Quattrociocchi and M. Piersanti thank the Italian Space Agency for the financial support under the contract "LIMADOU-2 fase B2/C/D/E1". The authors acknowledge the use of SuperDARN data. SuperDARN is a collection of radars funded by national scientific funding agencies of Australia, Canada, China, France, Italy, Japan, Norway, South Africa, United Kingdom and the United States of America.

Conflicts of Interest: The authors declare no conflict of interest.

Abbreviations

The following abbreviations are used in this manuscript:

1D	One dimensional
2D	Two dimensional
3D	Three dimensional
AE	Auroral Electrojet index
CSES-01	Chinese Seismo-Electromagnetic Satellite
EFD	Electric Field Detector
ESS	Extended Self-Similarity
GEO	Geographical Reference System
HPM	High Precision Magnetometer
K41	Kolmogorov 1941
KL	Kullback-Leibler
MHD	Magnetohydrodynamic
PDF	Probability Density Function
PSD	Power Spectral Density
SCM	Search-Coil Magnetometer
SH	She-Leveque
SuperDARN	Super Dual Auroral Radar Network

References

1. Kintner, P.M., Jr. Observations of velocity shear driven plasma turbulence. *J. Geophys. Res.* **1976**, *A28*, 5114–5122. [CrossRef]
2. Kintner, P.M.; Seyler, C.E. The status of observations and theory of high latitude ionospheric and magnetospheric plasma turbulence. *Space Sci. Rev.* **1985**, *41*, 1572–9672. [CrossRef]
3. Basu, S.; Basu, S.; MacKenzie, E.; Fougere, P.F.; Coley, W.R.; Maynard, N.C.; Winningham, J.D.; Sugiura, M.; Hanson, W.B.; Hoegy, W.R. Simultaneous density and electric field fluctuation spectra associated with velocity shears in the auroral oval. *J. Geophys. Res. Space Phys.* **1988**, *93*, 115–136. [CrossRef]
4. Tam, S.W.Y.; Chang, T.; Kintner, P.M.; Klatt, E. Intermittency analyses on the SIERRA measurements of the electric field fluctuations in the auroral zone. *Geophys. Res. Lett.* **2005**, *32*, L05109. [CrossRef]
5. Golovchanskaya, I.V.; Ostapenko, A.A.; Kozelov, B.V. Relationship between the high-latitude electric and magnetic turbulence and the Birkeland field-aligned currents. *J. Geophys. Res. (Space Phys.)* **2006**, *111*, A12301. [CrossRef]
6. Golovchanskaya, I.V.; Kozelov, B.V. On the origin of electric turbulence in the polar cap ionosphere. *J. Geophys. Res. (Space Phys.)* **2010**, *115*, A09321. [CrossRef]
7. Golovchanskaya, I.V.; Kozelov, B.V. Properties of electric turbulence in the polar cap ionosphere. *Geomagn. Aeron.* **2010**, *50*, 576–587. [CrossRef]
8. Kozelov, B.V.; Golovchanskaya, I.V. Scaling of electric field fluctuations associated with the aurora during northward IMF. *Geophys. Res. Lett.* **2006**, *33*, L20109. [CrossRef]
9. Kozelov, B.V.; Golovchanskaya, I.V.; Ostapenko, A.A.; Fedorenko, Y.V. Wavelet analysis of high-latitude electric and magnetic fluctuations observed by the Dynamic Explorer 2 satellite. *J. Geophys. Res. (Space Phys.)* **2008**, *113*, A03308. [CrossRef]
10. Kintner, P.M.; Franz, J.; Schuck, P.; Klatt, E. Interferometric coherency determination of wavelength or what are broadband ELF waves? *J. Geophys. Res.* **2000**, *105*, 21,237–21.250. [CrossRef]
11. Chang, T. Colloid-like Behavior and Topological Phase Transitions in Space Plasmas: Intermittent Low Frequency Turbulence in the Auroral Zone. *Phys. Scr. Vol. T* **2001**, *89*, 80–83. [CrossRef]
12. Chang, T.; Tam, S.W.Y.; Wu, C.C. Complexity induced anisotropic bimodal intermittent turbulence in space plasmas. *Phys. Plasmas* **2004**, *11*, 1287–1299. [CrossRef]
13. Tam, S.W.Y.; Chang, T.; Kintner, P.M.; Klatt, E.M. Rank-ordered multifractal analysis for intermittent fluctuations with global crossover behavior. *Phys. Rev. E* **2010**, *81*, 036414. [CrossRef]
14. Tam, S.W.Y.; Chang, T. Double rank-ordering technique of ROMA (Rank-Ordered Multifractal Analysis) for multifractal fluctuations featuring multiple regimes of scales. *Nonlinear Process. Geophys.* **2011**, *18*, 405–414. [CrossRef]
15. Fejer, B.G.; Kelley, M.C. Ionospheric irregularities. *Rev. Geophys. Space Phys.* **1980**, *18*, 401–454. [CrossRef]
16. Fejer, B.G. Low latitude electrodynamic plasma drifts—A review. *J. Atmos. Terr. Phys.* **1991**, *53*, 677–693. [CrossRef]
17. Anderson, D.; Anghel, A.; Yumoto, K.; Ishitsuka, M.; Kudeki, E. Estimating daytime vertical ExB drift velocities in the equatorial F-region using ground-based magnetometer observations. *Geophys. Res. Lett.* **2002**, *29*, 1596. [CrossRef]
18. Horvath, I.; Essex, E.A. Vertical E × B drift velocity variations and associated low-latitude ionospheric irregularities investigated with the TOPEX and GPS satellite data. *Ann. Geophys.* **2003**, *21*, 1017–1030. [CrossRef]
19. Paschmann, G.; Haaland, S.; Treumann, R. Auroral Plasma Physics. *Space Sci. Rev.* **2002**, *103*. [CrossRef]
20. Shen, X.; Zhang, X.; Yuan, S.; Wang, L.; Cao, J.; Huang, J.; Zhu, X.; Piergiorgio, P.; Dai, J. The state-of-the-art of the China Seismo-Electromagnetic Satellite mission. *Sci. China E Technol. Sci.* **2018**, *61*, 634–642. [CrossRef]

21. Shen, X.; Zong, Q.G.; Zhang, X. Introduction to special section on the China Seismo-Electromagnetic Satellite and initial results. *Earth Planet. Phys.* **2018**, *2*, 439. [CrossRef]
22. Huang, J.; Lei, J.; Li, S.; Zeren, Z.; Li, C.; Zhu, X.; Yu, W. The Electric Field Detector (EFD) onboard the ZH-1 satellite and first observational results. *Earth Planet. Phys.* **2018**, *2*, 469–478. [CrossRef]
23. Diego, P.; Huang, J.; Piersanti, M.; Badoni, D.; Zeren, Z.; Yan, R.; Rebustini, G.; Ammendola, R.; Candidi, M.; Guan, Y.B.; et al. The Electric Field Detector on Board the China Seismo Electromagnetic Satellite—In-Orbit Results and Validation. *Instruments* **2021**, *5*, 1. [CrossRef]
24. Cheng, B.; Zhou, B.; Magnes, W.; Lammegger, R.; Pollinger, A. High precision magnetometer for geomagnetic exploration onboard of the China Seismo-Electromagnetic Satellite. *Sci. China E Technol. Sci.* **2018**, *61*, 659–668. [CrossRef]
25. Consolini, G.; Quattrociocchi, V.; D'Angelo, G.; Alberti, T.; Piersanti, M.; Marcucci, M.F.; De Michelis, P. Electric Field Multifractal Features in the High-Latitude Ionosphere: CSES-01 Observations. *Atmosphere* **2021**, *12*, 646. [CrossRef]
26. Cicone, A.; Piersanti, M.; Consolini, G.; Materassi, M.; D'Angelo, G.; Diego, P.; Ubertini, P. Auroral oval layers detection by using CSES plasma and electric field data. *Il Nuovo Cimento* **2021**, *44*, 117. [CrossRef]
27. Greenwald, R.A.; Baker, K.B.; Dudeney, J.R.; Pinnock, M.; Jones, T.B.; Thomas, E.C.; Villain, J.P.; Cerisier, J.C.; Senior, C.; Hanuise, C.; et al. Darn/Superdarn: A Global View of the Dynamics of High-Lattitude Convection. *Space Sci. Rev.* **1995**, *71*, 761–796. [CrossRef]
28. Chisham, G.; Lester, M.; Milan, S.E.; Freeman, M.P.; Bristow, W.A.; Grocott, A.; McWilliams, K.A.; Ruohoniemi, J.M.; Yeoman, T.K.; Dyson, P.L.; et al. A decade of the Super Dual Auroral Radar Network (SuperDARN): scientific achievements, new techniques and future directions. *Surv. Geophys.* **2007**, *28*, 33–109. [CrossRef]
29. Nishitani, N.; Ruohoniemi, J.M.; Lester, M.; Baker, J.B.H.; Koustov, A.V.; Shepherd, S.G.; Chisham, G.; Hori, T.; Thomas, E.G.; Makarevich, R.A.; et al. Review of the accomplishments of mid-latitude Super Dual Auroral Radar Network (SuperDARN) HF radars. *Prog. Earth Planet. Sci.* **2019**, *6*, 27. [CrossRef]
30. Ruohoniemi, J.M.; Baker, K.B. Large-scale imaging of high-latitude convection with Super Dual Auroral Radar Network HF radar observations. *J. Geophys. Res.* **1998**, *103*, 20797–20811. [CrossRef]
31. Shepherd, S.G.; Ruohoniemi, J.M. Electrostatic potential patterns in the high-latitude ionosphere constrained by SuperDARN measurements. *J. Geophys. Res. Space Phys.* **2000**, *105*, 23005–23014. [CrossRef]
32. Thomas, E.G.; Shepherd, S.G. Statistical Patterns of Ionospheric Convection Derived From Mid-latitude, High-Latitude, and Polar SuperDARN HF Radar Observations. *J. Geophys. Res. (Space Phys.)* **2018**, *123*, 3196–3216. [CrossRef]
33. Group, S.D.A.W.; Thomas, E.G.; Ponomarenko, P.V.; Billett, D.D.; Bland, E.C.; Burrell, A.G.; Kotyk, K.; Reimer, A.S.; Schmidt, M.T.; Shepherd, S.G.; et al. SuperDARN Radar Software Toolkit (RST) 4.2. 2018. Available online: https://www.research.lancs.ac.uk/portal/en/publications/superdarn-radar-software-toolkit-rst-42(f02d5a1a-5e6b-40a7-88ea-35e2133a0d65).html (accessed on 11 February 2022).
34. Frisch, U. *Turbulence*; Cambridge University Press: Cambridge, UK, 1995.
35. Benzi, R.; Ciliberto, S.; Tripiccione, R.; Baudet, C.; Massaioli, F.; Succi, S. Extended self-similarity in turbulent flows. *Phys. Rev. E* **1993**, *48*, R29–R32. [CrossRef]
36. Biferale, L. A note on the fluctuation of dissipative scale in turbulence. *Phys. Fluids* **2008**, *20*, 031703. [CrossRef]
37. Benzi, R.; Biferale, L. Homogeneous and Isotropic Turbulence: A Short Survey on Recent Developments. *J. Stat. Phys.* **2015**, *161*, 1351–1365. [CrossRef]
38. Chevillard, L.; Roux, S.G.; Lévêque, E.; Mordant, N.; Pinton, J.F.; Arnéodo, A. Intermittency of Velocity Time Increments in Turbulence. *Phys. Rev. Lett.* **2005**, *95*, 064501. [CrossRef] [PubMed]
39. Benzi, R.; Ciliberto, S.; Baudet, C.; Ruiz Chavarria, G.; Tripiccione, R. Extended self-similarity in the dissipation range of fully developed turbulence. *Europhys. Lett.* **1993**, *24*, 275. [CrossRef]
40. Gruzinov, A.V.; Kukharkin, N.; Sudan, R.N. Two-Dimensional Convective Turbulence. *Phys. Rev. Lett.* **1996**, *76*, 1260–1263. [CrossRef]
41. Biskamp, D.; Bremer, U. Dynamics and statistics of inverse cascade processes in 2D magnetohydrodynamic turbulence. *Phys. Rev. Lett.* **1994**, *72*, 3819–3822. [CrossRef]
42. Tchen, C.M.; Pecseli, H.L.; Larsen, S.E. Strong turbulence in low-β plasmas. *Plasma Phys.* **1980**, *22*, 817–829. [CrossRef]
43. Basu, S.; MacKenzie, E.; Basu, S.; Coley, W.R.; Sharber, J.R.; Hoegy, W.R. Plasma structuring by the gradient drift instability at high latitudes and comparison with velocity shear driven processes. *J. Geophys. Res.* **1990**, *95*, 7799–7818. [CrossRef]
44. Meneveau, C.; Sreenivasan, K.R. Simple multifractal cascade model for fully developed turbulence. *Phys. Rev. Lett.* **1987**, *59*, 1424–1427. [CrossRef] [PubMed]
45. She, Z.S.; Leveque, E. Universal scaling laws in fully developed turbulence. *Phys. Rev. Lett.* **1994**, *72*, 336–339. [CrossRef] [PubMed]
46. Politano, H.; Pouquet, A. Model of intermittency in magnetohydrodynamic turbulence. *Phys. Rev. E* **1995**, *52*, 636–641. [CrossRef] [PubMed]
47. Biskamp, D.; Müller, W.C. Scaling properties of three-dimensional isotropic magnetohydrodynamic turbulence. *Phys. Plasmas* **2000**, *7*, 4889–4900. [CrossRef]
48. Ruiz-Chavarria, G.; Baudet, C.; Ciliberto, S. Scaling laws and dissipation scale of a passive scalar in fully developed turbulence. *Phys. D Nonlinear Phenom.* **1996**, *99*, 369–380. [CrossRef]

49. Benzi, R.; Massaioli, F.; Succi, S.; Tripiccione, R. Scaling Behaviour of the Velocity and Temperature Correlation Functions in 3D Convective Turbulence. *Europhys. Lett.* **1994**, *28*, 231–236. [CrossRef]
50. Biskamp, D.; Schwarz, E. On two-dimensional magnetohydrodynamic turbulence. *Phys. Plasmas* **2001**, *8*, 3282–3292. [CrossRef]
51. Giannattasio, F.; De Michelis, P.; Consolini, G.; Quattrociocchi, V.; Coco, I.; Tozzi, R. Characterising the electron density fluctuations in the high-latitude ionosphere at Swarm altitude in response to the geomagnetic activity. *Ann. Geophys.* **2019**, *62*, GM453. [CrossRef]
52. Kelley, M.C.; Kintner, P.M. Evidence for two-dimensional inertial turbulence in a cosmic-scale low-beta plasma. *Astrophys. J.* **1978**, *220*, 339–343. [CrossRef]
53. Cerisier, J.C.; Berthelier, J.J.; Beghin, C. Unstable density gradients in the high-latitude ionosphere. *Radio Sci.* **1985**, *20*, 755–761. [CrossRef]
54. Hulot, G.; Leger, J.M.; Clausen, L.B.N.; Deconinck, F.; Coïsson, P.; Vigneron, P.; Alken, P.; Chulliat, A.; Finlay, C.C.; Grayver, A.; et al. NanoMagSat, a 16U nanosatellite constellation high-precision magnetic project to initiate permanent low-cost monitoring of the Earth's magnetic field and ionospheric environment. In Proceedings of the Abstracts of the EGU21—The EGU General Assembly 2021, Vienne, Austria, 19–30 April 2021. [CrossRef]

remote sensing

MDPI

Article

A Preliminary Study on Ionospheric Scintillation Anomalies Detected Using GNSS-R Data from NASA CYGNSS Mission as Possible Earthquake Precursors

Carlos Molina [1,2,*], Badr-Eddine Boudriki Semlali [1], Hyuk Park [1,2] and Adriano Camps [1,2]

1 CommSensLab—UPC, Universitat Politècnica de Catalunya—BarcelonaTech, 08034 Barcelona, Spain; badr-eddine.boudriki.semlali@upc.edu (B.-E.B.S.); park.hyuk@upc.edu (H.P.); adriano.jose.camps@upc.edu (A.C.)

2 IEEC-Institut d'Estudis Espacials de Catalunya, 08034 Barcelona, Spain

* Correspondence: carlos.molina@upc.edu

Citation: Molina, C.; Boudriki Semlali, B.-E.; Park, H.; Camps, A. A Preliminary Study on Ionospheric Scintillation Anomalies Detected Using GNSS-R Data from NASA CYGNSS Mission as Possible Earthquake Precursors. *Remote Sens.* **2022**, *14*, 2555. https://doi.org/10.3390/rs14112555

Academic Editor: Fabio Giannattasio

Received: 23 March 2022
Accepted: 24 May 2022
Published: 26 May 2022

Abstract: Ionospheric perturbations affect the propagation of electromagnetic waves. These perturbations, besides being a problem for space communications, satellite navigation, and Earth observation techniques, could also be used as another Earth observation tool. Several recent studies showed correlations with earthquakes with ionospheric anomalies, but almost all of them use ground stations to measure the Total Electron Content (TEC) variations, and, in particular, the ones occurring after an earthquake. Here, a preliminary study is presented on how the ionospheric scintillation measured with GNSS-R instruments over oceanic regions shows a small, but detectable correlation with the occurrence of earthquakes, which in some cases occurs before the earthquakes. This study uses GNSS-R data from NASA CYGNSS Mission to measure the ionospheric amplitude scintillation (S_4) for 6 months from March 2019 to August 2019, applying a statistical analysis based on confusion matrixes, and the Receiver Operating Characteristic (ROC) curves to correlate S_4 anomalous variations to earthquakes. A small positive correlation is found between the ionospheric scintillation and the earthquakes during the six previous days. However, the study has some weakness because (a) a small number (~45) of large (M > 6) earthquakes over oceanic regions are studied, (b) the region studied is close to the geomagnetic equator, where ionospheric scintillations are usual, and (c) the overall correlation is small.

Keywords: ionospheric scintillation; GNSS-R; CYGNSS; earthquakes

1. Introduction

Ionospheric scintillation is one of the main concerns for satellite communications, in particular for Global Navigation Satellite Systems (GNSS) and low-frequency Earth observation missions, such as the upcoming P-band synthetic aperture radars and radar sounders. They may affect electromagnetic signals traversing the ionosphere, with rapid fluctuations in the intensity and/or phase of the signal. Several models [1] have been used in recent decades: Rino–Fremouw (1973) [2,3], Aarons (1985) [4], Franke and Liu (1985) [5], Iyer et al. (2006) [6], the WBMOD climatological model [7], or GISM [8].

Besides being a problem for satellite communications and Earth observation missions, ionospheric scintillation can also be used as a tool for Earth observation.

A novel technique, first introduced in [9], and continued in [10,11], uses GNSS Reflectometry (GNSS-R) technology to infer ionospheric amplitude scintillation over oceanic regions. This technique allows the creation of ionospheric scintillation maps that could help the calibration and improvement of the currently available models.

In the work presented here, the interaction between the ionosphere and the lithosphere is studied.

The lithosphere is the solid, outermost shell of the Earth, and it is subjected to slow movements. The crust is divided in tectonic plates that can move relative to each other because they lie on top of the viscous, upper part of the mantle. The upper mantle plastic deformations absorb the displacements of the crust on timescales of thousands of years, but the solid crust deforms elastically leading to earthquakes through brittle failure.

In the last decade, several studies revealed some coupling between the lithosphere, the atmosphere, and the ionosphere [12]. In this study, we will focus on the impact of the lithosphere on the ionosphere. According to [13–19], this coupling can be observed during earthquakes, in particular, before, during the so-called earthquake's preparation period, or immediately after them, with different physical explanations.

When the ionospheric perturbations are observed before the earthquake, one reason for this is the perturbation of the background electric field by an electric potential appearing on the Earth's surface during the earthquake's preparation period [14]. For example, microfractures near the epicenter can generate positively charged holes that can diffuse to the surface, and generate these electric fields, that in the end, disturb the ionosphere.

Furthermore, a thermal expansion of the atmosphere derived from the Land Surface Temperature (LST) increase prior to the earthquake occurrence, shown in several studies [15–17,20], can generate a small gravity wave disturbing the electron density profile of the ionosphere, inducing changes in the TEC.

The general interest in this field of study has increased in the last few years, as shown by the dedicated Chinese–Italian mission called China Seismo-Electromagnetic Satellite (CSES) to study the ionosphere-lithosphere coupling by monitoring electromagnetic and plasma density variations in the ionosphere. The satellite was launched to a 500-km altitude polar Sun-synchronous orbit in 2018, and it has already provided some interesting results of TEC variations related to several strong earthquakes in the Indonesian region [18].

Furthermore, some studies also detected ionospheric perturbations in the hours after large earthquakes, for example in the magnitude 9 Tohoku (Japan) earthquake in 2011. In this case, clear TEC deviations were measured following a concentric wave pattern around the epicenter [19]. Possible explanations for these perturbations are the acoustic waves generated by the sudden huge mass redistribution produced during the earthquake, that travel to the upper atmospheric layers or the ionosphere, inducing changes in the electron density. These waves can also be generated along the traveling Rayleigh waves propagating concentrically from the epicenter along the Earth's crust.

Most previous studies on this topic use the TEC anomalous variations as earthquake precursors [21–23]. Only a small fraction of the studies on ionospheric perturbations related to earthquakes use ionospheric scintillation to do their correlation [24]. These studies take GPS data from ground stations [25], or ground-based ionosondes [26] to measure the S_4 index and correlate it to earthquakes in the region. The novelty of this study is the use of the GNSS-R technique to obtain global oceanic maps of ionospheric scintillation and correlate them to earthquake precursors, allowing studying a large number of earthquakes globally distributed and making use of statistical tools such as the confusion matrixes and ROC.

Section 2 describes the data used, and the methods applied during this study; it is divided into four subsections. Section 2.1 describes how ionospheric scintillation is derived from GNSS-R CYGNSS measurements, Section 2.2 explains how changes in the scintillation index are calculated, Section 2.3 introduces the earthquake database used, and Section 2.4 describes how the statistical analysis is performed by means of the confusion matrix method. Section 3 presents the results of this study, and it is divided into two parts: the first shows the results for a particular case study in July 2019, and the second presents the statistical results of the complete analysis over a period of 6 months from March 2019 to August 2019. Finally, a discussion on the results is performed in Section 4, and the main conclusions are presented in Section 5.

2. Materials and Methods

2.1. Ionospheric Scintillation Measurement Using GNSS-R Data

In this study, ionospheric amplitude scintillation (S_4) data are inferred using GNSS-R products from the NASA Cyclone Global Navigation Satellite System (CYGNSS) mission, following the approach presented in [9]. This technique computes the scintillation in the signal's intensity using the following expression:

$$S_4 = \sqrt{\frac{\langle I^2 \rangle - \langle I \rangle^2}{\langle I \rangle^2}}, \tag{1}$$

in which I is the signal's intensity or, in our case, the magnitude of the Signal-to-Noise Ratio (SNR) of the Delay-Doppler Map (DDM) of the reflected signal over oceans in linear units, and its average ($\langle I \rangle$) is calculated over a 10 s moving window. This way, global S_4 maps can be elaborated on calm oceanic regions between 40°S and 40°N latitudes.

The signals emitted by the GNSS transmitters cross the ionosphere in the down-welling path, then coherently reflect on calm water surfaces, and reach the GNSS-R receiver, in this case, a CYGNSS satellite, which compares the signals with the direct ones. As the signals follow different paths, the reflected ones are more influenced by the ionosphere than the direct ones, therefore the ionospheric scintillation can be detected in the resulting DDM intensity.

The points where the scintillation values are geolocated correspond to the specular reflection point over oceans, which are provided in the CYGNSS L1 dataset. This can be seen as a problem because the actual scintillation is mainly induced at the altitude of the ionosphere, approximately 350 km, near the electron density peak. However, as the reflected signal crosses this layer two times, the middle point between these two pierce points nearly matches the specular reflection point. Furthermore, in later steps, a 100 km radius is used to aggregate data when searching for earthquakes around anomalies in the S_4 value, which helps to smooth this effect along with the uncertainty in the distance where earthquakes disturb the ionosphere.

The GNSS-R constellation used, CYGNSS, was originally designed for wind speed and cyclone monitoring. It consists of eight micro-satellites orbiting at around 520 km altitude, with an orbital inclination of 35°. Each satellite can track up to four GNSS signals reflected on the ground [27], allowing an effective coverage between latitudes −40° to 40°. The constellation was launched in December 2016, and the science data are available from March 2017, with a cadence of 1 DDM per channel per second, resulting in around $2.76 \cdot 10^6$ samples per day for the whole constellation. The DDM data rate was set to 2 Hz in July 2019.

After quality filtering, the total number of samples per day per pixel of $1° \times 1°$ varies from 100 in equatorial regions to 200 in ±30 to 35° North and South, as shown in Figure 1.

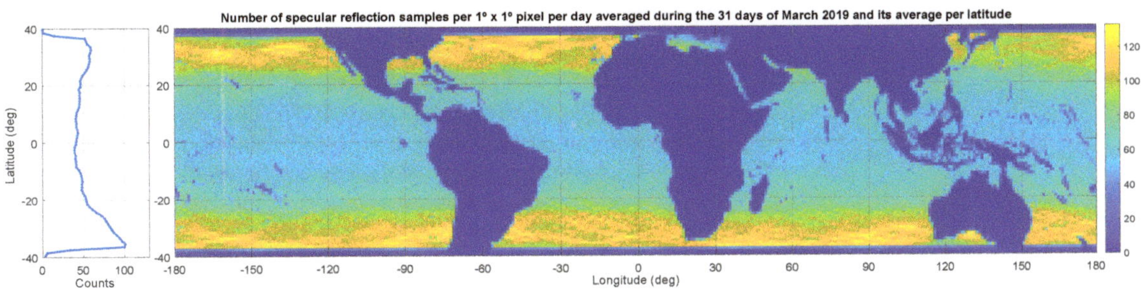

Figure 1. Map showing the average number of DDM samples per $1° \times 1°$ pixel per day, after filtering the land cover. The average is computed during the whole month of March 2019. On the left plot the computed average number of DDMs per latitude is presented.

CYGNSS L1 data are filtered out when the "poor_overall_quality" flag is true, which is the union of several flags described in the CYGNSS manual [27]. The most important flags refer to the land cover filter, which removes all the land points, and the ones that are less than 25 km from the coast. Furthermore, they remove the data when the satellite had suffer from attitude perturbations, or the DDM obtained has low confidence or a large noise figure, or when Radio-Frequency Interference (RFI) has been detected in the measurements.

2.2. Ionospheric Scintillation Changes Computation

Given that ionospheric scintillation is subjected to variations that follow, in the short term, a daily periodicity, and, in the long term, an annual, and 11-year solar cycle trend, data must be detrended to detect anomalous variations. As the daily variations are the short-term larger variations, and also because larger, more random fluctuations due to solar radiation are expected during the daytime, data has been filtered to keep only measurements from 0 h to 6h LT, during the night time, when the ionosphere is expected to be more stable, at least in equatorial regions where this study is conducted. This way we expect to better detect anomalous changes on a time-scale longer than 1 day, which are also more likely to be induced by Earth's internal sources rather than by solar or space-weather sources.

To detrend the monthly/annual variations, the average of two months has been computed, and then subtracted from a seven day window average, in which we want to compute the S_4 variation. More precisely, for a particular day D in which the relative variation is computed, the previous 61 day average is subtracted from the previous seven day average (both including the current day D). This is computed for every day, and every $1^\circ \times 1^\circ$ pixel in the region covered by CYGNSS (from 40°S to 40°N) using Equation (2).

$$\Delta S_4(x, y, D) = \frac{1}{7} \sum_{d=D-6}^{D} S_4(x, y, d) - \frac{1}{61} \sum_{d=D-60}^{D} S_4(x, y, d), \tag{2}$$

where x, y represent the longitude and latitude of each pixel in the map, and d is the day swept to compute the average. The S_4 variation (ΔS_4) is computed for every day D, at each pixel in the map. It usually exhibits values ranging ± 0.06, being positive when the S_4 in the last week has been larger than the typical value during the last two months.

The selection of a two month (61 day) average for detrending is made to remove the long term, annual evolution of the ionospheric activity in addition to the variability of the sea-surface conditions, while the 7 day short average helps to smooth fluctuations in this activity due to space-weather with a daily or weekly timescale, and also fits with the approximate timescales expected for the earthquakes ionospheric precursors to occur, according to the studies referenced in previous sections [12,13,18]. Additionally, both averaging windows are asymmetrical in time, including only values before the current day. The reason for that is that our study is focused on searching for precursory changes in ionospheric scintillation.

Due to the periodicity of the CYGNSS satellite's orbits, and the filter applied to keep only measurements from 0 h to 6 h LT, a time dependence was observed in the latitude of the regions covered by the study along the period studied.

Figure 2 shows the position of the median latitude of the valid data pixels for every day in the period studied (blue line). It also shows, the area between the first and third interquartiles of the available data pixels (green area), and the maximum and minimum latitudes covered (gray area). This plot shows the oscillating behavior of the region covered in the study.

This variation in the region covered will make that some regions away from the equator would not be included in the study during a certain time. Earthquakes taking place in these periods and regions would not be studied because of the lack of ionospheric scintillation data to correlate with. The method for discarding these data will be explained in the section regarding the creation of the confusion matrix.

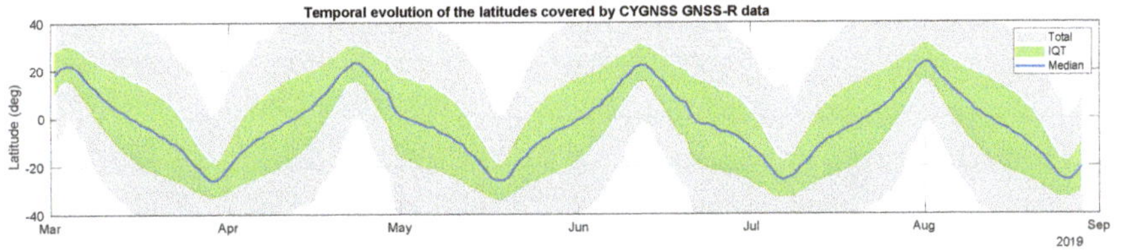

Figure 2. Temporal evolution of the region under study after applying the filters (LT and quality), and the detrending described in this section. The graph shows the latitudes covered at least by one pixel (gray), and the interquartile (green), and median (blue line) of these latitudes. The average period of this evolution is 50 days.

2.3. Earthquake Database

The second main data source is composed of the earthquakes database provided by the USGS containing the magnitude, UTC time, depth, and location of the epicenter, among other parameters. It is important to remark that the magnitude is a complex variable to describe, as there are different methods of computation and scales, which depend on the region of the Earth where earthquakes are studied, the composition of the lithosphere, or the measuring technique used.

Originally, the Richter scale (M_l) was developed for earthquakes in the California region by measuring the frequency of the seismic waves received at a seismographic station. Later on, when trying to use the same scale for other regions, it was found that the Richter method was not valid for all frequencies and distance ranges, in particular for larger magnitude earthquakes. Other methods to measure the magnitude were developed, designed to match the Richter scale in their range of validity [28,29]. The m_b, using the body wave, and the M_S, using the surface wave are two extensions of the Richter magnitude, but they were not valid for larger earthquakes. The moment magnitude scale M_W (usually noted as m_ww) is the one with a wider range of application, extending above M5, and in particular, for large earthquakes (M8 and greater events) is the most accurate one.

In the USGS database, each earthquake in the period studied is given by a single magnitude type [30], but, as stated before, they are all designed to be consistent. This is why, in our study, the magnitude provided is used for all the earthquakes, without taking into account the magnitude type. Table 1 shows the relative abundance of each magnitude type in the period studied:

Table 1. Magnitude types found in the USGS earthquakes database, indicating the counts and relative abundance for each of them.

Magnitude Type	Description [30,31]	Counts	Percentage
mb	Short-period body wave (m_b)	5268	86.25%
mww	Moment W-phase (M_W)	516	8.45%
mwr	Moment magnitude (regional)	127	2.08%
mlr	Revised local (Richter) magnitude	83	1.36%
mw	Moment magnitude	66	1.08%
ml	Richter (local) magnitude (M_l)	43	0.70%
md	Duration magnitude	4	0.07%
mb_lg	Short-period surface wave	1	0.02%

Figure 3 shows the distribution of all earthquakes in the period as a function of their depth and magnitude, and showing the magnitude type with different symbols and colors.

Figure 4 shows the position of all the earthquakes occurring in the six months studied, from March to August 2019. Please note that the color scale and size of the circles represent their magnitudes. It is important to remark that most of them happen in the middle of the oceans, near the faults between tectonic plates. With the method presented here, only these earthquakes in oceanic regions or near the shore can be monitored, because the land reflectivity is highly variable, and the possible ionospheric scintillation would become masked. For example, the largest earthquake in the period studied, with a magnitude of 8.0, on 26 May in Peru, was located on land, 500 km away from the closest seashore, which is larger than the maximum distance considered in this study.

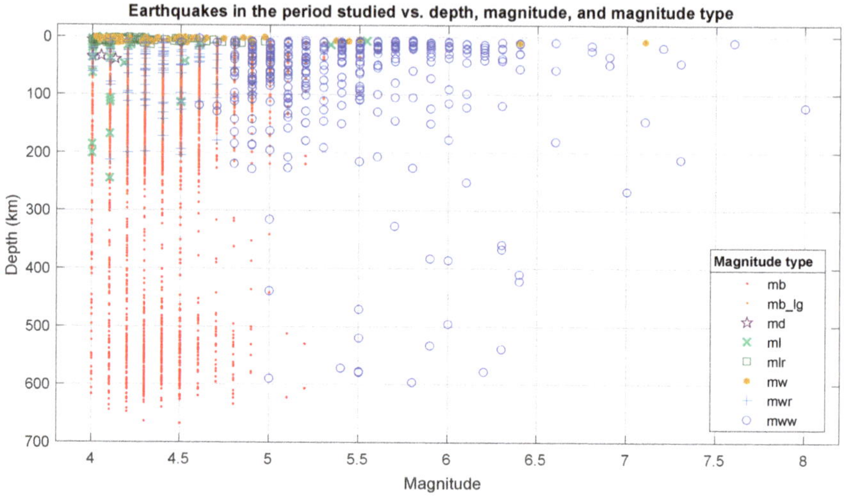

Figure 3. Distribution of earthquakes according their depth (km) and magnitude, indicating their magnitude type with different symbols and colors.

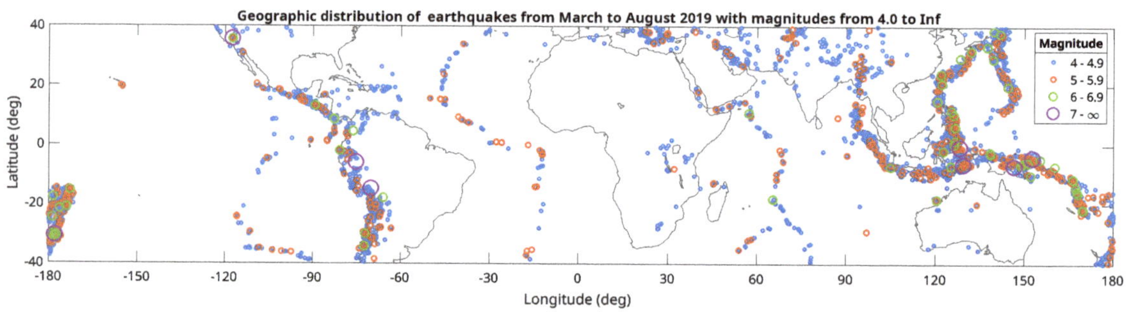

Figure 4. Geographic distribution of earthquakes occurring in the period from 1 March to 29 August 2019 with magnitudes larger to or equal than 4 within the region covered by CYGNSS. Magnitude is represented by the size and color of the circles.

The total number of earthquakes that took place in the period studied is 6108, and 29.4% of them are located on land, but many of them could happen on small islands or very near the shore so they are included. Figure 5 shows a histogram for all the earthquakes in the period (blue bars), and only the ones happening on land (in brown), as a function of their magnitudes. Please note that the vertical scale is in logarithmic scale to see the larger magnitude intervals, where only a few earthquakes occur.

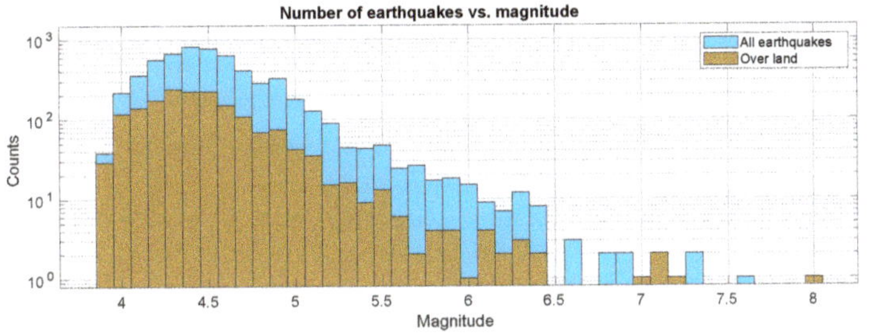

Figure 5. Histogram of earthquakes per magnitude during the period studied for all places (blue) and the ones happening on land (brown).

2.4. Confusion Matrix Calculation

Finally, the ionospheric scintillation variation data and the occurrence of earthquakes, are analyzed to assess the correlation of their mutual occurrence, and the confusion matrix is computed. The overall process is shown in the diagram in Figure 6.

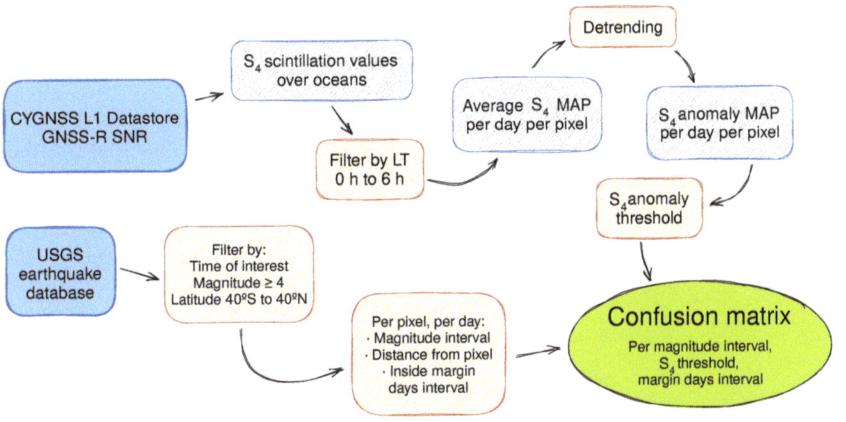

Figure 6. Diagram showing the complete data processing, from raw data sources to the CM inputs.

A confusion matrix (CM) is the usual way to study the effectiveness of an algorithm to detect an event. It is a 2 × 2 matrix classifying the events according to two binary classes: the predicted ionospheric scintillation signature, and the actual occurrence of the earthquake. This classification results in four possible cases, as depicted in Figure 7.

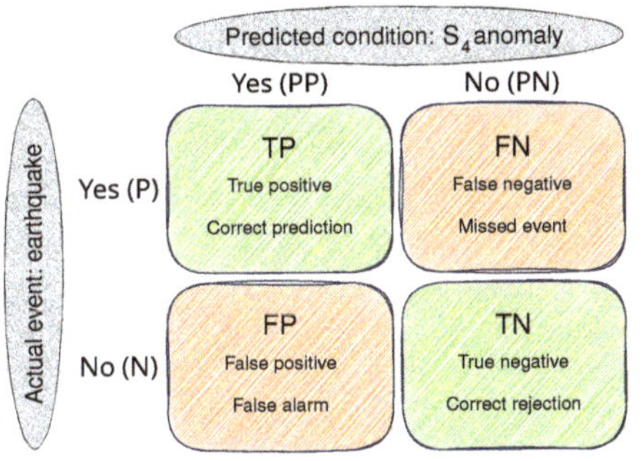

Figure 7. Confusion matrix diagram.

In particular, after the data pre-processing described in previous sections, the creation of the confusion matrix is described below, repeating this classification for each day d and each pixel (x, y) within the S_4 variations map:

- Predicted condition, S_4 positive increment (ΔS_4):
  - Yes (PP: Predicted Positive): If the ΔS_4 value in the current pixel is larger than a certain threshold *s4_th*.
  - No (PN: Predicted Negative): If the ΔS_4 value in the current pixel is smaller than the threshold.
- Actual event: earthquake:
  - Yes (P: Positive event): if at least one earthquake with a magnitude within the magnitude interval defined (*eq_mag*) has its epicenter within a certain distance (*radius*) from the center of the pixel (x, y), and it occurs within a certain time window $[d + d\_margin]$ (in days) from the studied day. Please note that this way, a positive *d_margin* means that the earthquake takes place after the ionospheric anomaly.
  - No (N: Negative event): if any earthquake matches the previous criteria.

The matrix accounts for the total number of events in each of the four categories. As planned, each event studied fit in one, and only one, category. The maximum number of events computed in a CM will be $N\_days \times N\_lats \times N\_lons$ events, where *N_days* is the number of days under study and $N\_lats \times N\_lons$ is the number of pixels in the map. In our case, using $1^\circ \times 1^\circ$ pixels, there are $80 \times 360 = 28,800$ points.

As explained in Section 2.2, in the S_4 variations maps there could be pixels with no data, either because they are completely located on land, and never fit a specular reflection point or because the LT filter applied removed all the days in the current small window used. This means that the total number of pixels computed (TOT) will be less than the total on the map, in particular, every day, a constant number referring to the land coverage will be subtracted, and then another quantity corresponding to the temporal evolution shown in Figure 2.

Several confusion matrixes are constructed after setting the different parameters mentioned before, which are detailed in Table 2:

Table 2. Description of the parameters swept to generate different confusion matrixes.

Parameter	Description	Values Swept	Units
radius	Maximum distance to look for earthquakes around the center of the pixels (x, y).	50, 100	km
eq_mag	Magnitude interval to filter earthquakes studied.	[4, 4.9], [5, 5.9], [6, ∞]	m_b, M_W, M_{Wr}
d_margin	Earthquakes occurring within the period.	[−6, −3], [−3, 0], [0, +3], [+3, +6]	day
s4_th	Inferior ΔS_4 threshold to consider the pixel as an anomalous change in S_4.	$0, 0.003, \cdots, 0.03$	-

The *d_margin* parameter is used to study whether there is a preferred time of appearance of the ionospheric anomalies before or after the earthquakes. Please note that when a CM is labeled as $d\_margin = +6$, it means that for each day D in which the anomaly is computed, earthquakes are searched within the days $(D + 3, D + 6)$ days, i.e., the anomalous change in S_4 is a precursor of the earthquake. For example, $d\_margin = -3$ would mean from $(D - 3, D + 0)$ days.

The *s4_th* parameter is used as a tuning parameter for the binary classification table. A smaller threshold would make the algorithm detect more ΔS_4 positives, which includes TP (correct predictions), but also FP (false alarms). The study will be used to optimize this threshold to maximize the effectiveness of the algorithm.

2.5. Statistical Parameters

From the confusion matrix, several metrics were extracted to quantify the goodness of the correlation. First of all, the normalized values of each category in the confusion table are computed according to the expressions in Equation (3)

$$\mathrm{TPR} = \frac{\mathrm{TP}}{\mathrm{P}}; \quad \mathrm{FPR} = \frac{\mathrm{FP}}{\mathrm{N}}; \quad \mathrm{FNR} = \frac{\mathrm{FN}}{\mathrm{P}}; \quad \mathrm{TNR} = \frac{\mathrm{TN}}{\mathrm{N}} \tag{3}$$

The simpler qualifier that can be extracted is the accuracy (ACC), which measures the ratio of correct predictions (TP and TN) among the total cases (TOT) according to Equation (4):

$$\mathrm{ACC} = \frac{\mathrm{TP} + \mathrm{TN}}{\mathrm{TP} + \mathrm{TN} + \mathrm{FP} + \mathrm{FN}} \tag{4}$$

The F_1 score is the harmonic mean between the precision $= \frac{\mathrm{TP}}{\mathrm{TP+FP}}$, and the recall or sensitivity $= \frac{\mathrm{TP}}{\mathrm{TP+FN}}$. F_1 is therefore calculated with Equation (5):

$$\mathrm{F}_1 = \frac{2\mathrm{TP}}{2\mathrm{TP} + \mathrm{FP} + \mathrm{FN}} \tag{5}$$

Another correlation parameter that performs better for unbalanced classifications, in which the number of positives and negatives is very different, is the Matthews Correlation Coefficient (MCC), also called phi-coefficient (ϕ), and it is given by the expression in Equation (6):

$$\mathrm{MCC} = \frac{\mathrm{TP} \times \mathrm{TN} - \mathrm{FP} \times \mathrm{FN}}{\sqrt{(\mathrm{TP} + \mathrm{FP})(\mathrm{TP} + \mathrm{FN})(\mathrm{TN} + \mathrm{FP})(\mathrm{TN} + \mathrm{FN})}} \tag{6}$$

Furthermore, finally, the Diagnostic Odds Ratio (DOR) is a measure of the effectiveness of a diagnostic test, and it is given by Equation (7). DOR has positive values larger than one when the test is useful and increases as the performance is better.

$$\mathrm{DOR} = \frac{\mathrm{TP}/\mathrm{FN}}{\mathrm{FP}/\mathrm{TN}} = \frac{\mathrm{TP} \cdot \mathrm{TN}}{\mathrm{FP} \cdot \mathrm{FN}} \tag{7}$$

Additionally, the Receiver Operating Characteristic (ROC) has been calculated. This is a method historically introduced by military radar receivers in the 1950s to diagnose the ability of a system to detect true positives with respect to false alarms. The ROC is the plot of the TPR vs. the FPR when moving a sensitivity parameter. In our case, the parameter is the *s4_th*, and the curve is studied for the rest of parameters indicated in Table 2.

3. Results

The results obtained using this technique cover a maximum period of 180 days (6 months) from March to August 2019. To compute the S_4 variations maps as explained in Section 2.2, it is required to obtain CYGNSS data from at least 60 days before the first day to be studied, so the average S_4 maps were computed from 1 January to 28 August.

Then, the S_4 variations (ΔS_4) maps were computed for every day between 1 March to 28 August, 180 days in total. As is shown in Figure 2, there is a temporal variation of the region covered after doing the LT filtering (from 0 h to 6 h) and the seasonal detrending. Figure 8 shows this map for one of the days with maximum coverage, almost ranging from −40° to +40°.

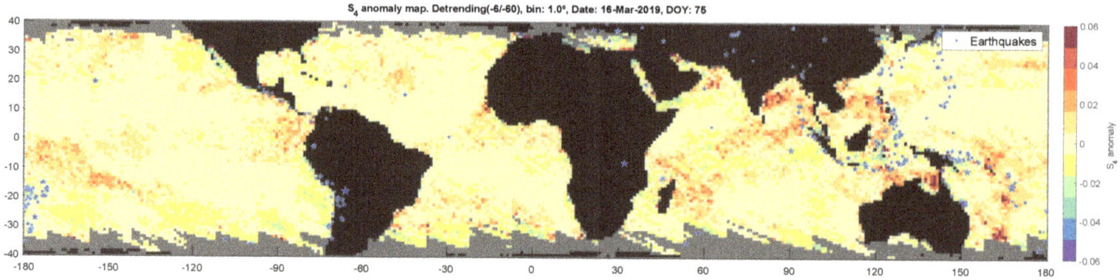

Figure 8. S_4 variations map computed from the seven days prior to 16 March 2019, after subtracting the preceding 60 day average over all oceanic regions between 40°S and 40°N latitudes. Blue stars indicate the earthquakes' epicenters happening from six days before to six days after.

The image shows the results of the detrended ΔS_4 values for the selected day over oceanic regions with the values ranging from −0.06 to 0.06 as indicated in the color scale. Blue stars mark the epicenter of all earthquakes happening during the period from −6 days to +6 days from the current day 16 March.

Figure 9 shows the S_4 variations map for a day in which the coverage is minimum, 30 March, in this case corresponding to the southern part. It can be observed that still several regions can be used to correlate ionospheric scintillation anomalous variations with earthquakes.

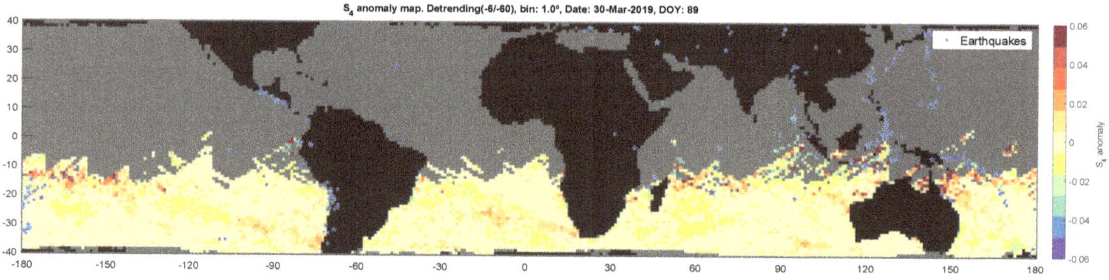

Figure 9. S_4 variations map for one of the minimum coverage days, from −40° to around −5° in average.

3.1. Case Studies

Before a more quantitative, global statistical analysis is made, some case studies are presented to visually check the results of the technique proposed. To do that, videos were created showing the daily ΔS_4 overlapped with the earthquakes. These videos are provided as supplementary materials. Video S1: anomaly_s4_180days_fullCoverage.m4v) shows the full region covered by CYGNSS for the whole period covered in the study, and Video S2: case_study.m4v show the period and region corresponding to the case study detailed here. As in the previous figures, in the videos, earthquakes are kept in each frame for six days before they occur until six days after, so it is possible to visually check whether there is an increasing S_4 anomalous deviation before the occurrence of an earthquake, and if it stays for some time after the earthquake happens.

A case in which some positive variations in the S_4 index are distributed around an earthquake cluster occurred at the beginning of July 2019, in the oceanic region of Vanuatu's archipelago. Vanuatu is a group of islands of volcanic origin that formed after the subduction of the Australian plate under the Pacific plate, which makes this region suffer from large recursive earthquakes [32]. Some of the largest earthquakes from the cluster mentioned are shown in Figure 10, indicating their magnitudes, UTC times, and depths. Please note that all of them are located eastward of the fault, where the subduction takes place. The circular curves around the earthquakes indicate the shakemap Modified

Mercalli Intensity (MMI) contours, which are defined from I to X measuring the intensity of the effects produced on the surface.

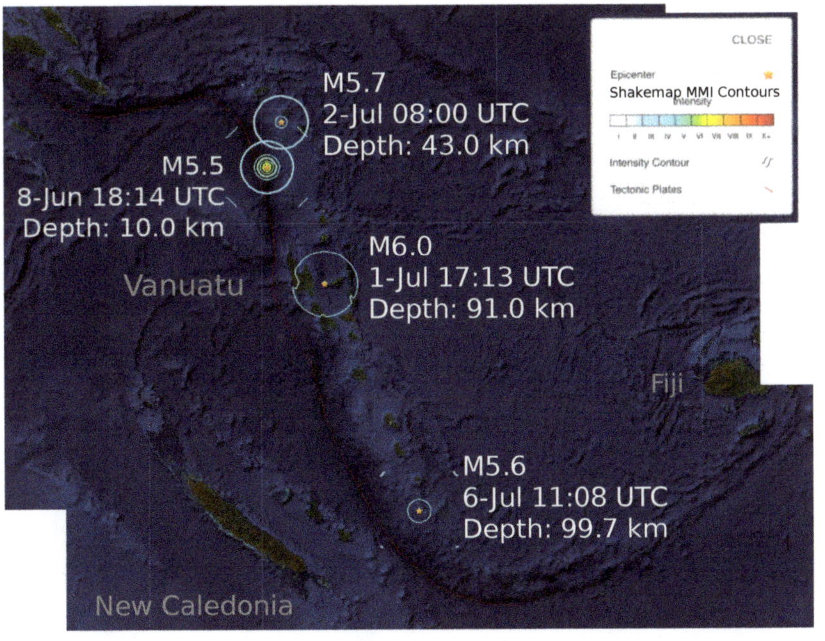

Figure 10. Some of the larger earthquakes around Vanuatu at the beginning of July 2017, showing their epicenters, time, depth and surface felt magnitude shakemaps. It can also be seen the submarine trench and the tectonic fault between Australian and Pacific plates.

During the last few days of June and the first week of July, several positive variations in the ionospheric scintillation values were found in this region. Moreover, the anomalies appear after a period of stabilization in the scintillation activity, indicated by close-to-zero ΔS_4 values in most of the area.

Figure 11a shows the map of the region around the Vanuatu archipelago for 30 June, a few days before the earthquakes shown in the previous image. Figure 11b, shows the map for 1 July, the day of the strongest earthquake (M6.0) in this cluster, and an increase is observed in the closest pixels to this earthquake, and also in the Southern New Caledonia region. Figure 11c corresponds to 4 July, and the peaks in S_4 variations are now clear in the whole area, and that can also be related to other earthquakes happening in 6 July in the South. It is important to remark that ΔS_4 is computed from the average of the previous 7 days, so, variations are shown for the same day D are produced before it, even if the day D is a few days after a particular earthquake.

Finally, Figure 11d shows 7 July, in which the positive values in the S_4 deviations persist although the northern region is starting to miss data due to the oscillation in the region covered shown before.

3.2. Temporal ΔS_4 Analysis around Earthquakes

In this section, the results of a more detailed temporal analysis of the S_4 changes are presented. The purpose is to study the temporal evolution of the S_4 changes around the earthquakes in the period under study. The way to measure this evolution is by averaging all the pixels in the S_4 deviations maps that fall inside a circle of a given radius around each earthquake and computing this average for each day during a period of some days before and after the earthquake day.

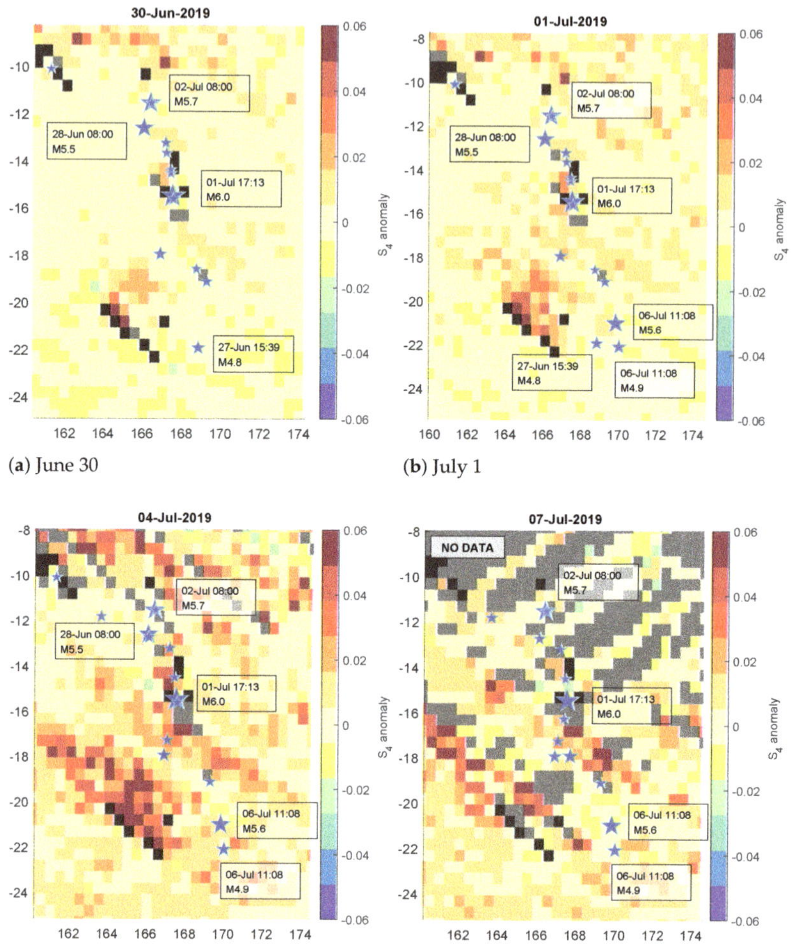

(**a**) June 30 (**b**) July 1

(**c**) July 4 (**d**) July 7

Figure 11. S_4 variations maps for selected days around Vanuatu islands at the beginning of July 2019, with a resolution of 0.5° per pixel. Stars represent the earthquakes' epicenters with magnitudes ≥ 4.5 in the period from -6 to +6 days from the current day.

Figure 12 presents different plots for different minimum earthquake magnitudes in rows, and different integration radii, in columns. Each plot shows the averaged ΔS_4 value of all the pixels inside a circle centered in the earthquake epicenter from 15 days before to 15 days after the event, and then averaged for all the earthquakes in the period from 1 March, to 29 August 2019. Only earthquakes that are surrounded by at least 40% water pixels are computed in this average. This filter is applied to remove false alarms due to earthquakes that occur on land, but close enough to the sea to have enough pixels with valid data. A low number of data points to average may produce unrealistic results. The total number of earthquakes in this period, and those remaining after filtering with the water coverage, are shown in each graph.

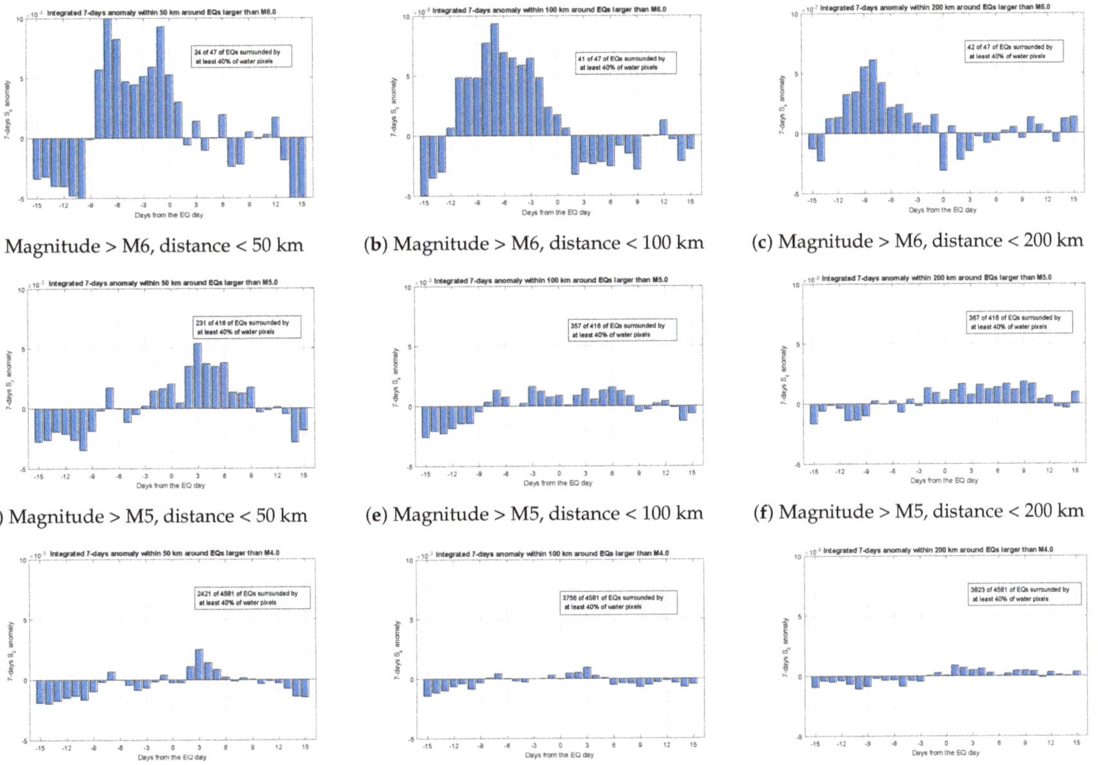

(**a**) Magnitude > M6, distance < 50 km (**b**) Magnitude > M6, distance < 100 km (**c**) Magnitude > M6, distance < 200 km

(**d**) Magnitude > M5, distance < 50 km (**e**) Magnitude > M5, distance < 100 km (**f**) Magnitude > M5, distance < 200 km

(**g**) Magnitude > M4, distance < 50 km (**h**) Magnitude > M4, distance < 100 km (**i**) Magnitude > M4, distance < 200 km

Figure 12. Integrated seven day precursory anomaly within a circle of the indicated radius averaged for all the earthquakes taking place in the period studied, with a magnitude larger than the one indicated in each plot. Only earthquakes surrounded by at least 40% of water are computed.

The results in Figure 12 are divided in rows according to the earthquake magnitude lower limit. For the largest magnitude interval, M6 and larger (Figure 12a–c), it is observed that for the 10 days before earthquakes, the value of seven days ΔS_4 is positive and larger in magnitude than the rest of the days, particularly, the ones after the earthquakes. Similar behavior is observed using different integration radii (50 km, 100 km, 200 km), only showing a small decrease in the peak when integrating larger radii. This is the expected response, as the larger the area to estimate scintillation, the more uncorrelated noise in the circle.

In the next figures for smaller magnitude intervals (from M5 and M4), a decrease in the peak amplitude is observed, which reduces with decreasing earthquakes' magnitude and larger radii. Furthermore, please note that the position of the peak has moved towards the right, indicating that the scintillation positive anomalous increments occur now closer to the event of earthquake. Since the seven day window introduces a 3.5 day, a positive ΔS_4 peak at $D = 3$ means that the precursor occurs just the day before the earthquake (Figure 12d).

3.3. Confusion Matrix Analysis

The qualitative results shown in the previous case studies are only some particular cases to illustrate the technique. A more robust and quantitative way to study if this technique is effective in detecting earthquake-related ionospheric anomalies is to perform a correlation test of the data by means of the confusion matrixes introduced in Section 2.4.

As explained, the S_4 variations maps from the full period of 180 days are computed pixel by pixel classifying them in the four categories depending on whether they are positive or negative cases for both the ΔS_4 value and the earthquake occurrence. With these results, it is possible to construct the confusion matrix, and its derived statistics for each set of parameters.

Table 3 shows a subset of the total number of CMs computed. All of them are calculated for 180 days of data using:

- *radius*: 100 km
- *eq_mag*: 6 to ∞
- *d_margin*: +3 d
- *s4_th*: $0, 0.003, \cdots, 0.030$

Table 3. Confusion matrixes for several values of the ΔS_4 threshold, showing the total number of events (TOT), Positives (P), Negatives (N), True Positives (TP), True Negatives (TN), False Positives (FP), and False Negatives (FN), using data from 180 days.

s4_th	TOT	P	N	TP	TN	FP	FN
0.000	2,603,715	363	2,603,352	199	1,366,209	1,237,143	164
0.003	2,603,715	363	2,603,352	151	1,906,229	697,123	212
0.006	2,603,715	363	2,603,352	117	2,177,099	426,253	246
0.009	2,603,715	363	2,603,352	97	2,312,815	290,537	266
0.012	2,603,715	363	2,603,352	74	2,393,857	209,495	289
0.015	2,603,715	363	2,603,352	64	2,447,989	155,363	299
0.018	2,603,715	363	2,603,352	51	2,486,012	117,340	312
0.021	2,603,715	363	2,603,352	35	2,513,181	90,171	328
0.024	2,603,715	363	2,603,352	29	2,533,037	70,315	334
0.027	2,603,715	363	2,603,352	21	2,547,932	55,420	342
0.030	2,603,715	363	2,603,352	19	2,559,017	44,335	344

Remember that the total number of events computed, marked as TOT is the total number of pixels with valid data during the period studied. Positives (P) and Negatives (N) refer to the pixels with earthquakes closer than the radius, and TP, TN, FP, and FN are the four categories of the CM.

The results of the statistical parameters described in Section 2.5 are presented here in Table 4 for each case studied, for the same previous cases used in Table 3.

Table 4. Statistical measurements extracted from the previous confusion matrix, using the same parameters: $radius = 100$ km, $eq\_mag \geq 6$, $d\_margin : +3d$, and sweeping the $s4\_th = 0, 0.003, \cdots, 0.030$.

s4_th	TPR (%)	FPR (%)	TNR (%)	FNR (%)	ACC	F_1 Score $\times 10^{-3}$	MCC $\times 10^{-3}$	DOR
0.000	54.8	47.5	52.5	45.2	0.525	0.32	1.73	1.340
0.003	41.6	26.8	73.2	58.4	0.732	0.43	3.95	1.948
0.006	32.2	16.4	83.6	67.8	0.836	0.55	5.06	2.429
0.009	26.7	11.2	88.8	73.3	0.888	0.67	5.83	2.903
0.012	20.4	8.0	92.0	79.6	0.919	0.70	5.36	2.926
0.015	17.6	6.0	94.0	82.4	0.940	0.82	5.81	3.373
0.018	14.0	4.5	95.5	86.0	0.955	0.87	5.43	3.463
0.021	9.6	3.5	96.5	90.4	0.965	0.77	3.99	2.974
0.024	8.0	2.7	97.3	92.0	0.973	0.82	3.85	3.128
0.027	5.8	2.1	97.9	94.2	0.979	0.75	2.99	2.823
0.030	5.2	1.7	98.3	94.8	0.983	0.85	3.22	3.188

As introduced in Section 2.4, the Receiver Operating Characteristic (ROC) curves are also a good way to identify better performance of a classification algorithm and to tune the threshold parameter used, in this case *s4_th*. The ROC curve is constructed by plotting the TPR vs. the FPR, so it is able to compare simultaneously the performance of the algorithm to correctly associate positive changes in the S_4 value with earthquakes against the rate of false alarms produced when the algorithm detects an positive change in the scintillation, but there is no earthquake in this region and period.

ROC curves for the three magnitude intervals are presented in each of the plots in Figure 13. Four curves per graph are shown with different colors indicating the time interval from the anomaly to the occurrence of the earthquakes, being positive when it is prior to the earthquake so it could be considered as a precursor.

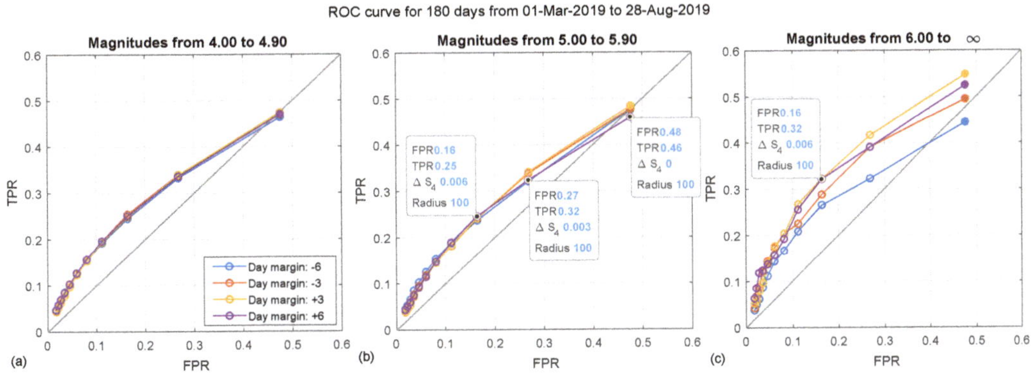

Figure 13. ROC curves for three magnitude intervals using $radius = 100$ km. Each curve indicated in the legend is using a different d_margin parameter, being positive for S_4 positive increments prior to earthquakes. Points within each curve are sweeping the value of *s4_th* each 0.003 from zero in the top right point.

Each point in the curve is for a different value of the threshold parameter $s4\_th$ shown as rows in previous Table 3. Please note that the point closer to $(0,0)$ is the one the with largest $s4\_th = 0.03$, which makes sense because larger threshold are filtering more true detections, and also reducing the false alarm rate. Decreasing the threshold in 0.003 intervals, both parameters increases until the last point (the one closest to $(1,1)$) for $s4\_th = 0$.

ROC curves indicate better performance when they are more separated from the $y = x$ line, which would be the result of a completely random detection algorithm. ROC-AUC (ROC Area Under the Curve), is a good indicator of this performance, and it will be discussed in the next section.

Figures 14 and 15 show the ROC curves for the three magnitude intervals using a radius of 50 km and 100 km, respectively. In these plots, *s4_th* has been swept from 0 to 0.04 in intervals of 0.01.

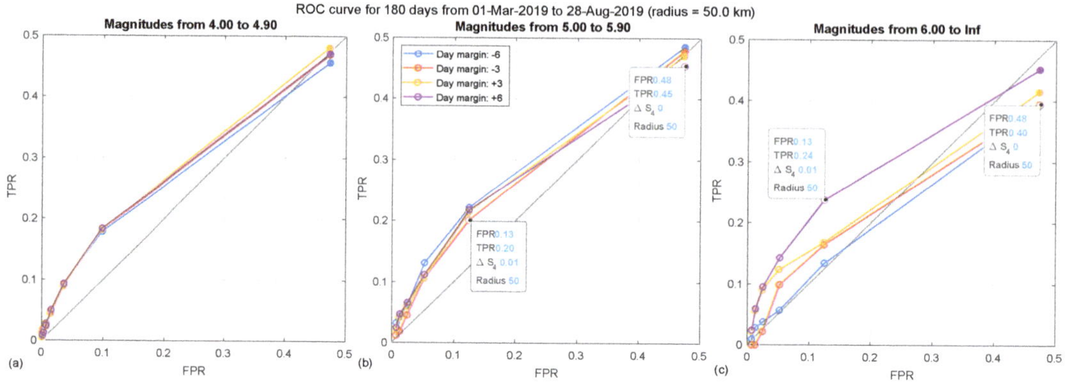

Figure 14. ROC curves for three magnitude intervals using *radius* = 50 km.

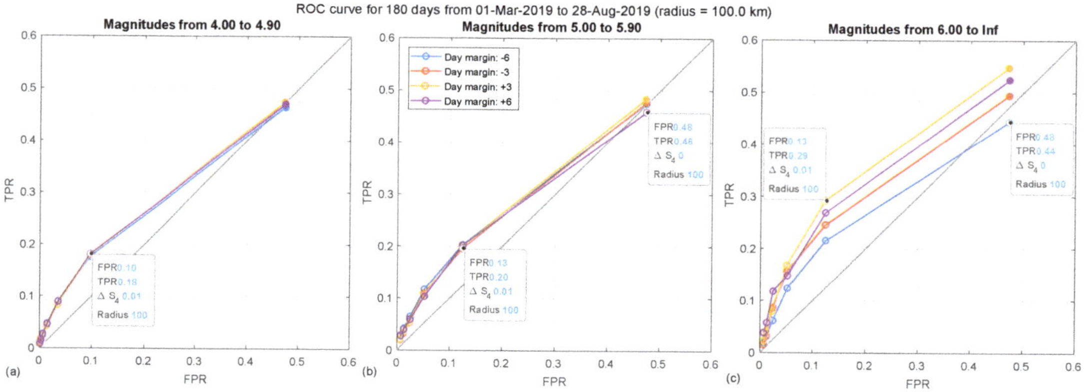

Figure 15. ROC curves for three magnitude intervals using *radius* = 100 km.

4. Discussion

The results presented here are showing a small, but positive correlation between the occurrence of earthquakes and the anomalous variations of ionospheric scintillation. The confusion matrix shows the values for the True Positives, True Negatives, False Negatives, and False Positives integrated for all the days in the period studied. The positives (P) show the number of pixels with at least one earthquake within the radius distance, and it is only varying when this parameter changes. For example, in Table 3, is 363 for all cases, as the radius is always 100 km.

It is observed that when the ΔS_4 threshold ($s4\_th$) increases, the number of true positives (TP) and false positives (FP) both decrease. The relative ratio of change is studied with the ROC curves. Larger values of the TPR with respect to the FPR for the same threshold mean that is more probable that a positive in the predicted condition (ionospheric anomaly) is associated with an earthquake event, than it is not, and they will be represented by curves above the diagonal.

Figure 13 shows the ROC for the three magnitude intervals. All of them show a small, but positive indicator, with curves above the main diagonal for almost all the threshold values selected. What is also very interesting is that, as the magnitude of the earthquakes increases, the curves are further from the diagonal, and also they separate from each other. In the last graph (Figure 13c), for magnitudes larger than 6, curves for $d\_margin = +3$ and $d\_margin = +6$ are above their corresponding negatives values. This means that for the largest earthquakes a better correlation is observed when the anomalies are recorded a few days (3 to 6) before the earthquakes.

Similar results show the ROC curves in Figures 14 and 15, which have been drawn using a coarser *s4_th* interval. In this case, for the 50 km radius and magnitudes larger than 6 (Figure 14c), we observe that only the one corresponding to the $d\_margin = +6$ is separating from the main diagonal. Here we can see that the correlation is less when the radius decreases from 100 km to 50 km.

The values of the ROC-AUC are also a good indicator of the performance of the correlation. Please note that in all the ROC curves shown before in Figure 13, none of their values reach the point (1,1) even that the sensitivity parameter, *s4_th*, has been swept from 0 to 0.03 or 0.04. $s4\_th = 0$ is the one closer to the (1,1), but decreasing this value under 0 makes no sense as it would mean considering a detection when the average S_4 during the last 7 days has been larger than a value lower than the last 2-month average. Then, the ROC-AUC has been computed using the data points shown in the curves, and adding the points (0,0) and (1,1), doing a linear extrapolation. The resulting areas are shown in Tables 5 and 6.

Table 5. ROC-AUC values for $radius = 100$ km .

ROC-AUC 100 km	***d_margin* (Days)**			
Magnitudes	**[−6,−3]**	**[−3,0]**	**[0,+3]**	**[+3,+6]**
4–4.9	0.5204	0.5249	0.5256	0.5242
5–5.9	0.5223	0.5250	0.5278	0.5168
≥ 6	0.5148	0.5487	0.5766	0.5635

Table 6. ROC-AUC values for $radius = 50$ km .

ROC-AUC 50 km	***d_margin* (Days)**			
Magnitudes	**[−6,−3]**	**[−3,0]**	**[0,+3]**	**[+3,+6]**
4–4.9	0.5117	0.5180	0.5232	0.5193
5–5.9	0.5297	0.5202	0.5201	0.5141
≥ 6	0.4670	0.4753	0.4881	0.5199

Table 5 for $radius = 100$ km identifies the magnitude larger than 6, as the set of curves with larger ROC-AUC values in general, being the largest one, 0.5766, when the interval for *d_margin* is $[0, +3]$, closely followed by the $d\_margin = [+3, +6]$. The rest of the table exhibits values around 0.52, very similar among them, remarking the similarity among the curves for the lowest two magnitude intervals in Figure 13.

Table 6 shows the results of the ROC-AUC for $radius = 50$ km. In the case of the largest magnitude interval (above 6), as observed in Figure 14 the ROC lines are closer to the main diagonal, or even under it, resulting in ROC-AUC values under 0.5. In this case, the better results in terms of ROC-AUC values are for the lower magnitude intervals, [5–5.9] and [4–4.9]. This may be an indicator that smaller earthquakes may produce perturbations at smaller distances from the epicenter.

Another way to analyze the correlation between both events is through the statistics extracted from the CM. In particular, the values of the DOR are shown in Figure 16. The DOR, given by Equation (7), is an indicator of the diagnostic performance that takes into account the whole parameters in the CM and it ranges from 0 to infinity, being larger than one for positive correlations, and higher for better performance. In the plots, it is shown that it goes from 1 to 3 or 4 (depending on the case) as *s4_th* increases. Figure 16a, for the magnitudes from 4 to 4.9 we can see that all the curves for different *d_margin* values are very similar, and they separarte for the largest magnitude intervals, as in Figure 16c, where the largest DOR is almost 5 for a $d\_margin = +6$ and $s4\_th = 0.024$.

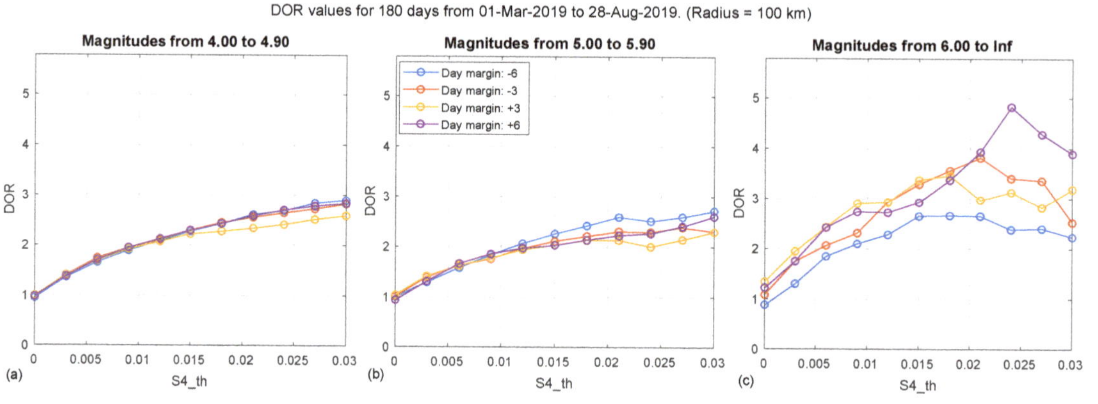

Figure 16. DOR values for two magnitude intervals using $radius = 100$ km with respect to the ΔS_4 threshold: $s4\_th$.

Another metrics computed is the accuracy (ACC) using Equation (4), which is shown in Figure 17 for the 100 km radius case. The problem with this indicator for very unbalanced classifications such as this is that they will give unrealistically high values because they are only accounting for the TP and TN counts with respect to the total TOT. It is observed in the plot that it goes from 50% accuracy to almost 100%, and all the curves are overlapping in the one shown here, for all *d_margin*, and all magnitude intervals. In this case, as the number of Positives (P) and Negatives (N), i.e., earthquakes vs. no-earthquakes, is very unbalanced, the accuracy is not a good indicator of correlation.

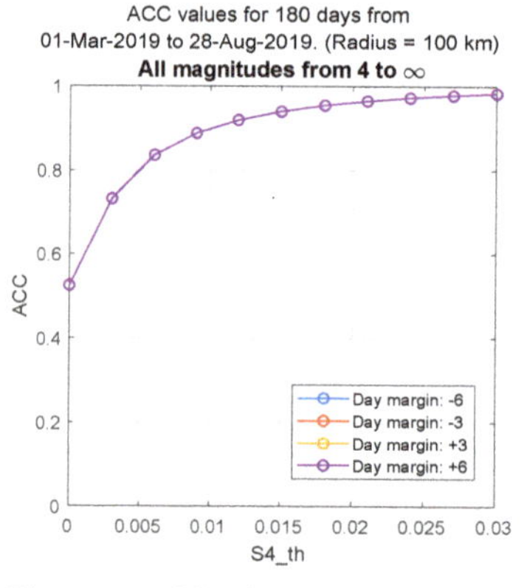

Figure 17. ACC values for two magnitude intervals using $radius = 100$ km with respect to the ΔS_4 threshold: $s4\_th$.

To complete the analysis, curves for the F_1 score and Matthews Correlation Coefficient (MCC) are shown in Figures 18 and 19, respectively. F_1 score is an improvement of the ACC parameter, but still, it is less indicative than the DOR or the MCC for this case. Remember from Equation (5), that the numerator is only accounting for the true positive (TP) cases, which for this study are decreasing a lot for large magnitudes intervals, which explains the low values in the plot in Figure 18c.

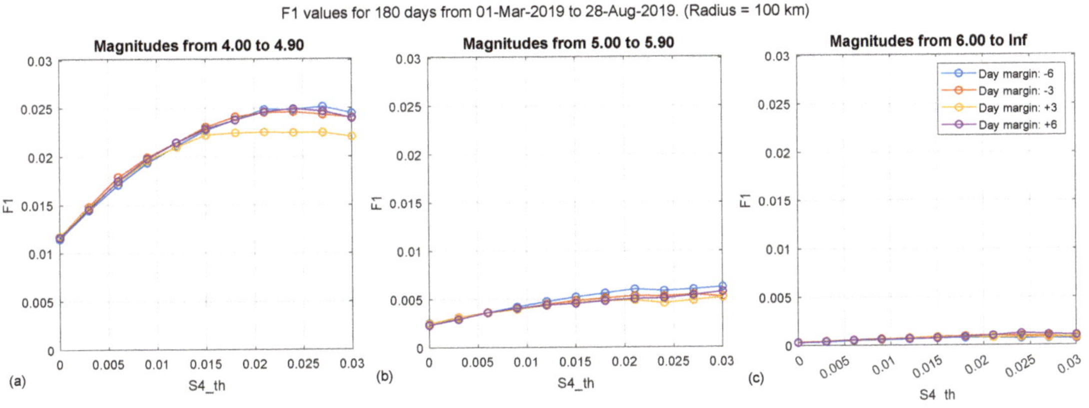

Figure 18. F_1 score values for two magnitude intervals using *radius* = 100 km with respect to the ΔS_4 threshold: *s4_th*.

Finally, the MCC is shown in Figure 19a–c and it indicates positive values for all magnitude intervals. MCC ranges from −1 to +1, being zero when there is no correlation. In the same way as other parameters studied here, MCC is indicating positive values although they are small. As the F_1 score, the MCC shows smaller values for the largest earthquake magnitudes, the height of the peak is decreasing from Figure19a to Figure19c.

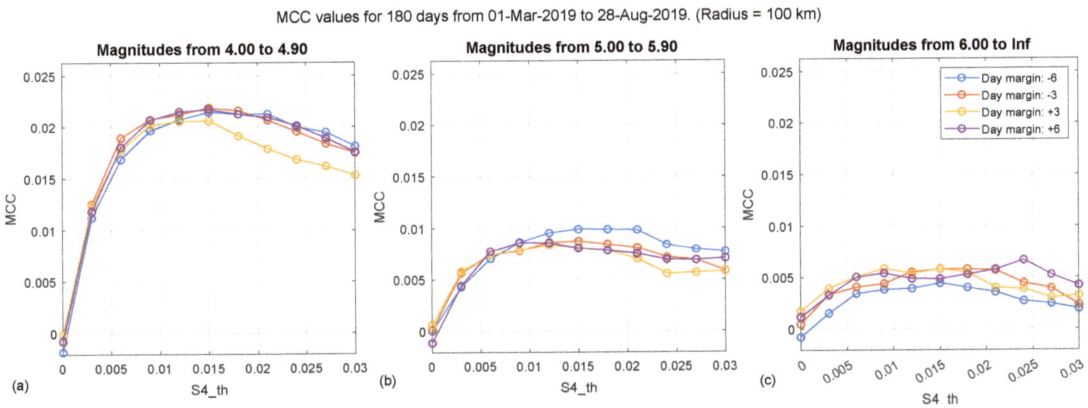

Figure 19. MCC values for two magnitude intervals using *radius* = 100 km with respect to the ΔS_4 threshold: *s4_th*.

An interesting conclusion that can be extracted from the two most significative statistical parameters (DOR and MCC), is the value of the ionospheric scintillation threshold (*s4_th*) parameter that is giving the best performance when correlating earthquakes with S_4 positive variations. As observed in Figure 16a,b for small magnitude intervals (from 4 to 6), using the DOR metrics, *s4_th* is larger than 0.03 (the largest value swept), but it is close to 0.02 for magnitudes larger than 6, in Figure 16c. For example, the values previously shown in Table 4 correspond to the yellow line (*d_margin*: +3 d) in Figure 16c, which has a peak DOR = 3.463 for a *s4_th* = 0.018.

In the case of the MCC it also exhibits a peak in the *s4_th* interval studied, which falls around 0.01–0.015 for lower magnitude intervals, from 4 to 6, and it is a bit wider for large earthquakes, ranging from around 0.006 to 0.020.

The position of both peaks in DOR and MCC also matches the values of *s4_th* in which the ROC curves are further away from the main diagonal, as found in Figures 13–15.

5. Conclusions

This study used six months of data from the NASA CYGNSS GNSS-R mission from March to August 2019 to estimate ionospheric amplitude scintillation to try to analyze possible correlations between the occurrence of earthquakes and the associated changes in the ionospheric amplitude scintillation index S_4, and also try to determine if these changes are produced with higher probabilities before (as a precursor) or after the earthquakes. The technique used computes the confusion matrix (CM), a usual technique in diagnostic tests, e.g., in medical sciences or machine learning algorithms, to quantify the goodness of a diagnostic test.

The results presented in this work show a small, but detectable, positive correlation for all earthquakes with magnitudes above 4, with better results as the magnitude increases. A slightly better correlation was also observed when the positive increments in the S_4 index are observed between 6 and 3 days before the earthquakes, with respect to the ones observed after them. These results have been extracted from the analysis of ROC curves for different magnitude intervals showing the relation between the probability of finding a correct prediction vs. a false alarm. In the best cases, the correct prediction probability is around 32% and the false alarm probability is 16%.

It should be also pointed out, that the probability of detection is small for all the cases. In our opinion, despite these results showing possible evidence of ionospheric scintillation increments as precursors of earthquakes, mostly for the largest magnitude ones, the signature is still small, and it should not be regarded as an early warning system for earthquakes. Moreover, the study was made using earthquakes during a 6 month period, when only a small number of large earthquakes ($M \geq 6$) took place, specifically 68. From them, only 47 were in oceanic regions. Moreover, the region covered by CYGNSS is always close to the geomagnetic equator, where ionospheric scintillation is also more probable to appear.

The novelty of this study lies in the use of GNSS-R global oceanic data to measure amplitude scintillation S_4 positive increments as a possible precursor for earthquakes. Other studies detected anomalies in the ionosphere, but they were mostly related to changes in the Total Electron Content (TEC) after the earthquakes, which can be produced by the mechanical energy transmission from the sudden crust movement through the atmosphere to the ionosphere. Furthermore, few of the previous studies used satellite data, allowing the possibility to perform global studies such as this one, and they were focused on particular cases using local GNSS ground stations.

Future work will extend this study using longer periods of time, and larger Earth coverage. CYGNSS constellation used to obtain the GNSS-R data used here only covers from 40°N to 40°S, which is good to have a better revisit time, but it is missing earthquakes occurring at higher latitudes. Furthermore, this technique, using GNSS-R is only valid for calm oceanic regions, which loses around 30% of the earthquakes. So, a possible way to study ionospheric scintillation on land would be to use GNSS Radio-Occultation (GNSS-RO) data as well. Using this technique the overall performance may be even better as it does not suffer from the ocean surface conditions.

Supplementary Materials: The following supporting information can be downloaded at: https://www.mdpi.com/article/10.3390/rs14112555/s1, Video S1: anomaly_s4_180days_fullCoverage.m4v, Video S2: case_study.m4v.

Author Contributions: Conceptualization, C.M. and A.C.; methodology, C.M. and A.C.; software, C.M.; validation, C.M., B.-E.B.S., H.P., and A.C.; formal analysis, C.M.; investigation, C.M., B.-E.B.S., H.P., and A.C.; resources, A.C.; data curation, C.M.; writing—original draft preparation, C.M.; writing—review and editing, C.M., B.-E.B.S., H.P., and A.C.; visualization, C.M.; supervision, A.C.; project administration, A.C.; funding acquisition, A.C. All authors have read and agreed to the published version of the manuscript.

Funding: This work was supported by the Spanish Ministry of Science, Innovation and Universities and EFRD, "Sensing with Pioneering Opportunistic Techniques" SPOT, grant RTI2018-099008-BC21/AEI/10.13039/501100011033, and by the Unidad de Excelencia Maria de Maeztu MDM-2016-0600 and, Grant RYC-2016-20918 financed by MCIN/AEI /10.13039/501100011033 and by ESF Investing in your future.

Data Availability Statement: GNSS-R data used in this study comes from the open NASA CYGNSS L1 data storage Physical Oceanography DAAC (PO.DAAC) available at https://podaac.jpl.nasa.gov/dataset/CYGNSS_L1_V2.1 [24 January 2022] and earthquake data comes from the open USGS earthquake database available at https://earthquake.usgs.gov/earthquakes/search/ [3 February 2022].

Acknowledgments: We want to acknowledge the rest of the team at CommSensLab for their help on setting the computing infrastructure to process the amount of data used in the study, and their recommendations and suggestions in the way to perform this study.

Conflicts of Interest: The authors declare no conflict of interest.

Abbreviations

The following abbreviations are used in this manuscript:

ACC	Accuracy
CYGNSS	Cyclone Global Navigation Satellite System
DAAC	Distributed Active Archive Center
DDM	Delay-Doppler Map
DOR	Diagnostic Odds Ratio
FN	False Negative
FP	False Positive
FPR	False Positive Rate
GNSS	Global-Navigation Satellite System
GNSS-R	GNSS Reflectometry
GNNS-RO	GNSS Radio-Occultation
LST	Land Surface Temperature
LT	Local Time
MCC	Matthews Correlation Coefficient
MMI	Modified Mercalli Intensity
ROC	Receiver Operating Characteristic
ROC-AUC	Receiver Operating Characteristic Area Under the Curve
SNR	Signal-to-Noise Ratio
TEC	Total Electron Content
TN	True Negative
TP	True Positive
TPR	True Positive Rate
USGS	United States Geological Survey
UTC	Universal Time Coordinated

References

1. Priyadarshi, S. A review of ionospheric scintillation models. *Surv. Geophys.* **2015**, *36*, 295–324. [CrossRef] [PubMed]
2. Fremouw, E.; Rino, C. An empirical model for average F-layer scintillation at VHF/UHF. *Radio Sci.* **1973**, *8*, 213–222. [CrossRef]
3. Rino, C.L. A power law phase screen model for ionospheric scintillation: 1. Weak scatter. *Radio Sci.* **1979**, *14*, 1135–1145. [CrossRef]
4. Aarons, J. Construction of a model of equatorial scintillation intensity. *Radio Sci.* **1985**, *20*, 397–402. [CrossRef]
5. Franke, S.; Liu, C.H. Modeling of equatorial multifrequency scintillation. *Radio Sci.* **1985**, *20*, 403–415. [CrossRef]
6. Iyer, K.; Souza, J.; Pathan, B.; Abdu, M.; Jivani, M.; Joshi, H. A model of equatorial and low latitude VHF scintillation in India. *Indian J. Radio Space Phys.* **2006**, *35*, 98–104.
7. WBMOD Ionospheric Scintillation Model. 2022. Available online: https://ccmc.gsfc.nasa.gov/modelweb/ionos/wbmod.html (accessed on 10 March 2022).
8. Béniguel, Y. Global ionospheric propagation model (GIM): A propagation model for scintillations of transmitted signals. *Radio Sci.* **2002**, *37*, 1–14. [CrossRef]

9. Camps, A.; Park, H.; Foti, G.; Gommenginger, C. Ionospheric Effects in GNSS-Reflectometry From Space. *IEEE J. Sel. Top. Appl. Earth Obs. Remote. Sens.* **2016**, *9*, 5851–5861. [CrossRef]
10. Molina, C.; Camps, A. First Evidences of Ionospheric Plasma Depletions Observations Using GNSS-R Data from CYGNSS. *Remote. Sens.* **2020**, *12*, 3782. [CrossRef]
11. Molina, C.; Boudriki Semlali, B.E.; Park, H.; Camps, A. Possible Evidence of Earthquake Precursors Observed in Ionospheric Scintillation Events Observed from Spaceborne GNSS-R Data. In Proceedings of the 2021 IEEE International Geoscience and Remote Sensing Symposium IGARSS, Brussels, Belgium, 11–16 July 2021; pp. 8680–8683. [CrossRef]
12. Pulinets, S.A.; Krankowski, A.; Hernandez-Pajares, M.; Marra, S.; Cherniak, I.; Zakharenkova, I.; Rothkaehl, H.; Kotulak, K.; Davidenko, D.; Blaszkiewicz, L.; et al. Ionosphere Sounding for Pre-seismic anomalies identification (INSPIRE): Results of the project and Perspectives for the short-term earthquake forecast. *Front. Earth Sci.* **2021**, *9*, 131. [CrossRef]
13. Kamogawa, M. Preseismic lithosphere-atmosphere-ionosphere coupling. *Eos Trans. Am. Geophys. Union* **2006**, *87*, 417–424. [CrossRef]
14. Namgaladze, A.A.; Zolotov, O.V.; Karpov, M.I.; Romanovskaya, Y.V. Manifestations of the earthquake preparations in the ionosphere total electron content variations. *Nat. Sci.* **2012**, *4*, 848–855. [CrossRef]
15. Pavlidou, E.; Van der Meijde, M.; Van der Werff, H.; Hecker, C. Time Series Analysis of Land Surface Temperatures in 20 Earthquake Cases Worldwide. *Remote Sens.* **2019**, *11*, 61. [CrossRef]
16. Ma, J.; Chen, S.; Hu, X.; Liu, P.; Liu, L. Spatial-temporal variation of the land surface temperature field and present-day tectonic activity. *Geosci. Front.* **2010**, *1*, 57–67. [CrossRef]
17. Blackett, M.; Wooster, M.J.; Malamud, B.D. Exploring land surface temperature earthquake precursors: A focus on the Gujarat (India) earthquake of 2001. *Geophys. Res. Lett.* **2011**, *38* , [CrossRef]
18. Song, R.; Hattori, K.; Zhang, X.; Sanaka, S. Seismic-ionospheric effects prior to four earthquakes in Indonesia detected by the China seismo-electromagnetic satellite. *J. Atmos.-Sol.-Terr. Phys.* **2020**, *205*, 105291. [CrossRef]
19. Tsugawa, T.; Saito, A.; Otsuka, Y.; Nishioka, M.; Maruyama, T.; Kato, H.; Nagatsuma, T.; Murata, K.T. Ionospheric disturbances detected by GPS total electron content observation after the 2011 off the Pacific coast of Tohoku Earthquake. *Earth Planets Space* **2011**, *63*, 875–879. [CrossRef]
20. Hassini, A.; Belbachir, A.H. Thermal infrared geostationary satellite sensor data application for prediction and monitoring earthquake in Algeria. *Int. J. Inf. Technol. Manag.* **2016**, *15*, 293–312.
21. Shah, M. Earthquake ionospheric and atmospheric anomalies from GNSS TEC and other satellites. In *Computers in Earth and Environmental Sciences*; Elsevier: Amsterdam, The Netherlands, 2022; pp. 387–399.
22. Oikonomou, C.; Haralambous, H.; Pulinets, S.; Khadka, A.; Paudel, S.R.; Barta, V.; Muslim, B.; Kourtidis, K.; Karagioras, A.; İnyurt, S. Investigation of Pre-Earthquake Ionospheric and Atmospheric Disturbances for Three Large Earthquakes in Mexico. *Geosciences* **2021**, *11*, 16. [CrossRef]
23. Sotomayor-Beltran, C. Positive and negative ionospheric disturbances prior to the 2016 christmas earthquake in Chile. *Geomat. Nat. Hazards Risk* **2019**, *10*, 622–632. [CrossRef]
24. Kandalyan, R.A.; AlQuran, M.K. Ionosphere scintillation and earthquakes. *Jordan J. Phys.* **2010**, *3*, 69–76.
25. Guha Bose, A.; Das, A.; Chowdhury, S.; Deb, A. Studies of scintillations and TEC variations with GPS satellite links together with soil radon anomalies preceding Nepal earthquakes of April–May 2015. *Nat. Hazards* **2022**, *112*, 1137–1163. [CrossRef]
26. Nayak, C.; Tsai, L.C.; Su, S.Y.; Caton, R.G.; Groves, K.M. Can earthquakes affect ionospheric scintillation? In Proceedings of the AGU Fall Meeting Abstracts, San Francisco, CA, USA, 12–16 December 2016; Volume 2016, p. NH51A–1919.
27. CYGNSS Level 1 Science PODAAC Data Store. V3.0 netCDF Data Dictionary. Available online: https://podaac-tools.jpl.nasa.gov/drive/files/allData/cygnss/L1/docs/148-0346-8_L1_v3.0_netCDF_Data_Dictionary.xlsx (accessed on 10 February 2022).
28. USGS Magnitude Types. 2022. Available online: https://www.usgs.gov/faqs/moment-magnitude-richter-scale-what-are-different-magnitude-scales-and-why-are-there-so-many (accessed on 16 March 2022).
29. Michigan Tech: The Richter Scale. 2022. Available online: https://www.mtu.edu/geo/community/seismology/learn/earthquake-measure/ (accessed on 16 March 2022).
30. USGS Magnitude Types. U.S. Geological Survey. 2022. Available online: https://www.usgs.gov/programs/earthquake-hazards/magnitude-types (accessed on 20 February 2022).
31. SCSN Catalog Change History. Southern California Earthquake Data Center. 2022. Available online: https://scedc.caltech.edu/eq-catalogs/change-history.html (accessed on 22 April 2022).
32. Cleveland, K.M.; Ammon, C.J.; Lay, T. Large earthquake processes in the northern Vanuatu subduction zone. *J. Geophys. Res. Solid Earth* **2014**, *119*, 8866–8883. [CrossRef]

remote sensing

MDPI

Article

Assessment of Polar Ionospheric Observations by VIPIR/Dynasonde at Jang Bogo Station, Antarctica: Part 1—Ionospheric Densities

Eunsol Kim [1], Geonhwa Jee [1,2,*], Young-Bae Ham [1,2], Nikolay Zabotin [3], Changsup Lee [1,2], Hyuck-Jin Kwon [1], Junseok Hong [4], Jeong-Han Kim [1] and Terence Bullett [5]

1 Division of Atmospheric Sciences, Korea Polar Research Institute, Incheon 21990, Korea; eunsol518@kopri.re.kr (E.K.); astro422@kopri.re.kr (Y.-B.H.); cslee@kopri.re.kr (C.L.); hjkwon@kopri.re.kr (H.-J.K.); jhkim@kopri.re.kr (J.-H.K.)
2 Department of Polar Science, Korea University of Science and Technology, Daejeon 34113, Korea
3 Department of Electrical, Energy and Computer Engineering, University of Colorado, Boulder, CO 80309, USA; nikolay.zabotin@colorado.edu
4 Korea Astronomy and Space Science Institute, Daejeon 34055, Korea; junseok@kasi.re.kr
5 Cooperative Institute for Research in Environmental Sciences, University of Colorado, Boulder, CO 80309, USA; terry.bullett@noaa.gov
* Correspondence: ghjee@kopri.re.kr; Tel.: +82-32-760-5306

Abstract: Vertical incidence pulsed ionospheric radar (VIPIR) has been operated to observe the polar ionosphere with Dynasonde analysis software at Jang Bogo Station (JBS), Antarctica, since 2017. The JBS-VIPIR-Dynasonde (JVD) provides ionospheric parameters such as the height profile of electron density with NmF2 and hmF2, the ion drift, and the ionospheric tilt in the bottomside ionosphere. The JBS (74.6°S, 164.2°E) is located in the polar cap, cusp, or auroral region depending on the geomagnetic activity and local time. In the present study, an initial assessment of JVD ionospheric densities is attempted by the comparison with GPS TEC measurements which are simultaneously obtained from the GPS receiver at JBS during the solar minimum period from 2017 to 2019. It is found that the JVD NmF2 and bottomside TEC (bTEC) show a generally good correlation with GPS TEC for geomagnetically quiet conditions. However, the bTEC seems to be less correlated with the GPS TEC with slightly larger spreads especially during the daytime and in summer, which seems to be associated with the characteristics of the polar ionosphere such as energetic particle precipitations and large density irregularities. It is also found that the Dynasonde analysis seems to show some limitations to handle these characteristics of the polar ionosphere and needs to be improved to produce more accurate ionospheric density profiles especially during disturbed conditions.

Keywords: polar ionosphere; VIPIR; Dynasonde; Jang Bogo Station (JBS); Antarctica

Citation: Kim, E.; Jee, G.; Ham, Y.-B.; Zabotin, N.; Lee, C.; Kwon, H.-J.; Hong, J.; Kim, J.-H.; Bullett, T. Assessment of Polar Ionospheric Observations by VIPIR/Dynasonde at Jang Bogo Station, Antarctica: Part 1—Ionospheric Densities. *Remote Sens.* **2022**, *14*, 2785. https://doi.org/10.3390/rs14122785

Academic Editor: Fabio Giannattasio

Received: 6 May 2022
Accepted: 7 June 2022
Published: 10 June 2022

1. Introduction

The ionospheric density is principally governed by solar EUV radiation, but the polar ionospheric density exhibits various characteristic features due to the additional magnetospheric forcings such as electric fields and energetic particles as well as the unique geometry of nearly vertical geomagnetic field line (e.g., [1,2]). The energetic particles precipitate into the polar upper atmosphere and produce additional ionization mainly in the auroral region but also in the polar cap region. The soft electron precipitation also produces the F-region ionization in the cusp region. The polar ionospheric density is further redistributed by the plasma convection induced by the magnetospheric electric field, which transports the dayside plasma to the night side to produce the characteristic features of the polar cap ionosphere, such as the tongue of ionization (TOI) and the polar cap patch (e.g., [3,4]). The ionospheric density distributions in the polar region are closely associated with coupling processes between the ionosphere and the magnetosphere, being strongly affected by solar wind conditions, which makes it difficult to understand and requires routine monitoring of the states of the polar ionosphere. Another important factor of the polar ionosphere's dynamics is its atmospheric wave activity.

The world-wide network of ionosondes has a relatively good coverage over the globe, but it is sparse at high latitudes, particularly in the southern hemisphere. Other ground-based observations for the polar ionospheric density are also mostly located in the Arctic: for example, incoherent scatter radars (ISRs) at Poker Flat (Alaska), Resolute Bay (Canada), Sondrestrom (Greenland), Kiruna (Sweden), Tromsø and Svalbard (Norway), and Sodankylä (Finland). On the other hand, there are only a few observation sites for the ionospheric density in the southern polar region. Only one ISR has been operational in Syowa station, and a few digisondes are operational, for example, at Zhongshan station and Casey station. Recently, an ionospheric sounding system was installed at Jang Bogo Station (JBS), Antarctica, and started operating in 2017 to collect ionospheric parameters in the southern polar region. The sounding system is called the Vertical Incidence Pulsed Ionospheric Radar (VIPIR), and it utilizes the Dynasonde mode of operation and the Dynasonde analysis software to conduct echo recognitions and ionogram inversions to produce ionospheric parameters such as bottomside ionospheric electron density profiles with error bars, the F-region peak density (NmF2) and the peak height (hmF2), estimates for the ion drifts, and ionospheric tilts [5–8]. The JBS-VIPIR-Dynasonde (JVD) is distinguished from a conventional digital ionosonde, for example, the digisonde series from Lowell Digisonde International, which is one of the most widely operated digital ionospheric sounding systems around the globe (e.g., see https://www.digisonde.com/index.html, accessed on 3 March 2022). The data acquisition and analysis procedures in the Dynasonde are performed with minimized assumptions and no data pre-processing such as Fourier transform is applied to reduce the loss of precision in the phase-based physical parameters of the radio echoes such as the line-of-sight Doppler, range resolution, and angles of arrival of received signals. This approach also allows the application of sophisticated upper-level analysis techniques producing parameters of the Traveling Ionospheric Disturbances (TIDs), of kilometer-scale irregularities, and vector velocities of the isodensity ionization contours, all from a single standard ionogram mode. All of this data processing is performed autonomously and in real time. Historically, there have been a few Dynasondes in the polar regions such as at the EISCAT Tromsø and Svalbard observatories, at the IRF Kiruna and Lycksele stations in the northern polar region, and at the Halley base in Antarctica. However, those at Lycksele and at the Halley base are no longer in operation. The JVD is currently the only available Dynasonde for the ionospheric observation in the southern polar region. In order to evaluate the overall quality of the ionospheric data obtained from JVD, we compare the JVD bottomside ionospheric densities with the independent measurements of total electron content (TEC) from a Global Positioning System (GPS) receiver simultaneously operated at JBS. We fully understand that parameters measured by the two instruments are not exactly the same. However, the GPS TEC is the only available measurement related to the ionospheric density to be compared with JVD data at the moment. There have been a few comparison studies between ionosonde and GPS TEC measurements. However, these are mostly conducted at low and middle latitudes [9–11], and no comparison has been performed in the polar region. As for the other ionospheric parameter such as ion drift and irregularity, there will be a separate study on the comparison with the SuperDARN radar observations around JBS. In the following sections, we briefly introduce the observations of JVD densities, as well as GPS TEC, and the results of the comparison will be presented.

2. Data

2.1. The Observation of Ionospheric Density from JVD

The VIPIR was installed at JBS (74.62° S, 164.23° E geographic coordinates and 79.87° S geomagnetic latitude), Antarctica, in 2015, but the ionospheric data with a high temporal resolution of 2 min were not available until 2017, when the installation and subsequent test operations were complete. The Dynasonde data analysis procedure estimates the ionospheric quantities by using the echo information, including two angles of arrival, phase-based group range, line-of-sight Doppler, polarization, and amplitude. Every sounding session produces a long list (up to a few thousand) of detected radio echoes. The echo information is inverted to the height profiles of ionospheric density, of the vertical Doppler, and of the two tilt components via the NeXtYZ inversion procedure within the Dynasonde analysis software [8] to produce the ionospheric parameters. More details of the VIPIR hardware and Dynasonde software can be found in Ham et al. [7].

In principle, the ground-based ionospheric sounding technique is optimized to measure the height profiles of the bottomside ionosphere when the density monotonically increases to the F-region peak with the increase in height. Note that it is not intended for TEC measurements since no direct information of the topside ionosphere is available from the ground sounding technique. It is also important to notice that the ionospheric density profiles can be severely disturbed, especially in the polar region during magnetic storm/substorm or sporadic E. For example, when the E-region peak density (NmE) exceeds the NmF2 during auroral events, the ionosonde can theoretically observe the density only up to the E-region peak height (hmE) around 110–120 km and is unable to observe the F-region ionosphere. The ionospheric density with so-called E-layer-dominated ionosphere (ELDI) has been frequently observed in the auroral region, for example, in Tromsø and Svalbard in the winter for the solar minimum period [12]. During the ELDI events, the radar signals from the ionosonde are blocked and cannot reach above the E-layer peak heights, and the F-region density profiles can only be observed by other observation techniques such as ISRs [2,12] and GPS radio occultation from COSMIC or CHAMP satellites [13,14]. Since the JBS is mostly located in the polar cap during the nighttime, at the vicinity of the poleward boundary of the auroral oval at dawn and dusk, and in the cusp region near magnetic local noon [15], the ionospheric density profiles can be affected by geomagnetic activities. It is found that the JVD observations can be erroneous when the ionosphere is severely disturbed, as will be shown later in the paper. In order to minimize the effects of the severely disturbed ionosphere as well as the ELDI events from JVD observations, we perform the assessment study only for geomagnetically quiet times (Kp < 2), during which about 80% of the data were utilized. The ionospheric density parameters including NmF2 and bottomside TEC (bTEC) from the JVD observations are compared with the GPS TEC measurements during the solar minimum years of 2017 to 2019. The mean F10.7 index was about 72 solar flux units (sfu) for the three-year period. The JVD observations for disturbed times will be briefly discussed in Section 4.

2.2. The Observation of GPS TEC at JBS

The National Geographic Information Institute (NGII) has been operating a dual-frequency GPS receiver at JBS for geodetic survey, providing a slant TEC (STEC) along a ray path between the GPS satellites and the receiver. The STECs were derived by the Space Geodesy group at the Korea Astronomy and Space Science Institute (KASI) with a 30 s time interval using a geometry-free combination as below:

$$\mathrm{STEC_P} = \frac{1}{40.3}\frac{f_1^2 f_2^2}{f_1^2 - f_2^2}(P_2 - P_1), \tag{1}$$

$$\mathrm{STEC_L} = \frac{1}{40.3}\frac{f_1^2 f_2^2}{f_1^2 - f_2^2}(\phi_1\lambda_1 - \phi_2\lambda_2), \tag{2}$$

where f and λ are the signal frequency and wavelength, respectively, and P and ϕ are pseudorange measurements and carrier phase measurements, respectively. The unit of TEC is TECU (1 TECU = 10^{16} electrons/m^2). The $\mathrm{STEC_P}$ is absolute but noisy, while $\mathrm{STEC_L}$ is more precise but includes ambiguities. Therefore, the final STECs are derived through a phase leveling from both code and carrier-phase-based STECs within an arc of each satellite [16]. Then, the STECs can be converted into vertical TECs (VTECs) using a single layer approximation as follows:

$$\mathrm{VTEC} = \sqrt{1 - \left(\frac{R_E}{R_E + h_{ipp}}\cos(el)\right)^2} \times (\mathrm{STEC} - b_r - b_s), \quad (3)$$

where R_E and el are the Earth's radius and the elevation angle of the satellite, respectively, and h_{ipp} is the height of the ionospheric pierce point (IPP), which is set to be 350 km in altitude in this study. The b_s and b_r are the differential code biases (DCBs) of the satellites and the receiver, respectively. The satellite DCBs are provided by the Center for Orbit Determination in Europe of Astronomical Institute University of Bern, and the receiver DCBs are calculated from a single-receiver method [17].

The STEC to VTEC conversion is inherently ambiguous. As noted above, it invokes a thin-shell model of the ionosphere. This model is easy to handle, but it is very far from being realistic, as it compresses electron content from the altitude range extending up to about 20,000 km into the single point at about 350 km (the effective altitude of the shell). The unrealistic character of the thin-shell model may introduce substantial uncertainty into the VTEC estimates. Even though the phase leveling is applied for more precise TEC and both satellite and receiver DCBs are corrected, it is still challenging to estimate TEC in Antarctica. Extremely low electron density in high latitude regions can sensitize the TEC estimation because even a small amount of TEC changes in the phase-leveling process could be critical. Furthermore, strong ionospheric density irregularities causing scintillations or loss of signal lock are frequent at JBS during the December solstice [18], and they can also affect the phase-leveling accuracy, owing to the changes in ambiguities. This means that using VTEC data as a truth for assessing an ionospheric sounding data has a certain limitation.

A lack of the TEC measurements with high elevation angles greater than about 70° due to the inclination of GPS satellites causes large displacements among the IPPs within a specific time window. At the same time, there are various ionospheric density structures in the polar region, causing large density variations. For example, the density levels of TOI or the polar cap patch in the polar ionosphere can be at least two times greater than the ambient electron density [19]. Hence, the resulting large density gradients with large displacements of IPPs in the GPS TEC measurements should be considered in the comparison with the JVD densities. Even though a cut-off elevation angle of 15°~30° is typically applied to avoid multipath effects [20,21], it may not be appropriate for representing the VTECs over the JBS. Therefore, we used the larger elevation cut-off angle of 50° to increase the accuracy of the averaged VTECs at the zenith of JBS during the times when the JVD ionospheric density measurements were available. Figure 1a shows the spatial distribution of IPPs around the JBS for the GPS TEC measurements with the elevation angles greater than 50° for a day (DOY 291, 2018) as an example. Note that the spatial distributions of IPPs are mostly located at the lower latitudes of the JBS, considering the 55° inclination of GPS satellites. They are mostly located within less than 2.5° around the JBS (i.e., within about 250 km of the JBS). This spatial difference from the JBS may still be large and may not be negligible for the comparison between JVD electron densities and GPS TEC measurements, so it needs to be considered to interpret the results of the comparison. There are about 1~4 GPS satellites located around the JBS for a 10 min time interval, during which the JVD densities and GPS TEC measurements are averaged for the comparison. Figure 1b shows the traces of the GPS satellite paths for 10 min in the azimuth-elevation coordinate.

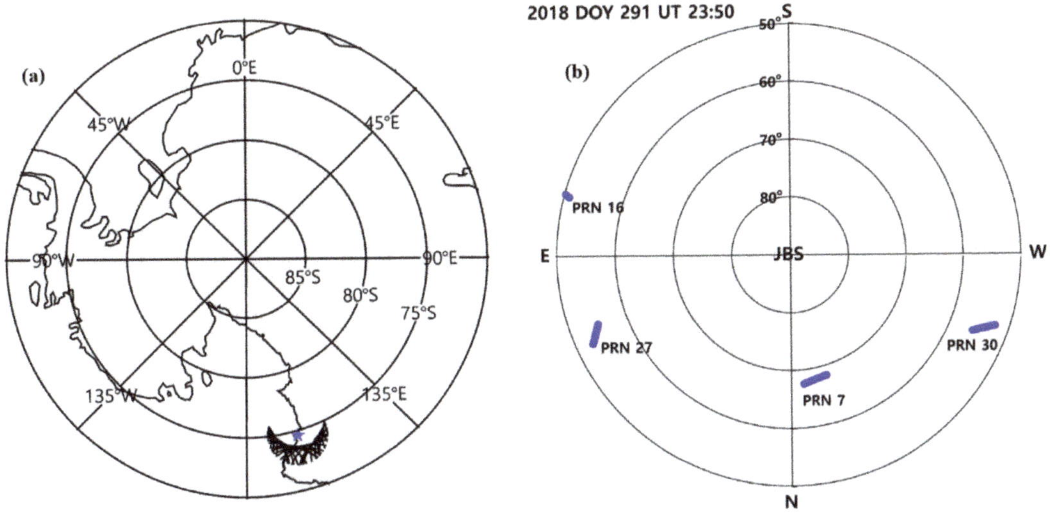

Figure 1. Spatial distribution of the ionospheric pierce points (IPPs) (black dots) for the TEC measurements over the JBS (blue star) from the GPS satellites with the elevation angles greater than 50° for DOY 291, 2018 (**a**) and four GPS satellite paths for 10 min from 23:50 to 23:59 UT in azimuth-elevation coordinate (**b**) over JBS.

3. Comparisons between JVD-Observed Densities and GPS TEC Measurements

Both JVD and GPS measurements are valuable additions to the sparse ionospheric observations in the southern polar region. Bearing in mind all limitations of such approach, we initially compared the JVD NmF2 and bTEC with VTEC measurements from a co-located GPS receiver at JBS for geomagnetically quiet times during solar minimum years. Figure 2 shows the scatter plot of 10 min averaged GPS TEC vs. JVD NmF2 (left) and bTEC (right) from 2017 to 2019 from the top to bottom panels. The linear Pearson correlation coefficients for each case are shown at the upper-right corner of each panel. Note that both NmF2 and bTEC are supposed to be somewhat correlated to the GPS TEC, considering how much the F-region peak density contributes to the GPS TEC (e.g., [9,22]). It was found in Figure 2 that the JVD NmF2 is highly correlated with GPS TEC, but the JVD bTEC shows slightly lower correlations, which seems to imply that the JVD measurement of the F-region peak density is more accurate than the measurement of bTEC calculated from the density profile of the bottomside ionosphere. However, we should remember that bTEC has the more complex nature compared to NmF2: it is computed from the density profiles that are estimated by the NeXtYZ inversion procedure using not only the observed information but also the physics-based and empirical models for the daytime D-region (or nighttime E-region) ionization and for the E-F valley region [8], which may deviate from the true ionosphere. The bottomside TEC involves an effective thickness H_b of the bottomside ionosphere (bTEC = NmF2 × H_b), and the VTEC involves an effective thickness H_t of the entire ionosphere (VTEC = NmF2 × H_t). These two TECs may not correlate well with each other, especially in the polar region. Nonetheless, the correlations between the two independent measurements are fairly strong.

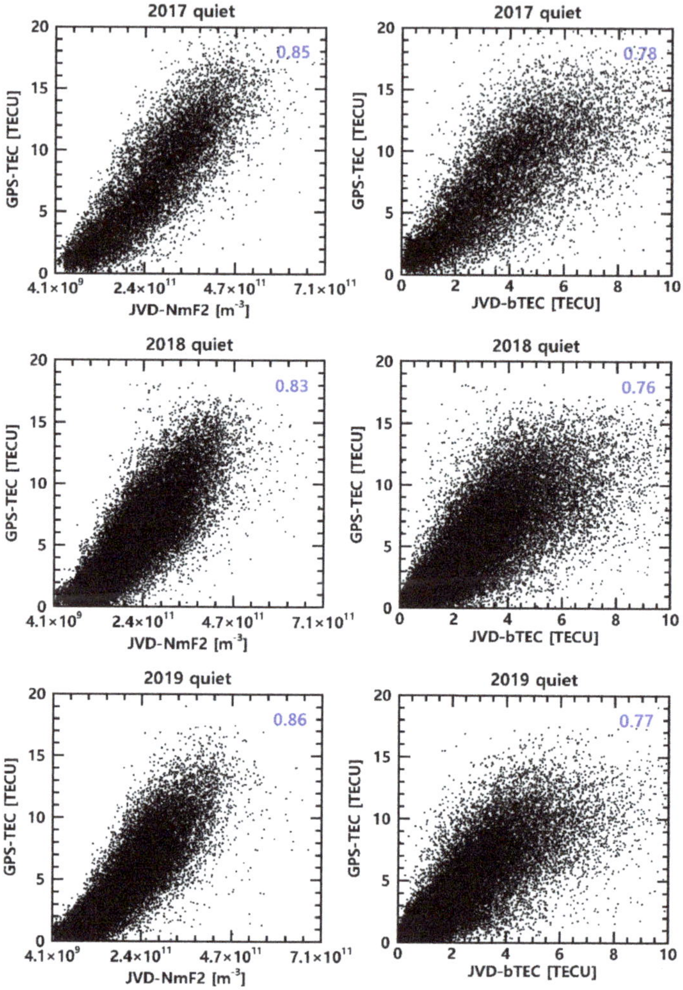

Figure 2. The scatter plot of JVD NmF2 (**left**) and bottom bTEC (**right**) vs. GPS TEC averaged for 10 min during quiet time from 2017 (top) to 2019 (bottom). The linear correlation coefficients for each case are shown at the upper right-corner of the panels.

The next comparisons with the GPS TEC measurements were performed for the international quiet days (IQDs) during the study period. Figure 3 shows the diurnal variations in 10 min. averaged JVD bTEC and NmF2 and GPS TEC for eight IQD cases (mean Kp~0.4). We chose the cases when the geomagnetically quiet condition persists for at least three consecutive days from the lists of the five quietest days for each month provided by the World Data Center for Geomagnetism, Kyoto. The diurnal variations in JVD NmF2 very closely follow the variations in GPS TEC, although the IPPs of the GPS satellites mostly exist at lower latitudes. The JVD bTECs also show similar diurnal variations from the JVD NmF2 and GPS TEC but with a somewhat larger spread. The larger spread in the JVD measurements may be associated with the ionospheric structures causing spread F on the ionogram. Since the spread F ionograms show multiple refractive scatterings, it may complicate the analysis of the ionogram data [23]. The TID activity caused by atmospheric gravity waves is the most common mechanism of the spread F ionogram. Shimazaki [24] reported that the spread F can also be caused by charged particles precipitating into the

F-region ionosphere at high latitudes with lower energy than auroral particles. According to their study, the high-latitude spread F appears even in the sunlit conditions, while it mainly occurs during nighttime at low and middle latitudes. When the spread echoes appear on the ionogram, it may be a challenge for automatic scaling to produce the realistic ionospheric density profiles. In addition, the ionospheric irregularities have been reported to frequently occur especially in summer in the southern polar ionosphere [18]. Figure 3 also shows a tendency of a larger spread (particularly in bTEC) during the daytime in the summer season (see Figure 3f). As will be discussed in the next section, the Dynasonde analysis procedure seems to show some limitations to estimate the ionospheric density profiles, in particular when the characteristic features of the polar ionosphere exist: for example, energetic particle precipitations and ionospheric density irregularities.

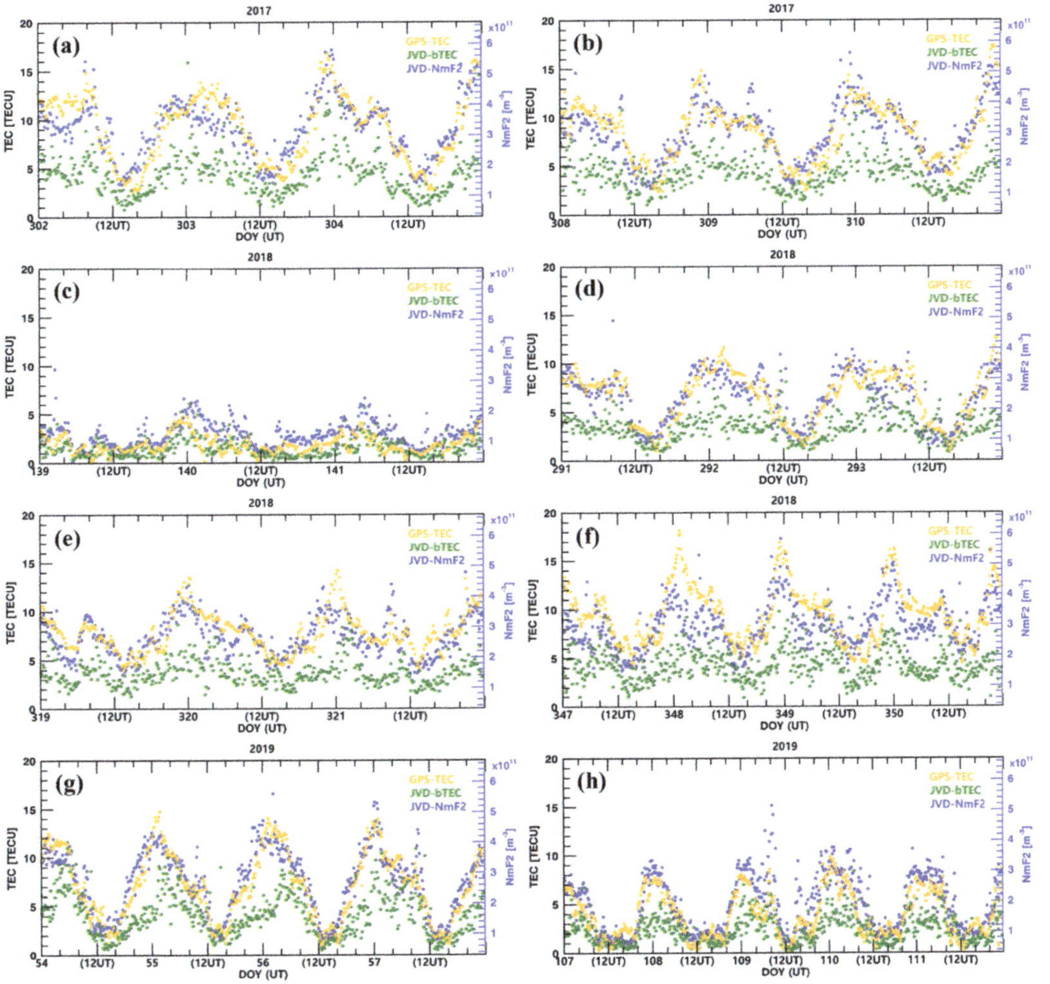

Figure 3. The diurnal variations of 10 min averaged JVD bTEC (green), NmF2 (blue), and GPS TEC (yellow) for eight IQDs. Note that LT = UT + 11 at JBS. The eight IQD cases include DOY 302–304, 2017 (**a**); DOY 308–310, 2017 (**b**); DOY 139–141, 2018 (**c**); DOY 291–293, 2018 (**d**); DOY 319–321, 2018 (**e**); DOY 347–350, 2018 (**f**); DOY 54–57, 2019 (**g**); DOY 107–111, 2019 (**h**).

When the solar-production is negligibly small near the polar winter (see Figure 3c), the magnitudes of JVD bTECs are very close to the GPS VTECs for DOY 139–141 (early winter), 2018. In summer season, however, the differences between the two TECs become large, especially during daytime (see Figure 3f). Typically, the bottomside ionospheric TEC is known to contribute to GPS TEC by about 10~40% in the low and middle latitude ionospheres [10]. Figure 4 shows the annual variations in the ratios of 10 min averaged JVD bTECs to GPS VTECs for quiet conditions during the study period of 2017 to 2019. Each pixel in the figure indicates the total amount of data for three years within a bin of a day and a 0.1 ratio interval. The ratios are mostly less than about 0.5 in the austral summer season, peaking at about 0.35, but they tend to be slightly enhanced up to about 0.7 in the austral winter season. The daily medians, as depicted by the yellow line, are the average values per each DOY for the three-year period. However, the ratios are sometimes greater than 1, indicating that the JVD bTECs occasionally exceed the GPS VTECs, particularly in the winter season. Note that the enhanced contribution of bTEC to GPS TEC may be related with the E-region density enhancements by stronger energetic particle precipitations in winter season [25–27]. Moreover, the solar production mainly responsible for the F-region density is nearly absent in polar winter, which reduces the contribution of the F-region density to GPS TEC. This aspect of the ratio is a unique characteristic of the polar ionosphere. Nonetheless, occasions of the large ratio (greater than 1), while statistically insignificant, clearly imply that there might be some quality issues in measurements from both GPS receiver and JVD.

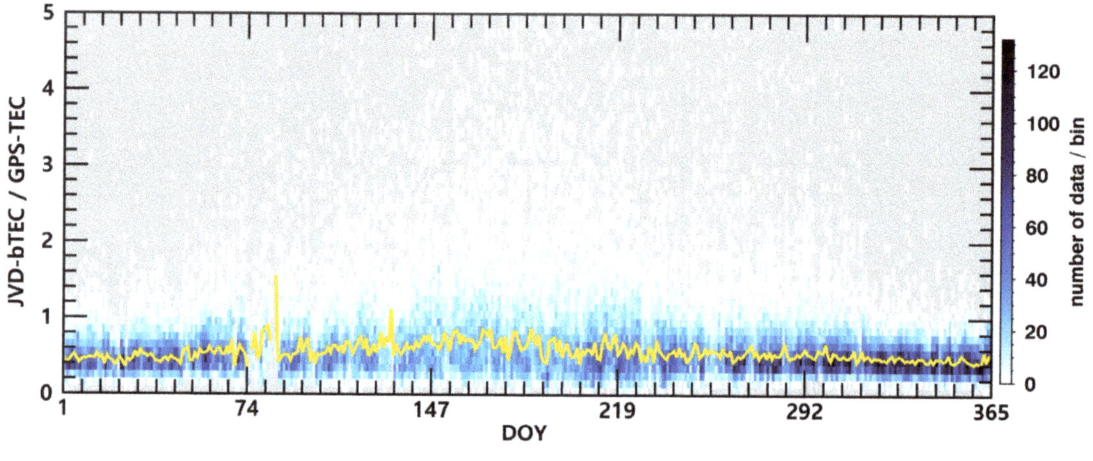

Figure 4. The annual variation in the ratios of JVD bTECs to GPS VTECs averaged for 10 min during the period of 2017 to 2019. The yellow line indicates daily median values, and the bluish colors indicate the numbers of data in each bin.

Figure 5 shows the diurnal variations in the hourly mean GPS VTEC (yellow) and JVD bTEC (green) for each month from 10 min averaged VTEC measurements with standard deviations (error bars) during quiet times for the period from 2017 (top) to 2019 (bottom). The GPS TEC data were not available in April and May in 2017. Note that LT = UT + 11 at JBS. The mean values of the bottomside ionospheric TEC are mostly well below the GPS TEC measurements. However, the differences between GPS TEC and JVD bTEC become smaller at night and especially in winter when the solar production is nearly absent but the additional production by auroral precipitation exists in the polar nighttime E-region ionosphere. As will be discussed later in the next section, however, the JVD seems to have some issues with regard to the estimation of the ionospheric density profiles during auroral events, which may be the reason for slightly larger standard deviations in JVD bTECs. Nonetheless, the climatological characteristics of JVD bTEC are generally consistent with the GPS TEC measurements in the polar region: the ionospheric densities are greater during daytime than nighttime and in summer than in winter during solar minimum years. Both measurements also show the variations with the solar activity, decreasing toward the solar minimum from 2017 to 2019.

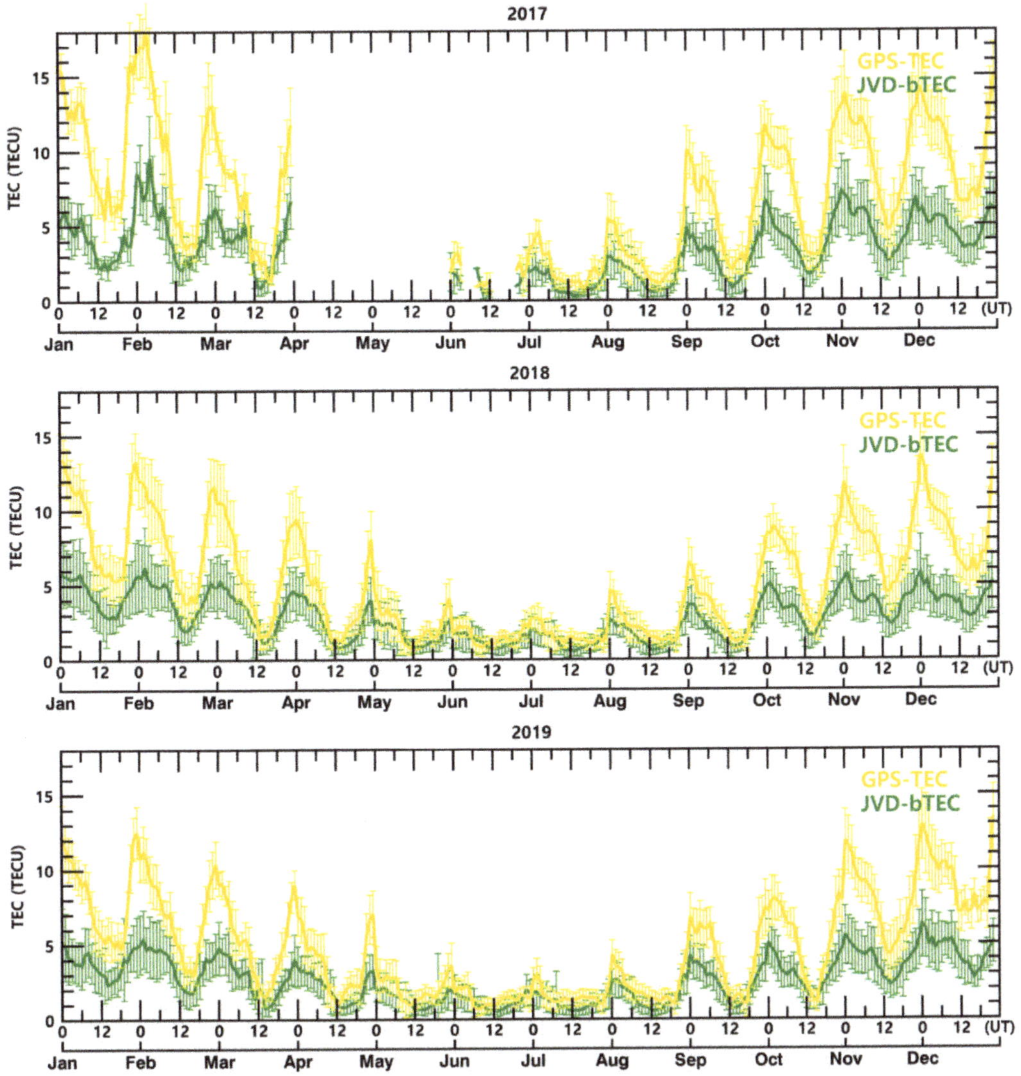

Figure 5. The diurnal variations in mean GPS TEC (yellow) and JVD bTEC (green) from 10 min-averaged values for each month are presented for a three-year period. Note that LT = UT + 11 at JBS.

4. Common Type of Misestimated Ionospheric Density Profiles from the Dynasonde

One of the characteristic features of the polar ionosphere is that the E-region peak density can be equivalent or even greater than the F-region peak density, due to the additional production by energetic particle precipitations. The so-called E-layer dominated ionosphere (ELDI) frequently occurs at high latitude in winter for solar minimum and geomagnetically disturbed times when the F-region density is minimized due to the reduced solar production, but the E-region density is increased by the enhanced energetic particle precipitations (e.g., [2,12]). Since the ionospheric sounding technique cannot observe the ionosphere above the density peak height whether it occurs in the F-region or in the E-region, the observed ionospheric densities from the JVD should be carefully examined, in particular when the energetic particle precipitations exist. During disturbed times, it is well known that the signals can be blocked by enhanced E-region density (e.g., ELDI), attenuated by increased D-region densities (e.g., polar cap absorption), and experience scintillations by density irregularities [18,28,29]. When these happen, the measurements from the ionospheric sounding must be affected by them, and the resulting ionospheric density profiles may not represent the state of the ionosphere well. Figure 6 shows an example of the ionogram produced by JVD at around 21.5 LT on 6 May 2018 (F10.7 = 68.4 sfu, Kp = 3), at which there was an auroral event over JBS. The ionogram and resulting density profile in Figure 6 imply that the echoes are reflected by the enhanced E-region density caused by auroral particle precipitation at around 100–120 km altitude: that is, the F-region could not be observed by JVD since the F-region density is probably smaller than the E-region density, which is indicated by there being no reflected signal above the E-region.

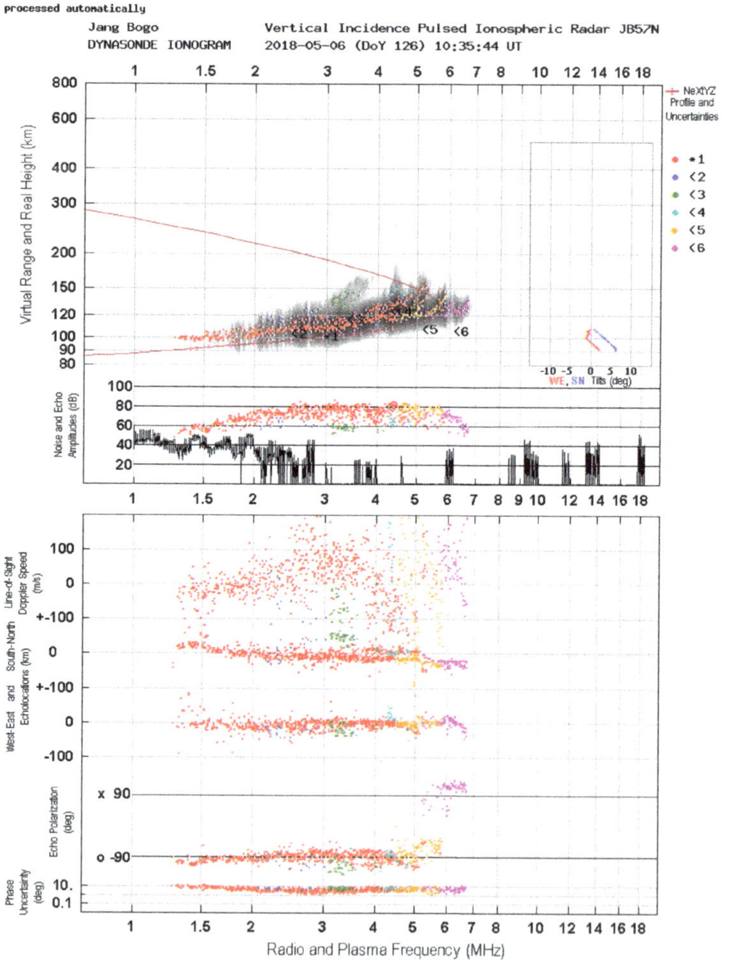

Figure 6. An example of the ionogram and resulting density profile obtained by JVD in winter at night on 6 May 2018 when the F10.7 was 68.4 sfu and the Kp index was 3.

Note that this is an example of absolutely correct processing of available ionogram data. The Dynasonde analysis detected 998 radio echoes, calculated their physical parameters and classified them into 6 traces (5 of which were ordinary polarization, reflecting complex structure of the disturbed E region, and 1, #6, was extraordinary). The autonomous analysis has chosen trace #1 for the profile inversion and successfully obtained the E-region profile with reasonable error bars (The topside part of this profile is not an actual inversion, it is just a Chapman-model-based extrapolation, which, in this case, has little to do with the real upper ionosphere).

The described procedure is completely based on objective physical parameters contained in the list of detected radio echoes. Note that it does not use the poorly defined notion of "the leading edge" of the ionospheric reflections. Most of the time, it works very well. Sometimes, however, when the ionospheric structure is particularly complex, the autonomous analysis makes mistakes in trace selection. Figure 7 shows an example of the erroneous density profile estimated at around 03 LT on 15 August 2018 with an auroral event over JBS during low solar and moderately disturbed geomagnetic conditions (F10.7 = 70.6 sfu, Kp = 3). The autonomous analysis software successfully identified 4130 radio echoes and classified them into 22 traces. It should have selected traces #1 and #15 for further analysis. However, wrong traces #4, #7, and #14 were chosen instead of the trace #1 for the profile inversion and this resulted in an unusually thin density profile peaking around 250 km in height, which does not represent the actual ionospheric density profile. This erroneous density profile belongs to the type of the misestimated density profiles by Dynasonde analysis, mostly occurring during the auroral events. This example indicates that the trace selection within the autonomous Dynasonde analysis software does not work dependably when the aurora occurs. This is not to say that such problems are unique to JVD, or to Dynasondes in general. This kind of misestimation has been observed at other high-latitude Dynasonde locations and with other amplitude-based ionospheric sounding techniques. The phase-based approach, with multiple physical parameters of the radio echoes readily available for the analysis, definitely has more diverse tools for improving results of this sort, and this must be one of the directions of future work. The current version of the software includes the expert reprocessing capability when an experienced operator is able to correct the trace selection mistakes.

During such disturbed conditions as auroral events in the polar ionosphere, the ionospheric density profiles often deviate from typical mid-latitude ionospheric density profiles with an F-region peak at around 300 km and a smaller E-region density peak at around 120 km or no E-region at night. The disturbed density profiles in the polar ionosphere may still be a challenge in the ionospheric sounding techniques such as the Dynasonde, as well as the more conventional digisonde. In conclusion, the electron density profiles estimated by the Dynasonde are mostly in reasonable agreement with the independent GPS TEC measurements during undisturbed conditions, but caution is required when using some of the analysis products during disturbed conditions in the polar ionosphere.

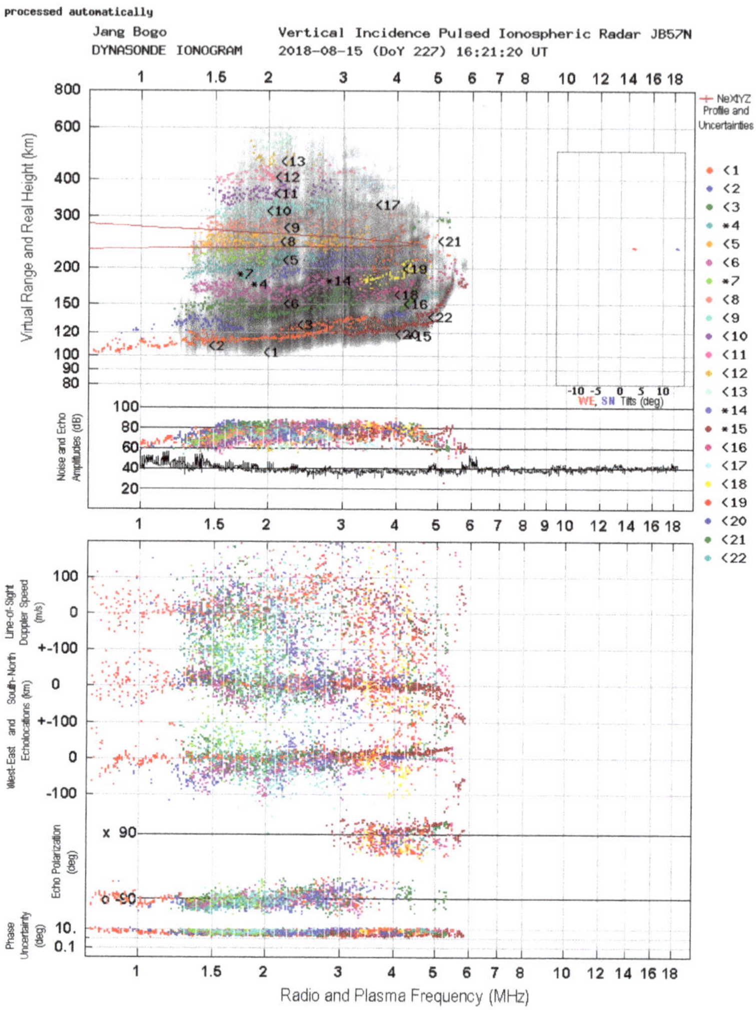

Figure 7. An ionogram and the resulting density profile obtained by autonomous Dynasonde analysis on 15 August 2018 (F10.7 = 70.6 sfu, Kp = 3).

5. Summary and Conclusions

The ionospheric densities obtained from the JVD are assessed using GPS VTEC data, the only available independent measurement of the ionospheric densities at JBS, Antarctica. We fully understand that the two instruments do not exactly measure the same parameters. Moreover, the VTEC measurements at high latitudes are very difficult by themselves, and their results cannot be considered as the ground truth. This study was performed mostly for geomagnetically quiet times during solar minimum years from 2017 to 2019. The JVD NmF2 is well correlated with GPS VTEC measurements but the JVD bTEC shows slightly less correlations. The more detailed comparisons between JVD densities and GPS VTEC have been performed for international quiet days. Both JVD NmF2 and bTEC are generally in a good agreement with GPS TEC but the JVD bTECs show relatively larger spread, which may be associated with the diverse characteristic features of the polar ionosphere such as energetic particle precipitations and large density irregularities. Those same features create additional difficulties for estimates of proper density profiles. The median ratios of JVD bTEC to GPS VTEC are about 0.5 on average but tend to be larger in winter. It indicates that the solar production is nearly absent in polar winter, but the effect of particle precipitation enhances the contribution of bottomside ionosphere to the total electron content in the polar region. Finally, it was found that the autonomous Dynasonde estimation of the ionospheric density profiles seems to have issues with trace selection during disturbed conditions, which need to be addressed in future versions of the Dynasonde analysis software.

Author Contributions: Conceptualization, G.J. and E.K.; methodology, E.K. and G.J.; software, E.K.; validation, G.J. and N.Z.; formal analysis, E.K.; investigation, E.K., G.J., N.Z., Y.-B.H., C.L. and H.-J.K.; resources, Y.-B.H., C.L., J.H., H.-J.K., J.-H.K., N.Z. and T.B.; data curation, E.K., Y.-B.H., C.L., J.H. and H.-J.K.; writing—original draft preparation, E.K.; writing—review and editing, E.K. and G.J.; visualization, E.K.; supervision, G.J. All authors have read and agreed to the published version of the manuscript.

Funding: This research was funded by the grant PE22020 from Korea Polar Research Institute (KOPRI).

Acknowledgments: Junseok Hong was supported by a basic research fund from the Korea Astronomy and Space Science Institute (KASI) (2022185010).

Conflicts of Interest: The authors declare no conflict of interest.

References

1. Heelis, R.A.; Lowell, J.K.; Spiro, R.W. A model of the high latitude ionosphere convection pattern. *J. Geophys. Res.* **1982**, *82*, 6339–6345. [CrossRef]
2. Kim, E.; Jee, G.; Ji, E.-Y.; Kim, Y.H.; Lee, C.; Kwak, Y.-S.; Shim, J.-S. Climatology of polar ionospheric density profiles in comparison with mid-latitude ionosphere from long-term observations of incoherent scatter radars: A review. *J. Atmos. Solar-Terr. Phys.* **2020**, *211*, 105449. [CrossRef]
3. Sojka, J.J.; Bowline, M.D.; Schunk, R.W. Patches in the polar ionosphere: UT and seasonal dependence. *J. Geophys. Res.* **1994**, *99*, 14959–14970. [CrossRef]
4. David, M.; Sojka, J.J.; Schunk, R.W.; Coster, A.J. Polar cap patches and the tongue of ionization: A survey of GPS TEC maps from 2009 to 2015. *Geophys. Res. Lett.* **2016**, *43*, 2422–2428. [CrossRef]
5. Rietveld, M.T.; Wright, J.W.; Zabotin, N.; Pitteway, M.L.V. The Tromsø dynasonde. *Polar Sci.* **2008**, *2*, 55–71. [CrossRef]
6. Kwon, H.-J.; Lee, C.; Jee, G.; Ham, Y.-B.; Kim, J.-H.; Kim, Y.H.; Kim, K.-H.; Wu, Q.; Bullett, T.; Oh, S.; et al. Ground-based observations of the polar region space environment at the Jang Bogo Station, Antarctica. *J. Astron. Space Sci.* **2018**, *35*, 185–193. [CrossRef]
7. Ham, Y.-B.; Jee, G.; Lee, C.; Kwon, H.-J.; Kim, J.-H.; Zabotin, N.; Bullett, T. Observations of the polar ionosphere by the Vertical Incidence Pulsed Ionospheric Radar at Jang Bogo station, Antarctica. *J. Astron. Space Sci.* **2020**, *37*, 143–156. [CrossRef]
8. Zabotin, N.A.; Wright, J.W.; Zhbankov, G.A. NeXtYZ: Three-dimensional electron density inversion for Dynasonde ionograms. *Radio Sci.* **2006**, *41*, 1–12. [CrossRef]
9. Belehaki, A.; Jakowski, N.; Reinisch, B.W. Comparison of ionospheric ionization measurements over Athens using ground ionosonde and GPS-derived TEC values. *Radio Sci.* **2003**, *38*, 1105. [CrossRef]
10. McKinnell, L.-A.; Opperman, B.; Cilliers, P.J. GPS TEC and ionosonde TEC over Grahamstown, South Africa: First comparisons. *Adv. Space Res.* **2007**, *39*, 816–820. [CrossRef]
11. Zhu, Q.; Lei, J.; Luan, X.; Dou, X. Contribution of the topside and bottomside ionosphere to the total electron content during two strong geomagnetic storms. *J. Geophys. Res.* **2016**, *121*, 2475–2488. [CrossRef]
12. Cai, H.; Li, F.; Shen, G.; Zhan, W.; Zhou, K.; McCrea, I.W.; Ma, S. E layer dominated ionosphere observed by EISCAT/ESR radars during solar minimum. *Ann. Geophys.* **2014**, *32*, 1223–1231. [CrossRef]
13. Mayer, C.; Jakowski, N. Enhanced E-layer ionization in the auroral zones observed by radio occultation measurements onboard CHAMP and Formosat-3/COSMIC. *Ann. Geophys.* **2009**, *27*, 1207–1212. [CrossRef]
14. Kamal, S.; Jakowski, N.; Hoque, M.M.; Wickert, J. Evaluation of E layer dominated ionosphere events using COSMIC/FORMOSAT-3 and CHAMP ionospheric radio occultation data. *Remote Sens.* **2020**, *12*, 333. [CrossRef]
15. Jee, G.; Ham, Y.-B.; Choi, Y.; Kim, E.; Lee, C.; Kwon, H.; Trondsen, T.S.; Kim, J.E.; Kim, J.-H. Observations of the aurora by visible all-sky camera at Jang Bogo Station, Antarctica. *J. Astron. Space Sci.* **2021**, *38*, 203–215. [CrossRef]
16. Rungraengwajiake, S.; Supnithi, P.; Saito, S.; Siansawasdi, N.; Saekow, A. Ionospheric delay gradient monitoring for GBAS by GPS stations near Suvarnabhumi airport, Thailand. *Radio Sci.* **2015**, *50*, 1076–1085. [CrossRef]
17. Arikan, F.; Nayir, H.; Sezen, U.; Arikan, O. Estimation of single station interfrequency receiver bias using GPS-TEC. *Radio Sci.* **2008**, *43*, 1–13. [CrossRef]
18. Hong, J.; Chung, J.; Kim, Y.H.; Park, J.; Kwon, H.; Kim, J. Characteristics of Ionospheric Irregularities Using GNSS Scintillation Indices Measured at Jang Bogo Station, Antarctica (74.62°S, 164.22°E). *Space Weather* **2020**, *18*, e2020SW002536. [CrossRef]

19. Crowley, G.; Ridley, A.J.; Deist, D.; Wing, S.; Knipp, D.J.; Emery, B.A. The transformation of high-latitude ionospheric F- region patches into blobs during the March 21, 1990 storm. *J. Geophy. Res.* **2000**, *105*, 5215–5230. [CrossRef]
20. Otsuka, Y.; Ogawa, T.; Saito, A.; Tsugawa, T.; Fukao, S. A new technique for mapping of total electron content using GPS network in Japan. *Earth Planets Space* **2002**, *54*, 63–70. [CrossRef]
21. Panda, S.K.; Gedam, S. Robustness of Elevation Cut-off in Estimating Ionospheric Total Electron Content from GPS Observation Data. In Proceedings of the 8th International Conference on Microwaves, Antenna, Propagation & Remote Sensing (ICMARS-2012), Jodhpur, India, 11–15 December 2012.
22. Hoque, M.M.; Jakowski, N.; Prol, F.S. A new climatological electron density model for supporting space weather services. *J. Space Weather Space Climate* **2022**, *12*, 1. [CrossRef]
23. Scotto, C.; Pezzopane, M. Automatic scaling of polar ionograms. *Antarctic Sci.* **2011**, *24*, 88–94. [CrossRef]
24. Shimazaki, T. A statistical study of occurrence probability of spread F at high latitudes. *J. Geophys. Res.* **1962**, *67*, 4617–4634. [CrossRef]
25. Newell, P.T.; Meng, C.I.; Lyons, K.M. Suppression of discrete aurorae by sunlight. *Nature* **1996**, *381*, 766–767. [CrossRef]
26. Borovsky, J.E. Still in the dark. *Nature* **1998**, *393*, 312–313. [CrossRef]
27. Barth, C.A.; Baker, D.N.; Bailey, S.M. Seasonal variation of auroral electron precipitation. *Geophys. Res. Lett.* **2004**, *31*, L04809. [CrossRef]
28. Bates, H.F. HF propagation through the auroral curtain. *J. Geophys. Res.* **1970**, *75*, 143–151. [CrossRef]
29. Kintner, P.M.; Ledvina, B.M.; Paula, E.R. GPS and ionospheric scintillations: Gps and ionospheric scintillations. *Space Weather* **2007**, *5*. [CrossRef]

remote sensing

Article

Ionospheric Behavior during the 10 June 2021 Annular Solar Eclipse and Its Impact on GNSS Precise Point Positioning

Juan Carlos Valdés-Abreu [1,2,†], Marcos A. Díaz [1,2,*,†], Manuel Bravo [3,4,†], Juan Carlos Báez [5,†] and Yohadne Stable-Sánchez [1,†]

1 Space and Planetary Exploration Laboratory (SPEL), Faculty of Physical and Mathematical Sciences, University of Chile, Av. Tupper 2007, Santiago 8370451, Chile; juanvaldes@ug.uchile.cl (J.C.V.-A.); spel@ing.uchile.cl (Y.S.-S.)
2 Department of Electrical Engineering, Faculty of Physical and Mathematical Sciences, University of Chile, Av. Tupper 2007, Santiago 8370451, Chile
3 Centro de Instrumentación Científica, Universidad Adventista de Chile, km 12 Camino a Tanilvoro, Chillan 3780000, Chile; manuelbravo@unach.cl
4 Department of Geophysics, Faculty of Physical and Mathematical Sciences, University of Concepcion, Av. Esteban Iturra s/n–Barrio Universitario, Concepcion 4091124, Chile
5 National Seismological Center, Faculty of Physical and Mathematical Sciences, University of Chile, Av. Beaucheff 1225, Santiago 8370583, Chile; jcbaez@csn.uchile.cl
* Correspondence: mdiazq@ing.uchile.cl
† These authors contributed equally to this work.

Abstract: The main effects of the 10 June 2021 annular solar eclipse on GNSS position estimation accuracy are presented. The analysis is based on TEC measurements made by 2337 GNSS stations around the world. TEC perturbations were obtained by comparing results 2 days prior to and after the day of the event. For the analysis, global TEC maps were created using ordinary Kriging interpolation. From TEC changes, the apparent position variation was obtained using the post-processing kinematic precise point positioning with ambiguity resolution (PPP-AR) mode. We validated the TEC measurements by contrasting them with data from the Swarm-A satellite and four digiosondes in Central/South America. The TEC maps show a noticeable TEC depletion ($<-60\%$) under the moon's shadow. Important variations of TEC were also observed in both crests of the Equatorial Ionization Anomaly (EIA) region over the Caribbean and South America. The effects on GNSS precision were perceived not only close to the area of the eclipse but also as far as the west coast of South America (Chile) and North America (California). The number of stations with positioning errors of over 10 cm almost doubled during the event in these regions. The effects were sustained longer ($\sim$10 h) than usually assumed.

Keywords: solar eclipse; ionosphere; precise point positioning; GNSS; total electron content; rate of total electron content index; Swarm satellite measurements; ionosonde; electron density

Citation: Valdés-Abreu, J.C.; Diaz, M.A.; Bravo, M.; Báez, J.C.; Stable-Sánchez, Y. Ionospheric Behavior during the 10 June 2021 Annular Solar Eclipse and Its Impact on GNSS Precise Point Positioning. *Remote Sens.* **2022**, *14*, 3119. https://doi.org/10.3390/rs14133119

Academic Editor: Fabio Giannattasio

Received: 21 April 2022
Accepted: 13 June 2022
Published: 29 June 2022

1. Introduction

A solar eclipse is a natural phenomenon that occurs when the Moon moves in the way between the Sun and Earth, totally or partially blocking the Sun, casting a shadow over the Earth. Since the Sun is one of the major drivers of atmospheric effects, such as its ionization at high altitudes, its blocking produces several disturbances. The atmospheric effects of a solar eclipse have been the subject of extensive research, mainly in meteorological parameters, total column ozone, photochemistry, gravity waves, and ionospheric parameters [1]. Despite the large number of studies concerning eclipses, the event of a solar eclipse is still unique since it happens at different seasons, different times of the day, different locations, and under different synoptic and geomagnetic conditions [1–3]. In addition, with every new eclipse, the scientific community gains larger numbers and a variety of instruments, which allow us to revisit the proposed conclusions from previous eclipses.

The ionosphere is directly affected since this atmospheric layer is produced by solar radiation. The total electron content (TEC) is a measure of the electron density in the ionosphere integrated along the line of sight, thus, an indication of its ionization. TEC can be obtained using a radio link between a satellite and the ground. Nowadays, the most common system delivering TEC measurements is the Global Navigation Satellite Systems (GNSS), which requires TEC measurements to improve the precision of position estimation. TEC is expressed in TEC units (TECu), where 1 TECu $= 10^{16}$ e/m^2. The perturbation of the ionosphere can be analyzed through the variations of TEC. The main parameters of the TEC variations during the eclipses are the delay value (τ) relative to the maximum phase of the eclipse; its amplitude (A), which generally is a decrease; and the duration (ΔT) of the perturbation [4]. Since the Moon's shadow moves rapidly from west to east across the Earth at supersonic speed, the total eclipse lasts just a few minutes anywhere [5–7]. Previous works have reported the depletion of TEC after the onset of the partial solar eclipse and have presented values of A in percent (A[%]) that can reach up to −64% with τ from −30 to 180 min [8–11]. This delay has been interpreted as an indicator of the combined effect of the photochemical processes and plasma dynamics [1,12]. Some works have reported ΔT from 50 to 240 min [4,13]. However, some studies have reported even longer effects [14,15]. A historical summary of ionospheric responses to solar eclipses since 1920 can be found in the Appendix in Bravo et al. [16].

Recent studies have shown that the effects of a solar eclipse on the ionosphere are not only local but can affect other geographic regions outside the umbra/penumbra of the eclipse [15,17–20]. These effects may be due to transport between hemispheric magnetic conjugates, alteration of the equatorial fountain effect, generation of a disturbed dynamo, and/or Atmospheric Gravity Waves (AGWs) that generate Traveling Atmospheric Disturbances (TADs) and/or Traveling Ionospheric Disturbances (TIDs).

An annular solar eclipse took place on 10 June 2021. The first external contact (P1 time) and the last external contact (P4 time) of the solar eclipse were at 08:12:22 UT and 13:11:22 UT, respectively. The partial solar eclipse was seen from the following geographic regions: in parts of the eastern United States and northern Alaska, Canada and parts of the Caribbean, Europe, Asia, and northern Africa. The annular eclipse was visible from parts of northeastern Canada, Greenland, and the Arctic Ocean, passing through the North Pole, and ending in Russian territory. It maximum magnitude was 0.944: this is the fraction of the angular diameter of a celestial body being eclipsed. This magnitude value was reached at geographic coordinates 80.815°N, and 66.78°W, at 10:41:57 UT (Greatest Eclipse time, GE time, http://xjubier.free.fr/en/site_pages/solar_eclipses/ASE_2021_GoogleMapFull.html, last accessed on 15 June 2022). The paths at ground level and at 350 km of altitude of the annular eclipse are shown in Figure 1 (see Supplementary Materials, Video S1). Due to the specific geometry of each eclipse, the paths differ both geographically and temporally according to the height considered, which could be significant when analyzing them [21].

Solar eclipses are rare events and, particularly, the 10 June 2021 event is an excellent opportunity to study the eclipse-induced effects on the polar ionosphere. Since the ionospheric variations can perturb GNSS, the eclipse can be used to study the positioning errors in these regions. There are some studies on the effects of the ionosphere during solar eclipses over the northern polar region. One of the first reported ones was the total solar eclipse that occurred on 9 March 1997 over Kazakhstan, Mongolia, eastern Siberia, and the Arctic Ocean (τ = −26 to 180 min and A = −5 TECu) [4,22]. Another reported one is the total solar eclipse that occurred on 1 August 2008 over Canada, northern Greenland, the Arctic Ocean, central Russia, Mongolia, and China (τ = −27 to 44 min, and A[%] = −40 to −11%) [10]. The most recent one is the eclipse that occurred on 20 March 2015 that covered the North Atlantic, Faroe Islands, and Svalbard (A[%] = −50 to −10%) [5,23]. This last one happened during the recovery phase of the most intense geomagnetic storm during Solar Cycle 24, the so-called St. Patrick's Day Storm. Due to the limited availability of GNSS stations around the globe at the time of these previous studies, they were focused on a regional scale. The increasing number of accessible GNSS stations around the world allows a study on a global scale, facilitating the search of potential interactions between regions. This can show how spreadable GNSS disturbances are. In particular, the poles are of interest since several ionospheric disturbances can start from there during geomagnetic storms.

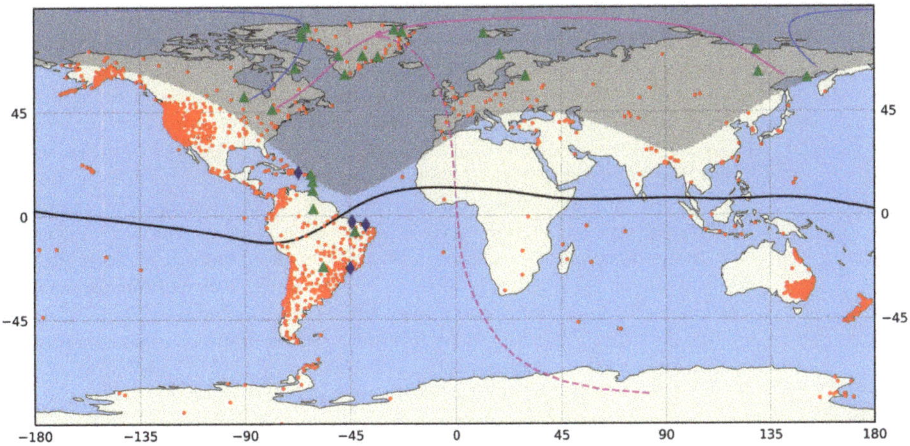

Figure 1. Instruments used in present work: 2337 GNSS stations (red dots), 4 digisondes (blue rhombuses), and 24 selected GNSS stations (green triangles). The magnetic equator (black line) and the annular eclipse path at ground level (blue line) are shown. Eclipse obscuration mask from P1–P4 time (shaded region), the annular solar eclipse path (magenta line), the maximum obscuration (magenta dot) at 350 km altitude are presented. The magenta dashed line starts from the maximum obscuration of the solar eclipse to its conjugate location in the Southern Hemisphere.

In GNSS receivers, TEC is estimated simultaneously from several satellites of the network, which serves to study the ionosphere. (e.g., [4,8,24], among many others). During the eclipses, the ionization decreases, producing a depletion in TEC. Although a decrease in electron concentration during a solar eclipse could produce an improvement in the positioning precision, it actually generates positional errors [25,26]. Few authors have analyzed the GNSS positioning errors caused by the influence of solar eclipses. The eclipses that occurred over Croatia on 11 August 1999 [27], over China on 22 July 2009 [28], and over the United States on 21 August 2017 [26] are some of the studies that analyzed GNSS positioning errors.

For the 1999 solar eclipse, Filjar, et al. [27] used a single frequency receiver located in the north of Croatia with ∼95% of maximum percentage of obscuration (MPO). In this work the authors did not relate ionospheric disturbances with positioning variations. The authors collected the horizontal positioning at the eclipse's maximum obscuration time (MOT). They calculated an average positioning error of ∼34 m on horizontal Global Positioning System (GPS) accuracy for that time. These horizontal values could be due to the use of a single-frequency, the number of receiver channels, and the possible influence of Selective Availability (until May 2000).

Jia-Chun et al. [28] used eight GPS stations to study the TEC changes and their effect on the positioning during the 22 July 2009 solar eclipse. They possessed a real-time point positioning and real-time precision of single baselines. The measurements were affected by a geomagnetic storm (Dst peak = -80 nT and Kp-index = 5^{+}), which made it difficult to separate the influence of the eclipse from the storm one.

In the case of the 2017 eclipse, Park et al. [26] computed and compared the rate of change of the TEC (ROT) with respect to the day before and the day after the eclipse; and with a time window of 3 h, from 16 UT. They determined the means of positioning errors at four GNSS stations (localized in Oregon, United States) within the path of the total solar eclipse which reached ~32 cm. However, on reference days, the means of positioning errors were between 7–14 cm. The authors used the average length of the eight baselines, which was ~270 km. On the eclipse day, the means of positioning results were −4 to 324% over the day before and the day after the eclipse. Yuan et al. [25] established the ionospheric eclipse factor method (IEFM) to model the ionospheric delay searching for the improvement of the GNSS positioning estimation. In this contest, the paper introduces the concept of the ionospheric eclipse factor method for the IPP for relatively precise separation of daytime from nighttime for the ionosphere. Although the ionospheric eclipse factor is not related to a solar eclipse as an astronomical phenomenon that occurs when the Moon obscures the Sun from Earth, this method could be used in future studies related to the impact of solar eclipses on GNSS positioning.

In our case, we obtain the apparent position variation using the post-processing kinematic precise point positioning (PPP) with ambiguity resolution (PPP-AR) mode. We chose this method because PPP demonstrates a high ability to improve position estimation. PPP is used for calculating the coordinates of a single receiver without the need for a reference station nearby as a control station. In addition, we can find some free PPP services available online [29]. PPP-AR is an enhanced version of the PPP technique that resolves the carrier phase ambiguities, improving the PPP accuracy [29–31]. Katsigianni et al. [30] recently presented a comparison between PPP and PPP-AR. In order to offer the community the possibility of evaluating our analysis, we used an online service. Thus, we selected the CSRS-PPP service for this work because it is one of the most commonly used PPP online services in the field. We also applied the common noise filter to more than 2300 GNSS stations, to correct the time series of the North, East, and Up components of the GNSS receivers, as described in [29].

The regular ionospheric effects of solar eclipses are not yet fully understood. Studies of the eclipse-induced effects on the ionosphere are important because they provide a better understanding of the processes that control the ionosphere and that can cause GNSS positioning errors. In the present paper, we present the impact of the 10 June 2021 annular solar eclipse on ionospheric variations that also cause errors in GNSS positioning. Therefore, we first analyze the ionospheric behavior at a global scale based on 2337 dual-frequency (DF) GNSS stations, Swarm-A satellite, and four ionospheric stations. We used GNSS stations distributed around the world since they will allow us to evaluate the effects beyond the northern polar region with a higher spatial resolution than ever before. Unlike previous studies about the GNSS positioning errors caused by the influence of solar eclipses, our study is focused on a global scale. This allowed us to find other locations in the world that could be affected by a perturbation in the north pole and how that perturbation propagates to those potential locations.

2. Materials and Methods

The methodology used in the 10 June 2021 annular eclipse is mainly based on the one described in Valdés-Abreu et al. [29]. However, in this work we incorporate the processing of ionospheric data from a Low Earth Orbit (LEO) satellite. The procedure of this work also includes the analysis of geophysical and geomagnetic conditions close to the date of the eclipse (10 June 2021). How we use this new set of data is detailed below.

2.1. Estimation of the Ionospheric Total Electron Content

The inherent space-time variability of the ionosphere can be observed through TEC that can be obtained using GNSS stations [32]. Then, GNSS measurements based on dual-frequency signals f_1 and f_2, were used to obtain the vertical TEC (VTEC) data. The ground-based dual-frequency GNSS (DF-GNSS) receiver continually records two types of delay: the pseudoranges and the carrier phases of the two signals. The obtained data was used to estimate the slant TEC (STEC) and to calculate the VTEC. STEC and VTEC were calculated from Receiver Independent Exchange Format (RINEX) files by using the GPS-TEC analysis software (GPS-TEC program Ver 2.9.5, developed by Dr. Gopi Seemala, https://seemala.

blogspot.com/2017/09/gps-tec-program-ver-295.html, last accessed on 17 April 2022) [24]. VTEC values were estimated with a satellite cut-off elevation angle of 30° at an altitude of 350 km to reduce possible errors. The TEC values were released every 30 s and were corrected for the satellite and receiver bias using the data obtained from the AIUB Data Center of Bern University in Switzerland (ftp://ftp.aiub.unibe.ch/CODE/, last accessed on 17 January 2022).

For the final selection of the RINEX files of each GNSS station, we took into account several aspects. First, we considered the quality of the files during the selected period of days (DoYs 159–163). Second, we verified that there were no errors or data-gap after TEC estimation and post-processing PPP-AR. This aspect is critical to relating TEC and/or ROTI with positioning variations. We used data from all available stations in the polar regions. We also tried to cover regions such as Africa, Australia, and Asia. The International GNSS Service (IGS) stations (http://www.igs.org, last accessed on 17 January 2022) [33]; the Chilean network of GNSS receivers operated by the National Seismological Center at University of Chile (CSN in Spanish); University NAVSTAR Consortium (UNAVCO); the Argentine Continuous Satellite Monitoring Network (RAMSAC in Spanish) [34]; the Brazilian Network for Continuous Monitoring of the Institute of Brazilian Geography and Statistics (IBGE in Portuguese); the Geoscience Australia; the Low-Latitude Ionospheric Sensor Network (LISN, http://lisn.igp.gob.pe/, accessed on 26 July 2021); and the African Geodetic Reference Frame (AFREF) provided RINEX files of 2337 GNSS stations that met the requirements we imposed (see Figure 1).

Additionally, the differential VTEC (DVTEC) in TECu and the percentage changes of DVTEC (DVTEC[%]) were used. These parameters are studied in the analysis of ionospheric irregularities, defined as the relative variation of VTEC, epoch by epoch, with respect to the mean value (in time) of $\overline{VTEC}$ as shown in Equations (1) and (2) [35].

$$DVTEC_t = VTEC_t - \overline{VTEC_t} \tag{1}$$

$$DVTEC[\%]_t = \frac{DVTEC_t}{\overline{VTEC_t}} \cdot 100 \tag{2}$$

where t represents the epoch, and $\overline{VTEC_t}$ is calculated by averaging the values of VTEC at the same time of the day, t, for the reference DoYs 159, 160, 162 and 163 which correspond to 2 days before and 2 days after the day of the eclipse (DoY 161).

According to the methodology [29], we used the ordinary Kriging interpolation method to produce the TEC maps at each ionospheric pierce point (IPP). With this method, we filled in the spatial gaps of the global ionosphere TEC maps, minimizing the effects of the inhomogeneous distribution of GNSS receivers. Before interpolating, we selected a spatial resolution of 2.5° × 2.5°. Then, we employed the Kriging package implemented in Python (https://github.com/ERSSLE/ordinary_kriging, last accessed on 17 January 2022).

2.2. ROT and ROTI

In order to detect, investigate and characterize the occurrence of ionospheric irregularities, we have used the Rate of change of the TEC Index (ROTI). The ROT and ROTI values are usually expressed in TECu/min. ROTI is defined as the standard deviation of the rate of TEC (ROT), and it is estimated by dual-frequency GNSS data with the time interval of 5 min by using Equation (3) [29]:

$$ROTI = \sqrt{\langle ROT^2 \rangle - \langle ROT \rangle^2} \tag{3}$$

where $\langle \cdot \rangle$ represents the temporal average. ROT is defined as the TEC variation rate of two successive epochs as stated in Equation (4) [29]:

$$ROT = \frac{STEC_t^i - STEC_{t-1}^i}{k_t - k_{t-1}} \tag{4}$$

where i indicates the observed satellite and t denotes the time of epoch. Hence, $k_t - k_{t-1}$ is the time interval between the subsequent epochs.

Depending on the ROTI value, the activity level can be classified in ranges such as: weak (if $0.25 \leqslant$ ROTI < 0.5); moderate (if $0.5 \leqslant$ ROTI < 1); and strong (if ROTI $\geqslant 1$), according to Liu et al. [36].

2.3. *Low Earth Orbit Satellite Measurements and Ionospheric Data*

Additionally, we analyze ionospheric measurements provided by a LEO satellite, the European Space Agency's Swarm mission. This mission is a constellation of three LEO satellites that were successfully launched on 22 November 2013, and are still operating. This constellation is designed to provide measurements of the Earth's magnetosphere and ionosphere, studying the impact of the solar wind on the dynamics of the upper atmosphere [37,38]. The Swarm-Alpha (A), Bravo (B), and Charlie (C) are three identical satellites that share the same design and payloads.

All three satellites were put into a circular near-polar orbit with a low eccentricity. Swarm-A/C pair have the same orbit configuration (inclination of 87.35°, altitude of ~450 km, east–west separation of about 1–1.5° in longitude), while Swarm-B has a different one (inclination of 87.75°, altitude of ~510 km). These satellites fly above the F-layer peak (the peak altitude of the ionospheric electron density). In addition, Swarm-A/C fly in tandem, while Swarm-B moves away from the couple Swarm-A/C by covering different local times [37,38].

The Swarm spacecraft were equipped with different payloads, including GPS receivers and Langmuir Probes (LP), among others. We considered the ionospheric VTEC values associated with the point where the link path between GPS and Swarm-A satellite pierces the spherical thin shell located 400 km above the Swarm-A orbit. We also used in-situ electron density (N_e) measurements by LP at ~450 km [38,39] for each of the five selected DoYs in June 2021 (https://Earth.esa.int/web/guest/swarm/data-access, last accessed on 17 January 2022).

We used the Swarm Level 2 (L2) TEC (TECxTMS_2F) data product, which contains time series of slant and vertical (absolute and relative) TEC for each GPS satellite in view (at most eight due to instrumentation design). The cadence of the ionospheric TEC data is 1 Hz since it was changed from 10 s (0.1 Hz) to 1 s (1 Hz) on 14 July 2014 [38,40].

We also used the Swarm LP data, which is part of the EFI package (EFIX_LP_1B plasma data). LP provides measurements of in situ N_e and electron temperature with a 2 Hz sampling rate [38,39].

2.4. *Apparent Position Variation Using Kinematic Precise Point Positioning*

The RINEX files of 2337 GNSS stations were processed using the Canadian Spatial Reference System (CSRS-PPP) online service (https://webapp.geod.nrcan.gc.ca/geod/tools-outils/ppp.php, last accessed on 6 January 2022) [41] with ambiguity resolution (PPP-AR) mode. The CSRS-PPP provides centimeter-level estimations with converged float solutions [41–44].

Usually, the CSRS-PPP report can provide a different reference start value for different days. To facilitate the evaluation of the apparent position variation time series, we process the data to have an equal reference for all the used data. At each of the 2337 stations, we applied the common noise filter to correct the time series of the North, East, and Up components, using the equation [29]:

$$CAPdoy_t = APdoy_t - RP_t \tag{5}$$

where t is the epoch, $CAPdoy_t$ is the corrected apparent position, $APdoy_t$ is the apparent uncorrected position, and RP_t is the reference position. We use the average of AP, $\overline{AP}$, from the same reference days mentioned in Equation (1) to calculate RP_t.

At each of the 2337 stations, the maximum error was obtained within the selected five days. Subsequently, the error of each station per component was classified by intervals, counting the percentage of the total number of stations that fell into each interval. In addition, a 3D position error was calculated as:

$$3D_t = \sqrt{East_t^2 + North_t^2 + Up_t^2} \quad (6)$$

We use two threshold values. First, we selected threshold values for maximum 3D positioning error greater than or equal to 10 cm (3D $\geqslant$ 10 cm) since according to data on quiet days, over 90% of the GNSS stations had 3D errors of less than 10 cm, while during ionospheric disturbances, only ~40% kept this level of accuracy [29]. Second, we applied the Equation (6) to the horizontal and the vertical components presented in [29,30], obtaining the threshold of the 3D positioning error root mean square (3D-RMS) greater than or equal to 3 cm (3D-RMS $\geqslant$ 3 cm).

2.5. Geomagnetic and Geophysical Conditions

The geomagnetic data downloaded from OMNIWeb Plus Data (https://omniweb.gsfc.nasa.gov, last accessed on 11 May 2022) for the 10 June 2021 annular solar eclipse indicates a period of low activity. Except in DoY 163, where it was 4^- between 3 and 6 UT, the estimated 3-hour planetary index (Kp) was $\leqslant 3^-$. The disturbance storm time index (Dst) peak was >-17 nT, except after DoY 162 where a minimum of -37 nT was reached at 11 UT. The interplanetary magnetic field (IMF) Bz component in GSM coordinate peak was >-7.4 nT after DoY 162 and the solar wind speed (Vsw) was 330–520 km/s during 8–13 June 2021 (see Figure 2).

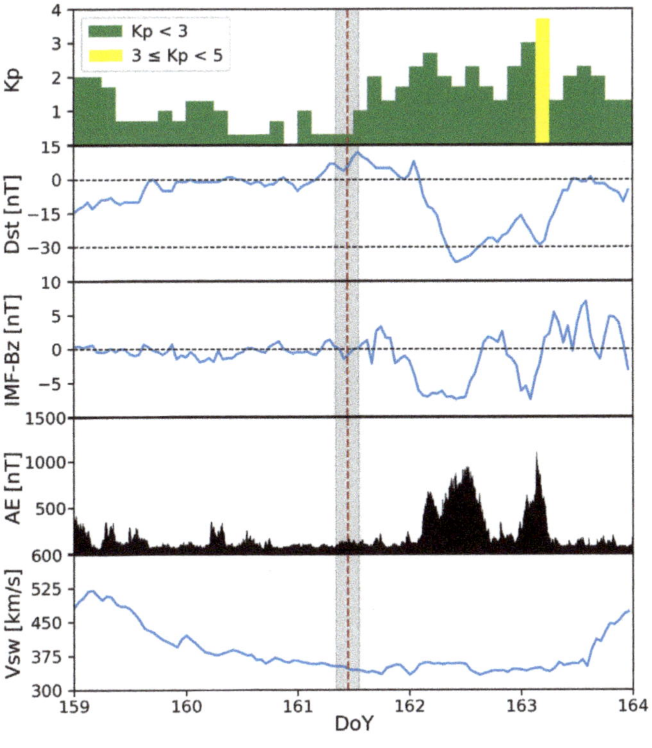

Figure 2. Variations of 3-hourly Kp, Dst, IMF-Bz, AE, and Vsw indices that characterize the geomagnetic conditions on 8–13 June 2021. P1–P4 time (light grey bar) and GE time (red dashed line) are also represented.

The Auroral Electrojet index (AE) is a good proxy of the geomagnetic activity level at mid/high latitudes [29,45]. Following De Michelis et al. [46], we selected two distinct datasets corresponding, respectively, to geomagnetically quiet (AE < 50 nT) and active (AE > 300 nT) periods. Figure 2 also illustrates that the AE index was over 500 nT between 5 and 15 UT on DoY 162, and between 1 and 5 UT on DoY 163; so these time periods showed some activity in the auroral regions. Since these days are used by comparison with the day of the eclipse, these periods of time were treated with care to avoid interfering with the eclipse analysis.

Therefore, the geomagnetic conditions were generally quiet, except on DoY 162 where a weak geomagnetic storm took place between 8 and 16 UT. DoY 162 did not cause problems in the ionospheric TEC background to our results for the eclipse day. However, the geomagnetic activity the day after the eclipse had significant effects on GNSS positioning errors comparable to the positioning errors caused by the annular solar eclipse. These effects will be presented in more detail in the coming section.

Earthquake Occurrence

We also reviewed the occurrence of earthquakes (EQs) around the world, with a moment magnitude greater than 5 Mw and a depth of over 70 km on 8–13 June 2021. This review is important because EQs are sources of TEC disturbances and thus positioning errors. In the period analyzed, 15 moderate EQs of less than 5.7 Mw occurred (https://earthquake.usgs.gov, accessed on 17 January 2022). However, none of them produced noticeable effects on TEC or on the position estimation on the GNSS receivers during the analyzed period of days.

3. Results

In this section, we present the main results obtained after applying the methodology described in the previous section. The results obtained in this work can be divided into two main parts: (1) the analysis of the TEC maps that present the effects on the ionosphere at a global scale; and (2) the calculation of the positioning errors that these ionospheric effects generate.

3.1. Ionospheric Behavior and TEC Maps

From the data of each station, we can estimate the VTEC for each station during the selected period of days. By using the ordinary Kriging interpolation, as described in Section 2.1, it is possible to obtain VTEC maps. Figure 3 shows a summary of the TEC maps by contrasting the eclipse (VTECe) and control ($\overline{VTEC}$) days. We present some particular hours: 09.15 UT, 10.70 UT (GE time), 12.00 UT, 13.19 UT (P4 time), 13.72 UT (P4 time + $\sim$0.5 h), and 17.66 UT. Figure 3 also shows the eclipse masks from 20% obscuration and with intervals of 20%, at 350 km altitude (white line). From these figures, it is possible to notice that the Greenland and South American sectors are two of the most affected in terms of VTEC depletion. VTECe in IPP and DVTEC[TECu] in Figure 3 use the Kriging interpolation method and are shown in equidistant cylindrical projection and Northern Hemisphere polar plots (see Supplementary Materials, Figure S1).

The 09.15 UT, 12.00 UT, and 13.72 UT maps were chosen in particular because they show the greatest apparent position variations during the eclipse time window. The 17.66 UT map was chosen because the ionosphere was roughly recovered by that time. For a better visualization of the eclipse effects, a third column has been incorporated where the VTEC differential in TEC units (DVTEC[TECu]) is shown.

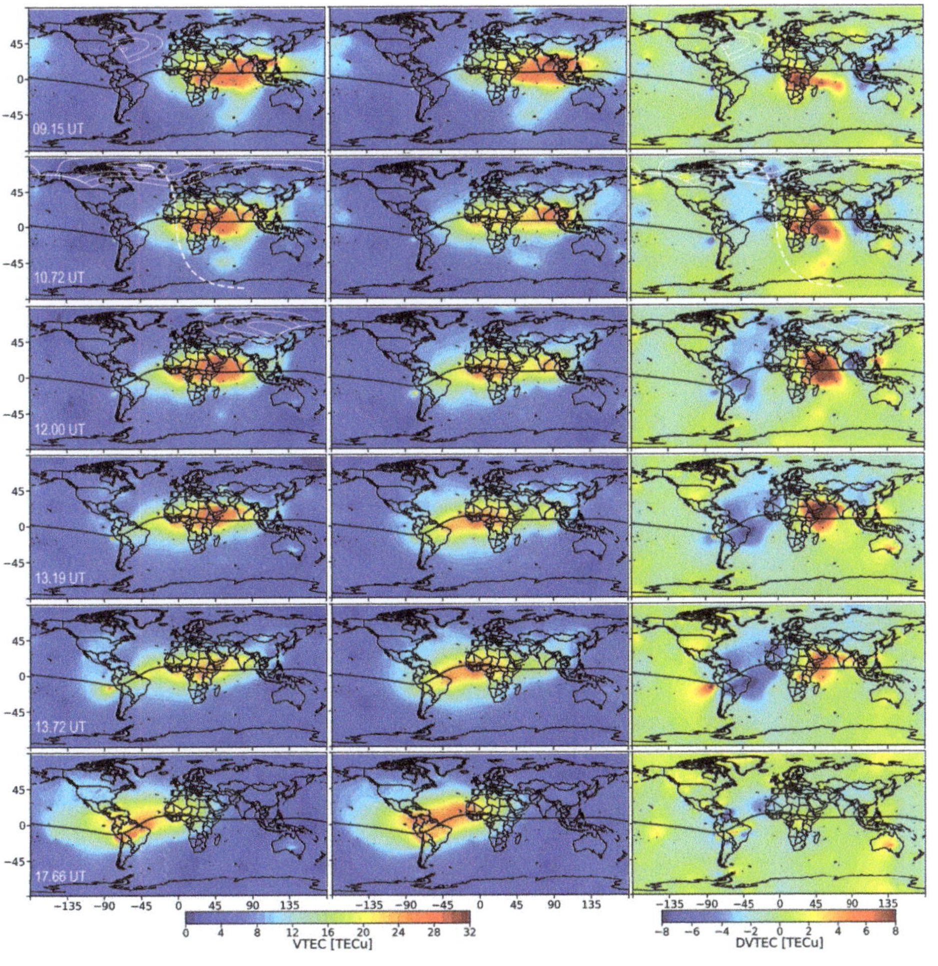

Figure 3. Ionospheric TEC maps during the 10 June 2021 Annular Solar eclipse using the Kriging interpolation method. From left to right panels: VTECe, $\overline{VTEC}$, and DVTEC[TECu]. From top to bottom panels: 09.15 UT, 10.70 UT (GE time), 12.00 UT, 13.19 UT (P4 time), 13.72 UT, and 17.66 UT. Eclipse obscuration masks from 20% obscuration and with intervals of 20%, at 350 km altitude (white line) are shown. The white dashed line starts from the maximum obscuration of the solar eclipse to its conjugate location in the Southern Hemisphere at 350 km altitude.

DVTEC had values of around −20% (−2 TECu) over the oceanic sectors when the eclipse began (P1-time). However, at these locations, the GNSS receivers are scarce, which can cause less reliable interpolation. This value could be considered as part of the non-significant variations in DVTEC. A similar problem is identified over Central Africa, where there is a value of 50% (7 TECu), possibly also due to the few receivers in this area (see Figure 1). The anomalies in these areas were observed more than 5 h before the eclipse. We will focus mainly on changes generated over the continental areas of America and Europe, while the other areas will not be considered for this analysis.

When the ionospheric TEC effects due to eclipse have already begun, the 09.15 UT maps show a slight depletion of −30% (−1 TECu) across eastern Canada under the shadow of the eclipse. At 10.70 UT (GE-time), these changes expand beyond the shadow area of the eclipse (see second row of Figure 3). The 10.70 UT maps show that ionospheric TEC depletion did not only occur across the obscuration region over the Northern Hemisphere. DVTEC[%] had values of around −60% (−4 to −2.5 TECu) over the South and East coasts of Greenland, −50% (−3 TECu) eastern Canada, and around −50% (∼−3 TECu) over the Lesser Antilles.

The 12.00 UT, 13.19 UT, and 13.72 UT maps show DVTEC[%] had values of around −30% (−3 to −1.5 TECu) over Russia after GE time. Figure 3) also illustrates how the TEC disturbance moved from West to East over the Northern Hemisphere, following the path of the annular solar eclipse. At 12 UT, there is a recovery of the ionospheric TEC over Canada and Greenland regions, but DVTEC[%] had values of less than −50% in East coast of Greenland. Ionospheric TEC depletion had values of around −60% (∼−5 TECu) over the Lesser Antilles near the north crest of Equatorial Ionization Anomaly (EIA) and less than −30% (∼−5 TECu) appeared over South America near the south crest of EIA. The 13.19 UT and 13.72 UT maps show another slight recovery of the ionospheric TEC in the North Atlantic and Greenland, as well as a TEC depletion over Russia. Moreover, TEC depletion was accentuated in the EIA crests over South America, where DVTEC[%] had values of less than −60% (<−11 TECu). It is shown that the effects lasted beyond the end of the eclipse.

The 17.66 UT maps present the global recovery of the ionosphere a few hours after the end of the eclipse. These maps show a slight DTEC[%] enhancement in the center of the EIA form ∼−16% to ∼10%. But the TEC depletion was ∼−20% (−5 to −3 TECu) in the EIA crest over South America. The TEC behavior in the EIA crests was maintained until after 19.66 UT.

On the other hand, DoY 162 had geomagnetic activity (see Section 2.5). Therefore, we checked if the DVTEC changes that we observed for the day of the eclipse were due to using DoY 162 as one of the reference days. We compute a new $DVTEC_t$ ($DVTECnew_t$, Equation (1)), and a new $\overline{VTEC_t}$ by averaging the VTEC values at the same time of the day, t, for the reference days, DoYs 159, 160, and 163. For each map, we used the map algebra ($DVTEC_t - DVTECnew_t$). The mean value ranged between −0.5 and 0.1 TECu, with a standard deviation of less than 0.7 TECu (see Supplementary Materials, Figure S2, Table S1). Therefore, geomagnetic activity during the DoY 162 did not cause problems in the background to our ionospheric TEC results for the eclipse day.

Ionospheric Behavior Using Swarm Satellite Measurements

We also present ionospheric behavior using Swarm-A measurements (see Figure 4). We illustrate VTEC at 850 km altitude on DoY 161 compared to DoY 159 during three ascending passes of the Swarm-A satellite (⩾45°S, see Figure 4 (upper panels)). We selected the three Swarm-A satellite passes that best fit the eclipse region and eclipse time window. The first satellite pass (∼8.50–9.20 UT) occurred after P1 time. The greatest ionospheric TEC degradation was −30°S–30°N (∼−1.7 TECu, −35%). The second pass (∼10.10–10.80 UT) was close to the GE time. As latitude increases, TEC decrease. The third satellite pass (∼11.70–12.40 UT) was performed prior to P4 time (∼−1.9 TECu, −37%). It is possible to see that TEC decrement was concentrated between 10 and 30°N (VTEC was close to ∼−1 TECu, and ∼−30%). Ionospheric TEC depletion was greatest in the 15–75°N region (∼−2 TECu, −45%).

Figure 4 (bottom panels) depict in situ N_e measurements made by Swarm-A Langmuir probe. Figure 4 (upper panels) show that VTEC behaves similarly to N_e. We observed that N_e decrease was −46% (-0.26×10^5 e/cm^3) at ∼8.86 UT in ∼1°N; −55% (-0.39×10^5 e/cm^3) at ∼10.62 UT in ∼50°N; and −55% (-0.72×10^5 e/cm^3) at ∼12.05 UT in ∼20°N.

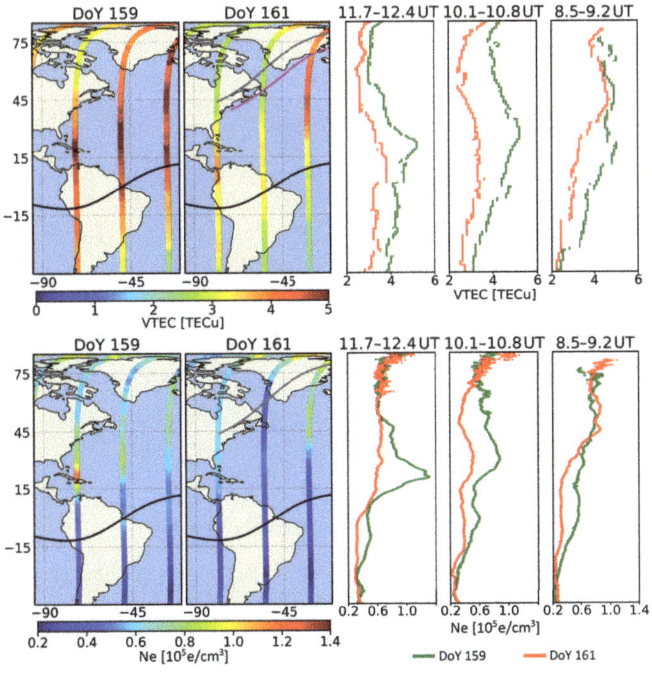

Figure 4. Ionospheric behavior using Swarm-A measurements. (**Upper panels**) present ionospheric TEC data taken by Swarm-A satellite at 850 km (at 400 km above the Swarm-A). The TEC gathered through the satellite orbit is presented over an Earth map (left panel) and as a profile with data obtained in one of the comparison days (write panel). The three satellite passes are from 45°S to ∼90°N, from left to right, between ∼11.70–12.40, ∼10.10–10.80, and ∼8.50–9.20 UT during 2 days before eclipse day (DoY 159), and eclipse day (DoY 161). The annular eclipse path at 450 km (grey line) and 850 km altitude (magenta line) are also shown on DoY 161. VTEC on DoY 161 (red dots) compared to DoY 159 (green dots) between ∼11.70–12.40 UT, ∼10.10–10.80 UT, and ∼8.50–9.20 UT. (**Bottom panels**) depict Swarm-A in situ electron density (N_e) presented in the same way than the VTEC data.

3.2. ROTI and GNSS Precise Point Positioning Accuracy Maps

We estimated the ionospheric TEC, ROTI, and positioning for the full 5 days but only show 6 hours per day. On the eclipse day, we observe the largest positioning variations during this time window (around P1–P4 time, see Figure 5). This study focuses on the positioning accuracy of the stations during the 10 June 2021 annular solar eclipse, during the time between 8 and 14 UT. We estimated the PPP-AR using the CSRS-PPP service, as described in Section 2.4. Comparing the eclipse day with respect to the DoYs 159 and 160, we can see that the percentage of GNSS stations that exceeded maximum 3D positioning error ⩾10 cm and 3D-RMS ⩾3 cm (positioning thresholds) jumped from ∼180 (∼8%) to 333 (∼14%) and from ∼170 (∼7%) to 210 (∼9%), respectively. In addition, the ROTI threshold ⩾0.25 TECu/min was taken according to Liu et al. [36], and used in the methodology [29]. Figure 6 shows maximum 3D positioning errors, 3D-RMS of the apparent position, and ROTI maps, for each of the five selected DoYs.

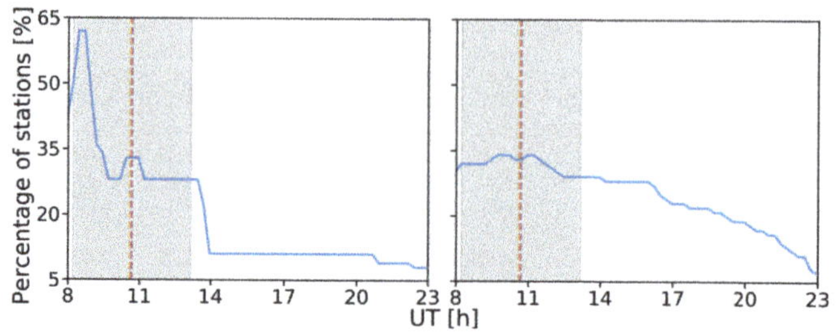

Figure 5. Time variations of the percentage of stations with 3D positioning error greater than 10 cm on DoY 161. The (**left panel**) shows the 36 GNSS stations localized in Greenland. The (**right panel**) presents the 335 GNSS stations that are situated in South America.

ROTI was calculated as described in Section 2.2 to study the relationship between the variation of TEC and the positioning error for the eclipse. The images in the Figure 6 (right panel) show five maps of ROTI, each representing the stations that had ROTI greater than 0.25 TECu/min. Each map represents a different day, but at the same time as that of the solar eclipse, DoYs 160–163 between 8 and 14 UT. We do not show the maps for DoY 159 because they are similar to those for DoY 160.

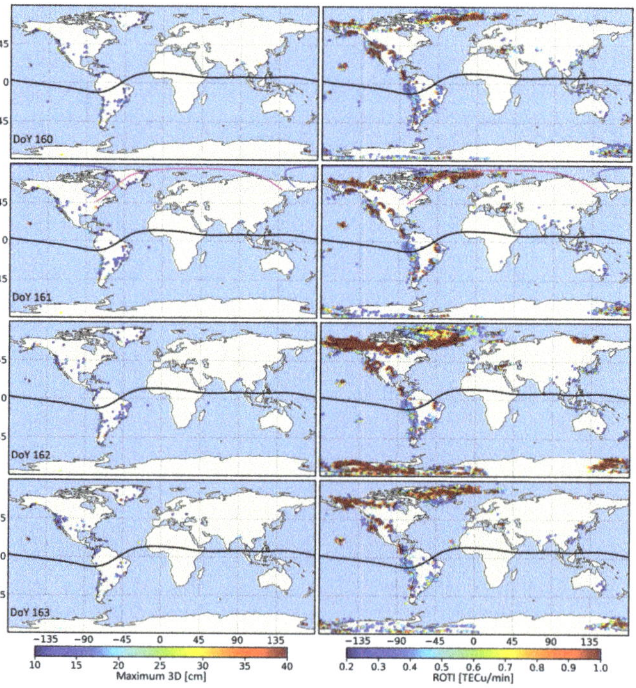

Figure 6. Maximum 3D positioning errors ⩾10 cm (**left panels**); and ROTI ⩾0.25 TECu/min (**right panels**) between 8 and 14 UT. From top to bottom shows DoYs 160–163. Annular eclipse path at ground level (blue line), and at 350 km of altitude (magenta line) are also shown.

The TEC data (e.g., Figure 3) and the derived PPP-AR (see Figure 6 (left panels)), and Figure 7) show that there are two regions where the errors are more severe during the solar eclipse (DoY 161, between 8 and 14 UT). These regions are Greenland and South America. For this reason, we focused this study on these sectors.

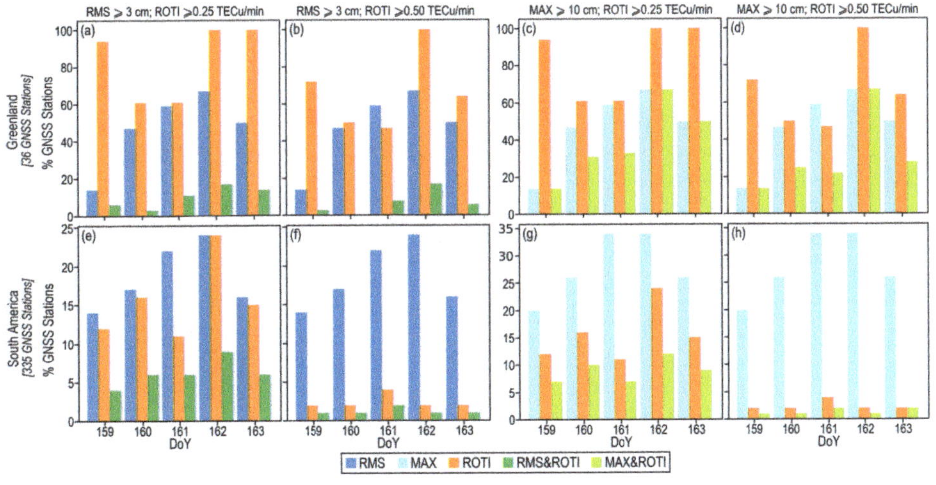

Figure 7. Behavior of the maximum 3D positioning error, 3D-RMS, and ROTI in Greenland and South America sectors, where 36 GNSS stations are localized in Greenland (**upper panels**) and 335 GNSS stations are situated in South America (**bottom panels**). Percentage of GNSS stations where (**a**,**b**,**e**,**f**) 3D-RMS $\geqslant$ 3 cm; (**c**,**d**,**g**,**h**) maximum 3D position $\geqslant$ 10 cm; (**a**,**c**,**e**,**g**) ROTI $\geqslant$ 0.25 TECu/min; and (**b**,**d**,**f**,**h**) ROTI $\geqslant$ 0.50 TECu/min. Percentage of GNSS stations meeting 3D-RMS (blue bars); maximum 3D positioning error (light blue bars); ROTI (orange bars); 3D-RMS and ROTI (green bars); and maximum 3D positioning error and ROTI (light green bars) values.

In the Greenland and South America regions, we can find 36 (~2%) and 335 (~14%) of the 2337 total available GNSS stations, respectively. We determined the percentage of stations localized in both regions that had errors that fell at certain intervals during the time period of the annular eclipse (between 8 and 14 UT). Figure 7 shows the percentages of stations that meet the thresholds of maximum 3D positioning error, 3D-RMS, and ROTI. We determined the number of GNSS stations based on ROTI activity and positioning values (see Supplementary Materials, Tables S2–S4, where each column in these tables represents the percentage of stations with maximum 3D position error, 3D-RMS, and ROTI at certain intervals in the selected period).

Figure 5 shows in more detail the percentage of stations with maximum 3D errors $\geqslant$10 cm on eclipse day. In Greenland, during the inicial period of the eclipse (~P1 time), the percentage of stations with maximum 3D positioning error rises to 60%. Subsequently, the value remains at ~28% until it increases to ~33% between 10.5 and 11.25 UT (around GE time). The value then returns to ~28% until 14 UT (after P4 time), when it drops to ~11% of stations. In South America, the percentage of stations with maximum 3D positioning error had a maximum of ~34% between 9.75 and 11.5 UT (around GE time). We also were able to observe a decrease in stations that exceeded the threshold maximum 3D positioning error from ~28% to ~22% between 16 and 17 UT.

3.3. Ionospheric Behavior and GNSS Positioning Errors by Region

To study the effects of the eclipse, we selected 24 stations from among the 2337 GNSS stations (see Figure 1). We chose five GNSS stations close to the annular solar eclipse (KMOR, KAGZ, MARG, IQAL, PICL). There were six GNSS stations located in the partial eclipse region (CN00, TRO1, SVTL, TIXI, MAG0, YAKT). Furthermore, we used five stations located in the Caribbean and South America (LMMF, BOAV, PIFL, MSBL). The sunrise (in PICL and CN00 stations) and the sunset (in MAG0 and YAKT stations) happened during the eclipse time window at ground level but did not take place at the ionospheric height of 350 km. More details about the GNSS stations and eclipse conditions (with respect to the ionospheric height of 350 km) can be found in Table 1.

Table 1. Detail about the GNSS stations analyzed, their location, eclipse condition, magnitude and change in VTEC at each station. Eclipse characteristics (time and maximum obscuration) for the 24 selected GNSS stations. Further, we have also estimated eclipse conditions at the ionospheric height of 350 km by the method suggested by Verhulst et al. [21]. Ionosphere, 3D eclipse, and 3D non-eclipse values are between 8 and 14 UT. In 3D non-eclipse: MAX refers to the maximum 3D positioning error of the reference days, and we calculate RMS by taking all the values of the reference days between 8 and 14 UT.

GNSS Station	Geographic		Eclipse					Ionosphere			3D Non Eclipse		3D Eclipse	
	Lat [°N]	Lon [°E]	C1 [UT]	MOT [UT]	C4 [UT]	MPO [%]	τ [min]	A [%]	A [TECu]	ROTI [$\frac{TECu}{min}$]	MAX [cm]	RMS [cm]	MAX [cm]	RMS [cm]
KMOR	81.25	−63.53	9.63	10.72	11.83	84	32	−44	−1.8	0.5	9.9	1.8	12.7	2.4
KAGZ	79.13	−65.85	9.57	10.63	11.75	84	23	−43	−1.6	0.5	10.1	2	11	2
MARG	77.19	−65.69	9.5	10.57	11.68	84	7	−30	−1.0	1.1	9.5	1.8	9.4	2.1
IQAL	63.75	−68.51	9.1	10.12	11.18	85	12	−56	−3.0	0.3	4.2	1.1	5.2	1.4
PICL	51.48	−90.16	9.03	9.92	10.87	81	1	−29	−1.3	0.1	3	0.9	4.2	1.8
ALGO	45.96	−78.07	8.75	9.68	10.63	89	17	−40	−1.6	0.1	3.5	1.1	3.4	1.6
LEFN	80.46	−26.29	9.62	10.8	10.08	90	15	−55	−2.8	1.8	7.5	1.9	9.8	2.2
BLAS	79.54	−22.97	9.6	10.77	11.95	90	22	−59	−3.1	2.2	9.5	1.8	11.4	2
KAGA	69.22	−49.81	9.18	10.32	11.47	90	18	−43	−1.8	0.2	7	1.6	8.9	1.8
GLS2	69.09	−39.65	9.18	10.32	11.53	88	29	−61	−3.3	0.2	9.9	2.2	16.7	2.6
KUAQ	68.59	−33.05	9.18	10.33	11.57	84	19	−61	−3.5	0.1	7.6	1.5	12.8	1.8
SENU	61.07	−47.14	8.92	10	11.18	86	17	−49	−3.3	0.1	6.6	1.4	10.6	2.5
CN00	17.67	−61.79	8.2	8.9	9.65	26	237	−45	−5.3	0.2	11.5	2.7	7.4	2.6
LMMF	14.59	−61	8.22	8.85	9.52	19	141	−61	−4.3	0.2	7.1	2.2	6.5	3
CN57	10.84	−60.94	8.25	8.78	9.37	11	161	−50	−4.4	0.1	6.1	1.8	8	2.7
BOAV	2.85	−60.7	8.58	8.7	8.82	0.1	288	−28	−5.1	0.1	6.7	1.7	8.1	3
PIFL	−6.79	−43.04	–	–	–	–	110 [1]	−65	−11.8	1.9	24.4	6.4	74.2	8.9
MSBL	−22.11	−56.53	–	–	–	–	126 [1]	−35	−4.2	0.1	8.5	2.9	11.6	4.2
NYA1	78.93	11.87	9.77	10.97	12.18	83	6	−42	−1.9	1.4	4	1.2	4.9	1.5
TRO1	69.66	18.94	9.72	11	12.28	59	37	−52	−2.5	0.1	4.7	1.3	12.3	2.7
SVTL	60.53	29.78	9.97	11.23	12.45	33	28	−52	−3.6	0.2	4.1	1.3	4.6	1.5
TIXI	71.63	128.87	10.55	11.57	12.57	89	100	−30	−1.4	0.2	6.5	1.8	8.3	2.1
MAG0	59.58	150.76	10.72	11.63	12.52	88	110	−27	−1.5	0.1	4.2	1.2	4.9	3.9
YAKT	62.03	129.68	10.82	11.8	12.73	87	62	−38	−2.8	0.1	4.1	1.2	4.9	1.5

[1] τ refers to GE time, not to MOT.

Figure 8 presents the results of ionospheric TEC of 12 stations from among the 24 selected GNSS stations for the eclipse day (VTECe), the reference days ($\overline{VTEC}$), and the final results of DVTEC [%]. The vertical blue shaded region between P1 time and P4 time, with GE time (brown dotted line). The vertical yellow shaded region between C1 time and C4 time, with MOT (black dotted line). Each plot is shown between 5 and 23 UT. The maximum reduction values of TEC for each station are indicated in Table 1 ($\tau = 1$ to 288 min, and A[%]= −65 to −27%). This eclipse occurred during the morning at most of the selected stations but took place in the afternoon at five stations (NYA1, TRO1, SVTL, TIXI, MAG0, YAKT). The Sun's activity became stronger around noon and the clear TEC reduction during the eclipse can be observed. The stations that are near the path of the annular eclipse at 350 km altitude and the east coast of Greenland reached lower values of DVTEC [%] ∼−55% than the stations with the annular eclipse at the surface level DVTEC [%] ∼−40%.

Figure 8 shows the ionospheric TEC changes for 12 of the 24 GNSS stations presented in Table 1. The TEC disturbance lasted longer at the GNSS stations located in South America and the Lesser Antilles (CN00, LMMF, CN57, BOAV, PIFL, MSBL). In these GNSS stations, the ionospheric effect caused by the eclipse started at ∼8.5–9 UT and ended ∼18–21 UT ($\Delta T > 10$ h). The ionospheric response is similar in BOAV and MSBL stations where A[%] ∼−30%. In MAG0, TIXI, and YAKT stations, we observe a TEC depletion during the eclipse time window, but it is not as noticeable as in the other cases.

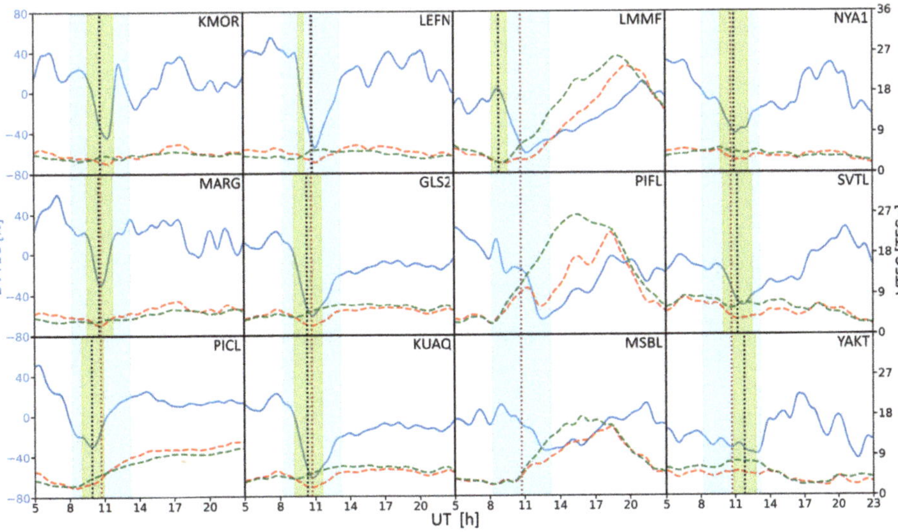

Figure 8. The behavior of the ionospheric TEC during the 10 June 2021 Annular Solar Eclipse in 12 of the 24 selected GNSS stations. DVTEC[%] (blue line), VTECe (red dashed line) and $\overline{VTEC}$ (green dashed line). The GNSS stations are ordered by latitude and then by longitude. P1–P4 time is represented by the light blue bar, C1–C4 time by the yellow bar, MOT by black dotted line, and GE time by the red dotted line.

Additionally, the number of GNSS stations according to ROTI activity was: 5 strong (BLAS, PIFL, LEFN, NYA1, and MARG stations), 2 moderate (KMOR and KAGZ stations) and 17 without activity (see Table 1).

In the same way, we presented the results of PPP-AR of 24 DF-GNSS stations during eclipse day. The time series were corrected for the common noise filter of the East, North, and Up components. The stations had variations in position within the time window of the eclipse (between 8 and 14 UT). The station with the highest positioning errors in the East, North, and Up components was PIFL stations. KAGA, GLS2, SENU, and MSBL stations also showed position variations between 5 and 8 UT.

Equation (6) is used to obtain the 3D results. Then, the maximum 3D positioning error and 3D-RMS values (between 8 and 14 UT) for each station are indicated in Table 1. We note that the GNSS stations can be separated according to the percentage of maximum 3D positioning error and 3D-RMS, with respect to the maximum values of reference days.

In the case of maximum 3D positioning error, four stations presented values below 0% (MARG, ALGO, CN00, and LMMF); eight 0–25% (KAGZ, IQAL, BLAS, BOAV, NYA1, SVTL, MAG0, and YAKT); seven GNSS stations were between 25 and 50%; three stations were 60–70% (GLS2, KUAQ, and SENU); and two stations > 100% (PIFL and TRO1).

On the other hand, for 3D-RMS in a percentage, 1 station was < 0% (CN00); 11 stations were 0–25% (KAGZ, MARG, LEFN, BLAS, KAGA, GLS2, KUAQ, NYA1, SVTL, TIXI, and YAKT); 7 stations were 25–50% (KMOR, IQAL, ALGO, LMMF, CN57, PIFL, and MSBL); 3 stations were 75–100% (SENU, PICL, and BOAV); and 2 stations > 100% (TRO1 and MAG0).

A Case Study

We will describe in more detail the results obtained with PIFL GNSS station (6.79°S, 43.04°W). PILF had the largest ionospheric disturbances and GNSS positioning errors (see Table 1). TEC depletion had values around −65% (−11.8 TECu) at 110 min after GE time (see Figure 8). Figure 9 presents the ionospheric behavior (ROT, ROTI) and kinematic

DF-GNSS PPP-AR mode during DoYs 160–163 between 5 and 23 UT. We do not show DoY 159 because it does not differ significantly from DoY 160.

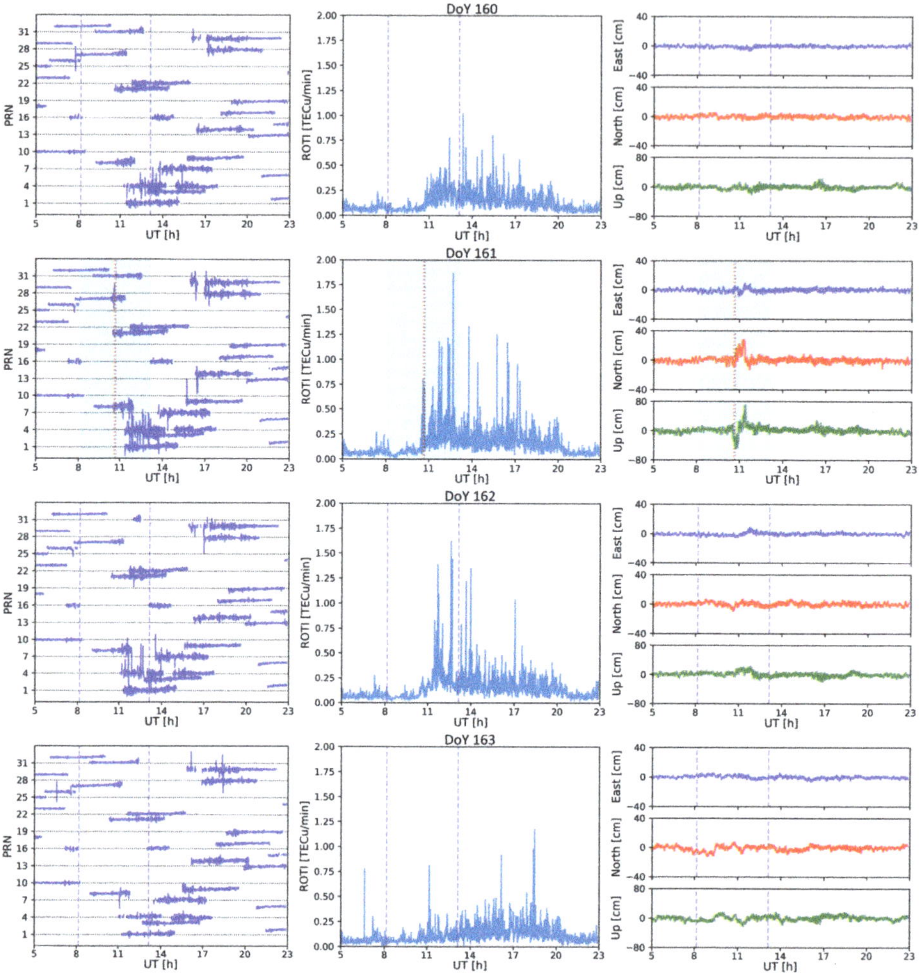

Figure 9. Ionospheric behavior and apparent position variation of the PIFL GNSS station. From top to bottom show DoYs 160–163. ROT (TECu/m) (**left panels**), ROTI (TECu/m) (**middle panels**), DF-GNSS PPP-AR (cm): East [cm] (blue line), North [cm] (red line), and Up [cm] (green line) (**right panels**). P1–P4 time on non-eclipse day (blue dashed lines), P1–P4 time on eclipse day (light blue bar), GE time (red dotted line).

Figure 9 (left, middle panels) illustrates examples of GPS ROT and GPS ROTI variations along with all visible GPS satellites. On eclipse day, we can observe a |ROT| > 1.5 TEC/min in eight Pseudo Random Noises (PRN-4, 7, 8, 9, 14, 27, 28, 30). The |ROT| value was exceeded by 3–4 PRNs during the reference DoYs 159, 160, and 163. In contrast, the |ROT| was exceeded by seven PRNs on DoY 162 (see Figure 9 (left panels)). On DoY 161 between 10.66 and 17.39 UT (6.73 h), we could note 22 and 8 ROTI values >0.5 and >1 TECu/min, respectively (see Figure 9 (middle panels)). The ROTI peak was 1.9 TECu/min at 12.75 UT, estimated from the PRN-4. On this day, nine PRNs (PRN-4, 7, 8, 9, 10, 14, 27, 28, 30) presented a moderate and/or strong ROTI activity. Regarding DoYs 159, 160, 162 and 163, we observed 12, 8, 20 and 6 values with moderate and/or strong ROTI activity. Then, this station showed strong TEC activity during each of the five DoYs. The ROTI value was higher on eclipse day 1.9 TECu/min at 12.76 UT (∼25 min after P4 time).

Figure 9 (right panels) show the apparent position variation of kinematic DF-GNSS PPP-AR mode in the East, North, and Up components for the PIFL GNSS station. These time series has been corrected for the common noise filter. On DoY 161, the apparent peak ground displacement in the East, North, and Up components were 18, 40, and 119.8 cm, respectively. Moreover, the maximum 3D positioning error ⩾10 cm ∼9.80 UT by ∼3.30 h. The Up, North, and East components are ordered from highest to lowest errors. Then, positioning errors in the three components and their results were clear during the eclipse time window (after GE time), relative to the reference days.

4. Discussion

In this section, we discuss the main findings regarding the 10 June 2021 annular solar eclipse. The main goal is to study the positioning errors of GNSS receivers caused by this solar eclipse. In order to verify our findings, we compare our ionospheric values with results presented for other solar eclipses over the northern polar region (9 March 1997 [4,22]; 1 August 2008 [10]; and 20 March 2015 [5,23]).

There are several free-to-use software available for single-station TEC estimation methods [47,48]. We selected GPS-TEC software because it is a widely used method by the scientific community to study phenomena such as geomagnetic storms [29] and solar eclipses [49], among others. GPS-TEC software is fundamentally based on the assumption that ionospheric density depends on altitude to determine VTEC from STEC.

4.1. Ionospheric Behavior

The present analysis aims to show, as best as currently possible, the effects that the solar eclipse generates both in the ionosphere under the moon's shadow as well as in the global ionosphere. The relevance of this event is that there are few of them that occur in polar regions, in this case, in the Arctic.

We have used interpolated global maps from TEC and we have calculated the difference between eclipse and reference days. The results of the TEC maps show a significant reduction under the moon's shadow, except at the CN00 station that has similar behavior to the LMMF, CN57, BOAV, and MSBL stations (see Figures 3 and 8, Table 1); the GNSS stations located in the region of the eclipse reaching a maximum of τ = 1 to 288 min, A peak ∼−5 TECu, A[%] = −61 to −27%. Table 1 details these parameters for GNSS stations of some selected regions (see Figure 1). The values of these parameters agree with those obtained for the solar eclipses of 9 March 1997 solar eclipse [4,22], 1 August 2008 [10], and 20 March 2015 [5,23].

TEC depletion was not as pronounced in the MAG0, TIXI, and YAKT stations (A[%] = −38 to −30%) compared to others GNSS stations (A[%] = −61 to −40%) with a similar percentage of obscuration (∼88%). This could be due to the fact that the sunset in MAG0, TIXI, and YAKT stations happened during the time-window of the eclipse at ground level. Moreover, the other stations were closer to the greatest eclipse (see Figures 1 and 8, Table 1).

In addition to the decrease of TEC in the ionosphere under the Moon's shadow, we have observed interesting and significant effects far from that region. This is the case of a significant decrease in TEC that seems to move southward from the shadow, passing through the North Atlantic, and remaining stationary for several hours over the Caribbean and the north of Brazil, at the stations CN00, LMMF, CN57, BOAV, PIFL, MSBL (see Figures 3 and 8, and Table 1). The delay value relative to GE time was between ∼30 and ∼168 min, A peak ∼−11 TECu, A[%] = −65 to −28%, ΔT > 10 h. This area coincides with the location of the crests of EIA. The TEC variations were more intense north and south of the magnetic equator, where they were similar to those obtained at the GNSS stations located in the eclipse region.

On the other hand, applying $DVTEC_t - DVTECnew_t$, we observed that the mean was between −0.5 and 0.1 TECu; and standard deviation was less than 0.7 TECu. Therefore,

TEC changes due to the weak geomagnetic activity during the DoY 162, did not cause problems in the ionospheric TEC background to our presented results for the eclipse day (see Supplementary Materials, Figure S2, Table S1).

In order to verify the negative disturbance in TEC on EIA crests, we have compared them with ionosonde observations of the sector involved (Figure 10). The Ramey (RA, 18.5° N, 67.1° W) station on the Caribbean side, and Sao Luis (SL, 2.6° S, 44.2° W), Fortaleza (FZ, 3.9° S, 38° W) and Cachoeira Paulista (CP, 22.7° S, 45.0° W) stations on the Brazil side were selected. The geographic locations of these stations are indicated with blue rhombuses in Figure 1. The data is obtained from the Digital Ionogram Data Base (http://giro.uml.edu/didbase/scaled.php, accessed on 4 August 2021) [50].

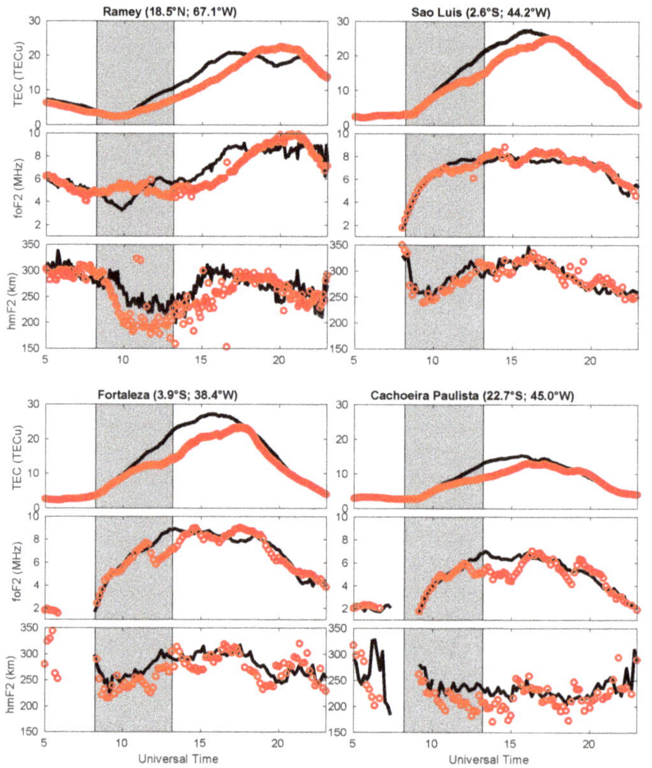

Figure 10. Comparison between TEC differences and differences in the critical frequency of the plasma (foF2) and its height (hmF2). Observations (red circles), reference variation (black line), and P1–P4 time (shaded interval) are shown.

As a result, Figure 10 shows coherence between TEC and the critical frequency of the plasma (foF2) of each station. That is, the electron concentration after the eclipse maximum (~11 UT) decreases (red circles) with respect to the reference curve (black line) calculated as indicated in Section 2.1. These same changes can be seen in the height of maximum electron concentration (hmF2). The decrease in foF2 and hmF2 is notorious at stations near the anomaly's crest (RA, FZ, CP); however, it is not very significant in the stations at the magnetic equator (SL). Moreover, similar ionospheric effects were seen in distant regions in the moon's shadow [16–20,51–56]. Differences between foF2 and TEC may be due to the fact that foF2 was the result of the original autoscaled records, and also that TEC was calculated from a spatial average.

A possible explanation for this phenomenon is that the eclipse could alter the thermospheric neutral wind regime and thus generate a ionospheric disturbance dynamo, which could be observed at the equator as a counter-electrojet. This counter-electrojet could be observed in the vertical drift of the plasma, for instance, the one measured by the Jicamarca incoherent radar. However, there are no measurements at Jicamarca for this period. Another way to observe is to calculate the difference in the horizontal component between an equatorial magnetometer and another in low latitude [57], or in the temporal variation of the same horizontal component of an equatorial magnetometer. In this case, neither the difference between Jicamarca (12.0°S, 76.8°W, I = 1°)—Piura (5.2°S, 80.6°W, I = 11°; available at http://lisn.igp.gob.pe/, last accessed on 22 May 2022), in the west coast of South America, nor the variation of the magnetometer of Kourou (5.2°N, 52.7°W; I = 13°; available at https://intermagnet.github.io, last accessed on 22 May 2022), in the east coast of South America shows significant variations during the eclipse day with respect to the other days (figure not shown), which rejects this hypothesis. Another possible explanation could be that due to the fact that the partial eclipse begins at low latitudes (see Supplementary Materials, Video S1) the electron concentration never reaches normal values again. An eclipse also can cause effects on a global scale. Because the eclipse-induced abrupt cooling of the atmosphere can result in an instantaneous temperature shift and pressure differential, triggering AGWs, and associated TADs and/or TIDs. However, a detailed investigation of these causes is out of the scope of the current paper [17].

On the other hand, ionospheric effects in the magnetic conjugate of the eclipse (end of the white line in Figure 3, at 10.70 UT) are not possible to observe due to the lack of receivers in this region (see Figure 1).

On DoY 161, there was low ROTI activity in the western region of the United States of America, compared to the reference DoYs. The decrease in the percentage of GNSS stations in South America with weak ROTI activities caused the increase of stations without activity up to 89%. Then, the number of stations with strong ROTI activity only increased from 1% to 3% in this sector. However, the behavior of the ROTI activity in Greenland was less than the reference days (see Figures 6 and 7; Supplementary Materials, Table S4).

The behavior of ionospheric TEC and ROTI shows that electrons were less active in the ionosphere during the solar eclipse (see Table 1, and Figures 3, 4, 6–8 and 10). The behavior of the ROTI in the eclipse region was consistent with that indicated by Park et al. [26]. They found a significant reduction in the ROT during the eclipse. Furthermore, eclipse day was the least ROTI active in Greenland because we were able to observe a clear reduction in ROTI values compared to the other four DoYs (see Figure 6 (left panels), Figure 7 (upper panels)).

On the other hand, as a consequence of the geomagnetic activity (AE-index > 500 nT) in the polar regions from 5–15 UT on DoY 162; we can see an increase in ROTI activity (ROTI $\geqslant$ 0.25 TECu/min) that starts at the northern polar region, propagating later the increment toward the equator ($\sim$50°N), which agrees with previous studies [29,58,59] (see Figure 6 (left panels)). In South America, the percentage of stations with ROTI activity increased from $\sim$12% to 24% (see Figure 6 (left panels), Figure 7 (upper panels), and Supplementary Materials, Table S4).

The results obtained with the GNSS stations at 350 km (see Figures 3 and 8) were consistent with the ionospheric TEC behavior at 400 km above the Swarm-A (see Figure 4). At P1 time, we observe a TEC depletion ($\sim$−1.7 TECu, −35%) in the central Atlantic region, where the eclipse started and its conjugate. The greatest TEC reduction ($\sim$−2 TECu, −45%) occurred at GE time (see Figure 4 (upper panels)). This TEC value was similar to that reported by Cherniak and Zakharenkova (−2 to −1.5 TECu) [60]. From Figure 4 (buttom panels), we were also able to show that the disturbance remained in the North and South American regions (TEC $\sim$ −30%) even though the eclipse was already over the northern European and Asian regions. Moreover, we can observe a close similarity in the behavior of in situ Ne and VTEC (see Figure 4). Furthermore, the results of ionospheric plasma depletion using Swarm-A LP were consistent with the findings presented in [60].

4.2. Ionospheric Impacts on GNSS Positioning Errors

The manner in which we present the positioning errors in this work was through the statistics of perturbed stations around the world and, in particular, in the Greenland and South American sectors.

On the eclipse day, we could see a slight increase in the percentage of GNSS stations around the world that exceeded both positioning thresholds compared to previous days. The main increment suffered by the maximum 3D positioning error goes from ∼8% to ∼14%. Then, Greenland and the southern sector of America were within the regions that presented GNSS stations with the highest positioning errors during the eclipse time window. This positioning behavior in both regions was consistent with the global ionospheric TEC changes.

Contrary to what happens with the activity of electrons, the percentage of stations that exceed both positioning-error thresholds was greater on DoY 161 compared to DoYs 159, 160, and 163 (see Figures 6 and 7 (upper panels)). We could see similar behavior in both cases of the 3D-RMS and ROTI activity relationships (RMS&ROTI). The eclipse day was the second DoY with the highest percentage of stations that exceeded the positioning-error thresholds. The effects of the eclipse day were only exceeded by DoY 162 due to weak geomagnetic activity (AE-index >500 nT) in the polar regions from 5 UT to 15 UT. In South America, the behavior of maximum 3D positioning error ⩾10 cm (34%) and 3D-RMS ⩾3 cm (22%) on day 161 was similar to day 162. In Greenland, these parameters were also similar on days 161 and 162, where maximum 3D positioning error ⩾10 cm was >55% and 3D-RMS ⩾ 3 cm was ∼17%. However, the effects on positioning on DoY 162 were slightly higher (see Figures 6 and 7 (bottom panels)).

Our RMS position values for the quiet days were in accordance with those from previous results [29,30]. They showed that the precision of the post-processing kinematic PPP-AR method was ⩽0.8 and ⩽2 cm for the horizontal and vertical components, respectively. Moreover, our 3D-RMS results in percentage (⩽3 cm = −4 to 225%) are consistent with the −4 to 324% presented by Park et al. [26].

Unlike previous studies [61–64], the results presented by Valdés-Abreu et al. [29], suggested that positioning errors also occur, regardless of whether the ROTI has rapid variations, with or without ROTI activity, in this type of DF-GNSS stations with the use of PPP-AR. Moreover, our results confirm that ionospheric disturbance sources can cause degradation of the GNSS accuracy (maximum 3D positioning error ⩾ 10 cm and 3D-RMS ⩾ 3 cm) when ROTI ⩾ 0.25 TECu/min, ROTI ⩾ 0.5 TECu/min, and without ROTI activity (see Table 1, and Figures 6–9). In addition, not all GNSS stations that had ROTI activity presented position errors.

Further, the ROTI activity–positioning variation relationship would have been met if two necessary conditions had been observed on each day in Figure 7. First, the positioning bars (RMS and MAX) had to be greater than or equal to the ROTI activity bar. This condition ensures that any ROTI activity causes variations in GNSS positioning. Second, the positioning ROTI bars (RMS&ROTI and MAX&ROTI) had to be the same or similar to the ROTI activity bar.

In most GNSS stations, we can observe the positioning errors were around the beginning of the TEC reduction (∼P1 time), the TEC peak (∼MOT and GE time), and/or in the final phase of the TEC recovery (see Figure 8). Then, we can see from one to more than three time slots with positioning errors. In general, after P1 time, the behavior of the stations with maximum 3D positioning error ⩾10 cm is similar to the DVTEC [%] of the stations located in Greenland (see Figure 8 (left, center left panels), Figure 5 (left panels)) and South America (see Figure 8 (center right panels), Figure 5 (right panel)). For example, in the Greenland region, the ionospheric TEC depletion was significant until ∼14 UT, and the recovery also could be observed in the rapid decrease from 28% to 11% of stations that exceeded the threshold of maximum 3D positioning error ∼14 UT, where the persistence of the positioning errors provoked by the 10 June 2021 annular eclipse lasted ∼6 h. Although the TEC depletion in sectors of South America could be observed until

~19–21 UT, a ionospheric TEC enhancement was observed around 16–17 UT, similar to the behavior of the GNSS stations with a maximum 3D positioning error greater than the threshold of 10 cm (from 28% to 22% between 16 and 17 UT). Therefore, the persistence of the positioning errors provoked by the 10 June 2021 annular eclipse lasted ~10 h.

The annular eclipse in Greenland caused significant TEC changes (~−60%), although with low ROTI activity. However, the GNSS positioning errors are similar to those caused during a weak geomagnetic storm with high auroral activity.

From Table 1, Figures 8 and 9, we see that the stations (PIFL, TRO1, GLS2, KUAQ, SENU) presented maximum 3D position errors >60%, also had A[%] $\leqslant -49\%$, but without ROTI activity (<0.25 TECu/min). PIFL station was the only one with ROTI activity over the 5-day period under consideration. Additionally, not all the stations that had A[%] $\leqslant -49\%$ got maximum 3D erros >60% (BLAS, LEFN, IQAL, SVTL, LMMF, and CN57 stations). The LMMF station presented A[%] = −61%, but maximum 3D position errors = −8%, and ROTI = 0.2 TECu/min. Although the BLAS and LEFN stations had strong ROTI activity and the IQAL station had weak ROTI activity, the percentage of maximum 3D positioning error in these stations was between 24% and 31%.

Thus, the results suggest that when maximum 3D errors >60%, with respect to the maximum of the reference days, we can find A[%] $\leqslant -49\%$, but not the opposite. The results also reinforce the idea that ROTI activity is not a necessary condition to affect GNSS accuracy. We were not able to estimate the ionospheric effects on GNSS positioning in the magnetic conjugate region of MPO of the solar eclipse, due to the lack of GNSS stations in this region.

Our study showed that the ionospheric TEC disturbances due to the solar eclipse in the polar regions can produce disturbances in low and medium latitudes. Ionospheric changes can cause GNSS positioning errors. The estimation of these errors is critical in teleoperated and autonomous (ground, maritime, and aerial) applications and other high-precision activities. For example, mining, agriculture, and fishing are all key economic activities in Chile that are considering the use of more teleoperated or autonomous systems. If the positioning error in the GNSS receivers spikes in vehicles in these industries, it could impose a serious risk to people and infrastructure. For open-pit mines, a high error can generate a failure in the estimation of the terrace on which a vehicle is located, with the consequent risk of falling. Halting autonomous operations during some events such as eclipses can reduce potential risks, but they can be complex for these industries. Stopping the operation for even a short period of time, such as an hour, could be prohibitively expensive. Therefore, forecasting the impact should be precise in location and duration.

5. Conclusions

In this work, we analyzed the ionospheric behavior during the 10 June 2021 annular solar eclipse and its impact on DF-GNSS PPP-AR accuracy. We use a large global GNSS network located around the planet to estimate the effects on positioning. This solar eclipse had a trajectory over the northern polar region. We used global ionospheric TEC maps with data gathered by ground-based GNSS stations.

The TEC maps show a noticeable depletion under the moon's shadow, reaching A[%] < −60%. Furthermore, a significant TEC decrease (A[%] < −60%) can also be observed far from the ionosphere under the moon's shadow in regions close to the crests of the EIA over the Caribbean and South America, with a duration or ΔT over 10 h. Then, percentages of the ionospheric TEC over the Caribbean and South America were similar to those obtained for GNSS stations located in the region of the eclipse. Our study also confirms that there are cases and places where the disturbance can last much longer than previously expected.

We show that TEC enhancement caused by geomagnetic activity on the day after the eclipse did not cause problems in the ionospheric TEC background to our presented results for the eclipse day. We also validated the ionospheric variations estimated with GNSS receivers through measurements from other instruments such as the Swarm-A satellite (VTEC and in situ Ne), and four ionosondes (TEC, foF2, and hmF2). The ionospheric behavior clearly demonstrates that electrons are less active in that layer during the solar eclipse. Furthermore, our results are consistent with ionospheric effects reported in similar previous solar eclipses.

This study not only analyzes the eclipse's day but also compares the effects of the ionosphere and its impact on the positioning precision with those over 2 days previous and 2 days after the day of the eclipse. The day of the eclipse was the day with the second highest percentage of stations that exceeded the selected positioning thresholds (maximum 3D positioning error $\geqslant$ 10 cm, 3D-RMS $\geqslant$ 3 cm), only surpassed by the day after, which had geomagnetic activity. The data analysis shows that the eclipse had a significant effect on GNSS precision for a long time ($\sim$10 h). The Greenland and South America sectors are within the regions that presented GNSS stations with the highest positioning errors during the eclipse time window. Moreover, both regions had the greatest ionospheric TEC decrease ($\sim -60\%$).

The ROTI variations were not relevant. Thus, the results reinforce the idea that ROTI activity is not a necessary condition to affect DF-GNSS PPP-AR accuracy. Additionally, the results suggest that when maximum 3D errors are larger than 60%, the A[%] is much less than -49%. However, the opposite is not necessarily true.

Supplementary Materials: The following are available online at https://www.mdpi.com/article/10.3390/rs14133119/s1, Video S1: Eclipse obscuration mask from P1 to P4 time at 350 km altitude. Figure S1: Ionospheric TEC maps during the 10 June 2021 Annular Solar Eclipse: the world and Northern Hemisphere polar plots. Figure S2: DVTEC[TECu] maps using the Kriging interpolation method to the eclipse day. Table S1: Mean and standard deviation for each map of Figure S2. Table S2: 3D-RMS in Greenland and South America. Table S3: Maximum 3D positioning error in Greenland and South America. Table S4: ROTI in Greenland and South America.

Author Contributions: Conceptualization, J.C.V.-A., M.A.D. and M.B.; methodology, J.C.V.-A. and M.A.D.; software, J.C.V.-A., M.B. and Y.S.-S.; validation, J.C.V.-A., M.A.D., M.B., J.C.B. and Y.S.-S.; formal analysis, J.C.V.-A., M.B. and Y.S.-S.; investigation, J.C.V.-A., M.A.D., M.B., J.C.B. and Y.S.-S.; resources, J.C.V.-A., M.A.D., M.B., J.C.B. and Y.S.-S.; data curation, J.C.V.-A., M.B. and Y.S.-S.; writing—original draft preparation, J.C.V.-A., M.B. and Y.S.-S.; writing—review and editing, M.A.D. and J.C.B.; visualization, J.C.V.-A., M.A.D. and Y.S.-S.; supervision, M.A.D. and J.C.B.; project administration, M.A.D.; funding acquisition, M.A.D. All authors have read and agreed to the published version of the manuscript.

Funding: This research was funded by the Air Force Office of Scientific Research under award numbers FA9550-18-1-0249. This work was also partially funded by the ANID-FONDECYT 1211144 and the National Agency for Research and Development (ANID)/Scholarship Program/Doctorado Nacional/2018–21181599 (CONICYT Doctoral Grant Number 21181599). M.B. thanks the support of ANID-FONDECYT Regular 1211144 and FONDECYT Postdoctorado 3180742. J.C.B. was also supported by ANID PIA (ACT192169) and supported by Fondecyt project (N°1200779 ANID, Chile).

Data Availability Statement: The RINEX files were obtained from: IGS stations; the Chilean network of GNSS receivers operated by CSN; UNAVCO; RAMSAC; Brazilian Network for Continuous Monitoring of IBGE; the Geoscience Australia; LISN; and AFREF. The satellite and receiver bias the data were obtained from AIUB Data Center of Bern University in Switzerland. The geomagnetic data were downloaded from: WDC for Geomagnetism, Kyoto; and OMNIWeb Plus Data Documentation. EQ data were available in the USGS Comprehensive Catalog of Earthquakes. The ionospheric TEC and LP measurements of in situ electron density data provided by ESA Swarm mission.

Acknowledgments: The authors would sincerely thank Miguel Martínez-Ledesma from Universidad de Concepción for his advice on plotting the eclipse obscuration mask.

Conflicts of Interest: The authors declare no conflict of interest. The funders had no role in the design of the study; in the collection, analyses, or interpretation of data; in the writing of the manuscript, or in the decision to publish the results.

References

1. Kumar, S.; Singh, A. Changes in total electron content (TEC) during the annular solar eclipse of 15 January 2010. *Adv. Space Res.* **2012**, *49*, 75–82. [CrossRef]
2. Founda, D.; Melas, D.; Lykoudis, S.; Lisaridis, I.; Gerasopoulos, E.; Kouvarakis, G.; Petrakis, M.; Zerefos, C. The effect of the total solar eclipse of 29 March 2006 on meteorological variables in Greece. *Atmos. Chem. Phys.* **2007**, *7*, 5543–5553. [CrossRef]
3. Singh, A.K.; Singh, R.; Veenadhari, B.; Singh, A. Response of low latitude D-region ionosphere to the total solar eclipse of 22 July 2009 deduced from ELF/VLF analysis. *Adv. Space Res.* **2012**, *50*, 1352–1361. [CrossRef]
4. Afraimovich, E.L.; Astafyeva, E.I.; Demyanov, V.V.; Edemskiy, I.K.; Gavrilyuk, N.S.; Ishin, A.B.; Kosogorov, E.A.; Leonovich, L.A.; Lesyuta, O.S.; Palamartchouk, K.S.; et al. A review of GPS/GLONASS studies of the ionospheric response to natural and anthropogenic processes and phenomena. *J. Space Weather Space Clim.* **2013**, *3*, A27. [CrossRef]
5. Hoque, M.M.; Wenzel, D.; Jakowski, N.; Gerzen, T.; Berdermann, J.; Wilken, V.; Kriegel, M.; Sato, H.; Borries, C.; Minkwitz, D. Ionospheric response over Europe during the solar eclipse of March 20, 2015. *J. Space Weather Space Clim.* **2016**, *6*, A36. [CrossRef]
6. Sun, Y.Y.; Liu, J.Y.; Lin, C.C.H.; Lin, C.Y.; Shen, M.H.; Chen, C.H.; Chen, C.H.; Chou, M.Y. Ionospheric bow wave induced by the moon shadow ship over the continent of United States on 21 August 2017. *Geophys. Res. Lett.* **2018**, *45*, 538–544. [CrossRef]
7. Paul, A.; Das, T.; Ray, S.; Das, A.; Bhowmick, D.; DasGupta, A. Response of the equatorial ionosphere to the total solar eclipse of 22 July 2009 and annular eclipse of 15 January 2010 as observed from a network of stations situated in the Indian longitude sector. *Ann. Geophys.* **2011**, *29*, 1955–1965. [CrossRef]
8. An-hua, C.; Sheng-Bing, Y.; Ji-Sheng, X. Ionospheric responses to a total solar eclipse deduced by the GPS beacon observations. *Wuhan Univ. J. Nat. Sci.* **1999**, *4*, 439–444. [CrossRef]
9. Afraimovich, E.L.; Lesyuta, O.S. Ionospheric response to the total solar eclipse of 21 June 2001. *arXiv* **2002**, arXiv:0201047.
10. Momani, M.A.; Yatim, B.; Mohd Ali, M.A. Ionospheric and geomagnetic response to the total solar eclipse on 1 August 2008 over Northern Hemisphere. *J. Geophys. Res. Space Phys.* **2010**, *115*, A8. [CrossRef]
11. Panda, S.K.; Gedam, S.; Rajaram, G.; Sripathi, S.; Bhaskar, A. Impact of 15 Jan 2010 annular solar eclipse on the equatorial and low latitude ionosphere over Indian region from Magnetometer, Ionosonde and GPS observations. *arXiv* **2015**, arXiv:1506.05245.
12. Sharma, S.; Dashora, N.; Galav, P.; Pandey, R. Total solar eclipse of 22 July 2009: Its impact on the total electron content and ionospheric electron density in the Indian zone. *J. Atmos. Sol. Terr. Phys.* **2010**, *72*, 1387–1392. [CrossRef]
13. Tsai, H.; Liu, J. Ionospheric total electron content response to solar eclipses. *J. Geophys. Res. Space Phys.* **1999**, *104*, 12657–12668. [CrossRef]
14. Farges, T.; Le Pichon, A.; Blanc, E.; Perez, S.; Alcoverro, B. Response of the lower atmosphere and the ionosphere to the eclipse of 11 August 1999. *J. Atmos. Sol. Terr. Phys.* **2003**, *65*, 717–726. [CrossRef]
15. Lei, J.; Dang, T.; Wang, W.; Burns, A.; Zhang, B.; Le, H. Long-Lasting Response of the Global Thermosphere and Ionosphere to the 21 August 2017 Solar Eclipse. *J. Geophys. Res. Space Phys.* **2018**, *123*, 4309–4316. [CrossRef]
16. Bravo, M.; Martínez-Ledesma, M.; Foppiano, A.; Urra, B.; Ovalle, E.; Villalobos, C.; Souza, J.; Carrasco, E.; Muñoz, P.R.; Tamblay, L.; et al. First Report of an Eclipse From Chilean Ionosonde Observations: Comparison with Total Electron Content Estimations and the Modeled Maximum Electron Concentration and Its Height. *J. Geophys. Res. Space Phys.* **2020**, *125*, e2020JA027923. [CrossRef]
17. Aa, E.; Zhang, S.R.; Shen, H.; Liu, S.; Li, J. Local and conjugate ionospheric total electron content variation during the 21 June 2020 solar eclipse. *Adv. Space Res.* **2021**, *68*, 3435–3454. [CrossRef]
18. He, L.; Heki, K.; Wu, L. Three-Dimensional and Trans-Hemispheric Changes in Ionospheric Electron Density Caused by the Great Solar Eclipse in North America on 21 August 2017. *Geophys. Res. Lett.* **2018**, *45*, 10933–10940. [CrossRef]
19. Le, H.; Liu, L.; Yue, X.; Wan, W. The ionospheric behavior in conjugate hemispheres during the 3 October 2005 solar eclipse. *Ann. Geophys.* **2009**, *27*, 179–184. [CrossRef]
20. Zhang, S.R.; Erickson, P.J.; Vierinen, J.; Aa, E.; Rideout, W.; Coster, A.J.; Goncharenko, L.P. Conjugate Ionospheric Perturbation During the 2017 Solar Eclipse. *J. Geophys. Res. Space Phys.* **2021**, *126*, e2020JA028531. [CrossRef]
21. Verhulst, T.G.W.; Stankov, S.M. Height Dependency of Solar Eclipse Effects: The Ionospheric Perspective. *J. Geophys. Res. Space Phys.* **2020**, *125*, e2020JA028088. [CrossRef]
22. Afraimovich, E.; Kosogorov, E.; Lesyuta, O. Effects of the 11 August 1999 total solar eclipse as deduced from total electron content measurements at the GPS network. *J. Atmos. Sol. Terr. Phys.* **2002**, *64*, 1933–1941. [CrossRef]
23. Stankov, S.M.; Bergeot, N.; Berghmans, D.; Bolsée, D.; Bruyninx, C.; Chevalier, J.M.; Clette, F.; De Backer, H.; De Keyser, J.; D'Huys, E.; et al. Multi-instrument observations of the solar eclipse on 20 March 2015 and its effects on the ionosphere over Belgium and Europe. *J. Space Weather Space Clim.* **2017**, *7*, A19. [CrossRef]
24. Seemala, G.K.; Valladares, C.E. Statistics of total electron content depletions observed over the South American continent for the year 2008. *Radio Sci.* **2011**, *46*, 1–14. [CrossRef]
25. Yuan, Y.; Tscherning, C.; Knudsen, P.; Xu, G.; Ou, J. The ionospheric eclipse factor method (IEFM) and its application to determining the ionospheric delay for GPS. *J. Geod.* **2008**, *82*, 1–8. [CrossRef]
26. Park, J.; Shahbazi, A.; Kim, S.K.; Oberg, R. Ionospheric Response to the Total Solar Eclipse of 21 August 2017, and Its Impact on GNSS Positioning. *J. Surv. Eng.* **2019**, *145*, 05019001. [CrossRef]
27. Filjar, R. Horizontal GPS positioning accuracy during the 1999 solar eclipse. *J. Navig.* **2001**, *54*, 293–296. [CrossRef]

28. Jia-Chun, A.; Ze-Min, W.; Dong-Chen, E.; Wei, S. Ionospheric Behavior During the Solar Eclipse of 22 July 2009 and Its Effect on Positioning. *Chin. J. Geophys.* **2010**, *53*, 731–739. [CrossRef]
29. Valdés-Abreu, J.C.; Díaz, M.A.; Báez, J.C.; Stable-Sánchez, Y. Effects of the 12 May 2021 Geomagnetic Storm on Georeferencing Precision. *Remote Sens.* **2022**, *14*, 38. [CrossRef]
30. Katsigianni, G.; Loyer, S.; Perosanz, F. PPP and PPP-AR Kinematic Post-Processed Performance of GPS-Only, Galileo-Only and Multi-GNSS. *Remote Sens.* **2019**, *11*, 2477. [CrossRef]
31. Abou-Galala, M.; Rabah, M.; Kaloop, M.; Zidan, Z.M. Assessment of the accuracy and convergence period of Precise Point Positioning. *Alex. Eng. J.* **2018**, *57*, 1721–1726. [CrossRef]
32. Turel, N.; Arikan, F. Probability density function estimation for characterizing hourly variability of ionospheric total electron content. *Radio Sci.* **2010**, *45*, 1–10. [CrossRef]
33. Dow, J.; Neilan, R.; Rizos, C. The International GNSS Service in a Changing Landscape of Global Navigation Satellite Systems. *J. Geod.* **2008**, *83*, 191–198. [CrossRef]
34. Piñón, D.; Gomez, D.; Smalley, B.; Cimbaro, S.; Lauría, E.; Bevis, M. The History, State, and Future of the Argentine Continuous Satellite Monitoring Network and Its Contributions to Geodesy in Latin America. *Seismol. Res. Lett.* **2018**, *89*, 475–482. [CrossRef]
35. Tsidu, G.M.; Abraha, G. Moderate geomagnetic storms of 22–25 January 2012 and their influences on the wave components in ionosphere and upper stratosphere-mesosphere regions. *Adv. Space Res.* **2014**, *54*, 1793–1812. [CrossRef]
36. Liu, X.; Yuan, Y.; Tan, B.; Li, M. Observational Analysis of Variation Characteristics of GPS-Based TEC Fluctuation over China. *ISPRS Int. J. Geo-Inf.* **2016**, *5*, 237. [CrossRef]
37. Friis-Christensen, E.; Lühr, H.; Hulot, G. Swarm: A constellation to study the Earth's magnetic field. *Earth Planets Space* **2006**, *58*, 351–358. [CrossRef]
38. Pignalberi, A. TITIPy: A Python tool for the calculation and mapping of topside ionosphere turbulence indices. *Comput. Geosci.* **2021**, *148*, 104675. [CrossRef]
39. Swarm L1b Product Definition. 2018. Available online: https://earth.esa.int/eogateway/documents/20142/37627/swarm-level-1b-product-definition-specification.pdf/12995649-fbcb-6ae2-5302-2269fecf5a08 (accessed on 15 March 2022).
40. Swarm L2 TEC Product Description. 2017. Available online: https://earth.esa.int/eogateway/documents/20142/37627/swarm-level-2-tec-product-description.pdf/8fe7fa04-6b4f-86a7-5e4c-99bb280ccc7e (accessed on 15 March 2022).
41. Abdallah, A.; Saifeldin, A.; Abomariam, A.; Ali, R. Efficiency of using GNSS-PPP for digital elevation model (DEM) production. *Artif. Satell.* **2020**, *55*, 17–28. [CrossRef]
42. Alkan, R.M.; Erol, S.; Ozulu, I.M.; Ilci, V. Accuracy comparison of post-processed PPP and real-time absolute positioning techniques. *Geomat. Nat. Hazards Risk* **2020**, *11*, 178–190. [CrossRef]
43. Yigit, C.O. Experimental assessment of post-processed kinematic Precise Point Positioning method for structural health monitoring. *Geomat. Nat. Hazards Risk* **2016**, *7*, 360–383. [CrossRef]
44. Mendez Astudillo, J.; Lau, L.; Tang, Y.T.; Moore, T. Analysing the Zenith Tropospheric Delay Estimates in On-line Precise Point Positioning (PPP) Services and PPP Software Packages. *Sensors* **2018**, *18*, 580. [CrossRef]
45. Davis, T.N.; Sugiura, M. Auroral electrojet activity index AE and its universal time variations. *J. Geophys. Res.* **1966**, *71*, 785–801. [CrossRef]
46. De Michelis, P.; Consolini, G.; Pignalberi, A.; Tozzi, R.; Coco, I.; Giannattasio, F.; Pezzopane, M.; Balasis, G. Looking for a proxy of the ionospheric turbulence with Swarm data. *Sci. Rep.* **2021**, *11*, 6183. [CrossRef]
47. Yasyukevich, Y.V.; Kiselev, A.V.; Zhivetiev, I.V.; Edemskiy, I.K.; Syrovatskii, S.V.; Maletckii, B.M.; Vesnin, A.M. SIMuRG: System for ionosphere monitoring and research from GNSS. *GPS Solut.* **2020**, *24*, 69. [CrossRef]
48. Pignalberi, A.; Pietrella, M.; Pezzopane, M.; Habarulema, J. Investigating different vTEC calibration methods for data assimilation in ionospheric empirical models. *Adv. Space Res.* **2021**, *68*, 2138–2151. [CrossRef]
49. Kundu, S.; Chowdhury, S.; Palit, S.; Mondal, S.K.; Sasmal, S. Variation of ionospheric plasma density during the annular solar eclipse on 26 December 2019. *Astrophys. Space Sci.* **2022**, *367*, 44. [CrossRef]
50. Reinisch, B.; Galkin, I. Global Ionospheric Radio Observatory (GIRO). *Earth Planets Space* **2011**, *63*, 377–381. [CrossRef]
51. Dang, T.; Lei, J.; Wang, W.; Zhang, B.; Burns, A.; Le, H.; Wu, Q.; Ruan, H.; Dou, X.; Wan, W. Global Responses of the Coupled Thermosphere and Ionosphere System to the August 2017 Great American Solar Eclipse. *J. Geophys. Res.* **2018**, *123*, 7040–7050. [CrossRef]
52. Huba, J.D.; Drob, D. SAMI3 prediction of the impact of the 21 August 2017 total solar eclipse on the ionosphere/plasmasphere system. *Geophys. Res. Lett.* **2017**, *44*, 5928–5935. [CrossRef]
53. Jonah, O.F.; Goncharenko, L.; Erickson, P.J.; Zhang, S.; Coster, A.; Chau, J.L.; de Paula, E.R.; Rideout, W. Anomalous Behavior of the Equatorial Ionization Anomaly During the 2 July 2019 Solar Eclipse. *J. Geophys. Res. Space Phys.* **2020**, *125*, e2020JA027909. [CrossRef]
54. Martínez-Ledesma, M.; Bravo, M.; Urra, B.; Souza, J.; Foppiano, A. Prediction of the Ionospheric Response to the 14 December 2020 Total Solar Eclipse Using SUPIM-INPE. *J. Geophys. Res. Space Phys.* **2020**, *125*, e2020JA028625. [CrossRef]
55. Sergeenko, N. Irregular Phenomena in Magnetically Conjugate Regions of the F2 Layer of the Ionosphere. *Geomagn. Aeron.* **2018**, *58*, 823–830. [CrossRef]
56. Reinisch, B.W.; Dandenault, P.B.; Galkin, I.A.; Hamel, R.; Richards, P.G. Investigation of the Electron Density Variation During the 21 August 2017 Solar Eclipse. *Geophys. Res. Lett.* **2018**, *45*, 1253–1261. [CrossRef]

57. Anderson, D.; Anghel, A.; Chau, J.; Veliz, O. Daytime vertical E × B drift velocities inferred from ground-based magnetometer observations at low latitudes. *Space Weather* **2004**, *2*, S11001. [CrossRef]
58. Kotulak, K.; Krankowski, A.; Froń, A.; Flisek, P.; Wang, N.; Li, Z.; Błaszkiewicz, L. Sub-Auroral and Mid-Latitude GNSS ROTI Performance during Solar Cycle 24 Geomagnetic Disturbed Periods: Towards Storm's Early Sensing. *Sensors* **2021**, *21*, 4325. [CrossRef]
59. Kotulak, K.; Zakharenkova, I.; Krankowski, A.; Cherniak, I.; Wang, N.; Fron, A. Climatology Characteristics of Ionospheric Irregularities Described with GNSS ROTI. *Remote Sens.* **2020**, *12*, 2634. [CrossRef]
60. Cherniak, I.; Zakharenkova, I. Ionospheric Total Electron Content Response to the Great American Solar Eclipse of 21 August 2017. *Geophys. Res. Lett.* **2018**, *45*, 1199–1208. [CrossRef]
61. Shagimuratov, I.; Chernouss, S.; Despirak, I.; Filatov, M.; Efishov, I.; Tepenitsyna, N.Y. Occurrence of TEC fluctuations and GPS positioning errors at different longitudes during auroral disturbances. *Sun Geosph.* **2018**, *13*, 89–94. [CrossRef]
62. Poniatowski, M.; Nykiel, G. Degradation of Kinematic PPP of GNSS Stations in Central Europe Caused by Medium-Scale Traveling Ionospheric Disturbances during the St. Patrick's Day 2015 Geomagnetic Storm. *Remote Sens.* **2020**, *12*, 3582. [CrossRef]
63. Luo, X.; Gu, S.; Lou, Y.; Xiong, C.; Chen, B.; Jin, X. Assessing the performance of GPS precise point positioning under different geomagnetic storm conditions during solar cycle 24. *Sensors* **2018**, *18*, 1784. [CrossRef]
64. Yang, Z.; Mrak, S.; Morton, Y.J. Geomagnetic Storm Induced Mid-latitude Ionospheric Plasma Irregularities and Their Implications for GPS Positioning over North America: A Case Study. In Proceedings of the 2020 IEEE/ION Position, Location and Navigation Symposium (PLANS), Portland, OR, USA, 20–23 April 2020; IEEE: Piscataway, NJ, USA, 2020; pp. 234–238. [CrossRef]

remote sensing

MDPI

Article

Advanced Classification of Ionospheric Troughs in the Morning and Evening Conditions

Alexander Karpachev

Pushkov Institute of Terrestrial Magnetism, Ionosphere, and Radiowave Propagation (IZMIRAN), 4, Kaluzhskoe Hwy, Troitsk, 108840 Moscow, Russia; karp@izmiran.ru; Tel.: +7-9164069331

Highlights:

- The separation of ionospheric troughs in the winter evening and morning ionosphere of the southern hemisphere was performed.
- The study is based on electron density measurements at CHAMP satellite altitudes of 405–465 km during 2000–2002.
- The main ionospheric trough was separated from poleward high latitude trough and from equaror-ward ring ionosperic trough.

Abstract: The separation and classification of ionospheric troughs in the winter evening and morning ionospheres of the southern hemisphere were performed using CHAMP satellite data for high solar activity (2000–2002). In the high-latitude ionosphere, the main ionospheric trough (MIT) was separated from the high-latitude trough (HLT). The separation was carried out using a thorough analysis of all the characteristic structures of the ionosphere in the framework of the auroral diffuse particle precipitation model. Two types of high-latitude troughs were identified: (1) a wide trough associated with zone II of diffuse precipitation on the poleward edge of the auroral oval and (2) a narrow trough of ionization, which is presumably associated with an electric field action. The poleward wall of MIT is as ever formed by diffuse precipitation in zone I on the equatorward edge of the auroral oval. The HLT and MIT separation is most difficult at the longitudes of the eastern hemisphere, where all structures are located at the highest latitudes and partially overlap. In the mid-latitude ionosphere, all the characteristic structures of the ionosphere were also identified and considered. MIT was separated from the ring ionospheric trough (RIT), which is formed by the decay processes of the magnetospheric ring current. The separation of MIT and RIT was performed based on an analysis of the prehistory of all geomagnetic disturbances during the period under study. In addition to the RIT, a decrease in the electron density equatorward of the MIT was found to be often formed at the America–Atlantic longitudes, which masks the MIT minimum. For completeness, all cases of a clearly defined polar cavity are also presented.

Keywords: main ionospheric trough; high latitude trough; ring ionospheric trough; low latitude trough; auroral diffuse precipitation

Citation: Karpachev, A. Advanced Classification of Ionospheric Troughs in the Morning and Evening Conditions. *Remote Sens.* **2022**, *14*, 4072. https://doi.org/10.3390/rs14164072

Academic Editor: Fabio Giannattasio

Received: 31 July 2022
Accepted: 18 August 2022
Published: 20 August 2022

1. Introduction

The ionization trough was discovered by Muldrew from the Alouette 1 data [1]. Muldrew identified it as the main ionospheric trough (MIT). MIT is located equatorward of the auroral oval. The results of numerous MIT studies are summarized in reviews [2–5]. Inside the auroral oval, another (high–latitude, HLT) trough was detected and studied in detail using the OGO 6 satellite data [6]. The so-called ring ionospheric trough (RIT) was later discovered equatorward at MIT [7]. RIT is formed by the decay process of the magnetospheric ring current and is typically observed in the recovery phase of a storm/substorm [8,9]. The multitudes of MIT and HLT, as well as MIT and RIT, partially overlap; therefore, the problem of trough separation arises in the region of the intersection. This problem was posed in prior studies [10,11] and elaborated in detail in a study by Karpachev [12]. However, this problem was not completely solved, so Liu and Xiong [13], processing a large dataset of CHAMP satellites, wrote: "There is no agreed quantitative definition of an MIT [14,15], as sometimes it is difficult to identify the individual mid-latitude trough from kinds of depletion structures extending along latitude, such as high-latitude trough and wave-like disturbances". An almost unambiguous solution to the problem of trough separation in the midnight ionosphere was obtained in a recent study by Karpachev [16]. Significant progress in the separation and classification of various structures of the midnight high-latitude and mid-latitude ionosphere has been achieved based on a thorough analysis of a large dataset of CHAMP satellites, which allows us to consider the phenomenon from different perspectives. As a result, two troughs were identified in the auroral ionosphere. A wide trough (HLT1) was identified in the framework of a simple visual model of diffuse auroral precipitation [17]. This model describes zone I of diffuse precipitation on the equatorward edge of the auroral oval and zone II on its poleward edge. It was determined that the precipitation in zone II forms the poleward wall of HLT1 and the precipitation in zone I forms the MIT poleward wall. This is the key point in separating MIT and HLT1.

The analysis is most effectively conducted in the framework of the longitudinal effect because the boundaries of both zones change with longitude by 2.5° in latitude [18], similar to the longitudinal variations in the MIT position. Analyzing the longitudinal variations of these structures, it was found that the problem of MIT and HLT1 separation is radically different in the western and eastern hemispheres. In the western hemisphere, MIT is located at lower latitudes than in the eastern hemisphere, and it is further removed from the auroral oval; therefore, it is quite simple to separate it from the HLT. In the eastern hemisphere, MIT shifts to high latitudes so that the region of its existence begins to overlap strongly with the statistical position of zone I of precipitation and the region of existence of HLT. Therefore, the separation of troughs in the eastern hemisphere was carried out according to the correspondence of the MIT poleward wall to the position of zone I and the HLT1 poleward wall to the position of zone II of precipitation. In addition, the trough minimum position relative to zone I was also monitored. The second trough (HLT2), described in [6,19], is associated with the action of local electric fields [19]; therefore, it is narrow in latitude (3–5°) and can be observed in any region of the auroral oval.

Because MIT is observed equatorward of the auroral oval, it is, by definition, a sub-auroral trough. The RIT is formed equatorward of MIT; therefore, it can be defined as a mid-latitude trough. RIT is formed even after a slight increase in geomagnetic activity and can be observed for a long time (sometimes two days) at latitudes of 53–57° GMLat [8,9]. Therefore, the frequency of RIT occurrence is determined by the degree of perturbation of the ionosphere, which is higher under high solar activity. The separation of MIT from RIT is also a difficult problem, but we can use the separation method carefully developed earlier [8,9]. It consists of the following: If MIT and RIT are observed simultaneously, then the equatorward trough is defined as the RIT. If there is one trough, its dynamics in prehistory must be considered. If the variations of the trough position correspond to the MIT model constructed in terms of the Kp index [20], then it is defined as MIT. If the trough is much more equatorward than the model position, then this is the RIT. However, there may be controversial cases that must be considered carefully. This will be shown below.

At longitudes of America and the Atlantic, a special structure is often recorded: a weak electron density minimum at the base of the MIT poleward wall and a deep minimum located much more equatorward. This structure is recorded in the evening more often than in the morning. The first minimum is located poleward of the average MIT position, whereas the second is, correspondingly, equatorward. Consequently, there is the problem of accurately determining MIT position, which is also discussed below.

Finally, to complete the pattern, the clearest cases of a polar cavity are also highlighted. However, a thorough analysis has not been conducted because, at most longitudes, the satellite did not reach the geomagnetic latitudes necessary for registration.

This work is an extension of the study of the irregular structure of the ionosphere in the early evening and early morning hours, based on the experience gained from the analysis of the midnight ionosphere [16]. In both studies, only the pattern in the southern hemisphere was considered. The southern hemisphere is much better suited for practicing this technique because there are a number of interesting effects in this hemisphere. This particularly applies to the longitudes of America.

2. Materials and Methods

The CHAMP satellite carried out in situ measurements of electron density *Ne* [21]. Variations in *Ne* are presented below in terms of plasma frequency *fp* ($Ne[\mathrm{cm}^{-3}] = 1.24{\cdot}10^4 fp^2$ [MHz]). The CHAMP altitude has changed from ~450 km to ~300 km, which is close to the height of the F2 layer maximum. It revolved on a nearly polar orbit with an inclination of 87°. The CHAMP data time resolution of 15 s is less than 1° of latitude, which allows accurate determination of the minimum trough position. The CHAMP data are available on the website http://op.gfz-potsdam.de/champ (accessed on 12 January 2015).

In this study, CHAMP data for local winter conditions in the southern hemisphere were used. The data only for high solar activity with F10.7~180 sfu for the period of 2000–2002 and the evening (17–19 LT) and morning (05–06 LT) conditions were considered. About 1500 CHAMP passes in the winter high- and mid-latitude ionosphere for $\mathrm{Kp} \leq 4$ were examined. The MIT is usually defined as a sufficiently deep decrease in electron density of at least ~30% relative to the top of the steep equatorward wall [3]. However, in the morning sector, the electron density usually monotonously falls to high latitudes without noticeable inflection on the equatorward wall, and the trough definition becomes uncertain. Therefore, we have not estimated the level of electron density decrease at the MIT minimum. The position of MIT was determined by the electron density minimum at several degrees equatorward of the base of the poleward wall [3]. In the morning, the poleward wall was almost always clearly defined. If the trough was poorly expressed or masked by ionospheric plasma irregularities on some satellite path, the position of its minimum was determined through coordination with well-expressed troughs on neighboring paths.

Stricter criteria were imposed on the selection of the HLT. The HLT is observed in the auroral oval, where the electron density is highly irregular and several density minima can be observed. Therefore, the HLT was recorded only in obvious cases wherein it was clearly structured and when its poleward wall did not extend beyond the poleward diffuse precipitation zone. Similarly, the polar hole was defined only as a broad minimum of the electron density at latitudes above the poleward precipitation zone. Finally, only pronounced troughs were recorded equatorward of MIT.

3. Structure of the Evening Ionosphere

Figure 1 shows the positions of the different structures in the winter evening (17–19 LT) ionosphere of the southern hemisphere in terms of geomagnetic latitude—geographical longitude. The following structures are presented in Figure 1: polar hole, HLT, MIT, RIT and specific equatorward minimum of electron density. To eliminate the dependence on geomagnetic activity, the positions of the MIT, RIT and HLT were reduced to Kp = 2 according to $\Lambda\mathrm{corr} = \Lambda\mathrm{c} - a(\mathrm{Kp}(\tau) - 2)$, where $\Lambda\mathrm{c}$ is the current position of the structure and the ***a*** factor is 2.0° for the MIT [20], 1.5° for RIT [9], and ~1.5° for HLT [6]. The Kp(τ) index was used because it considers the prehistory of geomagnetic activity development [22]. In the evening sector, the dependence of MIT position on local time is quite strong; it was also considered according to the model [20].

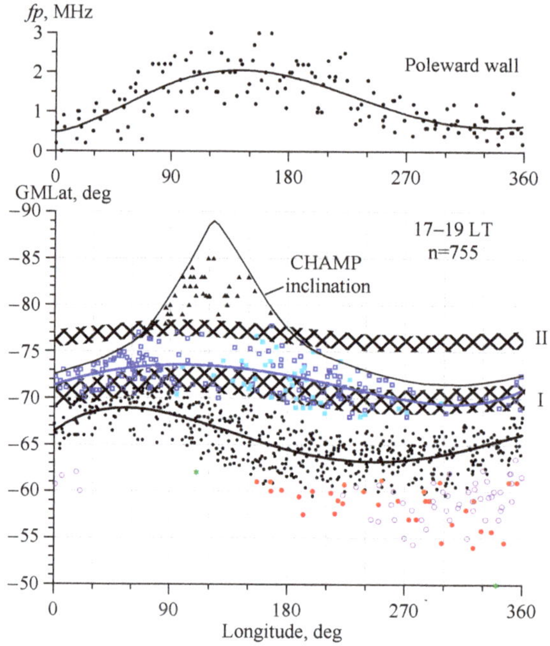

Figure 1. On the top: Longitudinal variations in the magnitude of the MIT poleward wall (dots and approximations). On the bottom: Longitudinal variations in the positions of main structures in the evening winter ionosphere of the southern hemisphere: polar hole (triangles), HLT1 (empty squares), HLT2 (filled squares), MIT (dots), RIT (red dots), equatorward density minima (purple circles), and low-latitude structures (green asterisks). The shaded latitude belts show diffuse auroral precipitation in zones I and II. The upper curve represents the CHAMP inclination equal to 87°.

An analysis of the structures of the high latitude ionosphere was conducted using a model of auroral particle precipitation constructed from DMSP satellite data recorded in both hemispheres [17]. The model agrees well with other statistical models of auroral electron precipitation in the nighttime conditions [23–25]. There are no data in the literature regarding the dependence of the position of the diffuse precipitation zones on longitude in the evening sector. However, because the position of MIT showed a pronounced longitudinal effect [12], we preserved the shape and amplitude of the longitudinal effect in the auroral precipitation in the same form as revealed in the interval of 21–03 MLT [18] and used it to analyze the structure of the midnight high-latitude ionosphere [16]. For the southern hemisphere, these boundaries are presented in Figure S2 (in Supporting Information) according to [18]. The equatorward and poleward boundaries of the auroral oval experience synchronous longitudinal variations with an amplitude of ~2.5°. In Figure 1, zones I and II of the diffuse precipitation taken from Figure S1 are shaded. The average (for all longitudes) position of the equatorward boundary of the auroral precipitation oval corresponds to model [17] for Kp = 2. The upper thin curve in Figure 1 depicts the CHAMP satellite inclination. The satellite inclination of 87° does not limit the observations of the discussed structures, except for the polar hole. However, polar hole cases are shown in Figure 1 solely for completeness of the pattern; only unambiguous cases were selected.

The longitudinal variations in the MIT position are quite confidently determined from the CHAMP data in the evening (correlation coefficient $r = 0.72$, standard deviation $\sigma = 1.8°$). The amplitude of the longitudinal effect reaches ~6°, which is greater than at midnight [16]. At longitudes of the eastern hemisphere between 0° and 120°E, the MIT is located farthest from the pole; thus, this region overlaps with zone I of diffuse precipitation, even when considering the assumed longitudinal effect in the position of this zone. Even more important is that the clearly expressed MITs in this region are superimposed on well-defined HLTs. Regardless, the problem of separating MIT and HLT in the evening sector is more complicated than in the midnight sector and requires detailed consideration. Figure 2 shows several examples of the structure of the evening ionosphere, which require careful analysis. The shaded bars in Figure 2 show zones I and II of diffuse precipitation, which positions correspond to the current values of the longitude and Kp index. In some cases, the zones are slightly shifted to better compliance of the structures on the grounds that CHAMP provided the current data, and the precipitation model is statistical.

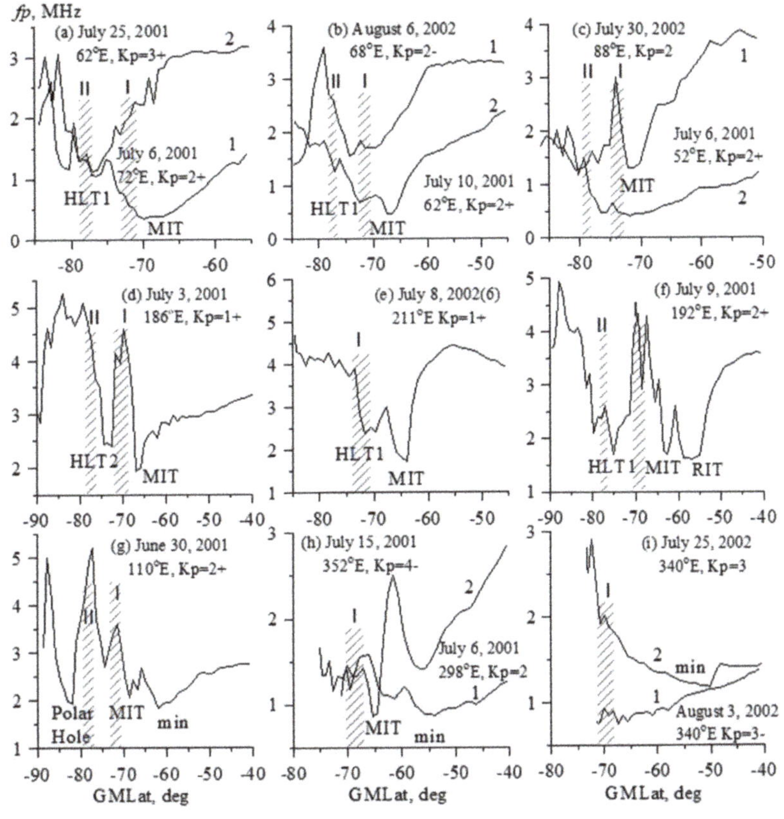

Figure 2. Characteristic examples of ionospheric troughs at different longitudes in the evening ionosphere. The shaded bars in (**a**) (zones I and II of precipitation) correspond to *fp* profile 1, i.e., Kp = 2+; the bar in Figure 2h corresponds to *fp* profile 1, i.e., Kp = 2.

The most difficult problem with the separation of MIT and HLT is observed in the eastern hemisphere at longitudes of 30–90°E, where the locations of MIT and HLT partially overlap. The weakly pronounced poleward wall of MIT at these longitudes (upper panel of Figure 1) complicated this problem. The data on the poleward wall were obtained during the quiet period from 10–17 July 2001.

The situation at problematic longitudes is considered below using individual characteristic examples. In Figure 2a, curve 1 depicts an example of a conventional MIT. Its minimum is located at quite a high latitude (−70°), but its poleward wall is definitely formed by diffuse precipitation in zone I. The MIT poleward wall is low, and the electron density minimum at −77° is apparently a sign of a high-latitude trough (HLT). Curve 2 depicts an example of a well-defined HLT with a minimum latitude of −76° (under Kp = 3+), i.e., deep inside the auroral oval. Its poleward wall is certainly associated with zone II of precipitation. This type of high-latitude trough has been defined as HLT1 [16]. The MIT on latitudinal *fp* profile 2 is completely filled, and only ionospheric plasma irregularities are observed at latitudes of its assumed minimum. This latitudinal *fp* profile is likely a consequence of the previous disturbance with Kp = 4+. Thus, the main criterion for the separation of MIT and HLT1 is the correspondence of the poleward wall of the trough to precipitation in zones I and II, respectively. The position of the trough minimum was also taken into consideration.

Figure 2b shows examples of the simultaneous existence of two electron density minima. Then, a lower-latitude minimum (<70°) corresponds to the MIT, and a higher-latitude one (>70°) refers to the HLT. The poleward walls of both troughs correspond to the related precipitation zones, although the MIT poleward wall is not pronounced in either case.

Curve 1 in Figure 2c represents a trough with a minimum at a very high latitude of −72° for Kp = 2. However, its poleward wall is clearly formed by precipitation in zone I, i.e., MIT. Of note that, in this case, MIT is located deep in the region of HLT existence. The latitudinal profile 2, as well as in Figure 2b, shows two *fp* minima. The equatorward minimum at latitude of −71.7° (under Kp = 2+) coincides with the MIT minimum on curve 1. However, its weak poleward wall is formed by only one *fp* value, i.e., it is not reliable. Therefore, this structure was attributed to the HLT, in which the poleward wall is located at latitudes of (−77–79°), i.e., at latitudes of zone II of diffuse precipitation. Thus, as the above examples show, the identification and separation of troughs in the challenge region at longitudes of 30–90°E was possible, although it required a thorough analysis of complex, controversial cases.

In Figure 2d, MIT and HLT are observed at the same time. However, in this case, the high-latitude trough is very narrow and deep. Such troughs were recorded onboard the OGO 6 satellite [6]. The authors of this study associated these troughs with the action of electric fields in the region of the high-latitude convection of ionospheric plasma. This trough can be defined as HLT2. In Figure 1, the approximating curve is determined for both HLT types. This curve shows the longitudinal effect in the HLT position, but it is distorted because, at most longitudes, the inclination of the satellite's orbit does not allow recording of the highest-latitude cases of HLT. (The HLT shown in Figure 1 sometimes goes beyond satellite inclination, which is associated with the reduction of the position of the troughs to Kp = 2).

In Figure 1, in the longitude range from 150° to 240°E, several HLTs are located at very low latitudes, equatorward of the polar oval. Such a case is presented in Figure 2e for HLT1. In this case, the MIT minimum and its poleward wall are much more equatorward than zone I of precipitation. We can deduce that zone I is located more equatorward at most longitudes than in Figure 1. However, this assumption cannot be verified without data on precipitation at all longitudes in the evening sector.

In Figure 2f, three troughs are observed simultaneously: HLT1, MIT and RIT. A well-expressed RIT was observed after a substorm with Kp = 5. The RITs in Figure 1 are indicated by red (filled) circles. The RIT is formed mainly in the western hemisphere. This is because the geomagnetic field in this hemisphere is weak, the precipitation of hot ions of the magnetospheric ring current is intense, and the trough is formed more often.

In Figure 2g, curve 1 shows the polar cavity at latitude of −82°, MIT at latitude of −68°, and the *fp* minimum at latitude of −61.5°. The polar cavity is observed at latitudes >78°; therefore it was recorded only at longitudes of the eastern hemisphere, where the satellite inclination allowed. The *fp* minimum is depicted by the green asterisk in Figure 1 at a longitude of 110°E. It clearly stands out from among other troughs, and the reason for its formation is unknown, although it is very clearly expressed.

Curve 1 in Figure 2h displays the structure often observed at the longitudes of America and the Atlantic, as discussed in the Introduction. This structure is characterized by a small *fp* minimum at latitude of −62° at the base of the MIT poleward wall and a deeper and wider minimum at latitude of −55°. This minimum makes it difficult to accurately determine the position of the MIT minimum. These equatorward electron density minima are marked in Figure 1 by purple circles. Figure 1 shows that both the RIT and the equatorward minimum are located in the same region, mainly in the western hemisphere. However, the RIT is formed after geomagnetic disturbance, and a structure with two electron density minima is formed under long-term quiet conditions. In addition, this structure has a specific shape.

Curve 2 in Figure 2h depicts a structure at a longitude of 352°E, formally similar to the structure represented by curve 1. However, the *fp* peak in this case is so pronounced that it represents a special independent structure. In other words, a question arises regarding the reasons for the formation of this peak. The structure in question was formed in a weakly disturbed ionosphere at Kp = 4−, but it is often observed at American longitudes under completely quiet geomagnetic conditions. The *fp* peak is located on the MIT equatorward wall so that the MIT minimum appears at a latitude of −65.3°. This peak forms a pronounced equatorward minimum at latitude of −57°, which can easily be mistaken for MIT with automatic and even manual data processing. However, even at Kp = 4−, this minimum is much more equatorward than the "normal" MIT, as shown in Figure 1.

The top panel in Figure 1 shows the longitudinal variations in the electron density on the top of the poleward wall derived with correlation coefficient r = 0.68, standard deviation σ = 0.43°, and amplitude A~1.5°. At small and especially large longitudes, the MIT poleward wall is surprisingly low. Another specific feature of the evening ionosphere at these longitudes is the slow decrease in electron density from middle to high latitudes, i.e., the small latitudinal gradient at the latitudes of MIT. This demonstrates curve 1 in Figure 2i. Latitudinal *fp* profile 1 was obtained on 3 August 2002 in the longitudinal sector of 340°E. The electron density slowly and almost monotonously decreases to high latitudes, showing neither a noticeable minimum nor a poleward wall. Such *fp* profiles are quite often observed at the discussed longitudes. Thus, in the evening, at small and large longitudes, the MIT as a structure is often not formed at all, neither its minimum nor the poleward wall. This is an unexpected fact that requires comprehension.

Latitudinal *fp* profile 2 in Figure 2i was obtained on 25 July 2002 under almost the same conditions in the same longitude sector. However, the latitudinal *fp* profile differs sharply from the usual profile at these longitudes. It is characterized by a high electron density at high latitudes, which quickly drops toward the equator. As a result, MIT is not observed, but a well-defined minimum is formed at an extremely low latitude of −50°. The whole structure of the ionosphere on this path looks strange. This structure has been recorded in the evening sector only once; a low latitude minimum is also marked by a green asterisk in Figure 1 at a longitude of 340°E and latitude of −50°. A possible explanation is associated with a strong particle precipitation equatorward of the "normal" boundary of diffuse precipitation, that is, at the latitudes of the MIT minimum, which is consequently filled with ionization [26]. A minimum latitude of −50° is probably a consequence of the formation of a subauroral polarization steam (SAPS) [27]. The study [27] provides an example of the formation of a trough at a latitude of 53°, according to the Millstone Hill radar and the DMSP F13 satellite measurements on 12 April 2001. The deep trough was formed under the action of the western component of the plasma drift, driven by a northward electric field. Thus, Figure 2h,i demonstrates the presence of problems in the identification of MIT at large longitudes in the western hemisphere.

As shown above, the main problem in the eastern hemisphere is the separation of high-latitude cases of the MIT from low-latitude cases of the HLT. The main criterion for separation is the correspondence of the MIT poleward wall to the precipitation in zone I and the HLT1 poleward wall to the precipitation in zone II. The solution to this problem was presented in sufficient detail above in the analysis of Figure 1. However, it is worth demonstrating this solution once again based on a very illustrative example obtained on 1 August 2001. Figure 3 on the left shows the longitudinal variations in the position of the troughs and both zones of diffuse precipitation. The Kp index varied in the period under review from 2− to 4−. To eliminate dependence on geomagnetic activity, the data were reduced to Kp = 2. Figure 3 shows that, in the western hemisphere, the trough minimum is located equatorward of the auroral oval of the precipitation and, by definition, is the MIT.

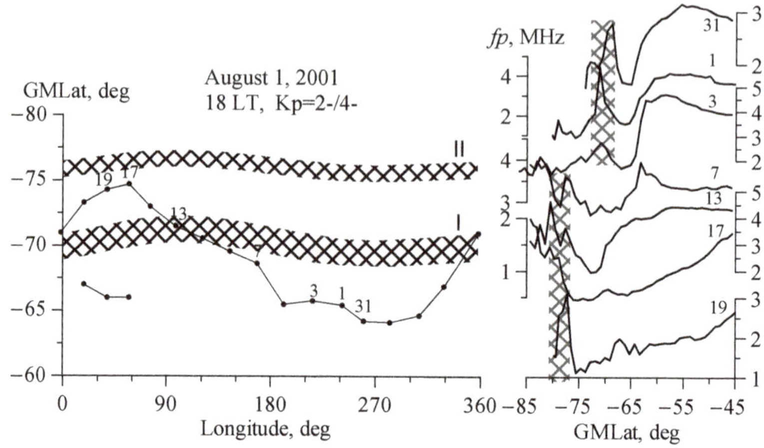

Figure 3. (**Left**): Longitudinal variations in the trough position on 1 August 2001. Precipitation zones I and II are hatched. (**Right**): Latitudinal *fp* profiles for the paths marked on the left.

Examples of the latitudinal cross-sections of the MIT are shown in Figure 3 on the right for paths 31, 1 and 3. The trough on these paths is clearly expressed; its poleward wall exactly corresponds to zone I of precipitation at a latitude of approximately −70°. In the eastern hemisphere, the troughs on paths 13–23 are located inside the auroral oval and by definition belong to the HLT. Examples of well-formed HLTs are shown on the right for paths 13, 17 and 19. Their poleward walls are located at latitudes of 76–80°, i.e., they are formed by precipitation in zone II. Path 7 is transitional; the poleward wall of the trough corresponds to zone II of precipitation, and the electron density minima at its wide bottom corresponds to both MIT and HLT. On paths 17 and 19, weak variations in ionospheric plasma are observed in the latitudinal region at approximately −66°, which can be assessed as a sign of MIT. On the other paths in the eastern hemisphere, there are no even weak manifestations, i.e., MIT has not formed in the eastern hemisphere. If the MITs in the western hemisphere combine with the HLTs in the eastern hemisphere into a single branch, there will be strong longitudinal variations in the trough position with an amplitude of ~11°. Note, however, the artificial character of the longitudinal effect in this case.

In Figure 2h, curve 2 depicts a specific structure, which is characterized by a shallow electron density minimum at the base of the poleward wall and a well-expressed minimum, which is much more equatorward than the "normal" minimum of the MIT. Because this structure is quite often formed at longitudes of America and the Atlantic (purple circles in Figure 1), it is worth considering its formation in detail. Figure 4 on the left shows the longitudinal variations in the trough position for quiet conditions (Kp varied from 1− to 2+) for the period of 4 July 2001. Both zones of diffuse precipitation are plotted in Figure 4. The thick curve shows the average position of MIT, which is highlighted in Figure 1. The numbers of characteristic satellite paths are indicated. The classic MIT is observed on path 14, and its poleward wall is formed by precipitation in zone I. In the eastern hemisphere, MIT is located slightly above the average position. MIT gradually enters zone I of precipitation, and on path 20, it transforms into HLT1, the poleward wall of which is already formed by precipitation in zone II. The MIT on path 20 did not manifest in any way, as in the previous event, at the same longitude (Figure 3).

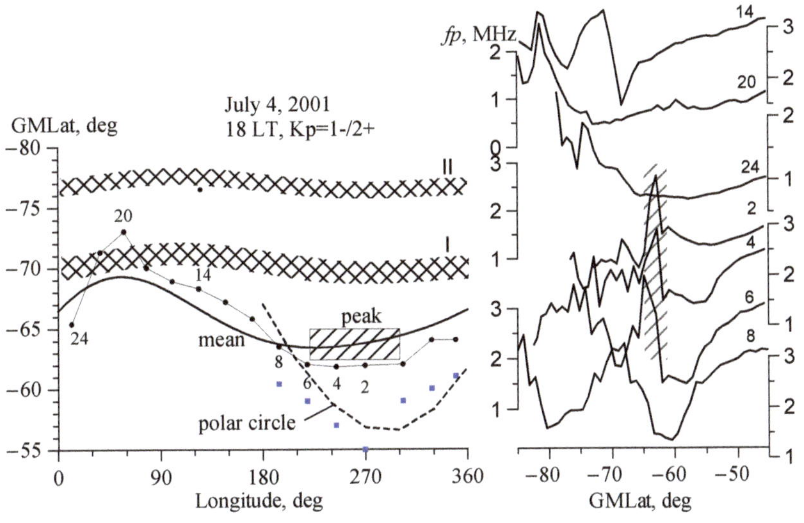

Figure 4. On the **left**: longitudinal variations on 4 July 2001 in the MIT position (black circles) and equatorward density minimum (blue squares). The average position of MIT (from Figure 1) is shown by a thick curve, and the polar circle is a dashed curve. Hatching shows the position of both zones of precipitation and the plasma density peak. On the **right**: *fp* latitudinal cross-sections for the paths indicated on the left.

A remarkable structure was formed in the western hemisphere. On paths 2, 4 and 6, a well-expressed ionospheric plasma peak is observed at latitudes of 62–65°, which is much more equatorward of the zone I precipitation. This peak defines the steep poleward wall of the trough. The appearance of plasma peaks distant from the auroral oval may be due to two reasons. Sufficiently strong particle precipitation sometimes occurs in the evening sector equatorward of the stable boundary of diffuse precipitation, i.e., already inside the trough [26]. The resulting peak of ionization can be amplified and shifted further to the equator by the zonal drift created by the SAPS [27], as the SAPS flows developed nearby. A small minimum *fp* is observed at the base of the poleward wall. Its position is plotted in Figure 4 as the MIT minimum because we have no other way to determine the position of the trough; it is masked by the well-defined equatorward minimum of *fp*. On path 2, this minimum is located at a latitude of −55°, which does not correspond to the minimum of the "normal" MIT at all. Then, this minimum shifts to high latitudes up to −61° on path 8. As a result, an ordinary MIT with a poleward wall formed by zone I precipitation was observed on path 8.

The dashed curve in Figure 4 shows the position of the polar circle at a geographic latitude of −66°. Thus, the observed minimum *fp* is obviously associated with the decay of electron density during the polar night. Automatic data processing of the MIT position has been frequently used recently. It is clear that any program will take the equatorward minimum as the minimum of MIT in Figure 4. Then, the longitudinal variations in the trough position, considering paths 20 and 22, will reach 18°, and the average position of MIT in the western hemisphere will be underestimated. There is a serious problem in determining the position of MIT because such a situation, as in Figure 4, is observed quite often. In the study by [2] on a large set of ISIS 1 and Injun 5 satellite data, it was determined that the minimum MIT was located 2–5° away from the auroral oval. This is true for the average position of the trough (the bold curve in Figure 4). However, in this case, even the base of the poleward wall is 6–7° away from the auroral oval. Thus, we were forced to take the minimum *fp* at the base of the poleward wall as the minimum of MIT.

4. Structure of the Morning Ionosphere

Figure 5 shows the longitudinal variations in the position of the main structures of the morning ionosphere. The designations are the same as in Figure 1. The CHAMP data for 05–06 LT were used. They were selected in the interval of Kp < 4 and were again reduced to Kp = 2. The longitudinal variations in the MIT position from the CHAMP data are less confidently determined in the morning (r = 0.55, σ = 2.1°) than in the evening because of the small amplitude A~4°. Longitudinal variations in the electron density on the MIT poleward wall (upper panel) were obtained during the quiet period from 3 July to 12 July 2001. These variations in the morning sector, as well as in the evening, are reliably revealed (r = 0.65, σ = 0.50°, A~1.5°).

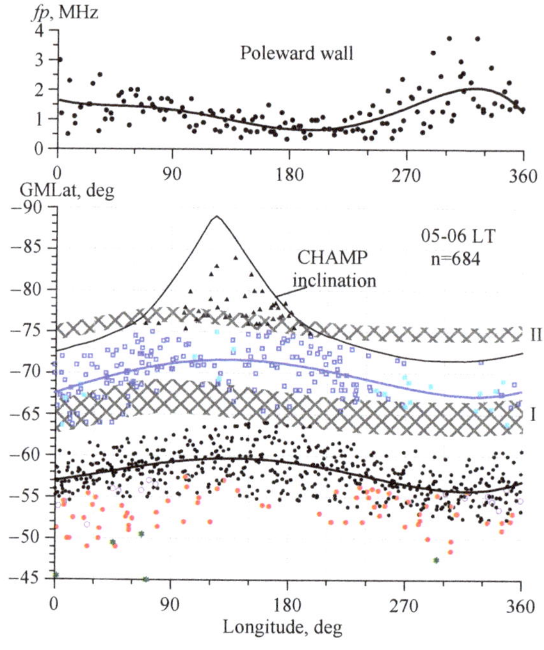

Figure 5. The same as in Figure 1, but for 05–06 LT.

Although in the position of the morning MIT, a weaker longitudinal effect was observed than in the evening; nevertheless, we retained the shape and amplitude of the longitudinal effect in precipitation in the same form as they were revealed for 21–03 MLT in [18]. The sharp difference between the morning and evening ionospheres is that MIT and HLT are separated by a large gap in the morning. Considering the results of the midnight ionosphere study [16], a trend can be identified: MIT is most closely located to the auroral oval in the evening, farther away from the oval at midnight, and even farther in the morning. Accordingly, the problem of separating MIT and HLT is most acute in the evening and simplest in the morning.

Examples of the most characteristic structures in the morning ionosphere are shown in Figure 6. Figure 6a shows an example of HLT1 recorded on 8 August 2002 at a longitude of 196°E. The poleward wall of the HLT is formed by precipitation at latitudes near −76°, which corresponds to zone II of precipitation. Only a slight decrease in electron density is observed at the latitude of MIT.

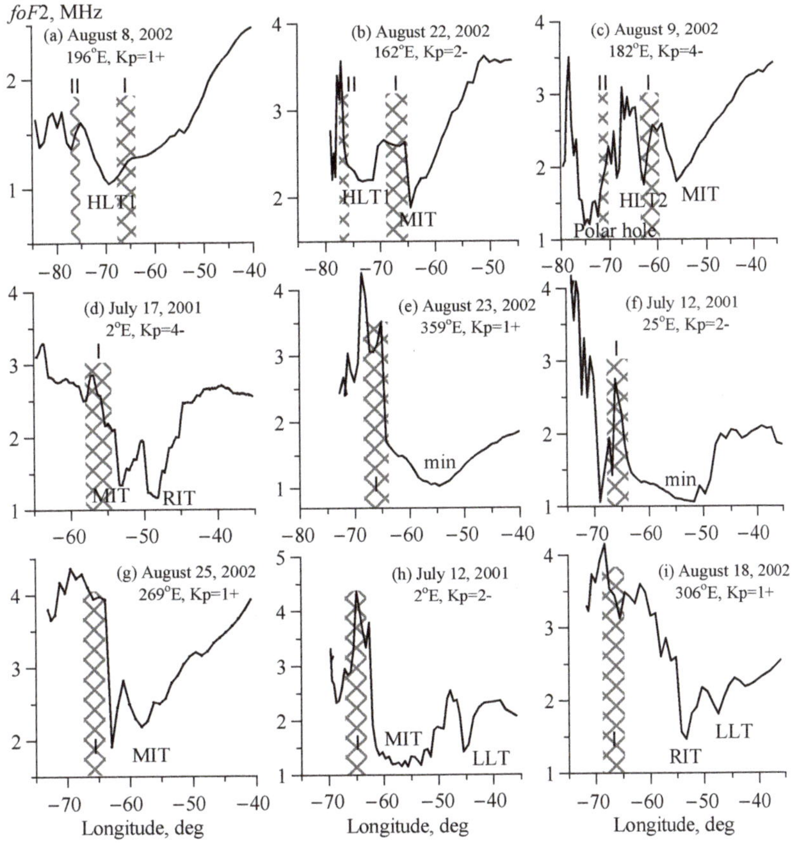

Figure 6. Characteristic examples of ionospheric troughs at different longitudes in the morning ionosphere.

At longitudes from 150° to 200°E, several MIT cases are located so high in latitude that their identification can be questioned. In Figure 6b, the minimum trough is located at a very high latitude of −64.6° for 5.0 LT and Kp = 2−. This is because the sharp minimum electron density corresponds to the base of a very steep poleward wall. However, the poleward wall is certainly formed by the precipitation in zone I, and this is MIT. Moreover, on this latitudinal *fp* profile, there is another high-latitude trough with the poleward wall located at a latitude of −78°, which corresponds to the precipitation in zone II.

Figure 6c clearly identifies three structures recorded on 9 August 2002 in the longitudinal sector of 182°E at Kp = 4−: the polar cavity, HLT and MIT. The high-latitude trough is quite narrow; therefore, it was assigned to HLT2. In Figure 5, it is located at the lowest possible latitude of −66° (reduced to Kp = 2), but it cannot be confused with MIT. Thus, even in the challenge longitudinal interval of 150–200°E, the MIT and HLT can be separated based on a thorough analysis. However, observing a trough path by path at longitudes of 120–210°E, the MIT can transfer into HLT and vice versa, and with a cursory analysis, it may not be noticed. This situation is demonstrated in Figure 3 for evening hours. The low poleward wall aggravates the problem, as well as in the evening sector.

The structures of the morning ionosphere are located at much lower latitudes than in the evening. As a result, the satellite inclination makes it possible to record the HLT at almost all longitudes. Therefore, the approximating curve for the HLT in Figure 5 reproduces the longitudinal variations in the HLT position more adequately than in Figure 1 for evening hours. The polar cavity in the morning is also much more equatorward than in the evening and is often superimposed on the precipitation in zone II. However, a detailed analysis of the polar cavity was not included in the objectives of this study.

On 17 July 2001, under Kp = 4, two troughs were also recorded: MIT and equatorward RIT (Figure 6d). The RIT is formed much more often in the morning than in the evening [9], which confirms the comparison of Figures 1 and 5. The morning RIT is, as in the evening, observed more often at longitudes with a weak geomagnetic field. Figure 6d shows the case in which MIT and RIT are separated very clearly. However, this is not always observed; therefore, we must separate them by analyzing the prehistory of the dynamics of both troughs during geomagnetic disturbance, starting with the main phase of the storm/substorm, as mentioned in the Introduction.

Figure 6e shows two examples of a structure that is most often recorded in the evening (Figure 2h, curve 1), but it is sometimes observed in the morning at longitudes in America and the Atlantic. It is characterized by a high poleward wall (compare with Figure 6b), shallow minimum or inflection of *fp* at its base and a deep minimum of *fp* far from the poleward wall. This structure is discussed in detail in Figure 4. In Figure 5, the equatorward minima of electron density is marked with purple circles. There are much fewer of these minima than in the evening. They are usually located far equatorward from the MIT average position. In any case, this minimum masks the MIT minimum and makes it difficult to determine the true MIT position.

The structure in Figure 6f looks like the structure in Figure 6e. However, in this case, the *fp* minimum at the latitude of -51.7° is more than 10° away from the base of the poleward wall and is accompanied by a sharp increase in *fp* to the equator. As a result, the structure in Figure 6f is similar to the structure that was observed in the evening in Figure 2i (curve 2) but is even more pronounced. In addition, in the evening, this structure is rarely observed; however, because the event repeats, a question arises about the mechanism of its formation.

In Figure 6g, two *fp* minima are observed in the longitudinal sector 269°E. The *fp* minimum at a latitude of -58.2° corresponds exactly to the average position of MIT for Kp = 2 in Figure 5.

The minimum latitude of -62.8° is located on the base of the steep poleward wall. The observed *fp* profile can be interpreted as an irregular structure at the bottom of the trough. However, the question remains as to why the deep minimum of the electron density is so often formed on the base of the sharp poleward wall of MIT.

The latitudinal *fp* profile in Figure 6h shows the deep and narrow low-latitude trough (LLT) on the equatorward wall of MIT at a latitude of -45.5°. Another such LLT was observed at a longitude of 72°E and latitude -45° (see Figure 5). Troughs at such low latitudes are observed quite rarely and probably belong to troughs associated with hot particle precipitation from the inner radiation belt [28] or with the penetration of electric fields deep into the plasmasphere [29].

The structure in Figure 6i also defies unambiguous identification. A deep *fp* minimum is observed at a latitude of -47.5°. This is 4–5° equatorward than observed on previous days at the same Kp and LT values. This minimum was formed after a weak disturbance with Kp = 3+. Even after such a weak disturbance, the RIT is often formed. However, in this case, it cannot be stated unambiguously. Nevertheless, in Figure 5, this trough is defined as the RIT. Equatorward of RIT at latitude of -47.5° another LLT is observed.

5. Discussion

The analysis of the localization of different structures of the high-latitude and mid-latitude ionospheres in terms of the longitudinal effect proved to be effective. This is because the positions of all structures, including the auroral oval of particle precipitation, experience quite strong variations with longitude. The MIT position changes with longitude by 6° in the evening and by 4° in the morning. The position of the boundaries of the auroral diffuse precipitation changes by 2.5°, according to the results, which was obtained so far only in one interval 21–03 MLT [18]. When variations in the position of the precipitation boundaries in the morning and evening sectors are revealed, the mutual location of the auroral oval and MIT can be clarified. However, we are confident that this will not fundamentally change the results of the analysis performed in this study.

In [2], the average distance of 2–5° between the auroral oval and MIT was determined. Now, we can refine this estimate. In the evening, MIT's average position is approximately 3° away from the equatorward border of diffuse precipitation, 4.5° at midnight [16], and as much as 6° in the morning.

Because MIT in the eastern hemisphere is located at a higher latitude than in the western hemisphere, the region of its existence in the eastern hemisphere is partly superimposed on zone I of precipitation and the location of the HLT. Therefore, the problem of MIT and HLT separation is most difficult at longitudes of 0–120°E in the evening and at midnight and 150–200°E in the morning. The problem is complicated by the fact that, at these longitudes, the poleward wall of the MIT is poorly expressed. Therefore, the main criterion for separating troughs is the correspondence of the MIT poleward wall to the precipitation in zone I and the HLT poleward wall to the precipitation in zone II. The position of the trough minimum also matters. All of this makes it possible to almost unambiguously separate MIT and HLT 1. However, to do this, we should carefully analyze each case because when observing path after path, situations arise when the MIT shifting to the pole transforms into HLT1 and vice versa. At the same time, only weak traces of MIT can be observed on the equatorward wall of HLT1, which is difficult to identify as a trough.

As for HLT2, the narrow trough described in [6] is most often observed at midnight, less often in the evening and even less often in the morning.

The polar cavity was recorded only where the inclination of the satellite allowed, i.e., at longitudes of 60–210°E. Analysis of the polar cavity was not a priority task; therefore, only obvious cases were recorded. The polar cavity was observed at latitudes above −76° in the morning and at those above −78° in the evening.

The RIT was also separated from MIT. For this purpose, the prehistory of the development of all, even weak disturbances, was considered. RIT is more often observed in the morning than in the evening. In the evening, it localizes exclusively at the longitudes of the western hemisphere, where the geomagnetic field is weak. In the morning, the RIT is observed at all longitudes, including in the eastern hemisphere.

In the evening, an extremely specific region of the ionosphere forms at the longitudes of America. First, in addition to the RIT, a special structure is often formed here (and partly over the Atlantic): a weak minimum of electron density is observed at the base of the poleward wall of the MIT, and a deep minimum is formed equatorward. This equatorward minimum is located at a much lower latitude than the statistical minimum of MIT. This seems to be associated with the decay of the electron density beyond the polar circle. This minimum masks the minimum of MIT, and with careless data processing, the true position of MIT can be greatly underestimated. Second, the poleward wall of MIT, probably driven by a horizontal drift, sometimes shifts to the equator, and then the entire structure of the high-latitude ionosphere is located much more equatorward than usual (see Figure 4). In this case, the amplitude of the longitudinal variations in the trough position can reach an incredible 18°. Third, the poleward wall of MIT at the longitudes of America is very low. Moreover, the electron density often monotonically falls to high latitudes without showing a noticeable minimum. As a result, the trough does not form at all at these longitudes.

6. Conclusions

The main result of the study is an almost unambiguous solution to the problem of separating different ionization troughs in the morning and evening ionosphere. The problem of the MIT and HLT separating is more complicated in the eastern hemisphere, the problem of the MIT and RIT separating is more complicated in the western hemisphere. Considering the results of [16], it can be argued that the problem of separation and classification of troughs in the nighttime ionosphere of the southern hemisphere is solved. This led to a decrease in the data scatter in the MIT position, the standard deviations are 1.8° in the evening, 1.85° at midnight, and 2.1° in the morning, which are less than in other statistical studies (see, for example, [10,30,31]). This is important for creating an accurate MIT model.

Supplementary Materials: The following supporting information can be downloaded at: https://www.mdpi.com/article/10.3390/rs14164072/s1. Figure S1: Model of auroral particle precipitation: diffuse auroral zone I equatorward of aurora (blue), structured auroral oval precipitation (auroral lights region or aurora, green), and soft diffuse precipitation zone II (orange) poleward of aurora. Figure S2: Longitudinal variations in the averaged auroral precipitation energy flux at 21–03 MLT under Kp = 2 for the June solstice (Jun.) in southern hemisphere [18].

Funding: This research received no external funding. The author would like to thank sponsors and operators of the CHAMP mission; Deutsches GeoForschungsZentrum (GFZ) Potsdam and German Aerospace Center (DLR).

Data Availability Statement: The CHAMP data are available on the website: http://op.gfz-potsdam.de/champ (accessed on 12 January 2015).

Conflicts of Interest: The author declares no conflict of interest.

References

1. Muldrew, D.B. F layer ionization troughs deduced from Alouette data. *J. Geophys. Res.* **1964**, *70*, 2635–2650. [CrossRef]
2. Ahmed, M.; Sagalyn, R.C.; Wildman, P.J.L.; Burke, W.J. Topside ionospheric trough morphology: Occurrence frequency and diurnal, seasonal and altitude variations. *J. Geophys. Res.* **1979**, *84*, 489–498. [CrossRef]
3. Moffett, R.J.; Quegan, S. The mid-latitude trough in the electron concentration of the ionospheric F-layer: A review of observations and modeling. *J. Atmos. Terr. Phys.* **1983**, *45*, 315–343. [CrossRef]
4. Rodger, A.S.; Moffett RJQuegan, S. The role of ion drift in the formation of ionisation troughs in the mid-and high-latitude ionosphere–a review. *J. Atmos. Terr. Phys.* **1992**, *54*, 1–30. [CrossRef]
5. Karpachev, A.T. Dependence of the MIT shape on longitude, altitude, season, local time, solar and magnetic activity. *Geomagn. Aeron.* **2003**, *43*, 256–269.
6. Grebowsky, J.M.; Tailor, H.A.; Lindsay, J.M. Location and source of ionospheric high latitude troughs. *Planet. Space Sci.* **1983**, *31*, 99–105. [CrossRef]
7. Karpachev, A.T. The characteristics of the ring ionospheric trough. *Geomagn. Aeron.* **2001**, *41*, 57–66. (In Russian)
8. Karpachev, A.T. Dynamics of main and ring ionospheric troughs at the recovery phase of storms/substorms. *J. Geophys. Res.* **2021**, *126*, e2020JA028079. [CrossRef]
9. Karpachev, A.T. Statistical analysis of ring ionospheric trough characteristics. *J. Geophys. Res.* **2021**, *126*, e2021JA029613. [CrossRef]
10. Werner, S.; Prolss, G.W. The position of the ionospheric trough as a function of local time and magnetic activity. *Adv. Space Res.* **1997**, *20*, 1717–1722. [CrossRef]
11. Karpachev, A.T.; Deminov, M.G.; Afonin, V.V. Two branches of day-time winter ionospheric trough according to the Kosmos-900 data at F2-layer heights. *Adv. Space Res.* **1998**, *22*, 877–882. [CrossRef]
12. Karpachev, A.T. Variations in the winter troughs' position with local time, longitude, and solar activity in the Northern and Southern hemispheres. *J. Geophys. Res.* **2019**, *124*, 8039–8055. [CrossRef]
13. Liu, Y.; Xiong, C. Morphology evolution of the mid-latitude ionospheric trough in nighttime under geomagnetic quiet conditions. *J. Geophys. Res.* **2020**, *125*, e2019JA027361. [CrossRef]
14. Prolss, G.W. The equatorward wall of the subauroral trough in the afternoon/evening sector. *Ann. Geophys.* **2007**, *25*, 645–659. [CrossRef]
15. Rodger, A.S.; Brace, L.H.; Hoegy, W.R.; Winningham, J.D. The poleward edge of the mid-latitude trough–its formation, orientation and dynamics. *J. Atmos. Terr. Phys.* **1986**, *48*, 715–728. [CrossRef]
16. Karpachev, A. Advanced separation and classification of ionospheric troughs in midnight conditions. *Sci. Rep.* **2022**, *12*, 13434. [CrossRef]
17. Vorobjev, V.G.; Yagodkina, O.I. Effect of magnetic activity on the global distribution of auroral precipitation zones. *Geomagn. Aeron.* **2005**, *45*, 467–473.
18. Luan, X.; Wang, W.; Burns, A.; Solomon, S.; Zhang YPaxton, L.J.; Xu, J. Longitudinal variations of nighttime electron auroral precipitation in both the Northern and Southern hemispheres from the TIMED global ultraviolet imager. *J. Geophys. Res.* **2011**, *116*, A03302. [CrossRef]
19. Williams, P.J.S.; Jain, A.R. Observations of the high latitude trough using EISCAT. *J. Atmos. Terr. Phys.* **1986**, *48*, 423–434. [CrossRef]
20. Karpachev, A.T.; Deminov, M.G.; Afonin, V.V. Model of the mid-latitude ionospheric trough on the base of Cosmos-900 and Intercosmos-19 satellites data. *Adv. Space Res.* **1996**, *18*, 221–230. [CrossRef]
21. Rother, M.; Michaelis, I. CH-ME-2-PLPT-CHAMP electron density and temperature time series in low time resolution (Level 2). *GFZ Data Serv.* **2019**. [CrossRef]

22. Deminov, M.G.; Shubin, V.N. Empirical model of the location of the main ionospheric trough. *Geomagn. Aeron.* **2018**, *36*, 45–52. [CrossRef]
23. Hardy, D.A.; Gussenhoven, M.S.; Holeman, E. A statistical model of auroral electron precipitation. *J. Geophys. Res.* **1985**, *90*, 4229–4248. [CrossRef]
24. Spiro, R.W.; Reiff, P.H.; Maher, L.J. Precipitating electron energy flux and auroral zone conductance–an empirical model. *J. Geophys. Res.* **1982**, *87*, 8215–8222. [CrossRef]
25. Sotirelis, T.; Newell, P.T. Boundary-oriented electron precipitation model. *J. Geophys. Res.* **2000**, *105*, 18655–18673. [CrossRef]
26. Khalipov, V.L.; Galperin, I.I.; Lisakov, I.V.; Krane, Z.; Nikolaenko, L.M.; Sinitsyn, V.M.; Sovo, Z.A. Diffuse auroral zone. II. Formation and dynamics of the poleward border of the subauroral ionospheric trough in the evening sector. *Space Res.* **1977**, *1*, 708–725. (In Russian)
27. Foster, J.C.; Burke, W.J. SAPS: A New Categorization for Sub-Auroral Electric Fields. *EOS* **2002**, *83*, 393–394. [CrossRef]
28. Karpachev, A.T. Sub-auroral, mid-latitude, and low-latitude troughs during severe geomagnetic storms. *Remote Sens.* **2021**, *13*, 534. [CrossRef]
29. Fejer, B.G.; Scherliess, L. Mid- and low-latitude prompt penetration ionospheric zonal plasma drifts. *Geophys. Res. Lett.* **1998**, *25*, 3071–3074. [CrossRef]
30. Kohnlein, W.; Raitt, W.J. Position of the mid-latitude trough in the topside ionosphere as deduced from ESRO-4 observations. *Planet. Space Sci.* **1977**, *25*, 600–602. [CrossRef]
31. Aa, E.; Zou, S.; Erickson, P.; Zhang, S.-R.; Liu, S. Statistical analysis of the main ionospheric trough using Swarm in situ measurements. *J. Geophys. Res.* **2020**, *125*, e2019JA027583. [CrossRef]

MDPI
St. Alban-Anlage 66
4052 Basel
Switzerland
Tel. +41 61 683 77 34
Fax +41 61 302 89 18
www.mdpi.com

Remote Sensing Editorial Office
E-mail: remotesensing@mdpi.com
www.mdpi.com/journal/remotesensing